20th Asian Pacific Confederation of Chemical Engineering Congress (APCChE 2023)

IOP Conference Series: Materials Science and Engineering
Volume 1318

Manila, Pasay City, Philippines
4-8 September 2023

ISBN: 979-8-3313-1004-2
ISSN: 1757-8981

Printed from e-media with permission by:

Curran Associates, Inc.
57 Morehouse Lane
Red Hook, NY 12571

Some format issues inherent in the e-media version may also appear in this print version.

Institute of Physics
Dirac House, Temple Back
Bristol BS1 6BE UK

Phone: 44 1 17 929 7481
Fax: 44 1 17 920 0979

techtracking@iop.org

Additional copies of this publication are available from:

Curran Associates, Inc.
57 Morehouse Lane
Red Hook, NY 12571 USA
Phone: 845-758-0400
Fax: 845-758-2633
Email: curran@proceedings.com
Web: www.proceedings.com

TABLE OF CONTENTS

IOP Conf. Series: Materials Science and Engineering **1318** (2024) 011001 doi:10.1088/1757-899X/1318/1/011001

PREFACE

The global landscape on industrial activities has changed with the COVID-19 pandemic. The drastic measures and activities (e.g., virtualization of gatherings, increased production of plastics and materials for face masks) that individuals must take leads to the realization of the persistence of the practices at the post-pandemic era. The idea of sustainable development now considers these drastic events in the future, in which chemical engineers can generate innovations that when the need arises, preparations will focus on large-scale development. Different aspects of sustainability, i.e., economic, environmental and social sustainability should be the backbone that guides the scientific community in developing new knowledge. For this volume, the editors have compiled the accepted papers submitted for the 20th Asian Pacific Confederation of Chemical Engineering (APCChE) Congress

APCChE is a bi-annual conference for academics and chemical industry experts in the Asian Pacific region to exchange research ideas and innovation in the field of chemical engineering and its related field. In 2023, the conference is done in conjunction with the 84th Philippine Institute of Chemical Engineers (PIChE) Convention. With its theme focusing on global partnership for industrial development, research papers from different themes are presented. This journal issue consolidates the accepted submissions of over 100 full papers from different themes. Submissions from the Biochemical and Biomedical Engineering theme feature the application of chemical engineering to innovate materials for biochemical processes and biomedical activities. Environment and sustainability research is a prominent aspect of the conference with research on different aspects where sustainability is needed such as water, air, and solid environments. The theme on water and wastewater engineering features research works on the application of chemical engineering knowledge for water security and wastewater treatment and minimization. Research work on energy management focuses on energy storage, energy sources and energy vectors while renewable energy is made into a separate theme in this conference to highlight the importance of renewable energy in the future. The applications of chemical engineering in food technology are presented in the conference to strengthen the approaches towards food security. Materials science and engineering are also highlighted in the conference through innovation in nanomaterials, and green chemistry. As the pharmaceutical industry is one of the main industries in the practice of chemical engineering, researchers have presented their findings in this field, with the focus on drug delivery. The core of chemical engineering knowledge is growing as evident in the presentations on process systems engineering, separation processes, reaction engineering and catalysis, and process safety and risk management. All these research works can make a big impact through partnerships for a common goal.

This volume is made possible with the efforts of the editors: **Dr. John Frederick Tapia**, De La Salle University; **Prof. Nathaniel P. Dugos, PhD**, De La Salle University; **Prof. Maria Natalia R. Dimaano, PhD**, University of Santo Tomas; **Prof. Allan N. Soriano, PhD**, De La Salle University; **Prof. Rizalinda L. de Leon, PhD**, University of the Philippines Diliman; **Prof. Kathleen B. Aviso, PhD**, De La Salle University; **Prof. Joey D. Ocon, PhD,** University of the Philippines Diliman; **Dr. Bonifacio T. Doma Jr.**, Mapua University; **Dr. Susan A. Roces**, De La Salle University; **Prof. Anchaleeporn W. Lothongkum, PhD**, King Mongkut's Institute of Technology Ladkrabang (KMITL); **Prof. Hidetoshi Sekiguchi, PhD**, Tokyo Institute of Technology; **Prof. Mohd Azlan, PhD**, University of Malaya; and **Prof. Muhammad Fahrurrozi, PhD**, Gadjah Mada University.

The International Scientific Committee of the 20th APCChE Congress is composed of **Prof. Maria Natalia R. Dimaano**, PhD, University of Santo Tomas; **Prof. David Shallcross**, University of Melbourne; **Prof. Xuhong Qian**, East China University of Science and Technology; **Prof. Shuichi Yamamoto**, Yamaguchi

University; **Prof. Woo-Sik Kim**, Kyung Hee University; **Prof. Dominic Foo,** University of Nottingham Malaysia; **Dr. Becky Macdonald**, National Cheng Kung University; Prof. **Dr. Anchaleeporn Lothongkum**, King Mongkut's Institute of Technology Ladkrabang; **Dr. Rizalinda De Leon**, University of the Philippines; **Dr. Kathleen Aviso**, De La Salle University; **Dr. Susan Roces,** De La Salle University; and **Dr. Bonifacio T. Doma, Jr.**, Mapua University.

The Organizing Committee of the 20th APCChE Congress is composed of **Engr. Maricris C. Vines** as Conference Chair, **Engr. Chamaigne Luz D. Jamandre**, **Dr. Nathaniel P. Dugos**, and **Engr. Ma. Elena T. Perlado** as Conference Co-Chair; **Dr. Araceli M. Monsada** as Treasurer; **Engr. Joshua Bon A. Roco** as Lead for Industry and Events Management; **Engr. Norman Giovanni M. Guevarra** as Lead for Academe and Technical Program; and **Dr. Nuna E. Almanzor, Prof. Alberto A. Laurito, Hon. Shirlyn C. Reyes, Hon. Cezar S. Dela Cruz,** and **Dr. Rex B. Demafelis** as advisers.

With the increasing trend to focus on holistic approach in research, the vision of the editors is a future where efficient ways to utilize materials will become possible, net-zero emissions can be realized, biodiversity is enriched, and people's lives are improved. The articles in this volume can be used to get closer to this ideality. Through partnerships of the academic, industry, and government sectors, the knowledge this volume contributes to what we know about the world can be transformed to sustainable technologies.

IOP Conf. Series: Materials Science and Engineering **1318** (2024) 011002 doi:10.1088/1757-899X/1318/1/011002

Peer Review Statement

All papers published in this volume have been reviewed through processes administered by the Editors. Reviews were conducted by expert referees to the professional and scientific standards expected of a proceedings journal published by IOP Publishing.

- **Type of peer review:** Double Anonymous
- **Conference submission management system:** Morressier
- **Number of submissions received:** 229
- **Number of submissions sent for review:** 132
- **Number of submissions accepted:** 138
- **Acceptance Rate (Submissions Accepted / Submissions Received × 100):** 60.3
- **Average number of reviews per paper:** 2.152173913043478
- **Total number of reviewers involved:** 34
- **Contact person for queries:**
  Name: John Frederick Tapia
  Email: apcche2023@gmail.com
  Affiliation: Department of Chemical Engineering, De La Salle University Manila

IOP Conf. Series: Materials Science and Engineering **1318** (2024) 012001 doi:10.1088/1757-899X/1318/1/012001

Hydrodynamic and mass transfer characterization of an internally illuminated airlift photobioreactor for growing Spirulina platensis

Butch G. Bataller[1] and Sergio C. Capareda[2]

[1]University of the Philippines Los Baños, Laguna, Philippines
[2] Texas A&M University, College Station, Texas, USA

bgbataller@up.edu.ph

Abstract. The dimensions of the of a photobioreactor play a major role in its hydrodynamic and mass transfer characteristics that also affect its growth performance. In this study, a bench-scale airlift photobioreactor (PBR) was designed and fabricated to have a riser-to-diameter ratio (A_r/A_d) equal to 0.42 and an aspect ratio (H/D) equal to 1.53. These dimension ratios are different from typical values wherein most airlift photobioreactors are designed to have A_r/A_d of at least 1.0 and H/D of at least 2.0. It was hypothesized, based on the results of Hwang and Cheng [1], the fabricated PBR in this study will have better hydrodynamic and mass transfer characteristics such as lesser mixing time, better liquid circulation and lesser gas entrainment that may lead to better growth performance. Due to the different dimension ratios from other studies, characterization of the PBR and testing of its performance are required. The hydrodynamic and mass transfer characteristics of the fabricated airlift photobioreactor in this study were measured at varying superficial gas velocities (U_{GR}) from 0.0017 ms^{-1} to 0.0124 ms^{-1}. The growth parameters of the *Spirulina platensis* were then tested at different aeration rates to relate the hydrodynamic and mass transfer characteristics to the photobioreactor performance. This study has proven that a photobioreactor that is designed and operated to have the best hydrodynamic and mass transfer characteristics will have the best growth performance at the same light intensity and initial optical density.

1. Introduction

Apart from the selection of optimal operating parameters such as light intensity, aeration or mixing rate, temperature, and pH among others, careful selection of the dimensions of a photobioreactor plays an important role in its performance. The photobioreactor dimensions can affect the reactor's hydrodynamic and mass transfer characteristics that determine the amount of light and nutrients made available to microalgal cells.

In the bench-scale cultivation of *Spirulina platensis*, the authors decided to fabricate an internally illuminated concentric tube airlift photobioreactor since its main body can be made from old fermenters available in the laboratory. Internal illumination was also provided to ensure that all incident light can be captured by the surrounding microalgal cells.

Airlift PBRs create circular mixing pattern where liquid passes continuously through dark and light phases, giving a flashing light effect to algal cells [2], leading to better photosynthetic efficiency [3].

They provide excellent heat and mass transfer between the liquid medium and aeration gas [4], homogeneous shear distribution, and high mixing with low energy consumption since it has no moving parts or internal stirring devices [5]. It also has higher biomass productivity than bubble column [6]. Airlift PBRs also offer large surface area to volume ratio, ease of control, and safe sterile operation [7].

The challenge is to design a photobioreactor that will promote the best growth performance. The dimensions of the airlift photobioreactor fabricated by the authors were based on the findings of Hwang and Cheng [1] who tested the effect of varying the ratio of the cross-sectional areas of the riser and downcomer region (A_r/A_d = 1.0 and 3.0) and aspect ratio (H/D = 3.7, 5.8, 7.9) on gas holdup, liquid velocity, and mixing time. They found that the PBR with A_r/A_d =1.0 has higher liquid velocity in the riser region and lesser bubble entrainment in the downcomer region than the PBR with A_r/A_d = 3.0. However, they also found that an A_r/A_d = 1.0 resulted in higher mixing time than A_r/A_d = 3.0. This means that the reactor with A_r/A_d =1.0 needs longer time to attain 95% homogeneity. The trade-off between mixing time, gas holdup and liquid velocity when using smaller A_r/A_d therefore needs further investigation. The airlift PBR in this study with A_r/A_d = 0.42 was investigated to determine if the expected higher liquid velocity in the riser could lead to lower mixing time.

Hwang and Cheng [1] also found higher gas entrainment in the downcomer region at higher H/D resulting in lesser liquid circulation. Column reactors are typically designed to have H/D that is greater than two [8]. But in this study, the airlift reactor was designed to have an effective H/D ratio of less than two to test if better liquid circulation can be obtained.

Since the fabricated airlift photobioreactor has dimension ratios different from typical values, there is a need to determine the mass transfer and hydrodynamic characteristics that may affect its performance [2],[9]. Characterization aids in optimizing operational conditions, incorporating design improvements, and in scaling up of the reactor [10],[11]. Operating at conditions with high mass transfer coefficient, carbon limitation is avoided since sufficient dissolved inorganic carbon in the medium is ensured [12]. Oxygen accumulation, that may lower the pH of the medium, is also prevented due to the high oxygen removal efficiency. Hydrodynamic forces may aid in liquid circulation but can also damage the cells, necessitating the determination of operating conditions that are optimal to cell growth [7].

It is then hypothesized that high growth performance can be obtained at the superficial gas velocity where the best liquid circulation is obtained, and gas holdup is minimized. Due to the high dependency of hydrodynamics and growth performance on reactor geometry, hydrodynamic and mass transfer characterization of the photobioreactor fabricated in this study is necessary.

This study fabricated and characterized a bench-scale internally illuminated concentric-tube airlift photobioreactor in terms of its hydrodynamic and mass transfer characteristics at different superficial gas velocities. The growth performance at the superficial gas velocity that gave the best hydrodynamic and mass transfer characteristics was then tested and compared to the performance at other superficial gas velocities.

2. Methodology

2.1. Design of the photobioreactor

A 3-L airlift photobioreactor was designed and fabricated wherein the main chamber and the light chamber were made of glass and the draft tube was made of clear acrylic pipe. The main chamber had a round bottom to promote better circulation of cells and to prevent accumulation of cells at the bottom if it had a sharp corner. The light chamber at the center of riser region also had a round bottom to promote smooth upward flow of liquid and cells that flow past it. Cool white light-emitting diodes (LEDs) with 12V DC input (3 Watts per foot) and 164 lumens per foot output were used as light source. The reactor was covered with a mirrored acrylic plate.

The liquid height in the reactor is kept at 80% of the total tank height to provide space for degassing and foam formation. The ratio of the cross-sectional area of the riser to the downcomer was

0.42 and the effective liquid height to diameter ratio was 1.53. The surface area to volume ratio was computed based on the reactor dimensions. A porous sparger was installed at the bottom of the reactor to promote air flow and liquid circulation.

Three reactors of the same dimensions were constructed to be used for replicate runs.

2.2. Liquid-phase physical and hydrodynamic characterization

2.2.1. Determination of physical characteristics of the medium. Five gallons of Zarrouk's medium was prepared to be used both in characterization and in cultivation of Spirulina. Twenty 100-ml samples were taken from the prepared medium and were used for physical characterization at 25 °C. The viscosity and density were measured using a Cannon–Fenske viscometer and pycnometer, respectively. The conductivity was measured using conductivity meter (VWR® SB90M5 SympHony Meter).

2.2.2. Mixing time (t_m) and circulation time (t_c) determination. Mixing time and circulation time determination followed the procedure of Fernandes et al. [10]. In this procedure, 5 mL saturated NaCl tracer solution was injected near the bottom of the riser area through a 1 mm stainless steel capillary. The conductivity of the liquid with the tracer was measured every 1-2 seconds using a conductivity probe (VWR® SB90M5 SympHony Meter) placed near the top of the riser at different superficial gas velocity (U_{GR}), which were varied from 0.0017 ms^{-1} to 0.0124 ms^{-1} (equivalent to 0.13 to 1.0 vvm air flow rate). Three replicate measurements were made per reactor. Mixing time was calculated as the time needed to reach 95% of the final conductivity or 95% concentration homogeneity. The circulation time was computed as the average time spans between maximum consecutive peaks in conductivity probe response curve.

2.2.3. Liquid circulation velocity (v_l) determination. Determination of the liquid circulation velocity in the riser region also followed the procedure of Fernandes et al. [10], which was conducted via a thermal tracer method. In this procedure, a pulse of 5 ml of hot water was injected to the flowing liquid and the time–temperature profile was plotted at two given points in the riser by means of two thermocouples. Two Type-K thermocouples were placed 13.7-cm apart in the riser area of the photobioreactor. Temperature was measured and logged every 0.5 sec using Campbell Scientific CR3000 Micrologger. The liquid linear velocity in the riser was obtained as the ratio of the distance between the two thermocouples and the differences in their response times.

2.3. Gas-phase hydrodynamic characterization

The overall gas holdup at different superficial gas velocities was determined using the volume expansion method [13]. In this procedure, the unaerated static liquid height (h_L) was first measured prior to each run. Then at the experimental superficial gas velocity, the air compressor was switched on and the aeration was continued until a constant expanded liquid height was achieved. The aerated expanded liquid height (h_D) was then measured. The overall gas holdup was computed as follows:

$$\varepsilon = \frac{h_D - h_L}{h_D} \quad (1)$$

Three replicate measurements per superficial gas velocity for each reactor were conducted.

2.4. Determination of overall volumetric mass transfer coefficient of oxygen ($K_La_{(O2)}$)

The overall volumetric mass transfer coefficient mass transfer coefficient (K_La) was determined using a modified general protocol of the American Society of Civil Engineers (ASCE) Standard 2-91 Measurement of Oxygen Transfer in Clean Water, presented in Babcock, Malda, and Radway [12]. The first step was sparging with nitrogen to deoxygenate the reactor contents. When the lowest or zero dissolved oxygen concentration was achieved, the nitrogen was turned off and air sparging at a certain superficial gas velocity was commenced. The oxygen concentration was measured by a dissolved oxygen meter (VWR® SB90M5 SympHony Meter) and logged every 10 s using HyperTerminal until a constant value was attained. The K_La was computed using the linearized form of equation (2):

$$C = C_\infty^* - (C_\infty^* - C_o)e^{K_L a \cdot t} \quad (2)$$

where:

$C =$	dissolved oxygen concentration (mgL^{-1}) at time, *t*
$C_\infty^* =$	saturation dissolved oxygen concentration (mgL^{-1})
$C_o =$	initial dissolved oxygen concentration (mgL^{-1})
$K_L a =$	overall mass transfer coefficient ($time^{-1}$)
$t =$	time

Three trials per reactor per superficial gas velocity were made.

2.5. *Determination of growth performance at varying superficial gas velocity*

Spirulina was grown at different aeration rates – 0.13, 0.30, and 0.80 vvm – at constant initial biomass concentration (0.05 gL^{-1}) and light intensity (166 μmol m^{-2} s^{-1}). Air, filtered using 0.2 μm air venting filter, was allowed to flow through the porous sparger at the reactor bottom then to the riser to promote liquid circulation. The initial height of the expanded liquid, due to aeration, was marked. A 60-ml sample was taken right after inoculation, after 5 h, and every 24 h thereafter for 8 days. Fresh medium was injected to the reactor up to the initial liquid height to maintain culture volume every after sampling. The sample's optical density was read at 680 nm using a UV-Vis spectrophotometer (VWR UV-Vis Scanning UV-3100). The protein, carbohydrate and lipid content were determined using the method developed by Bataller and Capareda [14]. The specific growth rate, doubling time, overall and daily volumetric biomass productivity, overall and daily volumetric protein productivity, and photosynthetic efficiency were then computed.

All culture runs were done in triplicate.

3. Results and Discussion

3.1. *Design of the photobioreactor*

The actual A_r/A_d and H/D of the internally illuminated concentric-tube airlift photobioreactor fabricated in this study were 0.42 and 1.53, respectively, and has a working volume of 3.0 liters. The dimensions of the reactor are shown in table 1 and the actual experimental setup is shown in figure 1.

Table 1. Dimensions of the 3-L short tube internally illuminated concentric-tube airlift photobioreactor.

	Main Chamber	Light Chamber	Draft Tube
D_o, cm (in.)	15.3 (6.02)	5.1 (2.01)	9.5 (3.74)
D_i, cm (in.)	14.6 (5.75)	4.7 (1.85)	8.8 (3.46)
H, cm (in.)	29.3 (11.54)	20.7 (8.15)	24.2 (9.53)
t, cm (in.)	0.35 (0.14)	0.2 (0.08)	0.35 (0.14)

This design of airlift photobioreactor has a high illuminated surface area-to-volume ratio (S/V = 67 m^2m^{-3}) that may result to high biomass productivity. A reactor is considered to have high surface-to-volume ratio if its value ranges from 25 m^2m^{-3} to 125 m^2m^{-3} [15].

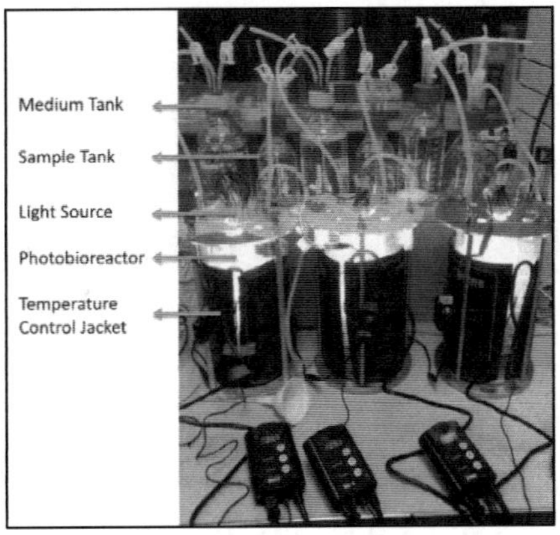

Figure 1. Actual experimental setup showing the various components of the 3-L internally illuminated airlift photobioreactor.

3.2. Liquid-phase physical and hydrodynamic characteristics

Being a dilute solution, the density and viscosity of the Zarrouk medium were found to be similar to water (ρ = 0.996 ± 0.006 g L-1 and μ = 0.997 ± 0.007 cP). But due to its higher ionic strength, the Zarrouk medium's conductivity is found to be significantly higher than tap water (σ_{medium} = 22.4 ± 0.333 mScm^{-1} compared to σ_{water} = 1.03 ± 0.046 mScm^{-1}).

Figure 2 shows the average mixing and circulation times at different superficial gas velocities (U_{GR} = 0.0017 – 0.0124 ms^{-1} corresponding to air flow rate of 0.13 to 1.0 vvm) in the three reactors. The small deviation among values imply that the reactors have similar hydrodynamic characteristics in terms of mixing time and circulation time.

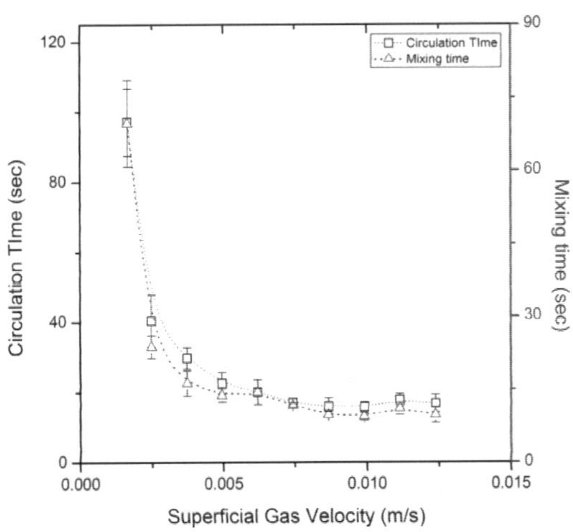

Figure 2. Liquid-phase hydrodynamic properties of the 3-l internally illuminated concentric-tube airlift photobioreactor at U_{GR} = 0.0017 – 0.0124 m s^{-1}.

As seen in figure 2, mixing time was found to be highly dependent on superficial gas velocity at low values. It decreased to 16.4 ± 2.62 s when the superficial gas velocity was increased to 0.0037 ms^{-1}. At this U_{GR}, it takes about 16.4 s to homogenize and distribute the reactor contents. Sufficient mixing, as in this case, can improve biomass productivity since the flashing light effect is promoted and the mass transfer of nutrients to the cells is increased.

This result is in agreement with the studies of Uyar, Ali, and Uyar [2], Banerjee et al. [9], and Konopacki et al. [11] that mixing time generally decrease with aeration rate. The increased iquid circulation velocity, liquid turbulence, and cycling frequency at higher aeration rate leads to better mixing and shorter mixing time in the photobioreactor [9].

At U_{GR} > 0.0037 ms^{-1}, mixing time and circulation time is almost independent of superficial gas velocity. This agrees with the results of Fernandes et al. [10] who found that mixing time was almost flow-independent for superficial gas velocities more than 0.003 ms^{-1}. This study, however, obtained lower mixing times, which is better, than the study of Fernandes et al. [10] even if both have similar volume capacities. This may be due to the different dimension ratios of the reactors resulting to different hydrodynamic characteristics. The split-cylinder airlift photobioreactor in Fernandes' study has a A_r/A_d = 1 and a H/D = 6.67 versus the A_r/A_d and H/D equal to 0.42 and 1.53, respectively, which supports the hypothesis that better hydrodynamic properties can be obtained at lower A_r/A_d and H/D. The airlift PBR fabricated in this study promoted better liquid circulation, as well, that may result to more frequent exposure of the cells to the light and dark cycles.

At the higher range of superficial gas velocity, mixing and circulation time only improved slightly and the flow is more turbulent. Cells can be potentially damaged due to the higher hydronamic stress. Working in this range, therefore, will not give additional significant benefit. For this reason, 0.0037 ms^{-1} superficial gas velocity or 0.3 vvm air flow rate was chosen as the best condition for Spirulina cultivation in the fabricated reactor. This assertion is tested in the performance evaluation phase of the study.

As for the liquid velocity measurements, the values obtained were high (0.091 to 0.274 ms^{-1}) even at the lowest air flow rate considered (0.13 vvm). The high liquid velocity obtained may be due to the smaller riser area than most reactors. According to Li et al. [5], smaller A_r/A_d causes a faster axial liquid flow in the riser, a higher liquid circulation velocity, a shorter circulation time, and smaller gas holdups in both the riser and downcomer. The A_r/A_d affects the liquid velocity since it modifies the resistance to flow by varying the fraction of the total volume contained in the riser and in the downcomer [10]. At lower A_r/A_d, the resistance to flow at the riser area is lower due to the lesser liquid volume than in the downcomer. The low A_r/A_d of the photobioreactor designed in this study led to the high liquid velocity in the riser area even at the lowest superficial gas velocity.

3.3. Gas-phase hydrodynamic and mass transfer characteristics

Gas holdup indicates the volume fraction and the mean residence time of the gas phase in aerated vessels. High gas holdup would mean longer residence time of entrained gases in the liquid phase. Since air is sparged to the reactor to be used for mixing, high gas holdup is not recommended for photosynthetic cultures. Oxygen in air may accumulate and may inhibit algal growth [16],[17].

The overall mass transfer coefficient of oxygen ($k_L a,_{O_2}$), on the other hand, represents the rate at which oxygen is dissolved in a reactor's liquid phase. Similar with gas holdup, high $k_L a,_{O_2}$ is undesired since oxygen accumulation may occur leading to lowering the pH of the culture medium that negatively impacts Spirulina that is known to grow in alkaline pH.

Figure 3 shows the gas hold up and overall volumetric mass transfer coefficient of oxygen, and their relationship, at various superficial gas velocities. It shows that gas holdup is constant at $6.76 \times 10^{-3} \pm 1.01 \times 10^{-5}$ at superficial gas velocities from 0.0017 ms^{-1} to 0.0025 ms^{-1}. Then, it rapidly increased to $3.40 \times 10^{-2} \pm 9.23 \times 10^{-4}$ at 0.0074 ms^{-1} superficial gas velocity. The gas holdup increased very slightly, thereafter. Similar relationship was observed between $k_L a,_{O_2}$ and superficial gas velocity. The $k_L a,_{O_2}$ increased very slightly from 0.057 ± 0.015 min^{-1} to 0.159 ± 0.020 min^{-1} from 0.0017 ms^{-1} to 0.0050 ms^{-1} superficial gas velocity. It then rapidly increased to 0.930 ± 0.065 min^{-1} at 0.0062 ms^{-1} superficial gas velocity. Then, the $k_L a,_{O_2}$ increased very slightly, thereafter.

In contrast, Xu, Liu, and Gu [18] and Zarei et al. [19] found that gas holdup increases almost linearly with increasing superficial gas velocity. The difference in the results may be due to the range of superficial gas velocities considered. In the study of Xu, Liu, and Gu [18], the U_{GR} they considered (0.0157 to 0.0262 ms^{-1}) were more than the highest U_{GR} in this study. While in the study of Zarei et al. [19], the difference may be due to the different H/D, which is equal to 13.3 in Zarei et al. study; while in this study, H/D is only 1.53. Aeration rate may have a greater influence on the gas holdup in reactors with smaller H/D.

It is desired to choose an air flow rate that lowers the gas holdup and $k_L a,_{O_2}$ but can still promote good mixing and circulation of reactor contents. The results from the gas-phase hydrodynamic

IOP Conf. Series: Materials Science and Engineering **1318** (2024) 012001 doi:10.1088/1757-899X/1318/1/012001

characterization supported the decision to choose 0.3 vvm (0.0037 ms^{-1} superficial gas velocity) as the best condition to grow Spirulina in the fabricated reactor since it is the upper limit of gas velocity where gas holdup and $k_L a,_{O_2}$ were minimum. At this value, the liquid hydrodynamic properties are desirable (e.g., low mixing and circulation time). It is expected that oxygen accumulation can be prevented at this gas velocity while still maintining a well-circulated algal culture.

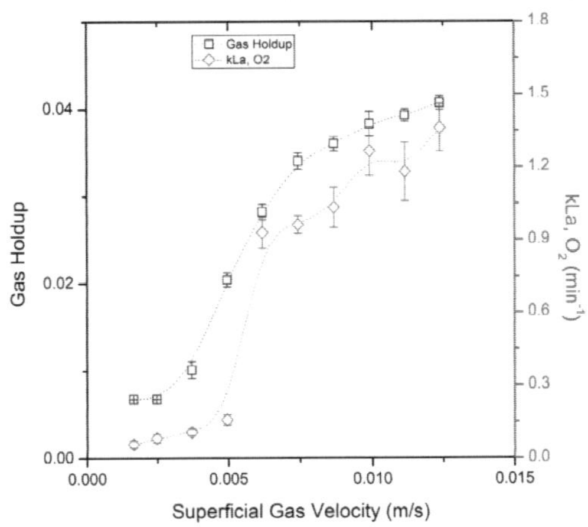

Figure 3. Gas-phase hydrodynamic and mass transfer properties of the 3-l internally illuminated concentric-tube airlift photobioreactor at U_{GR} = 0.0017 – 0.0124 m s^{-1}.

3.4. Growth performance at varying superficial gas velocity

From the previous sections, 0.30 vvm air flow rate is expected to give the best growth performance due to its desirable hydrodynamic and mass transfer characteristics. It is therefore chosen as the mid-level aeration rate in the performance tests. The other aeration rates chosen were 0.13 vvm and 0.80 vvm to test the growth performance at flow rates lower than and greater than 0.30 vvm. The results are summarized in table 2.

Table 2. Summary of growth parameters of *Spirulina platensis* cultivated at various air flow rates in a 3-l internally illuminated concentric-tube airlift photobioreactor. Light intensity was maintained at 166 μmol photons $m^{-2}s^{-1}$ and the initial biomass concentration was 0.05 gl^{-1}.

Growth Parameters	Air Flow Rate (vvm*)			
	0.13 (PBR-1)	0.13 (PBR-2)	0.3	0.8
Specific growth rate, μ (d^{-1})	0.823	0.364	0.664 ± 0.035	0.688 ± 0.006
Overall biomass productivity, Qx ($gl^{-1}d^{-1}$)	0.107	0.04	0.141 ±0.007	0.131 ± 0.007
Max. daily biomass productivity, ($gl^{-1}d^{-1}$)	0.252	0.135	0.217 ± 0.013	0.225 ± 0.018
Final biomass concentration, (g l^{-1})	0.906 ± 0.010	0.358 ± 0.002	1.172 ± 0.056	1.096 ± 0.056
Final biomolecule content (% w/w)				
Protein	59.70	58.5	57.3 ± 0.777	57.0 ± 0.156
Carbohydrate	36.20	38.2	38.8 ± 1.69	38.6 ± 1.47
Lipid	4.10	3.34	3.91 ± 0.910	4.42 ± 1.32
Overall product productivity ($gl^{-1}d^{-1}$)				
Protein	0.059	0.022	0.075 ± 0.004	0.0674 ± 0.0006
Carbohydrate	0.036	0.014	0.051 ± 6.02 × 10^{-7}	0.0459 ± 0.002
Lipid	0.0039	0.0011	0.0050 ± 0.0014	0.005 ± 0.002
Max. daily product productivity ($gl^{-1}d^{-1}$)				
Protein	0.11	0.041	0.185 ± 0.013	0.156 ± 0.033
Carbohydrate	0.066	0.029	0.108 ± 0.005	0.0934 ± 0.0177
Lipid	0.014	0.004	0.0146 ± 0.0001	0.0189 ± 0.0039
Photosynthetic Efficiency (%)	11.6	3.3	10.6 ± 0.121	11.0 ± 1.28

* Where, vvm is volume flow rate of air per volume of growth medium (L L^{-1} m^{-1})

It can be seen in table 2 that Spirulina grew best at the 0.30 vvm where the best hydrodynamic and mass transfer characteristics were also obtained. At this aeration rate the highest overall biomas productivity, and product productivity were obtained. The low mixing time at this flow rate ensures that the nutrients are made available to the cells at the lowest time possible. The low gas holdup and $k_L a_{,O_2}$ ensures that oxygen accumulation is reduced and better liquid circulation is obtained. The short circulation time also enhances the flashing light effect that enriches the photosynthetic activity of the cells.

At the lower air flow rate, liquid circulation is not enough that cells were observed to settle and accumulate at the bottom of the reactor. In this case, photolimitation was enhanced since the cells are not exposed to light. On the other hand, greater hydrodynamic stress may have experienced by the cells at the higher air flow rate that resulted to lower growth performance.

4. Summary and Conclusion

The different dimension ratios such as A_r/A_d and H/D of the airlift photobioreactor in this study necessitates hydrodynamic and mass transfer characterization. The aeration rate that will give the best mixing time, circulation time, liquid velocity, gas holdup, and overall mass transfer coefficient of oxygen was determined. It was found that at 0.30 vvm aeration rate, that is equivalent to 0.0037 ms^{-1} superficial gas velocity, the low mixing time and circulation time enhanced the flashing light effect resulting to better cell growth. Also U_{GR} = 0.0037 ms^{-1}, the gas holdup and $k_L a_{,O_2}$ are small ensuring that oxygen accumation is eliminated. The hypothesis that better growth performance can be obtained at the aeration rate where hydrodynamic and mass transfer characteristics are best is tested by determining the specific growth rate, biomass productivity, product productivity, and photosynthetic efficiency at three levels of aeration rates. This study has proven this hypothesis. Optimizing reactor hydrodynamic and mass transfer characteristics, therefore, is important in ensuring efficiency and performance. It can also be used in determining the optimum reactor operating conditions or in incorporating design improvements.

References

[1] Hwang S J and Cheng Y L 1997 *Chem. Eng. Sci.* **52** 3949–3960
[2] Uyar B, Ali M D, and Uyar G E O 2024 *Bioprocess Biosyst. Eng.* **47** 195-209
[3] Barbosa M J, Janssen M, Ham N, Tramper J and Wijffels R H 2003 *Biotechnol. Bioeng.* **82** 170–179
[4] Eriksen N T 2008 *Biotechnol. Lett.* **30** 1525–1536
[5] Li L, Xu X, Wang W, Lau R, and Wang C H 2022 *Bioresour. Technol.* **359** 1-18
[6] Oncel S and Sukan F V 2008 *Bioresour. Technol.* **99** 4755–4760
[7] Contreras A, García F, Molina E and Merchuk J C 1998 *Biotechnol. Bioeng.* **60** 317–325
[8] Singh R N and Sharma S 2012 *Renew. Sustain. Energy Rev.* **16** 2347–2353
[9] Banerjee S, Dasgupta S, Das D, and Atta A 2020 *Bioprocess Biosyst. Eng.* **43** 1487-1497
[10] Fernandes B D, Mota A, Ferreira A, Dragone G, Teixeira J A and Vicente A A 2014 *Chem. Eng. Sci.* **117** 445–454
[11] Konopacki M, Jablonska J, Dubrowska K, Augustyniak A, Grygorcewicz B, Glizniewicz M, Wroblewski E, Kordas M, Dolegowska B, and Rakoczy R 2023 *Microorganisms* **11** 1-14
[12] Babcock R W, Malda J, and Radway J A C 2002 *J. Appl. Phycol.* **14** 169–184
[13] Chisti MY 1989 *Airlift Bioreactors* (Essex, England: Elsevier Science Publishers, Ltd.)
[14] Bataller B G and Capareda S C 2018 *Algal Research* **32** 341-352
[15] Tredici M R, Carlozzi P, Chini Zittelli G and Materassi R 1991 *Bioresour. Technol.* **38** 153–159
[16] Torzillo G, Giovannetti L, Bocci F and Materassi R *Biotechnol. Bioeng.* **26** 1134–1135
[17] Marquez J, Sasaki K, Nishio N and Nagai S 1995 *Biotechnol. Lett.* **17** 225–228
[18] Xu C, Liu X, Gu X 2023 *J. Environ. Chem. Eng.* **11** 1-15
[19] Zarei Z, Malekshahi P, Trzcinski A P, Morowvat M H 2022 *Sustainability* **14** 1-15

IOP Conf. Series: Materials Science and Engineering **1318** (2024) 012002 doi:10.1088/1757-899X/1318/1/012002

Cellulose extraction from Cladophora rupestris for extraction of nanomaterials

M R F Persincula[1,2] **and C F Madrazo**[3]

[1] Chemical Engineering, College of Engineering, Batangas State University, Alangilan Campus, Golden Country Homes, Alangilan, Batangas City, Batangas 4200, Philippines
[2] Chemical Engineering Department, Gokongwei College of Engineering, De La Salle University, 2401 Taft Avenue, Manila 1004, Philippines
[3] Chemical Engineering Department, Gokongwei College of Engineering, De La Salle University, 2401 Taft Avenue, Manila 1004, Philippines

Email: maryrose.persincula@g.batstate-u.edu.ph, mary_rose_persincula@dlsu.edu.ph

Abstract. Cellulose from macroalgae is potential for the extraction of nanomaterials due to its availability and its great fraction in the biomass. *Cladophora rupestris* is a macroalgae and is a good material to which a cellulose may be extracted. The said macroalgae has no value-added products or uses yet. It has a lignocellulosic profile of: alpha-cellulose, 33.43 %; hemicellulose, 15.48%; holocellulose, 48.93; acid soluble lignin, 2.22%; acid insoluble lignin, 22.33%, extractives, 5.67% and total ash, 39.65%. Identifying the best method and conditions for the extraction of cellulose from the said material is a challenge that needs to be addressed. The process for the cellulose extraction considered variations on drying (oven, air, and sun drying), defatting (using solvents: methanol and hexane; and extraction time: 8 hrs and 4 days), solvent drying (oven and air drying), alkaline pre-treatment (0.1, 0.5, and 1M NaOH), fixed conditions on bleaching, bleached biomass oven drying. The biomass weight losses were monitored for some steps, attributing this on the removal of the lignin and extractives. The lignocellulosic profile of the crude cellulose extracted from the macroalgae using the best conditions was determined indicating an increased fraction of the cellulose components in the biomass and a transformation of the rigid biomass into a soft and paper-like texture. The functional groups present on the cellulose was determined. The extracted cellulose was used to produced cellulose nanofibril (CNF) with fiber diameter of the CNF ranges on 14.29 – 37.50 nm and crystallinity index of 92.48%. Attempts were done on extraction of cellulose nanocrystal (CNC) utilizing microwave-assisted acid hydrolysis (MAAH) and semi-batch acid hydrolysis (SBAH); and variation on the acid used (sulfuric acid, and oxalic acid), concentration of the acid, and the temperature conditions.

1. Introduction

A variety of biomass are abundant in our environment such as macroalgae but varies on their cellulose content and other chemical constituents to which it may comprise. Given the great application of cellulose as raw material for specialty chemicals and nanomaterials, exploring the use of macroalgae that can provide a great fraction of cellulose are those usually tapped for its extraction. *Cladophora rupestris* is a good material, with no economic use, and can provide up to 70% of cellulose. Those from algae have a high yield possessing high crystallinity [1,2] that enhances the mechanical properties when used for nanocomposite development [3]. In the Philippines, this green alga is found in the marine waters of Batanes and Paayas, Burgos, Ilocos Norte [4] and fresh waters of Taal Lake, Batangas [2,5]. It has dark green thallus which persists throughout the year especially during summer season [2] lasting until end of June. With the intention of coming up with a nanomaterial from this alga, the need to identify the conditions for every step of the cellulose extraction should be done to maximize its yield and attain a high degree of crystallinity on the extracted nanomaterials. This may be done by increasing the level of cellulose type Iα that can provide a high crystallinity over that of cellulose type Iβ in the matrix of the cellulose extracted.

2. Methodology

Cladophora rupestris were collected from seawater of Paayas, Burgos, Ilocos Norte. The chemicals used for the study were methanol, hexane (AR, Labscan), sodium hydroxide (NaOH) (AR, LobaChemie), hydrochloric acid, sodium acetate (AR, LobaChemie), glacial acetic acid (AR, Labscan), sodium hypochlorite, and oxalic acid.

2.1. Cellulose Extraction

Algae were washed thoroughly to remove any contamination from dirt, plants, and insects. It was either oven, air and sun dried) and weighed to obtain the total amount of dry biomass. The samples were defatted using either methanol and hexane as the solvent, at approximately a 1:7 v/v ratio of biomass and methanol for either 8 hours or 4 days. The defatted biomass was then filtered and subjected to oven drying at 60^0C until constant weight was reached. Consecutively, these samples were treated with (0.1, 0.5 and 1.0 M sodium hydroxide (NaOH) overnight, at 1:7 v/v ratio of defatted biomass and NaOH solution at 60^0C. The alkali treated biomass was filtered and washed until neutrality and oven dried again at 60^0C until constant weight. This was followed by the bleaching process in which the biomass was soaked in an acetate buffer solution (0.1M acetate solution: 0.1M acetic acid solution: 36g sodium hypochlorite) at 60^0C for 3 hours. The bleached biomass was filtered and washed until neutral pH and then oven dried at 60^0C.

2.2. Characterization of Cellulose

2.2.1. Lignocellulosic Profiling of Raw and Cellulose. The lignocellulosic profiling determined the holocellulose using TAPPI T212, hemicellulose by using alpha-, beta- and gamma-cellulose TAPPI T203, alpha-cellulose using TAPPI T429, acid soluble lignin using NREL/TP 510-42617, acid insoluble lignin by subtraction with the acid soluble lignin and alpha-cellulose. total extractives using NREL/TP 510-42617, and ash using TAPPI T211 [6]. The raw and cellulose after the bleaching process were determined for its lignocellulosic profile.

2.2.2. Functional Groups Determination. The Nicolet Nexus 6700 Fourier Transform Infrared Resonance (FTIR) instrument using the KBr pellet method was used for the generation of the FTIR Spectrum, absorbance and transmittance to determine the functional groups present in raw, defatted, alkaline treated, and bleached *Cladophora rupestris*.

2.2.3. Crystallinity Evaluation. The crystallinity was evaluated by computing the lateral order index, the hydrogen bond intensity, cellulose type Iα fraction, and the crystallinity. The lateral order index (LOI) can be used to interpret qualitative changes in cellulose crystallinity and computed using the ratio of absorbance at wavelength 1429 and 893 cm^-. As the LOI decreases, the crystallinity of the material also decreases. Hydrogen bond intensity (HBI) can be used interpret qualitative changes in crystallinity in cellulose and is the ratio of absorbance at wavelength 3336 and 1336 cm^-. As HBI increases, the crystallinity of the material decreases as it represents an increase in hydrogen bonding between certain hydroxyl functions in the cellulose [7]. The cellulose type Iα fraction was determined by using the equation:

IOP Conf. Series: Materials Science and Engineering **1318** (2024) 012002 doi:10.1088/1757-899X/1318/1/012002

$$I_\alpha = \frac{A_{750}}{A_{750} + 0.16A_{710}}$$

The A_{750} and A_{710} are the two absorbances that are the characteristic of Cellulose Iα and Cellulose Iβ, respectively [8].

2.3. Conversion of Cellulose-to-Cellulose Nanofibers (CNF)
The conversion of cellulose-to-cellulose nanofibers was based on the method described by Ramos et al. [9] using an ultrafine grinder. About 21 grams of cellulose were diluted to 3 liters of distilled water and fed in the ultrafine grinder (Masuko Supermass Colloider) with a speed of 1500 rpm for a total of 15 passes in varying clearances. The produced CNFs were filtered and oven dried at 60^0C until constant weight.

2.4. Characterization of Cellulose Nanofibers (CNF)

2.4.1. Average Fiber Diameter. This was assessed using field emission scanning electron microscopy (FE-SEM). The acceleration voltage of the Dual Beam Helios Nanolab 600i was set to 5.0 kV, and the beam current was set to 86 pA.

2.4.2. Crystallinity Index. X-ray diffraction (XRD) analysis was performed to determine the crystalline behavior of cellulose fibers. The samples were prepared as loose powders, and molded into a thin pellet with a pellet press before it was compressed in a glass sample holder. The XRD Program used in Shimadzu LabX XRD-6100 X-ray Diffractometer was θ Range – 5^0 to 50^0C with the rate of 20 per minute. The results were projected using the OriginLab Pro software. Crystallinity index (CI) was calculated as:

$$CI = \frac{Area\ of\ Crystalline\ Peaks}{Total\ Area} * 100$$

2.5. Sulfuric and Oxalic Acid Hydrolysis
The conversion of *cellulose-to-cellulose* nanocrystals relies on the effectiveness of the acid hydrolysis in stripping off the lignin and leaving a greater fraction of cellulose type Iα to attain a high crystallinity index. Attempts were made to extract cellulose nanocrystal by acid hydrolysis using sulfuric acid (H_2SO_4) at 5% (v/v) at 1:50 (w/v) ratio of biomass at 80-100^0C from 30-180 minutes and oxalic acid from 50-70% at 110^0C for 45 minutes. A browning of the material was seen indicating degradation of the cellulose components to glucose. Microwave-assisted acid hydrolysis and semi-batch acid hydrolysis utilizing both sulfuric and oxalic acid to shorten the contact time from 2-10 minutes.

3. Results and Discussion
Between the oven, air and sun drying, a good material as manifested by its color were obtained from the sun drying of *Cladophora rupestris.* Oven and air drying were both done in 3 days with a transition of color from green to brown. Solvent extraction utilizing hexane and methanol. No oil was extracted using hexane. The methanol manifested a green in color on the solvent, indicating an extraction of oil having a loss in mass of 21.75% which may be attributed to the presence of glycerides with affinity to polar solvents. The drying for every step was maintained at 60^0C due to manifestation on browning of the biomass beyond 80^0C. When the material after the sun drying up to alkaline treatment has a brown color, it did not respond well upon bleaching (a transition from green or light brown to white color should be seen). When the sodium hydroxide (NaOH) concentration was varied from 0.1, 0.5 and 1M; the losses at alkaline pretreatment, the losses at alkaline pretreatment plus washing, and the losses at alkaline pretreatment plus washing, plus bleaching were have no significant difference as indicated in the p-values of 0.7607, 0.8973, and 0.8984, respectively. This indicates that the percentage losses in biomass does not vary significantly within the alkaline range of 0.1 up to 1M of NaOH. Though the ones treated

at 1M NaOH had the highest percentage loss, the 0.5M NaOH was considered for economic reason of operation. About 33.46% cellulose yield was attained after bleaching of the macroalga.

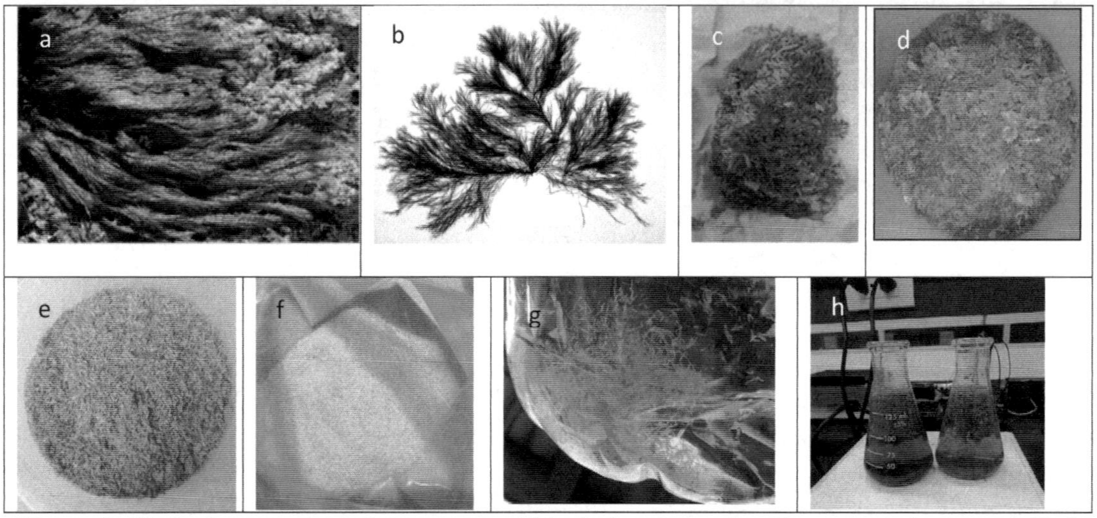

Figure 1. *Cladophora rupestris* from raw to acid hydrolyzed samples. From top to bottom, and left to right. (a) *Cladophora rupestris* at the marine site, (b) *Cladophora rupestris* with few fronds, (c) defatted using hexane, (d) defatted using methanol, (e) alkaline treated, (f) bleached, (g) sulfuric acid hydrolyzed cellulose, (h) oxalic acid hydrolyzed cellulose.

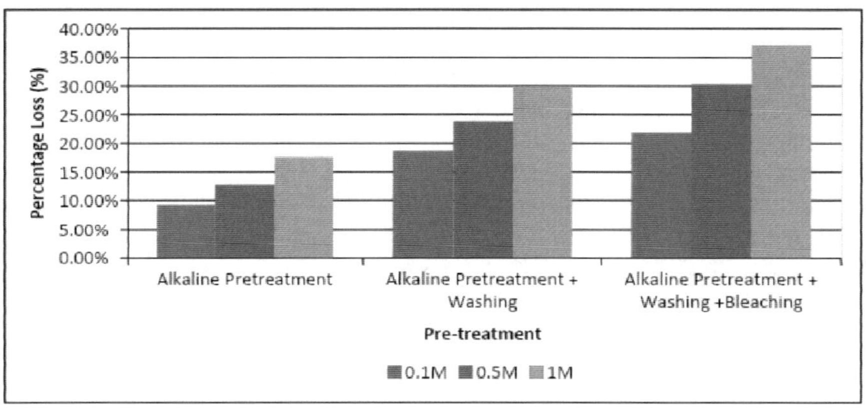

Figure 2. Percentage Weight Loss at every Cellulose Extraction Step and Varying Sodium Hydroxide Concentration

The lignocellulosic profile of *Cladophora rupestris* before and after cellulose extraction based on with defatting, alkaline treatment based on 0.5M NaOH and bleaching was summarized in Table 1.

Table 1. Lignocellulosic Profile of *Cladophora rupestris* Raw and Cellulose

Component	Raw (%)	Cellulose (%)
Alpha Cellulose	33.43	55.32
Hemicellulose	15.48	7.42
Holocellulose	48.93	62.74
Acid Soluble Lignin	2.22	3.28
Extractives	5.67	8.93
Total Ash	39.65	6.61

The *Cladophora rupestris* cellulose was comparably higher to the 20-30% cellulose component for Cladophora sp. [10] and with that of Cladophora glomerata amounting to 21.6% [11]. The alpha cellulose, extractives and total ash has p-values of 0.0279, 0.0215, and 0.0059, respectively implying the significant effect of cellulose extraction. For hemicellulose, holocellulose, acid soluble lignin and acid insoluble lignin, the p-values of 0.0751, 0.1895, 0.1222, and 0.0580, respectively were found indicating no significant effect of cellulose extraction on the mentioned components. This implies the need for acid hydrolysis to attain cellulose Iα that can provide a high crystallinity of the material and attain a nanoparticle at the same time.

Cellulose is a polysaccharide comprised of linear chains containing monomers of D-glucose linked with β(1-4) glycosidic linkages. The straight chains have side hydroxyls that are considered with weak bonds and can easily form bonds with other oxygen atoms within the same or neighboring chains forming a nanofibril [12]. When mixed with other compounds this hydroxyl bonds can readily detach to which functional groups of other compounds may attach. Based from the FTIR spectra, it was found that a common characteristic peak can be seen 710, 750, 1061, 1336, 1372, 1430, 1630, 2900, 3400 cm^{-} for raw, defattened, alkaline treated and bleached *Cladophora rupestris*. The mentioned wavelengths indicate the presence of -OH bonds, water adsorbed in the molecules, -CH bonds, $-CH_2$ bonds, -CO bonds, C-O-C stretch, and glycosidic linkages. The inherent characteristics for cellulose Iα and Iβ were manifested at peaks 710 and 750 cm^{-}, respectively [8]. It can be seen in Figure 3 the FTIR spectra and the characteristic peaks used for the crystallinity evaluation. The same peaks are common for raw, defatted, alkaline treated and bleached *Cladophora rupestris*.

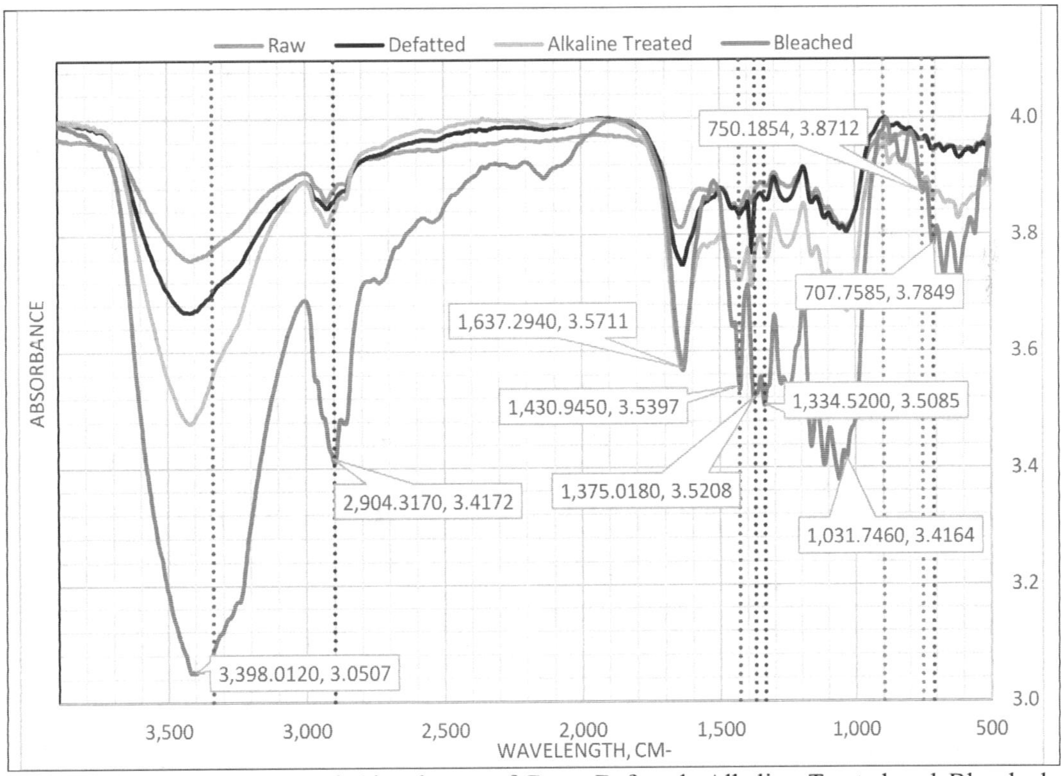

Figure 3. FTIR Spectra and Absorbance of Raw, Defatted, Alkaline Treated and Bleached *Cladophora rupestris*

The crystallinity was evaluated by determining the lateral order index and hydrogen bond intensity as a qualitative manifestation of increase or decrease in crystallinity. The fraction of cellulose Iα was determined indicating a greater presence over cellulose Iβ to qualitatively indicate high crystallinity as contributed by the cellulose Iα. The degree of crystallinity can be estimated using the infrared ratio of FTIR peaks at 1372 and 2900 cm⁻ [13]. The summary of the crystallinity evaluation results can be seen in Table 2. A decreasing LOI was seen as the macroalga undergone the cellulose indicating a decreasing crystallinity. The band towards 1430 cm- has a decreasing effect that indicates the increase in the amounts of cellulose II and amorphous cellulose with lower crystallinity. A decreasing effect was also manifested on the HBI indicating the decrease in hydrogen bonding between hydroxyl functions in the cellulose. Though a decrease was manifested on the HBI and LOI, the total crystallinity index and the cellulose Iα remains the same. This results also support the removal of the total extractives and ash as seen in the lignocellulosic profile of raw and bleached *Cladophora rupestris*

Table 2. Crystallinity Evaluation Summary

Material	Lateral Order Index (LOI) (A_{1429}/A_{A893}) [14]	Hydrogen Bond Intensity (HBI) (A_{3336}/A_{1336}) [15]	Cellulose Iα fractions [8]	Total Crystallinity Index [13]
Raw	0.97	0.97	0.86	1.00
Defatted	0.96	0.96	0.86	1.00
Alkaline Treated	0.94	0.94	0.86	0.98
Bleached	0.89	0.88	0.86	1.03

FE-SEM images (Figure 4a and b) showed that the CNF from Cladophora rupestris had a flexible filament shape conforming to the description of CNF features by Jawaid, M. et al. [16]. The CNF had a fiber diameter of 14.29-37.50 nm (Figure 4c and d). This filament form was caused by the defibrillation of cellulose during ultrafine grinding. The effect of ultrafine grinding with no further treatments to clean the fiber (using acid hydrolysis, washing, centrifugation and or dialysis) and the possible presence of some hemicelluloses which were not stripped off may cause the web-like aggregation of the fibers and sheet-like structures.

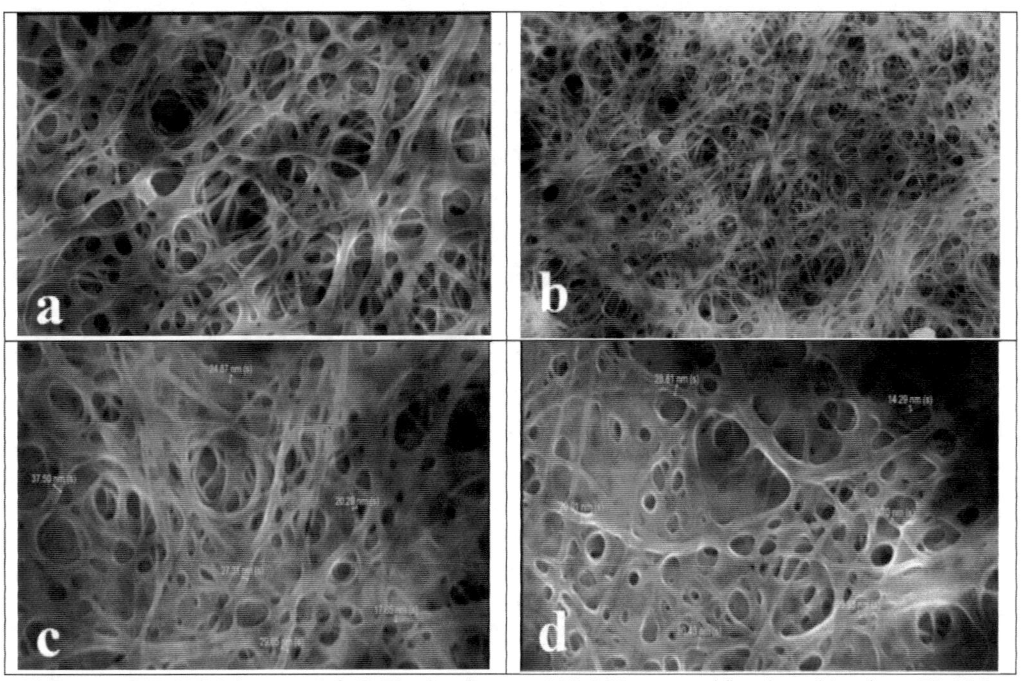

Figure 4. FE-SEM images of CNF, a) taken at 100,000x magnification, b) taken 50,000x magnification, c) containing the maximum value of fiber diameter, and d) containing the minimum value of fiber diameter

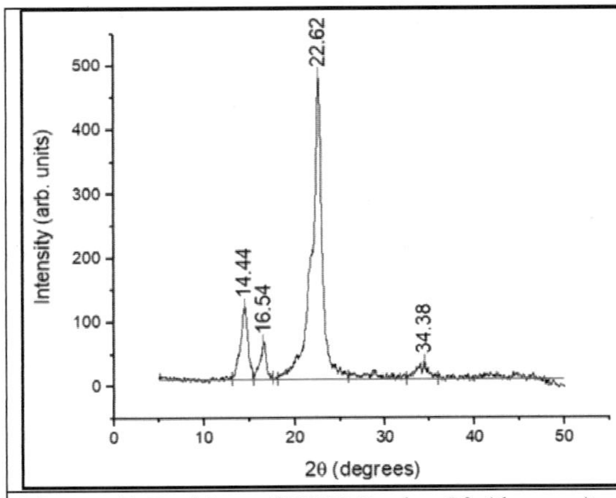

Figure 5. Position of XRD Peaks, 2θ (degrees)

Four peaks located at 14.44°, 16.54°, 22.62°, and 34.38° can be recognized from Figure 5, the first three being very well-resolved and narrow. The results of the individual peak areas calculated using the software at a baseline setting of 10 were 111.25, 48.99, 650.86, and 40, respectively. The overall area of the pattern (integrated from 5 to 50) was recorded to be 920.30. The crystallinity index of the CNF was 92.48% which agrees with the results of Le Bras et al. [17] confirming it to be comparably higher than other sources like wood. This result was comparably low compared with the study of Mihranyan [11] and Sucaldito and Camacho [2]. This implies the need for acid hydrolysis treatment first prior to mechanical defibrillation that can possibly enhanced its crystallinity.

Attempts were made to extract cellulose nanocrystal by acid hydrolysis using sulfuric acid (H_2SO_4) and oxalic acid. A browning of the material and the liquid was seen indicating degradation of the cellulose components to glucose. Microwave assisted acid hydrolysis was also used to shorten the period of acid hydrolysis, but the same was also seen. This may be attributed to the dehydration of the water molecules of the material upon microwave heating leading to cracking of the cellulose molecule to glucose. A semi-batch acid hydrolysis has a good response for cracking the cellulose into a nanocellulose with less browning of the material and the liquid.

4. Conclusion

Cellulose extraction involves the steps: drying, defattening, alkaline treatment, bleaching, and acid hydrolysis. The conditions on each step depends on the nature of the macroalga to be subjected to such treatments and the matrix of its interaction. To come up with a good nanomaterial from *Cladophora rupestris,* acid hydrolysis is necessary to increase its crystallinity index. The conditions for acid hydrolysis and the type of acid to be used will affect the surface functional groups that can affect the agglomeration of the nanomaterials. As much as possible, agglomeration should be broken apart prior to mixing to any other compound to come up with a good nanocomposite intended for certain applications. This characteristic may contribute to the mechanical properties of the composite to which the nanomaterial may be added.

Acknowledgments

The authors would like to extend gratitude to Mr. Mike Ajeros, Chemistry Instrumentation Laboratory of De La Salle University – Manila; and, Dr. Milagroza R. Martinez-Goss, Professor Emeritus at the Institute of Biological Sciences at University of the Philippines Los Baños and Dr. Zenaida M. Agngarayngay from Mariano Marcos State University in identifying the macroalgae collected from Burgos, Ilocos. To all colleagues and former students from Batangas State University namely: Naneth J. Aguado, Caryl Nicole D. Aleta, Matthew D. Culiat, Arvin Dave P. Dalisay, Zoren A. Del Mundo, Paola Angelu A. Untalan, Maricar B. Aguillon, Kiara Lainne J. Olivar, Reiane G. Tolentino, Grace Mary

Louise C. Butuan, Jay Jay B. Calingasan, and Jez Jaycris L. Hernandez; the authors would like to extend gratitude to all of you.

References

[1] Milledge J J, Smith B, Dyer W P, and Harvey P 2014 Macroalgae-derived biofuel: a review of methods of energy extraction from seaweed biomass *Energies* 7 (11) pp 7194-7222

[2] Camacho D H and Sucaldito M R 2017 Characteristics of unique HBr-hydrolyzed cellulose nanocrystals from freshwater green algae (Cladophora rupestris) and its reinforcement in starch-based film *Carbohydrate Polymer* 169 pp 315-323

[3] Abitbol T, Rivkin A, Cao Y, Nevo Y, Abraham E, Ben-Shalom T, Lapidot S, and Shoseyov, O 2016 Nanocellulose, a tiny fiber with huge applications. *Current Opinion in Biotechnology* 39 pp 76–88

[4] Silva, P C, Meñez E G and Moe R L 1987 *Catalog of the Benthic Marine Algae of the Philippines* (*Smithsonian Contributions to the Marine Science)* no 27 (Washington, D.C.: Smithsonian Institution Press) pp 98, 125

[5] Camacho D H, Gerongay S R C, and Macalinao J P C 2013 Cladophora Cellulose-Polyaniline Composite for Remediation of Toxic Chromium (VI) *Cellulose Chemistry and Technology* 47 (1-2) pp 125-132

[6] TAPPI 2023 Developing TAPPI Standards https://www.tappi.org/Get-Involved/Develop-Standards-Methods/develop-standards/develop-standards/.

[7] Kljun A, Benians TAS, Goubet F, Meulewaeter F, Knox JP, and Blackburn RS 2011 Comparative analysis of crystallinity changes in cellulose I polymers using ATR-FTIR, X-ray diffraction, and carbohydrate-binding module probes *Biomacromolecules* 12 pp 4121-4126

[8] Imai T and Sugiyama J 1998 Nanodomains of Iα and Iβ Cellulose in Algal Microfibrils *Macromolecules* 31 pp 6275-6279

[9] Ramos AR, Nacario R, Manalo M, and Razal R (n. d.) Production of Cellulose Nanofibrils (CNFs) ftom Kawayan Kiling (Bambusa Vulgaris Schrad) Pulp; Comparison of the Mechanically-Extracted CNFs From Various Chemically-Treated Pulp.

[10] Mihranyan A 2011 Cellulose from Cladophorales Green Algae: From Environmental Problem to High-Tech Composite Materials *Journal of Applied Polymer Science* 119 pp 2449–2460

[11] Xiang Z, Gao W, Chen L, Lan W, Zhu J Y, Runge T 2016 A comparison of cellulose nanofibrils produced from *Cladophora glomerata* algae and bleached eucalyptus pulp *Cellulose* 23 pp 493-503

[12] Zhou S, Nyholm L, Stromme M, and Wang Z 2019 Cladophora cellulose: unique biopolymer nanofibrils for emerging energy, environmental, and life sciences applications *Accounts of Chemical Research* 52 pp 2232-2243

[13] Nelson ML, O'Connor RT 1964 Relation of certain infrared bands to cellulose crystallinity and crystal lattice type. Part II. A new infrared ratio for estimation of crystallinity in celluloses I and II *Journal of Applied Polymer Science* 8 pp 1325–1341

[14] Hurtubise F G, Krässig H. 1960 Classification of Fine Structural Characteristics in Cellulose by Infared Spectroscopy. Use of Potassium Bromide Pellet Technique *Analytical Chemistry* 32 pp 177−181

[15] Nada A-A M A, Kamel S, El-Sakhawy M 2000 Physicomechanical properties of paper treated with polymers *Polymer Degradation Stability* 70 pp 347−355, *Restautor* pp 238-247

[16] Jawaid M, Mohammed T, and Naheed S 2018 Mechanical and Physical Testing of Biocomposites, Fibre-Reinforced Composites and Hybrid Composites *Woodhead Publishing* pp 49-50

IOP Conf. Series: Materials Science and Engineering **1318** (2024) 012003 doi:10.1088/1757-899X/1318/1/012003

Improvement of biodegradability prediction QSAR system through optimal combination of algorithms

Kazuhiro Takeda [1*], Kensuke Takeuchi [2], Yuki Sakuratani [2], and Kazuhide Kimbara [1]

[1] Department of applied chemistry and biochemical engineering, Shizuoka University, Hamamatsu, Japan
[2] Chemical Management Center, National Institute of Technology and Evaluation, Tokyo, Japan

*takeda.kazuhiro@shizuoka.ac.jp

Abstract. This paper introduces a novel quantitative structure-activity relationship (QSAR) system for predicting the biodegradability of chemicals. The system focuses on evaluating the ratio of biochemical oxygen demand (BOD) to theoretical oxygen demand (ThOD) (%BOD/ThOD) and the presence of degradation products to categorize chemicals as "readily biodegradable" (RB) or "not readily biodegradable" (NB). The conventional laboratory testing method for biodegradability assessment is time-consuming and expensive, thus highlighting the need for a quantitative structure-activity relationship (QSAR) system that can predict biodegradability without the need for extensive testing. The authors acquired a dataset of 3948 chemicals from the National Institute of Technology and Evaluation (NITE) database, which included information on molecular structure, %BOD/ThOD values, and the presence of degradation products. The QSAR system consists of two prediction models: a %BOD/ThOD prediction model and a discrimination prediction model. Multiple algorithms were employed in each model to enhance prediction accuracy. The %BOD/ThOD prediction model utilizes eight algorithms, including linear, nonlinear, and tree-structured models. The discrimination prediction model employs ten algorithms, and the outcomes are combined using majority voting. Explanatory variables for both models include molecular descriptors and spatial structure derived from the chemical's molecular representation. The performance of the QSAR system was evaluated based on training and validation datasets. Results showed that the combination of three %BOD/ThOD prediction algorithms improved the average correction rate compared to using all algorithms. Similarly, selecting the discrimination prediction algorithm further enhanced the average correction rate.

1. Introduction

For the assessment of new chemicals, it is necessary to consider risk management. In Japan, the evaluation of a chemical's biodegradability in the environment plays a crucial role in determining whether it is categorized as "readily biodegradable" (RB) or "not readily biodegradable" (NB) based on the ratio of biochemical oxygen demand (BOD) to theoretical oxygen demand (ThOD) (%BOD/ThOD) and the presence of degradation products. The conventional method for assessing biodegradability involves conducting laboratory tests following the Organization for Economic Co-operation and Development (OECD) [1] test guideline 301 (TG301) series, which can be both costly and time-consuming. Therefore, there is a growing need for a quantitative structure-activity relationship (QSAR) system that can predict biodegradability without the need for laboratory tests. Several QSAR systems have been proposed [2-4]. The authors present a novel QSAR system that includes a %BOD/ThOD prediction model and a discrimination prediction model [5]. The discrimination model categorizes chemicals into three classes: RD, "absence of degradation products for NB" (ANB), and "presence of degradation products for NB" (PNB). Both models utilize various algorithms. This paper focuses on optimizing the combination of these algorithms to achieve higher prediction performance.

2. Experimental data

Biodegradation test data for 3948 chemicals obtained under Chemical Substances Control Law in Japan were used to develop the QSAR models. A portion of the database was accessible through [6]. The dataset includes information on the molecular structure, %BOD/ThOD values, and the presence of degradation products for each chemical. The molecular structure was represented using the Simplified Molecular Input Line Entry System (SMILES) notation [7]. The distribution of %BOD/ThOD values is depicted in figure 1, which reveals that a significant number of chemicals exhibited poor decomposition, with a small peak observed around 80% of %BOD/ThOD. Among the 3948 chemicals, 960 were classified as readily biodegradable(RDs), while 2988 were classified as not readily biodegradable(NBs). In order to maintain the ratio of RDs to NBs, the chemicals were divided into training and untraining-verification sets, as illustrated in table 1.

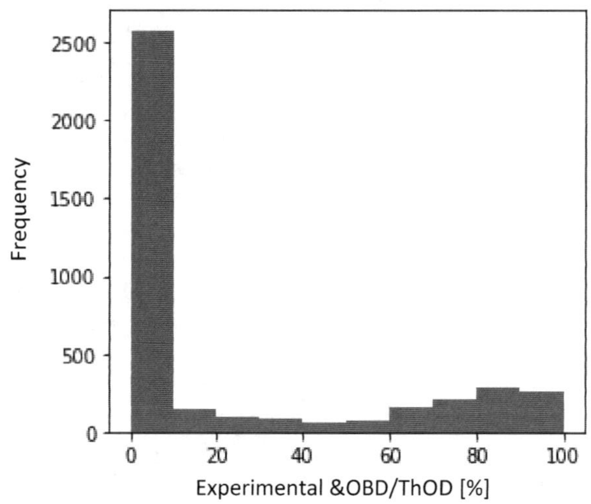

Figure 1. Frequency distribution of experimental %BOD/ThOD.

Table 1. Number of training and untraining-verification chemicals.

Class	Number of total chemicals (%)	Number of training chemicals (%)	Number of untraining-verification chemicals (%)
RD	960 (24.1 %)	768 (24.3 %)	192 (24.3 %)
NB	2988 (75.9 %)	2391 (75.7 %)	597 (75.7 %)
Total	3948 (100 %)	3159 (100 %)	789 (100 %)

IOP Conf. Series: Materials Science and Engineering **1318** (2024) 012003 doi:10.1088/1757-899X/1318/1/012003

3. QSAR system

Figure 2 presents the proposed QSAR system designed for predicting TG301 outcomes taking account of biodegradation products. In the initial step, the selection of chemicals was based on a %BOD/ThOD threshold of 60%. Subsequently, a discrimination prediction model was employed to classify the chemicals as RD, ANB, or PNB. Due to insufficient data for specific subdomains resulting from the %BOD/ThOD division, the discrimination prediction model was trained using all available data, regardless of the %BOD/ThOD values. Consequently, the discrimination prediction model encompassed the endpoints of RD, ANB, and PNB. Notably, the system was found to be not applicable to the domain defined as "%BOD/ThOD >= 60% and ANB," as well as the domain "%BOD/ThOD < 60% and RD."

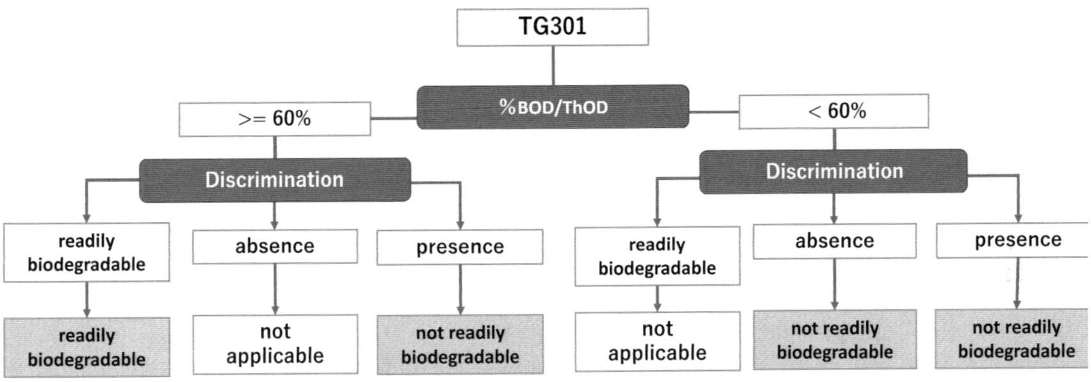

Figure 2. Schematic flow diagram for proposed QSAR system.

3.1. *Explanative variables*

The explanatory variables utilized in each model consisted of molecular descriptors and spatial structure, derived from the SMILES representations of the chemicals. The set of descriptors encompassed 3436 variables, generated using RDkit [8], Mordred [9], and alvaDesc [10] software. The spatial structure was obtained by employing Gaussian PM6 [11] to determine the most stable configuration, which was subsequently represented using voxels. Each voxel consisted of a grid of 64 cubes on each side. The assumption was made that atoms occupied space equivalent to their Van der Waals radius, and the voxel's atomic number was determined based on the occupied cube within it. However, in cases where adjacent atoms shared space, the closer atom took precedence, and the shared region was attributed the atomic number of the closer atom. To facilitate analysis, the atom coordinates were aligned along the principal component identified by PCA, serving as the x-axis direction, and adjusted to ensure a precise fit within the voxel.

3.2. *%BOD/ThOD prediction*

It is not the case that a single algorithm will provide the optimal prediction for all chemicals. Instead, the appropriate algorithm tends to vary depending on the specific chemical being considered. To address this challenge of algorithm selection, a range of algorithms including linear, nonlinear, and tree-structured models are available. By integrating the prediction results from these diverse algorithms, the overall accuracy can be robustly improved. In this study, eight algorithms were utilized, namely linear (Lasso), nonlinear continuous (Support Vector Machine (SVM), NN, Parallel, Serial, and SerialNarrow), and nonlinear discontinuous (eXtreme Gradient Boosting (XGB) and Random Forest (RF)). The NN algorithm solely utilizes descriptors as explanatory variables, while Parallel constructs two networks in parallel—one employing descriptors as explanatory variables, and the other employing molecular spatial structures as explanatory variables. The outputs of these parallel networks are then integrated to obtain the final prediction. Serial, on the other hand, consists of a series of networks—one using descriptors and the other utilizing the spatial structure of the molecule. SerialNarrow aggregates the three-dimensional structure into a single variable, which is subsequently used as an explanatory variable alongside descriptors.

The prediction endpoint in this study is a quantitative %BOD/ThOD value. To train each algorithm, the training data is grouped based on fitness [5], ensuring the models are tailored to the specific characteristics of the data.

3.3. Discrimination prediction

Similar to the prediction of %BOD/ThOD, multiple algorithms were employed and their prediction results were combined using majority voting to enhance stability and accuracy. A total of ten algorithms (XGB, k-Nearest Neighbor (kNN), Gaussian Process (GP), Multi Layer Perceptron (MLP), RF, Support Vector Classification (SVC), NN, Parallel, Serial, and SerialNallow) were utilized, and their outcomes were aggregated to obtain the final decision prediction results. The decision predictions focused on three classes of endpoints, namely RD, PNB, and ANB. To train each algorithm, the training data was grouped based on fitness criteria [5].

3.4. Classification

The final outcome of the %BOD/ThOD prediction and discrimination analysis is categorized into two labels: RD or NB according to the schematic flow diagram in Figure 2.

3.5. Model selection to improve prediction performance

Prediction of %BOD/ThOD and discrimination prediction employ multiple algorithms. In %BOD/ThOD prediction, the highest prediction among multiple algorithms is selected as the predicted value. For discrimination prediction, a majority vote among multiple algorithms is utilized to determine the prediction. While the combination of multiple algorithms aims to enhance robustness, it is important to consider that underperforming algorithms may potentially impact overall performance. To enhance prediction accuracy further, we systematically evaluate the performance of all algorithm combinations and search for the optimal combination.

4. Results and discussions

4.1. %BOD/ThOD prediction

As shown in figure 1, a significant number of training data points had %BOD/ThOD values close to 0. To train the NN, Parallel, Serial, and SerialNarrow algorithms, the data were split into two categories: those with %BOD/ThOD close to 0 and those with other values. An equal number of data points were randomly selected from each category for training purposes. Additionally, measures were taken to prevent overfitting to the training data. The same data was trained using early stopping techniques [12]. Subsequently, the prediction results from eight different algorithms were combined. Integration methods, including minimum, average, and maximum, were examined for comparison. The upper precision (*UP*) and lower precision (*LP*) are defined as follows:

$$UP = \frac{N_a}{N_a+N_c}, LP = \frac{N_d}{N_b+N_d} \quad (1)$$

where N_a, N_b, N_c, and N_d represent the counts of classes *a*, *b*, *c*, and *d* in the confusion matrix (table 2). The integration method that maximizes *LP* involves selecting the maximum value (table 2).

Table 2. Confusion matrix for %BOD/ThOD prediction.

		Predicted %BOD/ThOD	
		≧60%	<60%
Experimental %BOD/ThOD	≧60%	a	b
	<60%	c	d

Table 3. Prediction performance for % BOD/ThOD.

	Training data		Untraining-verification data	
	LP	UP	LP	UP
Max	0.99	0.58	0.96	0.45
Min	0.81	0.97	0.78	0.66
Mean	0.97	0.85	0.9	0.75

4.2. Discrimination prediction

As depicted in the table, the NB dataset contains approximately three times the amount of data compared to the RD dataset. To ensure balanced learning, the algorithms (NN, Parallel, Serial, SerialNallow) learned an equal number of randomly selected samples from both NB and RD. Moreover, to prevent overfitting to the training data, the same data was utilized with early stopping [12]. The evaluation metrics used include accuracy, precision for RD, PNB, and ANB, as well as recall for RD, PNB, and ANB, which are defined as follows:

$$Accuracy = \frac{N_a+N_e+N_i}{N_a+N_b+N_c+N_d+N_e+N_f+N_g+N_h+N_i} \quad (2)$$

$$precision\ of\ RD, PNB, and\ ANB = \frac{N_a}{N_a+N_b+N_c}, \frac{N_e}{N_d+N_e+N_f}, and\ \frac{N_i}{N_g+N_h+N_i} \quad (3)$$

$$recall\ of\ RD, PNB, and\ ANB = \frac{N_a}{N_a+N_d+N_g}, \frac{N_e}{N_b+N_e+N_h}, and\ \frac{N_i}{N_c+N_f+N_i} \quad (4)$$

where N_a, N_b, ..., and N_i are respectively the number of class a, b, ..., and i, respectively, as obtained from the confusion matrix (table 4). The prediction results are presented in the table 5.

Table 4. Confusion matrix for discrimination prediction.

		Predicted discrimination		
		RD	PNB	ANB
Experimental discrimination	RD	a	b	c
	PNB	d	e	f
	ANB	g	h	i

IOP Conf. Series: Materials Science and Engineering **1318** (2024) 012003 doi:10.1088/1757-899X/1318/1/012003

Table 5. Prediction performance for discrimination prediction.

	Accuracy	Precision			Recall		
		RD	PNB	ANB	RD	PNB	ANB
Training data	0.87	0.86	0.95	0.86	0.9	0.59	0.96
Untraining-verification data	0.76	0.69	0.76	0.79	0.8	0.39	0.88

4.3. classification prediction

The classification prediction results based on %BOD/ThOD prediction and discrimination are presented in tables 1 and 2. The prediction results includes 231 chemicals with %BOD/ThOD values greater than or equal to 60% and ANB. While this classification is not directly applicable, it comprises only approximately 10% of RD (readily biodegradable) chemicals, which is a minor practical concern. Only 27 chemicals were classified as RD with %BOD/ThOD below 60%. Further examination of these non-applicable chemicals individually is necessary in future studies. The analysis of these chemicals classified as non-applicable reveals that more than 90% of the chemicals in the training data and over 75% of the chemicals in the untrained data were suitable for their respective classifications.

Table 6. Classification prediction results for training data.

		Predicted class					
		%BOD/ThOD≧60			%BOD/ThOD<60		
		RD	ANB	PNB	RD	ANB	PNB
Experimental discrimination	ANB	48	153	3	8	1518	8
	PNB	57	50	192	4	168	192
	RD	678	38	9	15	28	0
	Total	783	231	204	27	1714	200
	Class	C1	C2	C3	C4	C5	C6

Table 7. Classification prediction results for untraining-verification data.

		Predicted class					
		%BOD/ThOD≧60			%BOD/ThOD<60		
		RD	ANB	PNB	RD	ANB	PNB
Experimental discrimination	ANB	30	64	6	9	315	5
	PNB	30	27	38	1	45	27
	RD	149	23	8	4	7	1
	Total	209	114	52	14	367	33
	Class	C1	C2	C3	C4	C5	C6

4.4. model selection results

For the prediction of %BOD/ThOD and discrimination, a diverse set of algorithms was employed. The maximum value was used to integrate the results for %BOD/ThOD prediction, while majority vote was used for discrimination prediction. These approaches yielded highly accurate prediction results. The selection of algorithms was further carried out to enhance accuracy. In the table, the classes are denoted as C1, C2, ..., C6 from left to right, with C1 being the favorable decomposition class and C3, C5, and C6 being the challenging decomposition classes. The correction rate for readily biodegradablechemicals (C_{RD}) and not readily biodegradablechemicals (C_{NB}) were defined by equation 5, respectively. The objective for the algorithm selection was to maximize the mean values of C_{RD} and C_{NB}.

$$C_{RD} = \frac{N_1}{N_{RD}}, C_{ND} = \frac{N_{356}}{N_{NB}} \tag{5}$$

where N_1 represents the number of chemicals in class C1, N_{356} represents the number of chemicals in classes C3, C5, and C6, N_{RD} represents the number of readily biodegradablechemicals, and N_{ND} represents the number of not readily biodegradablechemicals. The performance of all algorithm combinations was evaluated, and the results of the search for the optimal combination are presented in the table 8. By selecting three %BOD/ThOD prediction algorithms (XGB, RF, and Parallel), the average correct response rate improved to 0.903 compared to 0.890 when all algorithms were used. Moreover, by employing only XGB for discrimination prediction, the average correct response rate further improved to 0.952. When validated on untrained data, the average correct response rate improved to 0.761 by selecting the %BOD/ThOD prediction algorithm, compared to 0.757 with no selection, and it was further enhanced to 0.785 by selecting the discrimination prediction algorithm. Different combinations of algorithms can be selected based on various objective functions, providing flexibility according to the user's specific goals.

Table 8. Classification prediction results by selection optimization.

Used model		Balanced accuracy	
%BOD/ThOD prediction	Discrimination prediction	Training data	Untraining-verification data
All	All	0.890	0.757
XGB, RF, Parallel	All	0.903	0.761
XGB, RF, Parallel	XGB	0.952	0.785

5. Conclusions

In this study, a novel QSAR system was developed to predict the biodegradability of chemicals without the need for time-consuming and costly laboratory tests. The QSAR system included a %BOD/ThOD prediction model and a discrimination prediction model. The %BOD/ThOD prediction model utilized eight different algorithms, including linear, nonlinear continuous, and nonlinear discontinuous models, to optimize the prediction performance. The discrimination prediction model employed ten algorithms and employed a majority voting approach to classify chemicals into three categories: RD, ANB, and PNB.

The QSAR system was evaluated using a dataset of 3948 chemicals, and the results demonstrated promising prediction accuracy. The %BOD/ThOD prediction model achieved high precision by integrating the results from multiple algorithms, while the discrimination prediction model accurately classified chemicals into the appropriate categories using a majority voting approach.

Furthermore, the study explored the impact of algorithm selection on prediction performance. By systematically evaluating different algorithm combinations, the researchers identified optimal combinations that improved the overall prediction accuracy. The selection of three %BOD/ThOD prediction algorithms (XGB, RF, and Parallel) and one discrimination prediction algorithm (XGB) yielded the highest correct response rates, both on the trained and untrained data.

The findings of this study contribute to the development of a robust QSAR system for the prediction of biodegradability of chemicals. The optimized combination of algorithms enhances the accuracy and reliability of the predictions, reducing the need for extensive laboratory testing and supporting efficient risk management in the assessment of new chemicals. Future studies can build upon these results and further refine the QSAR system, considering additional factors and expanding the dataset to improve its applicability and generalizability.

6. References

[1] OECD (1992) http://www.oecd.org/chemicalsafety/risk-assessment/1948209.pdf
[2] J. Tunkel, P. H. Howard, R. S. Boethling, W. Stiteler, and H. Loonen, Predicting ready biodegradability in the Japanese ministry of international trade and industry test, Environ. Toxicol. Chem. 19, 10 (2000) pp.2478 -2485
[3] N. H. Dimitrova, I. A. Dermen, N. D. Todorova, K. G. Vasilev, S. D. Dimitrov, O. G. Mekenyan, Y. Ikenaga, T. Aoyagi, Y. Zaitsu and C. Hamaguchi, CATALOGIC 301C model - validation and improvement, SAR and QSAR in Environmental Research, 28, 6 (2017) pp.511-524, DOI: 10.1080/1062936X.2017.1343255
[4] K. Huang and H. Zhang, Classification and regression machine learning models for predicting aerobic ready and inhent biodegradation of organic chemicals in water, Environ. Sci. Technol., 56 (2022) pp.12755-12764
[5] K. Takeda, K. Takeuchi, Y. Sakuratani, and K. Kimbara, Development of a decision support system for biodegradation using chemical descriptors and 3D structures, SCEJ 88th Annual Meeting, Tokyo (2023)
[6] https://www.nite.go.jp/chem/jcheck/search.action?request_locale=en
[7] D. Weininger, SMILES, a chemical language and information system. 1. Introduction to methodology and encoding rules, Journal of Chemical Information and Computer Sciences, 28 (1988) pp.31-36.
[8] G. Landrum, P. Tosco, B. Kelley, S. Riniker, P. Gedeck, NadineSchneider, R. Vianello, A. Dalke, R. R. Schmidt, B. Cole, A. Savelyev, S. Turk, M. Swain, A. Vaucher, D. Nealschneider, M. Wojcikowski, A. Pahl, J.-P. Ebejer, F. Berenger, A. Stretton, N. O'Boyle, D. Cosgrove, P. Fuller, J. H. Jensen, G. Sforna, K. Leswing, S. Leung, and J. vanSanten, rdkit/rdkit: 2019 03 4 (Q1 2019) Release (2019); https://doi.org/10.5281/zenodo.3366468.
[9] Moriwaki, H., Tian, YS., Kawashita, N. et al. Mordred: a molecular descriptor calculator. J Cheminform 10, 4 (2018). https://doi.org/10.1186/s13321-018-0258-y
[10] Alvascience, alvaDesc (software for molecular descriptors calculation) version 2.0.12, (2022) https://www.alvascience.com
[11] Gaussian 16, Revision A.03, Frisch, M. J., et. al., Gaussian, Inc., Wallingford CT (2016).
[12] L. Prechelt, Early Stopping-But When?, Neural Networks (1996).

Acknowledgements

This study was conducted under the research project, "Research on the Introduction of Wight of Evidence Approach to the Evaluations of Biodegradation and Bioaccumulation of Chemicals (2022)", funded by Ministry of Economy, Trade and Industry.

IOP Conf. Series: Materials Science and Engineering **1318** (2024) 012004 doi:10.1088/1757-899X/1318/1/012004

Reflux extraction with saponification of crude wax from sugarcane filter cake for policosanol production

JMS Hernandez[1], JL Movillon[2], JA Capunitan[3], RE Arocena[4] and MG Borines[5]
[1] Chemical Engineer, [2] Professor Emeritus, [3]Associate Professor, [4]Assistant Professor, [5] Professor
Department of Chemical Engineering, CEAT, University of the Philippines Los Baños, College, Laguna, Philippines

E-mail: mgborines@up.edu.ph

Abstract. Filter cake, a by-product in the sugar industry, has significant amount of wax containing policosanol which can be extracted using proper method and solvent. Policosanol is a mixture of high molecular weight aliphatic primary alcohols and was reported to reduce cardiovascular diseases. This study explored the use of hot ethanol reflux extraction to recover the crude wax from the filter cake followed by saponification for policosanol purification. Crude wax yield of 84.5 mg crude wax/ g filter cake was obtained using reflux extraction at 60 ° C , 500 rpm and 12 mL absolute ethanol /g filter cake. The maximum crude wax yield obtained from Soxhlet extraction was 121.88 ± 0.46 mg crude wax/ g dry filter cake, which corresponds to 69.33% recovery. Using a two-level factorial design, the effects of sodium hydroxide (NaOH) concentration (0.1 M and 1 M), reaction time (30 minutes and 60 minutes) and reaction temperature (30 °C and 60 °C) on the policosanol yield during saponification were determined. Results showed that NaOH concentration and temperature have significant effects on policosanol yield while reaction time alone is insignificant. Increasing both the reaction time and NaOH concentration can lead to an increase in policosanol yield. The highest yield of 1.234 ± 0.322 mg policosanol/g crude wax after saponification was obtained at 60 °C for 60 minutes with 1 M NaOH. This is twice the amount of policosanol recovered as compared to 0.583 ± 0.026 mg policosanol/ g crude wax without saponification.

1. Introduction

About 65-70% of the world's sugar production is sourced from sugarcane with the Philippines as one of the main sugar producers [1]. The Philippines produced 1,609,861 metric tons of raw sugar and 608,039 metric tons of refined sugar during the crop year 2021-2022 [2].

Aside from sugar, the sugar industry also produces economically important by-products such as bagasse, molasses and filter cake. It has been reported that out of the sugarcane harvested, 82.4 % reaches a factory, 9.4% of it is sugarcane waste left in the field while the rest is sugarcane waste separated from the cleaning process. From the sugarcane sent to the factories approximately 13% is turned into sugar, 52% resulted to liquid waste, 28% to bagasse, 3% to molasses and 4% to filter cake [1]. With these values, the Philippine sugar industry generates significant amounts of by-products that have great potential if valorized. In the Philippines, molasses is currently used as the feedstock for bioethanol production while bagasse is used as fuel for power generation of the factory. Filter cake, on the other hand, is used mainly for soil amelioration and is given for free to the farmers.

On a wet basis, there are 1-7 kilograms of the filter cake produced for every 100 kilograms of sugar cane [3]. With the Philippines processing 18,524,684 metric tons of sugar cane in the crop year 2021-

2022, on the average, 730,187 metric tons of filter cake are produced [2]. This large amount of filter cake generated in the process poses a challenge in its disposal.

Some studies showed that filter cake has a considerable amount of a medically beneficial compound. Filter cake contains 10-14% of wax which significantly contains long-chain alcohols consisting of 18 carbon atoms or more called policosanol [4]. Policosanol is mainly composed of octacosanol, constituting 66-67% of the wax component while triacosanol and hexacosanol account for 12-14% and 7-8% of the wax, respectively [5, 6]. Policosanol was shown to lower total cholesterol and low- density lipoprotein (LDL), and also increased high-density lipoprotein (HDL), thereby reducing the risk for cardiovascular diseases [7]. Researchers from Cuba have found that 5 – 20 mg daily of policosanol was effective at improving serum lipid profiles [8]. Policosanol can reduce LDL cholesterol and increase HDL through inhibiting cholesterol synthesis and increased processing of LDL. Today, policosanol is used as nutraceuticals and as a food supplement [9].

Due to the health benefits of policosanol, the rise in its popularity has resulted in a demand for its sources other than beeswax, rice bran and flax straws. Thus, the recovery of policosanol from filter cake has potential commercial value in the health food industry. The utilization of filter cake as a source for policosanol could also reduce the amount of unused filter cake significantly. The development of efficient methods to extract and purify policosanol from filter cake has also drawn the attention of researchers.

Hot reflux extraction is a commonly used method in different policosanol extraction experiments. A study conducted on the optimization of hot reflux solvent extraction of crude wax using ethanol from sugarcane filter cake for policosanol production reported a recovery of 7.98% crude wax by weight from filter cake [10]. The study also showed that a high solvent-to-solid ratio, high extraction time, and moderate extraction temperature resulted in higher amount of crude wax yield.

To improve policosanol yield after extraction, further purification of the wax is needed. This is commonly done by physical techniques such as filtration, centrifugation or by chemical-aided technique such as saponification. Saponification, which involves the hydration reaction between metallic alkali (base, commonly KOH, NaOH and CaOH) and a fat or oil to form salts of fatty acid, commonly known as soap, and long-chain organic compounds such as aliphatic alcohols. It has been used for purification of waxes [11]. In the saponification of plant waxes, the soap is not the only product of interest but also the organic compound, or the unsaponified product, because of its nutritive substance content. In the case of sugarcane wax, it is the policosanol. Thus, this study aimed to explore the use of hot ethanol reflux extraction to recover the crude wax from the filter cake followed by saponification for policosanol purification. The effects of the saponification parameters such as NaOH concentration, temperature and reaction time on policosanol yield were explored in this study.

2. Materials and Methods

2.1 Materials and reagents

Filter cake was provided by the Universal Robina Corporation – Sugar and Renewables (Balayan, Batangas, Philippines). The filter cake was sun dried for two (2) days and then oven-dried at 105 °C for 6 hours, grounded and screened in a 20-mesh sieve.

Octacosanol (97.1%) was purchased from Sigma (St. Louis, MO) and was used as standard without further purification. All other chemical reagents were of analytical or HPLC grade.

2.2 Hot ethanol reflux extraction

Fifty grams of dried filter cake was suspended in a preheated 600 ml absolute ethanol in a 1-L flat bottom flask and was refluxed at 60 °C at 500 rpm for 1.64 hrs. After extraction, the mixtures were filtered using Whatman No. 1 filter paper. The filtrate was then transferred to centrifuge tubes and cooled at 4 °C. After cooling, the filtrate was centrifuged at 3000 rpm for 10 minutes and decanted to collect the sediments. It was then airdried to evaporate the remaining ethanol. The amount of extracted

IOP Conf. Series: Materials Science and Engineering **1318** (2024) 012004 doi:10.1088/1757-899X/1318/1/012004

crude wax was determined gravimetrically.

2.3 Soxhlet extraction

The maximum amount of crude wax that can be extracted from the filter cake was determined using Soxhlet extraction. Twenty grams of dried filter cake was packed in filter paper and was put in the Soxhlet chamber. A solvent-to-solid ratio of 12 mL absolute ethanol/ g dried filter cake was used in the study. The absolute ethanol was put in the flat bottom flask and was put in the hot plate operating at 80 °C for 24 hours. After extraction, the amount of crude wax was determined using the same method used in the hot ethanol reflux extraction method.

2.4 Purification of extracted crude wax

The extracted wax was subjected to saponification process based on the study on saponification at varying sodium hydroxide concentrations (0.1 M or 1 M), reaction temperature (30 °C or 60 ° C) and reaction time (30minutes or 60 minutes) [12]. The crude wax was allowed to react with sodium hydroxide and heated at the specified reaction temperature and time. After the reaction, the mixture was cooled at room temperature and separated using a simple filtration set up to collect the wax. The amount of wax was determined gravimetrically after drying. The octacosanol content of the wax was determined using Gas Chromatography.

2.5 Experimental Design

The study used a two-level factorial design with 3 factors having varying levels. The manipulated variables were NaOH concentration, reaction time and reaction temperature with policosanol yield as the response variable. Table 1 shows the high and low values of the parameters used in the study. A total of 16 experimental runs were conducted with two(2) replicates. The effects and interactions of each factor to the response variable were analyzed using the half- normal plot while Analysis of Variance (ANOVA) was used to further verify the results. The statistical model was evaluated at a level of significance of α=0.05.

Table 1. High and low values of parameters for the extraction of policosanol from filter cake.

Factor	Symbol	Level	
		Low (-1)	High (+1)
NAOH Concentration (M)	A	0.1	1
Reaction temperature (°C)	B	30	60
Reaction time (hrs)	C	0.5	1

2.6 Determination of octacosanol contents of extracted wax

Octacosanol ($C_{28}H_{58}O$) in the extracted wax was analyzed in a Shimadzu Gas Chromatograph using a calibration curve of octacosanol standard solutions with concentrations varying in the range of 0.2 to 0.8 g octacosanol /Li of hexane. About 0.01 grams of purified wax was dissolved in hexane and was filtered. Ten µL of each sample were accurately measured and injected to the Gas Chromatograph. Helium was used as the carrier gas at a flow rate of 1.3mL/min. The temperature of the injector was at 300°C. The oven condition was 50°C for one minute, which was then increased by 10°C/ min up to 250 °C and held for four minutes. The policosanol concentrations of the samples were determined from the measurements made in the GC. The yield of policosanol extracted from the filter cake was then computed and used as response variable in the statistical analysis of the data.

3. Results and Discussion

3.1 Crude wax extraction

The average amounts of crude wax extracted from sugarcane filter cake using Soxhlet extraction and hot reflux extraction with ethanol as solvent were found to be 121.875 ± 0.460 and 84.5 mg crude wax/ g filter cake, respectively. This corresponds to a 69.33% recovery of crude wax by ethanol reflux extraction leaving a significant amount of unextracted crude wax in the filter cake.

3.2 Effects of NaOH concentration, temperature, and reaction time on policosanol yield

The policosanol yields after subjecting the crude wax extracted from hot ethanol reflux extraction to saponification at different conditions are shown in Table 2.

Table 2. Policosanol amount in crude wax after saponification.

Parameters			Average policosanol amount* (mg/ g crude wax)
NaOH Conc. (M)	Reaction Temp. (°C)	Reaction Time (minutes)	
0.1	30	30	0.377 ± 0.107
0.1	30	60	0.523 ± 0.322
0.1	60	30	0.579 ± 0.066
0.1	60	60	0.576 ± 0.236
1	30	30	0.657 ± 0.076
1	30	60	0.838 ± 0.239
1	60	30	0.879 ± 0.019
1	60	60	1.234 ± 0.322

**Average of two(2) trials*

The lowest amount of policosanol produced after purification was computed to be 0.377 ± 0.107 mg policosanol/ g crude wax with saponification conditions of 0.1 M NaOH for 30 minutes at 30 °C while the highest amount of 1.234 ± 0.322 mg policosanol/ g crude wax was obtained at saponification conditions of 1 M NaOH for 60 minutes at 60 °C.

With the same NaOH concentration and reaction temperature, increasing the reaction time increases the amount of policosanol in the sample. This was observed for saponification conditions of 1 M NaOH both at 30 °C and 60 °C and at 0.1 M NaOH and 30 °C. However, at 0.1 M NaOH and 60 °C , there was a slight decrease in the policosanol content upon increasing the reaction time. In general, it was observed that increasing the reaction time would lead to an increase in policosanol extracted. During saponification, longer contact time between NaOH and crude wax would allow more esters in the crude wax to react with NaOH splitting into a carboxylate ion and an alcohol [13].

Increasing the concentration of NaOH from 0.1 M to 1 M also increased the amount of policosanol recovered at both levels of temperature and reaction time. This can be explained by the collision theory where an increase in the concentration of a reactant would increase the number of collisions between reacting species and in turn increases the reaction rate [14].

The amount of policosanol also increased with increasing reaction temperature at both levels of NaOH concentration and reaction time. The temperature of a substance is proportional to the kinetic energy of the particles in the reaction. An increase in the kinetic energy of the particles would lead to an increase in the intensity and frequency of the collision thus increasing the reaction rate of crude wax and NaOH during saponification.

IOP Conf. Series: Materials Science and Engineering **1318** (2024) 012004 doi:10.1088/1757-899X/1318/1/012004

The effect of saponification was determined by comparing the policosanol yields before and after saponification. The percent difference in policosanol yields was calculated and shown in Table 3. An increase in policosanol yield was only observed for 1 M NaOH at both levels of reaction time and temperature but not in lower NaOH concentration. A similar observation was also reported by some researchers [15]. The highest yield of 1.234 ± 0.322 mg policosanol/ g crude wax after saponification was obtained at 60 °C for 60 minutes with 1 M NaOH. This is twice the amount of policosanol recovered as compared to 0.583 ± 0.026 mg policosanol/ g crude wax without saponification.

The half- normal plot for policosanol extraction using hot ethanol reflux method (Figure 1) shows that NaOH concentration (A) and reaction temperature (B) showed significant effects on policosanol yield but not the interactions of these parameters. At 95% level of confidence, p-values of the NaOH concentration and temperature are less than 0.0500 indicating that these factors are significant while the reaction time is not significant to the model and shows little effect on the response.

Table 3. Percent difference of policosanol yield before and after saponification.

Parameters			
NaOH Conc. (M)	Reaction Temp. (°C)	Reaction Time (minutes)	Percent difference (%)
0.1	30	30	-35.335
0.1	30	60	-10.292
0.1	60	30	-0.686
0.1	60	60	-1.201
1	30	30	12.693
1	30	60	43.739
1	60	30	50.858
1	60	60	111.578

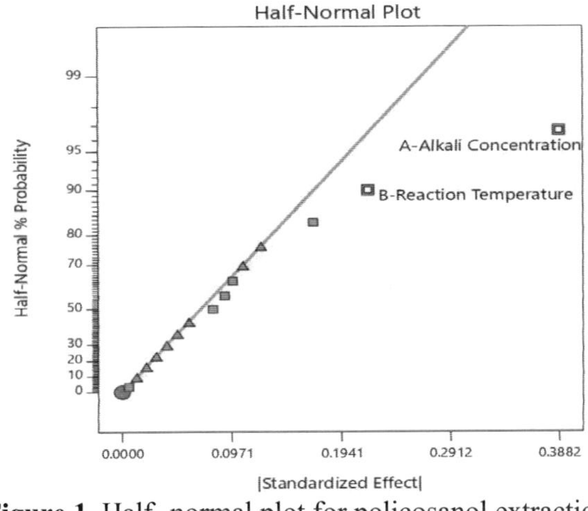

Figure 1. Half- normal plot for policosanol extraction using reflux method.

IOP Conf. Series: Materials Science and Engineering **1318** (2024) 012004 doi:10.1088/1757-899X/1318/1/012004

3.3 Effect of Alkali Concentration

The concentrations of policosanol after saponification with 0.1 M and 1 NaOH at 60°C and 60 minutes are presented in Figure 2. Increasing the concentration of the alkali will yield a higher amount of the product but this will only hold true up to the optimal concentration wherein the amount of product will be on its maximum possible value [11]. Saponification at 60 minutes and 60 °C with 0.1 M and 1 M NaOH yielded an average of 0.576 and 1.234 mg policosanol/ g crude wax, respectively. By increasing the concentration of the alkali, the purity of the policosanol in the crude wax also increased. This is due to the higher amount of hydroxide ions available that could hydrolyze the esters and fatty acids found in the wax. A patent that allowed production of higher primary aliphatic alcohols by saponifying and extracting sugarcane wax, reported that the concentration of the hydroxide ions must be such that the ratio in weight of the alkali with the wax must be above 1.05 [16]. This is to ensure that there is enough alkali that will react with the wax. The saponification process must remain for a period of at least 30 minutes with constant stirring. Increasing the reaction time further may result in minimal gains. Increasing the NaOH concentration also increased the amount of hydroxide ions available that will bond with the ester to form carboxylate ions and alcohols. This increased the amount of policosanol available in the wax.

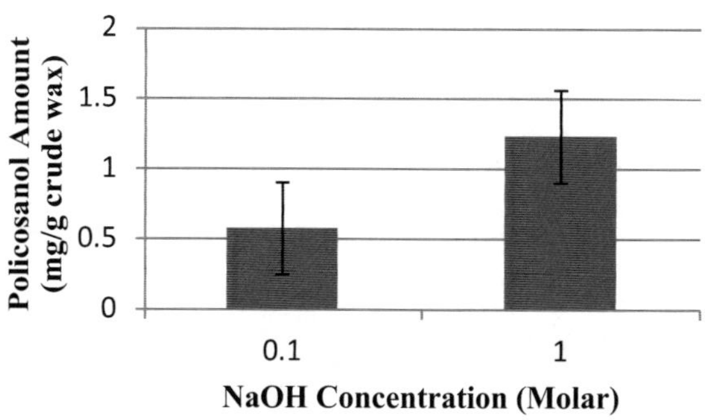

Figure 2. Policosanol amount after saponification with varying NaOH concentration at 60 °C for 60 minutes.

3.4 Effect of reaction temperature

The reaction temperature was found to have a significant effect on policosanol yield (Figure 3). Upon saponification at 30 °C and 60 °C using 1 M Sodium hydroxide for 60 minutes, the amount of policosanol was recorded at 0.838 and 1.234 mg policosanol/ g crude wax, respectively. By increasing the reaction temperature, the yield of the policosanol in the crude wax also increased. Higher temperatures will increase the efficiency of the purification process since the particles collide with each other at a higher rate [17].

3.5 Effect of reaction time

The amounts of policosanol extracted at varying reaction time with 1 M Sodium hydroxide at 60°C are shown in Figure 4. Based on the results, the effect of the reaction time to the response variable is relatively small and not significant to the amount of policosanol. A patent reported that after an hour of saponification, the recovery of sugarcane wax reaches its peak [16].

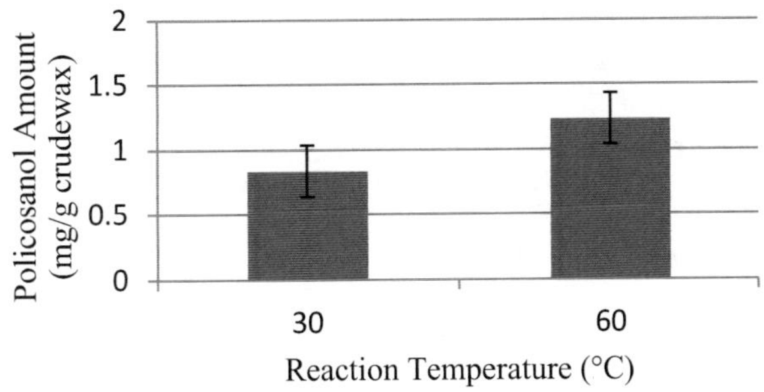

Figure 3. Policosanol amount after saponification with varying reaction temperature with 1M sodium hydroxide for 60 minutes.

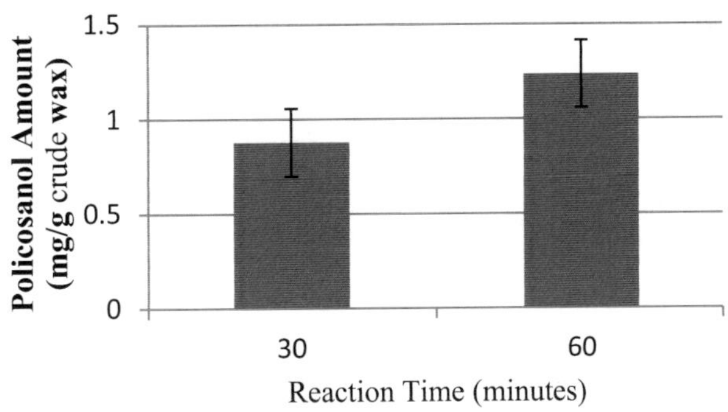

Figure 4. Policosanol amount after saponification with varying reaction time with 1M Sodium hydroxide at 60°C.

3.6 Effect of significant interactions of main factors

Increasing both the NaOH concentration and reaction temperature increased the amount of policosanol (Figure 5). The interaction between the two showed a positive increase in the amount of policosanol. The two parameters were not interfering with the effect of the other. While increasing the reaction time had a positive effect on the response, the effect was small to consider it as significant to the mathematical model.

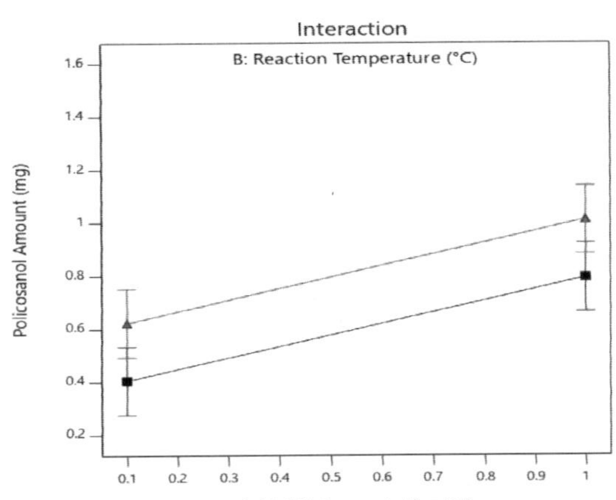

Figure 5. Effect of alkali concentration- reaction temperature interaction on the policosanol amount after saponification.

4. Conclusion

This study showed that hot ethanol reflux extraction with saponification was able to recover twice the amount of policosanol compared with hot reflux extraction alone. However, the crude wax recovery obtained in this study was only 69.33 %, leaving a significant amount of unextracted crude wax in the filter cake. Results showed that NaOH concentration and temperature have significant effects on policosanol yield while reaction time is insignificant. Increasing both the reaction time and NaOH concentration can lead to an increase in policosanol yield. While saponification as purification method increased the policosanol yield, extraction of crude wax from the filter cake can still be improved.

5. References

[1] Makul N and Sua-iam G 2016 Effect of incinerated sugarcane filter cake on the properties of self-healing concrete *Construction and Building Materials* pp 32-40
[2] Sugar Regulatory Administration 2022 Philippine Sugar Statistics: Crop Year 2021-2022
[3] Chopra A K and Kumar V 2016 Effects of sugarcane press mud on agronomical characteristics of hybrid cultivar of eggplant (*Solanum melongena* L.) under field conditions *International Journal of Recycling of Organic Waste in Agriculture* **5**(2) 149–162
[4] Albarelli J Q *et al.* 2018 Thermo-economic evaluation of a new approach to extract sugarcane wax integrated to a first and second generation biorefinery *Biomass and Bioenergy* **119** pp 69–7
[5] Keithley J and Swanson B 2009 Policosanol to manage dyslipidemia in older adults. *Complementary and Alternative Therapies and the Aging Population* pp 117–134
[6] Lockwood G B 2009 The plant nutraceuticals *Trease and Evans' Pharmacognosy* pp 459–470
[7] Berthold I G and Berthold H K 2002 Policosanol: Clinical pharmacology and therapeutic significance of a new lipid-lowering agent *American Heart Journal* **143** (2) pp 356-365
[8] Janikula M 2002 Policosanol: a new treatment for cardiovascular disease? *Alternate Medicine Review* pp 203-217
[9] Srisaipet A P 2017 Policosanol extraction from beeswax and improvement of purity *FluidsChE*
[10] Bagabay R 2016 Parametric and optimizaiton studies on crude wax extraction from sugarcane filtercake using ethanol for policosanol production Undergraduate thesis University of the Philippines Los Baños
[11] Salimon J *et al.* 2012 Saponification of *Jatropha curcas* seed oil: Optimization by D-Optimal design *International Journal of Chemical Engineering* **2012** Article ID 574780 p 6
[12] Huang K and Jia K 2004 Numerical analysis of microwave heating on saponification reaction. *In Wave Propagation, Scattering and Emission in Complex Media* pp 215-222
[13] Mcmurry J 2012 Organic Chemistry (8th ed.) Singapore: Cengage Learning Asia Pte
[14] Zumdahl S and Zumdahl S 2010 Chemistry (8th ed.) Singapore: Cengage Learning Asia Pte
[15] Ou S *et al.* 2012 Preparation of octacosanol from filter mud produced after sugar cane juice clarification China: Elsevier Scientific Publishing
[16] Granja A H 1993 Mixture of higher primary aliphatic alcohols, its obtention from sugarcane wax and its pharmaceutical uses. Patent No. US5663156A Cuba
[17] Bertholet R 1991 Obtaining compounds from wax saponification Patent No. US5159124A. Vevey Switzerland

IOP Conf. Series: Materials Science and Engineering **1318** (2024) 012005 doi:10.1088/1757-899X/1318/1/012005

Fabrication and experimental investigations on mechanical properties of stir cast aluminum MMC with the addition of ash

Pankaj Kumar Gupta[1], Mayank Verma[2], Tapas Bajpai[3]

[1, 2, 3] Malaviya National Institute of Technology, Jaipur, (Raj.)-302017-INDIA

E-mail: pankaj.mech@mnit.ac.in

Abstract. Aluminium alloys are lightweight and high strength materials which are extensively used in engineering applications. Metal matrix composites (MMCs) based on aluminium have significantly increased in the field of newer materials for high-performance mechanical uses. Aluminium metal matrix composites (Al-MMCs) are more involved in mining, aerospace, marine, and automobile applications. Aluminium liquid casting is typically used to treat metal matrix for composites. Such casting is carried out using the liquid casting method. In this article, using the stir casting technique, an effort has been made to create a metal matrix composite by choosing aluminium (Al6061) as matrix material and silicon carbide, cow dung and buffalo dung ash as reinforcement. This ash is biodegradable in nature. Four types of samples were prepared from this composite for all the tests which were named S1, S2, S3, and S4, with Al6061 as 100%, 84%, 82%, and 80%, respectively. Four mechanical properties, i.e., micro hardness, density, tensile strength and compressive strength, were investigated and the effect of hybrid ash was studied. Maximum density, microhardness, tensile strength and compressive strength were achieved for sample S2 due to the highest percentage of silicon carbide and the lowest percentage of hybrid ash. Addition of ash proved good alternative as a sustainable reinforcement.

Keywords: Stir casting, aluminium (Al6061), silicon carbide, Cow dung ash, buffalo dung ash,

1. Introduction

Materials made up of two or more chemically distinct elements separated by a different interface on a macro scale are known as composites. Metal matrix composites (MMCs) offer massively increased features such as high strength and toughness, high modulus, damping capability, and strong wear resistance when related to unreinforced alloys. [1] Aluminium metal matrix composites (Al-MMCs) has a high weight-to-strength ratio, extreme rigidity, low wear rate and minor abrasion. Excellent thermal conductivity, high creep corrosion-resistant, high stability in dimensions and ability to withstand at high temperatures [2]. Al-MMCs are used in critical domains of various industries such as defence and aerospace, automotive, and many other technology segments [3]. Very few investigations are underway on MMCs, because metals have several benefits over plastics and ceramics [4]. Various types of fabrication methods are available for producing MMCs. [5]. Stir casting is a feasible technology for developing metal matrix composite because of its cost-effectiveness and adaptability for large-volume production [6]. Some researchers have worked on the mechanical properties of Al-MMC, which are reviewed in the subsequent section.

Samal et al. found that when the quantity of silicon carbide (SiC) in the matrix increased, the micro hardness increased. It was found that 15% SiC in matrix material increased the micro hardness value up to 37% compared to the base material. Hardness increases due to SiC particles' hardness, making composite stiffer and more resistant to indentation [7]. Singh et al. found that the composite's hardness increased as the proportion of reinforcement in the matrix material increased. The increasing hardness of the composite is a barrier due to dislocations in the aluminium matrix [8]. Suresh and Moorthi used aluminium alloy (AA6061) as the matrix phase and titanium diboride (TiB2) as the reinforcing phase, in ratios of 0, 4, 8, and 12 percent by weight. The results revealed that the hardness, wear resistance, and micro-hardness were enhanced and weight loss was decreased [9]. Kumar and birr conducted a tensile test on a composite of aluminium matrix and bamboo leaf ash as reinforcement. The yield strength and ultimate tensile strength improved as the amount of bamboo leaf ash was increased. [10]. H.R. Ezatpour et al. experimented with the metal matrix AA6061 as the matrix material and Nano Alumina Al2O3 as the reinforcing phase. The mesh sizes were 50 microns and 40 nm, respectively. The manufacturing was done using the stir casting and extrusion process. The results demonstrated a high porosity. Porosity was increased with increased wt. % of reinforcement and decreased with extrusion. As-cast and extruded samples, ultimate yield and tensile strength increased as Al2O3 nanoparticles increased, but elongation decreased [11].

2. Materials and methods

2.1. Matrix material

The aluminium alloys include the AA6061, this is the most widely used matrix materials. It's a medium-to-high-grade ejected aluminium alloy that can be heat-treated. It is a potential material for manufacturing lightweight vehicles [12]. This alloy is preferred because of its high strength-to-weight ratio, better formability, weld-ability, resistance to corrosion and wear, and low cost [13].

2.2 Reinforcement material

The reinforcement could be utilized to increase, along with other things, its strength-to-weight ratio, creep and heat impact resistance, abrasion resistance, fatigue limit, and corrosion resistance [14]. Reinforcement used in the proposed work were (SiC), cow dung ash and buffalo dung ash. The cow dung and buffalo dung ash were chosen as reinforcement due to their organic and eco-friendly nature, easy availability and cost-effectiveness. The hybridization of ash particles contains many valuable elements such as silicon and carbon that enhance the composite's mechanical properties. Element analysis of buffalo dung ash and cow dung ash was done using a scanning electron microscope with EDS, as shown in Table 2. The amount of (SiC) percent as reinforcement remained constant at 10%. The amount of cow dung and buffalo dung ash percentage varied from 6%, 8% and 10%, with equal amounts as given in Table 3 for different samples. Figure 1 shows the SEM images of (SiC) cow dung ash and buffalo dung ash.

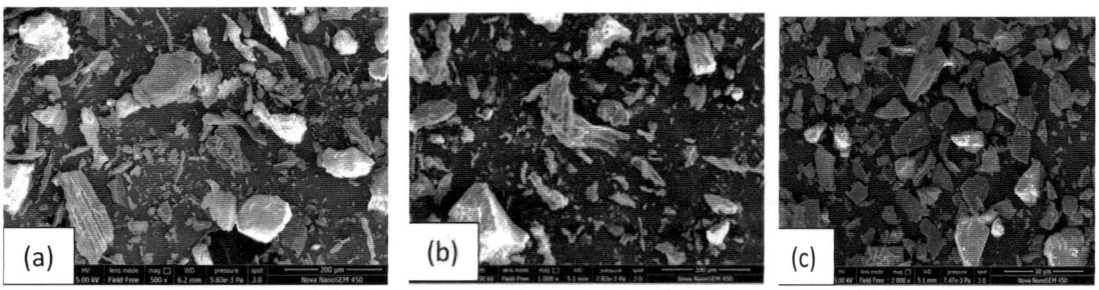

Figure 1. (a) SEM Image of Cow Dung (b) Buffalo Dung Ash (c) Silicon Carbide

Table 1. Chemical elements of cow dung ash and buffalo dung.

Element	Cow Dung Weight %	Buffalo Dung Weight %
Silicon	23.34	46.37
Carbon	12.74	18.63
Calcium	12.18	16.39
Sulfur	11.42	-
Aluminium	6.42	5.46
Magnesium	5.64	6.98
Oxygen	4.68	6.18
Sodium	4.53	-
Potassium	4.43	-
Iron	1.94	-

Table 2. Reinforcement percentage in samples

Sample	Al6061	Silicon carbide	Cow dung ash	Buffalo dung ash
S1	100%	0	0	0
S2	84%	10%	3%	3%
S3	82%	10%	4%	4%
S4	80%	10%	5%	5%

2.3 Fabrication

The stir casting method was used for the manufacturing of composite material. A schematic diagram of the stir casting setup is shown in Figure 2. First, matrix material was placed inside the crucible, maintained at 800°C. Matrix material was melted. Reinforcement and filler material simultaneously heated in preheater at 300°C. After melting the matrix material, reinforcement was added to the matrix material. Subsequently, the mixture was agitated at 400 RPM for 10 minutes for proper mixing of the mix, as shown in Figure 3. The mixture was discharged into the mold. The casted composite was allowed to solidify.

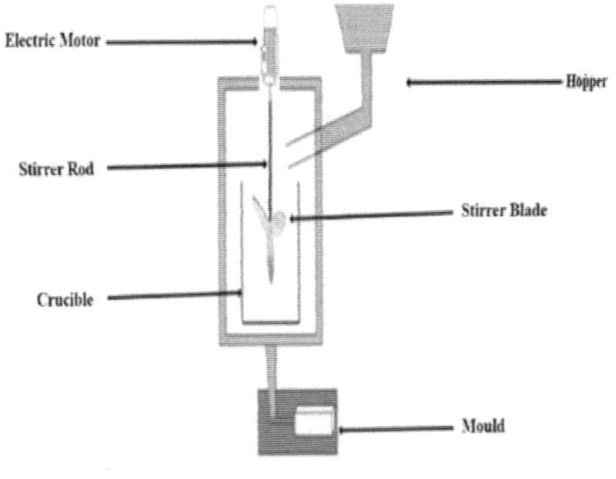

Figure 2 Schematic diagram of the stir casting

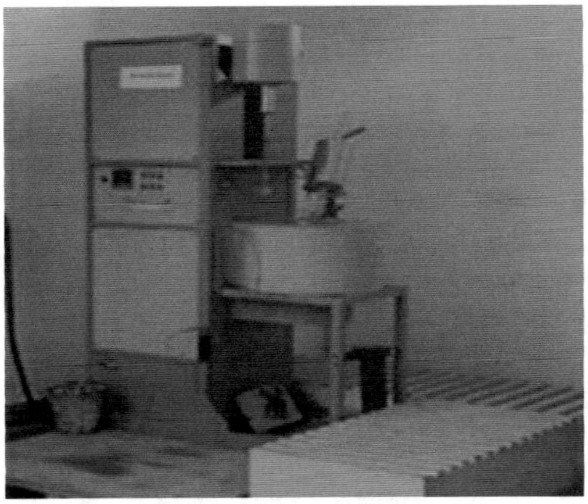

Figure 3 Stir casting setup

IOP Conf. Series: Materials Science and Engineering **1318** (2024) 012005 doi:10.1088/1757-899X/1318/1/012005

2.4 Density measurement

Density measurement test was conducted to determine the strength-to-weight ratio of composite material. Density was measured on a Contech high precision balance machine. First, measure the weight of the sample in the air. Then measured the weight in water. Finally, density was calculated using density balance.

2.5 Micro hardness

Micro hardness testing is functional when parts are too small for Rockwell or Brinell testing. It can be used to examine the effectiveness of composites. The Vickers hardness equipment (INNOVA TEST Europe BV 6222) was used to test the hardness of the composite material. The sample dimension was 10×10×10 mm^3. The specimen is shown above in Figure 4. The ASTM E10-07 standard was followed. Load applied was 500gf to the specimen for about 35 seconds. Five Indentations were done on the surface of each sample to take an average hardness reading.

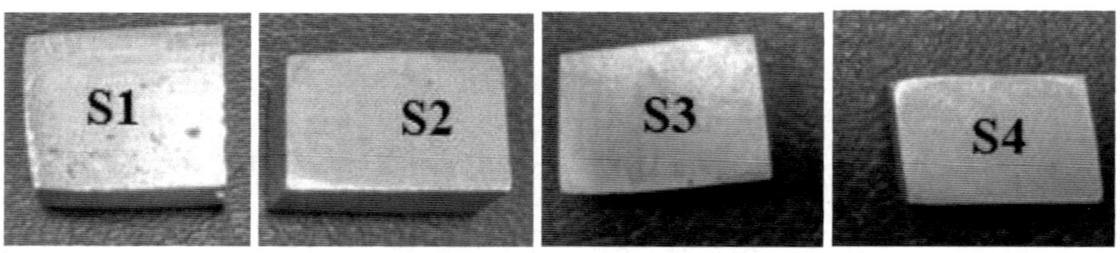

Figure 4 Samples for Micro Hardness

2.6 Compressive strength

This test's objective is to determine how a material will behave or respond to a compressive load by taking measurements of critical parameters like strain, stress, and deformation. A compression test was conducted on the Instron-5967 universal testing machine. Samples were made according to ASTM D1621. The load rate used was 2mm/min, and the maximum Load applicable for compression was 30 K.N.

2.7 Tensile strength

It determines elastic modulus, ultimate tensile, ductility, strain hardening and how far it can extend before breaking. Samples are prepared according to ASTM: E08. A tensile test was executed on (Tinius Olsen H25KL) universal testing machine. Strain rate were maintained at 6mm/min to 0.00048mm/min. for Load up to 12 K.N.

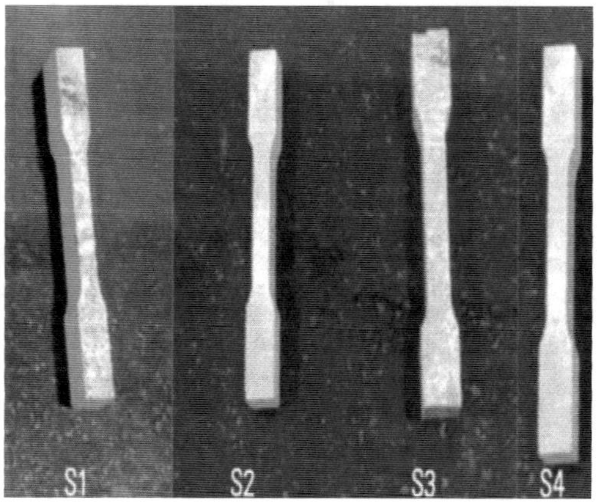

Figure 5 Samples for tensile test

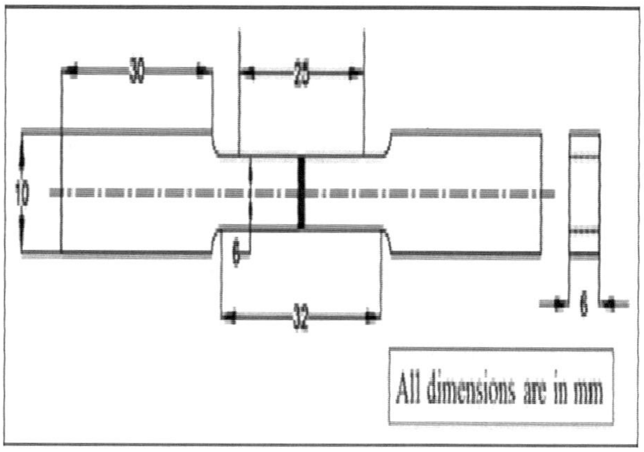

Figure 6 ASTM: E08 tensile strength test specimen.

IOP Conf. Series: Materials Science and Engineering **1318** (2024) 012005 doi:10.1088/1757-899X/1318/1/012005

3. Results and discussion

3.1 Density measurement

The density Al6061 matrix reinforced with cow dung ash, buffalo dung ash and SiC have been examined at room temperature. A simple water impression technique measured the density of the composite material. Density increased when the percentage of reinforcement increased for sample S2, and then it decreased for samples S3 and S4. The density enhancement is due to proper reinforcement mixing in matrix material for sample S2, but as ash reinforcement percentage was increased. Density was reduced due to the porous nature of ash particles that create cracks in a matrix material, resulting in the lightweight composite. Figure7 shows the variation of density with reinforcement percentage.

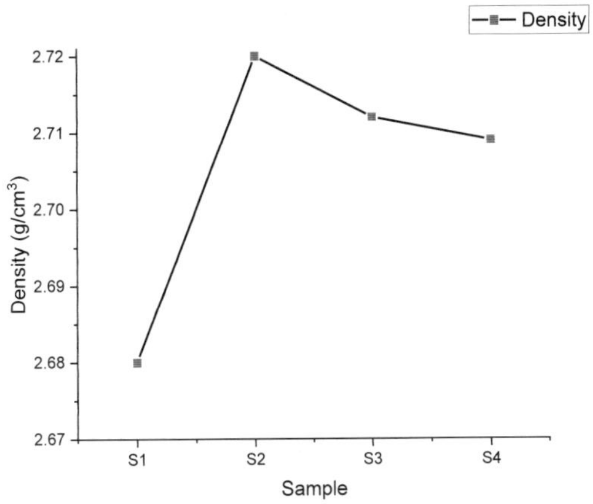

Figure 7 Effect Reinforcement percentage on the density of composite

3.2 Microhardness of composite material

Micro-Hardness tests were performed on a Vickers hardness scale with a load of 500gf. This test was performed using a hardness tester. Hardness was increased with increasing the percentage of reinforced particles. Increased hardness on composite was due to hardness of reinforcement particles, which was higher than matrix material. Further addition of ash particles reduced hardness, as shown in Figure 8. Porosity increases as the ash reinforcement's volume rise. This might have caused long processing times to increase hydrogen uptake from the air, weakening the composites' mechanical properties (micro hardness) [1].

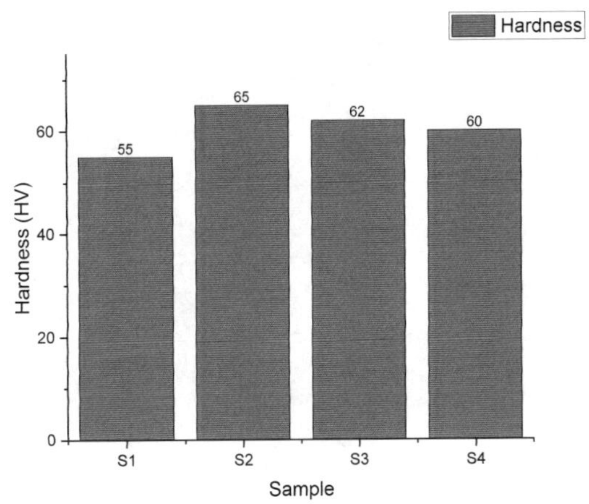

Figure 8 Effect Reinforcement percentage on micro hardness

IOP Conf. Series: Materials Science and Engineering **1318** (2024) 012005 doi:10.1088/1757-899X/1318/1/012005

3.3 Compressive strength of composite

The compression test was performed on a universal testing machine with a 2 mm/min load rate. Results conclude that compressive strength increased with an increased percentage of reinforcement particles. Further increment of ash as filler material reduces compressive strength, as shown in Figure 9. Agglomeration and ladle slag formation are responsible for this. These ladle slags were created by the improper incorporation of ash particles clumped together as a significant slag presence in the samples after failing to disperse evenly into the aluminium matrix during the stirring operation [1].

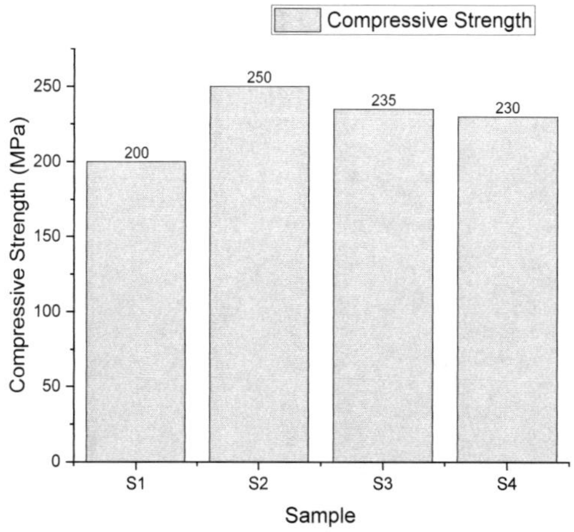

Figure 9 Effect of reinforcement % on compressive strength

3.4 Tensile strength of composite

A tensile test was performed on the universal testing machine at a load of 12KN with a load rate of 2mm/min. The result shows that sample S2 has maximum tensile stress compared to unreinforced, which was 125 MPa as shown in Figure 10. A decrease in tensile strength was observed after the further addition of ash particles. Trapped gases that were generated during the cooling of the cast, segregation, and aggregation of the reinforcements, produced the pores. This is thought to be the reason of the decrease in tensile strength [1]. Figure 11 shows the broken samples after the tensile test.

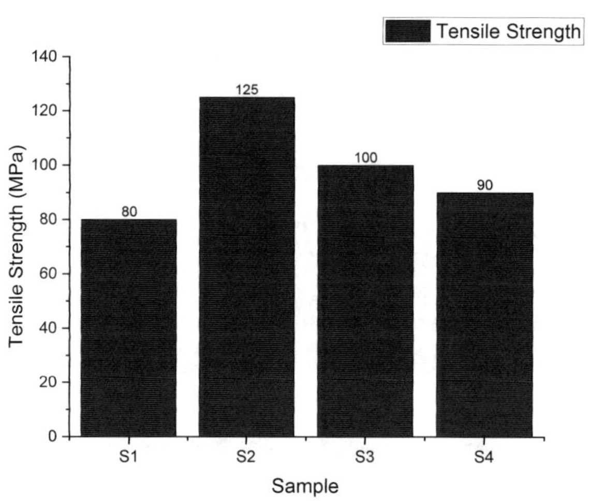

Figure 10 Effect of tensile strength with reinforcement %

Figure 11 Tensile test specimen after the test (S1, S2, S4, S4)

4. Conclusions

The conclusions that have been drawn from the experimental investigations on Al-MMC are following- The stir casting technique was used to fabricate metal matrix composite, due to its economic investment and adaptability for large volume production. A higher amount of SiC enhances the density because of the proper mixing of reinforcement in the matrix material. But the higher ash reinforcement percentage decreases the thickness due to the porous nature of ash particles that create cracks in a matrix material, resulting in light weight composite.

Micro hardness was increased with increased percentage of SiC reinforced particles. Further addition of ash particles reduced micro hardness. A rise in the volume percentage of the ash reinforcement gives rise to more porosity. This leads to an increase in the pickup of hydrogen from the atmosphere that causes composite weak for micro hardness.

Tensile and compressive strength improved with an increased percentage of SiC reinforcement particles. Further increment of ash as filler material reduces tensile strength and compressive strength due formation of ladle slags. These ladle slags originated from the poor disruption of ash particles into the aluminum matrix during the agitation process and grouped together as a large slag. This is an inclusion in the samples. Pores were also created due to entrapped gases formed during the cooling of the cast, segregation, and aggregation of the reinforcements. This has been attributed to being responsible for reducing the tensile strength.

Two-step stir casting can improve the mechanical properties at a higher magnitude than single-step stir casting. Heat treatment of casted reinforced composite can be done to improve the strength.

References

[1] Sun Y, Lyu Y, Jiang A 2014. Fabrication and characterization of aluminum matrix fly ash cenosphere composites using different stir casting routes. Journal of Materials Research **29**, 260–266.

[2] Gireesh Ch, Durga Prasad K.G, Ramji K 2018. Experimental Investigation on Mechanical Properties of an Al6061 Hybrid Metal Matrix Composite. Journal of Composites Science **2**, 49.

[3] Sharma V. K, Singh R.C, Chaudhary R 2007.Effect of fly ash particles with aluminium melt on the wear of aluminium metal matrix composites, Engineering Science and Technology, an International Journal **20**, 1318-1323.

[4] Balasubramanian I, Maheswaran R 2015.Effect of inclusion of SiC particulates on the mechanical resistance behavior of stir-cast AA6063/SiC composites, Materials & Design (1980-2015) **65**, 511-520.

IOP Conf. Series: Materials Science and Engineering **1318** (2024) 012005 doi:10.1088/1757-899X/1318/1/012005

[5] Selvam D.R. J, Smart R.D.S, Dinaharan I 2013.Synthesis and Characterization of Al6061-Fly Ash SiCp Composites by Stir Casting and Compocasting Methods, Energy Procedia **34**, 637-646.

[6] Umanath K, Palanikumar K, Selvamani S.T 2013.Analysis of dry sliding wear behavior of Al6061/SiC/Al2O3 hybrid metal matrix composites, Composites Part B: Engineering **53**, 159-168.

[7] Samal P, Surekha B, Vundavilli P.R 2022. Experimental Investigations on Microstructure, Mechanical Behavior and Tribological analysis of AA5154/SiC Composites by Stir Casting, Silicon **14**, 3317–3328.

[8] Singh J, Jawalkar C.S, Belokar R.M 2020.Analysis of Mechanical Properties of AMC Fabricated by Vacuum Stir Casting Process. Silicon **12**, 2433–2443.

[9] Suresh S, Gowd G.H, Deva Kumar M.L.S 2019. Mechanical Properties of AA 7075/Al2O3/SiC Nano-metal Matrix Composites by Stir-Casting Method. J. Inst. Eng. India Ser. D **100**, 43–53.

[10] Kumar B.P, Birru A.K 2017.Microstructure and mechanical properties of aluminium metal matrix composites with addition of bamboo leaf ash by stir casting method, Transactions of Nonferrous Metals Society of China **27**, 2555-2572.

[11] Ezatpour H.R, Torabi-Parizi M, SAJJADI S.A. 2013.Microstructure and mechanical properties of extruded Al/Al2O3 composites fabricated by stir-casting process, Transactions of Nonferrous Metals Society of China **23**, 1262-1268.

[12] Lakshmikanthan P, Prabu B 2016. Mechanical and tribological behavior of aluminium al6061-coconut shell ash composite using stir casting pellet method. Journal of the Balkan Tribological Association **22**, 4008–4018.

[13] Razak A, Kaladgi R, et al 2021. Fabrication characteristics and mechanical behavior of aluminium alloy reinforced with Al2O3 and coconut shell particles synthesized by stir casting. IOP Conf.Ser. Mater. Sci. Eng. **1057**, 012017.

[14] Suleiman I.Y, Salihu S.A, Mohammed T.A 2018. Investigation of mechanical, microstructure, and wear behaviors of Al-12%Si/reinforced with melon shell ash particulates. The international journal of advanced manufacturing technology volume **97**, 4137–4144.

IOP Conf. Series: Materials Science and Engineering **1318** (2024) 012006 doi:10.1088/1757-899X/1318/1/012006

Characterization and evaluation of gold mill tailings as an additive in cement-stabilized products

A Balbin[1,2*], I D Tabañag[1,3], M Lavapiez[2], R Balanay[2], R Varela[2], H Omping[2], K H McCarthy[2], M Mutia[2], M R Alvarado[2], I J Mirasol[2], J Jabien[2] and B Allih Tawantawan[2]

[1] School of Engineering, University of San Carlos, Talamban, Cebu City, 6000, Philippines
[2] Caraga State University, Butuan City,8600, Philippines
[3] Philippine Council for Industry, Energy, and Emerging Technology Research and Development, Department of Science and Technology (DOST-PCIEERD), Bicutan, Taguig City, 1631, Philippines

E-mail: albalbin@carsu.edu.ph

Abstract. Gold mill tailings (GMT) are waste from gold mill processing plants usually collected and stored in tailings storage facilities (TSF) and left for years in many abandoned mines in the world. GMT is in large amounts and contains high concentrations of chemicals and compounds that may harm human lives and the ecosystem if not well-maintained. A characterization study is conducted on GMT from a small-scale mill processing plant in Agusan del Sur, Philippines, and its potential utilization as an additive to sand and cement in making cement-stabilized products is also evaluated. The elemental and oxide composition, the presence of free and total cyanide, and particle size analysis were determined. Different proportions of gold mill tailings, cement, and sand were prepared and tested for compressive strength and water absorption. The results showed that the GMT contained some potentially valuable elements like titanium, zinc, and europium. GMT also contains potentially hazardous elements like lead (663ppm), arsenic (157ppm), and mercury (17ppm). The free cyanide in the effluent is also found at 1.9 ppm. These elements and compounds are potentially hazardous in case of tailings seepage or overflow to groundwater and neighboring agricultural lands. The evaluation of cement-stabilized samples shows that the GMT from the study area can be utilized as an additive to cement and sand in the production of cement-stabilized products however it must be properly treated and handled to ensure the safety of workers. The TCLP result showed that the product is non-toxic and non-reactive. The construction and proper maintenance of a TSF for small-scale gold milling plants in the study area is recommended to prevent environmental, health, and safety risks.

1. Introduction

The pressing environmental and climate crisis has prompted nations to work for the attainment of sustainable development goals to ensure a safe space for humanity. Essentially, this includes the management of waste generated by all industries [1]. An advocated paradigm to manage waste hopefully over a macro scale is the circular economy that has continuously gained increasing attention and traction worldwide. Circular economy is an approach that promotes close-loop operations and designing waste out, which is so aligned with sustainable production and consumption practices for ecological objectives [2]. One of the sectors that have explored and tried to incorporate circular economy operations is the gold mining sector because of its notoriety in producing hazardous waste [3-4]. The gold mining sector has been accounted for in a lot of environmental contamination and degradation issues [3], [5], [6]. One of the waste materials of the sector considered for the application of circular economy is the gold mill tailings (GMT) [6-7].

GMT is the waste produced in the extraction of gold. It contains predominantly gangue, which is the nonvaluable material in the ore. GMT is in large amounts and contains high concentrations of chemicals and compounds that may harm human lives and the ecosystem if not well-maintained [8]. GMT are generally contained in a Tailings Storage Facility (TSF) for large-scale mill processing plants or disposed of in poorly constructed containment facilities in some small-scale mill processing plants. These storage facilities are necessary for the containment of tailings but it creates environmental problems and safety hazards. Dust particles may be dispersed by strong winds which can contaminate the air and soil, and seepage from tailings storage facilities can contaminate the water bodies. Most radically, tailings storage facilities can break which may result in the discharge of waste to the environment and cause damage to the ecosystem or worse fatalities [9].

Caraga Region is dubbed as the mining capital of the Philippines. It is home to various mining and milling industries including large-scale and small-scale gold processing plants. Particularly, in Bayugan 3, Rosario, Agusan del Sur, Philippines is situated one large-scale and some small-scale gold mine and mill processing plants. To date, the estimated GMT from this place is 2 million metric tons (MMT) which covers hectares of land area. Tailing storage in the area, especially those coming from small-scale mills is already causing concerns to the local government and the community because of the poor management and maintenance of facilities.

In this study, a characterization of the GMT from a small-scale mine is conducted towards the possible utilization of the waste to reduce its volume in the facility. It is crucial to have a thorough understanding of the waste's physical and chemical properties to establish acceptable and efficient waste management strategies [10–12]. Additionally, prospective mine issues including reactivity, acid mine drainage, and metal leaching to groundwater can be predicted [13], [14]. In particular, the particle size, elemental and oxide composition, and free cyanide were determined. It is important to know the elemental composition to assess if there are possible valuable elements and hazardous elements like lead, and mercury. The determination of possible hazardous elements will lead us to the identification and recommendation of appropriate technology to be used to mitigate the risk of having to deal with those elements.

The evaluation and testing of cement-stabilized products using GMT mixed with sand is also conducted in this study. The use of GMT could lead to the reduction of volume in the storage facilities and the reduction of related hazards and risks. Moreover, the possible creation of new industries from the utilization of waste can boost the socioeconomic activities in the area while also providing alternative environmental solutions to the problem of mine waste.

2. Materials and Methods

The gold mine tailings were collected from the tailings pond of a small-scale gold milling plant in Rosario, Agusan del Sur. The tailings were collected from 30 different spots around the tailings pond at 2 meters distance. The samples were taken at 30cm depth using polyvinyl chloride (pvc) pipe to drill vertically downward and inserted into polyethylene terephthalate (PET) plastic then placed into an ice box and brought to the laboratory for analysis. The samples were filtered to separate the liquid and solid parts. The solid part was dried at 105°C for 36 hours.

2.1 Physical and chemical characterization

The summary of the analysis conducted is in Table 1. The particle size was determined by sieve analysis and using sieves with openings of 600, 300, 212, 150, and 75 μm. About 200 grams of dried samples were used and sieved at 100 rpm for 5 minutes. The elemental composition analysis was conducted with an X-ray fluorescence (XRF) Spectrometer (Malvern Panalytical, Epsilon 1). About 2 grams of the solids and 5 to 10 mL of the wastewater sample were used. The analysis was done in three replicates. A third-party laboratory was commissioned to analyze the free and total cyanide in the effluent, XRD of product, and toxic characteristic leaching procedure (TCLP).

Table 1. Summary of analysis and testing conducted in this study.

Parameter	Method/ Equipment	Reference
Particle Size	Sieve Analysis	[15]
Elemental and Oxide	XRF	[16], [17]
Free and Total Cyanide	Continuous flow analysis	EN ISO 14403-2
Compressive Strength	Unconfined	ASTM C67
Water Absorption	ASTM C140 method	ASTM C140, ASTM C67
Toxic Characteristic Leaching Procedure (TCLP)	AAS, ICP-OES	USEPA Method 1311

2.2 Cement-stabilized product (CSP) sample preparation

The mixing proportion of the CSP is presented in Table 2. The raw materials were manually mixed and added with water, weighing 50% of the total weight of the solid components. The mixture was then poured into a pre-cleaned mold of 100cm x 100cm x 60cm dimension and was allowed to dry for 24 hours at room temperature. The samples were then removed from the mold for testing.

Table 2. The percent weight composition of each raw material in the sample.

Formulation	Raw Materials (weight %)		
	Cement	GMT	Sand
A	50	30	20
B	30	20	50
C	30	40	30
D	20	30	50
E	20	50	30

2.3 Determination of compressive strength, water absorption, and TCLP

Water absorption was tested using ASTM C140, where test samples were oven-dried at 105°C - 115°C for at least 24 hours until a constant mass was achieved (W_d). The samples were then placed in a water bath for curing. After 24 hours of soaking, the water absorption was determined. Then, the samples were returned to the curing water until the desired number of curing days of 14, 21, and 28 days were reached. The samples were then immersed in a plastic basin container containing potable water for 24 hours at a temperature of 15.6°C to 26.7°C. The specimens were removed from the water after 24 hours and weighed after being cleaned with a paper towel; weighed and recorded as Ws (saturated weight). The average result of the three samples was reported as the water absorption for each formulation. The water absorption, A, expressed as a percentage of the relationship of the weight of water absorbed by the test samples to the weight of the dry test samples, is calculated as follows:

$$Water\ Absorption,\ A\ \% = [(Ws - W_d)] / W_d] \times 100 \quad (1)$$

Where; Ws = saturated weight of the test samples, g
W_d = oven-dry weight of the test samples, g

For compressive strength, samples were tested at 14, 21, and 28 days using a KENSAIN Analog E Series Compression Machine. 3 samples from each sample proportion were taken and tested. The average result of the samples was reported as compressive strength. After the desired curing day, samples were dried in an oven following the drying procedure described earlier. To determine if there are toxic elements that can leach out from the sample of CSP, a TCLP is conducted through an external laboratory.

3. Results and Discussions

3.1 Particle size distribution

The GMT particle size data presented in Table 3 indicates that the composition consists of approximately 80% fine sand and 20% silt or clay particles measuring less than 74 µm. According to the ASTM Unified Soil Classification System (2011), this composition qualifies as fine sandy loam.

Table 3. Particle size of GMT

Sieve Opening (µm)	Classification	GMT retained (g)	Mine Tailings (% wt. retained)	(% Cumulative retained)	(% Cumulative passing)
600	Medium Sand	0	0	0.00	100.00
300	Fine Sand	13.4	6.73	6.73	93.27
212		6.9	3.47	10.20	89.80
150		8.9	4.47	14.67	85.33
75		120.2	60.37	75.04	24.96
Bottom pan	Silt/ Clay	49.7	24.96	100.00	0.00

3.2 Elemental and oxide composition of gold mine tailings.

The elemental content of the gold mill tailings is shown in Table 4. Titanium, zinc, manganese, lead, and strontium are the top 5 elements present. The elements lead (663ppm), arsenic (157 ppm), and mercury (17ppm) were also found in high levels. The potential seepage of these elements to ground water and nearby agricultural areas may result to both environmental, and health and safety risk. The possible extraction of these elements for development of new products can be further studied. Moreover, the high level of mercury is believed to be coming from amalgamation process to separate the gold from other impurities. The use of mercury in small-scale mining has long been prohibited through executive order (EO 79 s. 2012) but some miners continue to use it and dispose it improperly with the tailings.

Table 4. Elemental Composition of GMT (solid)

Element	ppm	Element	ppm	Element	ppm	Element	ppm
Ti	2107	Sr	110	Ba	381	Rb	31
Zn	1827	Cl	62	Cu	314	Cr	23
Mn	932	Zr	57	Eu	264	Cd	17
Pb	663	V	53	As	157	Hg	17

The oxide analysis in Table 5 revealed that GMT contains silica (77%), alumina (9.34), iron oxide (5.9), potassium oxide (4.03), calcium oxide (1.03), and sulfur trioxide (2.23). Some of these oxides are essential materials for the utilization of mine tailings in glass making [6]. The GMT from Agusan del Sur may require some enhancement and treatment to qualify as a good material for glass making and this can be further worked out by the researchers. On the other hand, the presence of sulfur trioxide could indicate a possible problem in acid mine drainage since sulfuric oxide (SO_3) readily reacts to form sulfuric acid (H_2SO_4).

Table 5. Oxide analysis of gold mill tailings

Oxide	SiO_2	Al_2O_3	K_2O	Fe_2O_3	CaO	SO_3	Total
%	77.46	9.34	4.03	5.9	1.03	2.23	100

3.3 Presence of total and free CN in the effluent

The results showed that the effluent discharge for total cyanide was 42 mg/L and 1.9 mg/L for free cyanide. The free CN value both exceeds the Philippine and the US Environmental Protection Agency

(USEPA) water quality guideline. This calls for attention to the local government of Agusan del Sur for the stricter monitoring and implementation of existing laws.

Table 6. Total and free CN analysis

Parameter	Results	USEPA (water quality standard)	Philippines DAO 2016-08
Total CN	42 mg/L	if > 0.2 mg/L conduct free CN analysis	NS
Free CN	1.9 mg/L	<0.2 mg/L	*0.04-0.40 mg/L

NS – Not specified *depending on water body type

3.4 Water absorption and compressive strength

The results for water absorption and compressive strength are evaluated based on ASTM standards summarized in Table 7. Different type of bricks and masonry units requires different compressive strength and water absorption.

Table 7. ASTM Specifications for different bricks and concrete masonry units.

Specifications	Weathering Conditions	Code	Min. Compressive Strength (MPa)	Max. Water Absorption (%)	Reference
Building brick (BB)	SW	BB-SW	20.7	17	ASTM C62
	MW	BB-MW	17.2	22	
	NW	BB-NW	10.3	No limit	
Facing brick (FB)	SW	FB-SW	20.7	17	ASTM C216
	NW	FB-NW	17.2	22	
Pedestrian and light traffic paving brick (PB)	SW	PB-SW	55.2	8	ASTM C902
	MW	PB-MW	20.7	14	
	NW	PB-NW	20.7	No limit	
Load Bearing masonry units (LB)	-	LB	13.1	17	ASTM C90
Non-load-bearing masonry (NB)	-	NB	3.45	No limit	ASTM C129

SW- severe weathering MW - moderate weathering NW - negligible or no weathering

The result of the evaluation is summarized in Table 8. Formulation A with 50% cement, 30% GMT, and 20% sand can be used for all applications except for paving brick under severe weathering. Formulation B (30-20-50) can be applied for building brick under no weathering conditions, while C (30-40-30), and D (20-30-50) can be used for non-load bearing applications. Sample E with the highest GMT content of 50% did not pass the compressive strength requirement and thus may be considered for other purposes like non-load bearing decoratives such as miniatures, garden gnomes, and others.

Table 8. Water absorption and compressive strength of the samples.

Formulation				Characteristics				Possible Application (Reference: see Table 7)
				Water Absorption (%)	Compressive Strength (MPa)			
Code	C	GMT	S		14	21	28	
A	50	30	20	6.51	20.1	27.6	25.7	All except PB-SW
B	30	20	50	7.62	10.7	11.0	15.0	BB-NW
C	30	40	30	7.96	8.1	11.8	8.7	NB
D	20	30	50	14.16	3.2	4.1	3.5	NB
E	20	50	30	19.11	2.9	3.1	3.4	other purposes

The TCLP results in Table 9 indicate that no element in the bricks exceeded the regulatory level for TCLP both from the Philippines and USEPA guidelines. Moreover, the sample did not ignite when subjected to a flame test and did not exhibit the characteristic of a reactive substance.

Table 9. TCLP results of CSP

Parameters (Total)	Result (mg/L)	(DAO 2013-22) (mg/L)	USEPA (mg/L)	Parameters (Total)	Result (mg/L)	(DAO 2013-22) (mg/L)	USEPA (mg/L)
Mercury	< 0.001	0.1	0.2	Zinc	< 0.01		
Arsenic	0.002	1	5	Aluminum	0.6		
Cadmium	< 0.01	0.3	1	Barium	0.6		
Chromium	0.3	5	5	Cobalt	< 0.01	70	
Copper	< 0.05			Antimony	< 0.01		
Iron	< 0.05			Selenium	< 0.01	1	1
Magnesium	< 0.1			Silicon	7.7		
Manganese	< 0.05			Tin	< 0.01		
Nickel	< 0.1			Titanium	< 0.01		
Lead	< 0.05	1	5	pH	12.65@ 25°C		

4. Conclusion

The gold mill tailings contain elements of potential value as well as potentially hazardous elements which could contaminate nearby drinking water sources and agricultural land. The potentially valuable elements may be subject to further research for possible extraction from the waste. Treatment of the waste to remove the hazardous elements and proper handling are necessary if the use of GMT as an additive in cement-stabilized products is decided to ensure the safety of the workers. The effluent showed a high concentration of free cyanide which exceeds regulatory requirements and requires immediate action from the local government and the industry.

Based on the evaluation of cement-stabilized samples CSP containing GMT are non-toxic and non-reactive. GMT can be used as an additive for applications in almost all types of bricks if the cement amount is at 50%. But considering the cost, the most practical application is to limit cement at 30% and the amount of GMT at 20-40% for application in building brick under no-weathering conditions and non-load-bearing masonry. The use of GMT for other products such as ceramic and glass can be further studied.

Acknowledgment

The authors would like to express appreciation to the Philippines Department of Science and Technology CRADLE Program for funding this research and DOST-ERDT for the scholarship of the main author.

References

[1] Castro-León G, Baquero-Quinteros E, Loor B J, Alvear J, Montesdeoca Espín D E, De La Rosa A and Montero-Calderon C 2020 Waste to catalyst: synthesis of catalysts from sewage sludge of the mining, steel, and petroleum industries *Sustainability* **12** pp 9849,

[2] Halog A, Balanay R, Anieke S and Yu T Y 2021 Circular economy across Australia: Taking stock of progress and lessons *Circ. Econ. Sustain.* **1** pp 283–301

[3] Kinnunen P and Kaksonen A 2019 Towards circular economy in mining: Opportunities and bottlenecks for tailings valorization *J. Clean. Prod.* **228**, pp 153–60

[4] Upadhyay A, Laing T, Kumar V and Dora M 2021 Exploring barriers and drivers to the implementation of circular economy practices in the mining industry *Resour. Policy*, **72** p

102037 doi: 10.1016/j.resourpol.2021.102037.

[5] Lebre E, Corder G and Golev A 2017 The Role of the Mining Industry in Circular Economy *Wiley Online Library*

[6] Uchenna O, Makhatha M, Mekuto L and Mavumengwana V 2020 Gold Mine Tailings: A Potential Source of Silica Sand for Glass Making *Minerals* **10** doi: 10.3390/min10050448

[7] Balegamire C, Nkuba B and Dable P 2022 Production of gold mine tailings based concrete pavers by substitution of natural river sand in Misisi, Eastern Congo *Clean. Eng. Technol.* **7**, p 100427

[8] Hudson-Edwards K, Jamieson H and Lottermoser B 2011 Mine wastes: Past, present, future *Elements* **7** pp. 375–80 doi: 10.2113/gselements.7.6.375.

[9] Brown B 2002 Management of Tailings Disposal on Land *Mineral Processing Plant Design, Practice, and Control* pp 1809–27

[10] Jamieson H 2011 Geochemistry and mineralogy of solid mine waste: Essential knowledge for predicting environmental impact *Elements* **7** pp 381–6 doi: 10.2113/gselements.7.6.381

[11] Bian Z, Miao X, Lei S, Chen S, Wang W and Struthers S 2012 The challenges of reusing mining and mineral-processing wastes *Science (80-.)* **337** pp 702-3 doi:10.1126/science.1224757

[12] Chileshe M, Syampungani S, Festin E S, Tigabu M, Daneshvar A and Odén P C 2020 Physico-chemical characteristics and heavy metal concentrations of copper mine wastes in Zambia: implications for pollution risk and restoration *J. For. Res.* **31** pp. 1283–93 doi:10.1007/s11676-019-00921-0

[13] Anderson N et al. 1999 Phytomining for nickel, thallium and gold *J. Geochemical Explor.* **67** pp 407–15 doi: 10.1016/S0375-6742(99)00055-2

[14] van der Ent A, Baker A J, van Balgooy M M and Tjoa A 2013 Ultramafic nickel laterites in Indonesia (Sulawesi, Halmahera): Mining, nickel hyperaccumulators and opportunities for phytomining *J. Geochemical Explor.* **128** pp 72–9 doi: 10.1016/J.GEXPLO.2013.01.009

[15] Gee G and Or D 2004 Methods of soil analysis part 4. Physical Methods *Vadose Zo. J.* **3** pp 722–3 doi: 10.2136/vzj2004.0722

[16] Farrokhpay S, Cathelineau M, Blancher S, Laugier O and Filippov L 2019 Characterization of Weda Bay nickel laterite ore from Indonesia *J. Geochemical Explor.* **196** pp 270-81 doi: 10.1016/j.gexplo.2018.11.002

[17] Li G *et al.* 2018 Selective leaching of nickel and cobalt from limonitic laterite using phosphoric acid: An alternative for value-added processing of laterite *J. Clean. Prod.* **189** pp 620–6 doi:10.1016/j.jclepro.2018.04.083.

IOP Conf. Series: Materials Science and Engineering **1318** (2024) 012007 doi:10.1088/1757-899X/1318/1/012007

Production of nitrogen-free syngas (H_2 + CO) from CO_2 and NH_3

Martin Keller

National Institute of Advanced Industrial Science and Technology, Global Zero Emission Research Center, 16-1, Onogawa, Tsukuba, Ibaraki, Japan

keller-m@aist.go.jp

Abstract. Towards the goal of carbon neutrality, future chemical production could utilize captured CO_2 as a carbon source instead of fossil carbon from petroleum or natural gas. However, the production of chemicals from CO_2 is highly energy-intense. The required energy could be imported to Japan in the form of hydrogen. However, the long-distance transport of molecular hydrogen is challenging. NH_3 has recently emerged as the most promising molecule for long-distance hydrogen transport and storage. The use of imported NH_3 as an energy vector to realize carbon recycling in the Japanese chemical industry is promising. Syngas (H_2+CO) is an important intermediate in the production of chemicals and fuels from CO_2. Hydrogen is required not only as a constituent of syngas, but also as a reductant of CO_2. If hydrogen is imported in the form of NH_3, conventionally, NH_3 would be cracked to H_2 and N_2, the product gases and unconverted NH_3 separated, and CO/syngas then be produced through the reverse water gas shift reaction (RWGS). This conventional pathway is energy-intense and requires many unit operations. Here a novel process to produce nitrogen-free syngas directly, without producing molecular H_2 first and without any dedicated gas separation steps is presented. This is realized by a using a metal oxide bed material that is active for catalytic cracking of NH_3, as well as able to transport oxygen through a redox reaction in a newly designed process. This leads to an inherent separation of N_2. This process, referred to herein as "NH3-RWGS" has the potential to decrease costs and increase efficiency of syngas production from CO_2 and NH_3.

1. Introduction

Towards the goal of carbon neutrality, future chemical production could utilize captured CO_2 as a carbon source instead of fossil carbon from petroleum or natural gas. However, the production of chemicals from CO_2 is highly energy-intense. The required energy could be imported to Japan in the form of hydrogen. However, the long-distance transport of molecular hydrogen is challenging. NH_3 has recently emerged as the most promising molecule for long-distance hydrogen transport and storage. The use of imported NH_3 as an energy vector to realize carbon recycling in the Japanese chemical industry is illustrated in Figure 1.

IOP Conf. Series: Materials Science and Engineering **1318** (2024) 012007 doi:10.1088/1757-899X/1318/1/012007

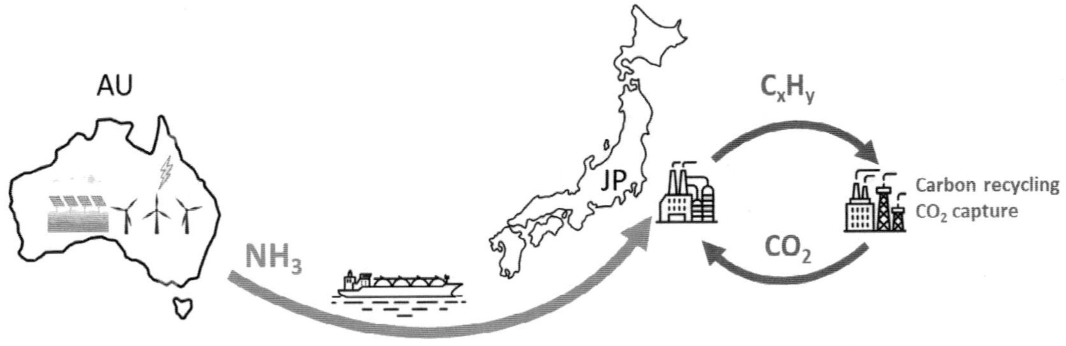

Figure 1. Illustration of chemical production from imported NH_3.

Syngas (H_2+CO) is an important intermediate in the production of chemicals and fuels from CO_2, for example in methanol synthesis or the Fischer-Tropsch process. Hydrogen is required not only as a constituent of syngas, but also as a reductant of CO_2. If hydrogen is imported in the form of NH_3, conventionally, NH_3 would be cracked to H_2 and N_2, the product gases and unconverted NH_3 separated, and CO/syngas then be produced through the reverse water gas shift reaction (RWGS).

$$CO_2 + H_2 \rightarrow CO + H_2O \qquad (1)$$

This conventional pathway is energy-intense and requires many unit operations. Here, a novel process is proposed that instead produces syngas in a two-reactor setup directly from NH_3, integrating the NH_3 cracking, inherent N_2 separation and RWGS reaction. This can be achieved by utilizing the chemical looping principle [1–5], in which oxygen is transferred between the two reactors through lattice oxygen of a solid, reducible metal oxide bed material. Because NH_3 cracking and the RWGS reaction are carried out in a single, integrated system, this novel proposed process is referred to as "NH3-RWGS", cf. Figure 2.

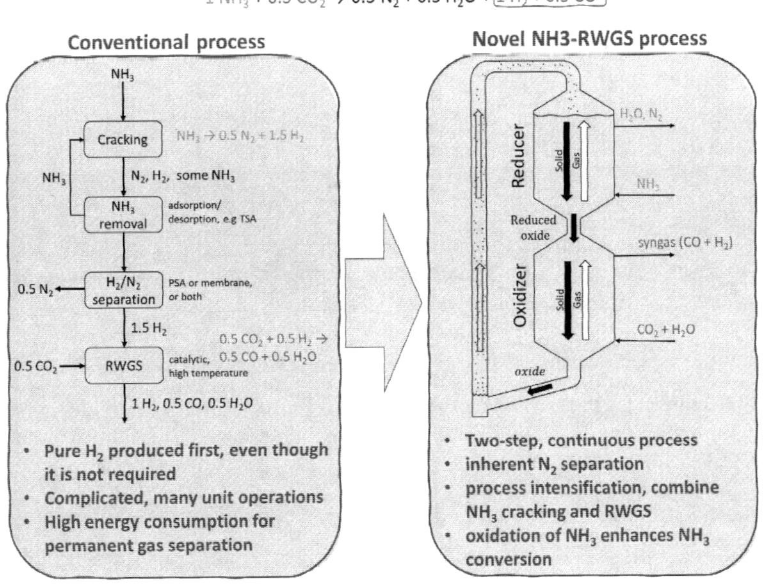

Figure 2. Comparison of a conventional approach for syngas production from CO_2 and NH_3 and the proposed NH3-RWGS process.

In the "Reducer" reactor, NH_3 is cracked (eq. 2) and the metal oxide is reduced by the H_2 produced in the cracking step (eq. 3), producing H_2O and N_2 as offgases.

$$NH_3 \rightarrow 1.5\ H_2 + 0.5\ N_2 \quad (2)$$

$$1.5\ H_2 + MeO_{ox} \rightarrow 1.5\ H_2O + MeO_{red} \quad (3)$$

The reduced bed material, MeO_{red}, is transferred to the "Oxidizer reactor, where CO_2, or a mixture of CO_2 and H_2O is introduced to produce the desired CO and H_2.

$$1\ H_2O + MeO_{red} \rightarrow 1\ H_2 + MeO_{ox} \quad (4)$$

$$0.5\ CO_2 + MeO_{red} \rightarrow 0.5\ CO + MeO_{ox} \quad (5)$$

This idealized process is constrained by the thermodynamic equilibrium of the overall RWGS reaction [6]. Here, the approach to overcome this limitation is to operate both reactors as counter-current gas-solid moving beds; this conceptual design is discussed in the next section.

With respect to the sources of CO_2 for syngas production, two scenarios are plausible. First, CO_2 could be obtained from industrial processes such as cement production, waste incineration or direct air capture (DAC). Second, the source of CO_2 could be CO_2 as a constituent of raw producer gas from the gasification of biomass or waste. In conventional biomass/waste-to-liquid processes, in which fuel or wax is produced through gasification and subsequent Fischer-Tropsch process, the carbon efficiency (that is the fraction of carbon in the liquid product of the total carbon in the biomass/waste feed) is usually around 30-50% [7–9]. The remainder of the carbon is emitted in the form of CO_2, mostly from scrubbing CO_2 produced in the gasifier from the syngas. The carbon efficiency could be significantly enhanced by reconverting the CO_2 back to CO through the RWGS reaction. This is illustrated in Figure 3b. In the context of sustainable aviation fuel (SAF) production, this is often considered as a hybrid between biofuels and e-fuels.

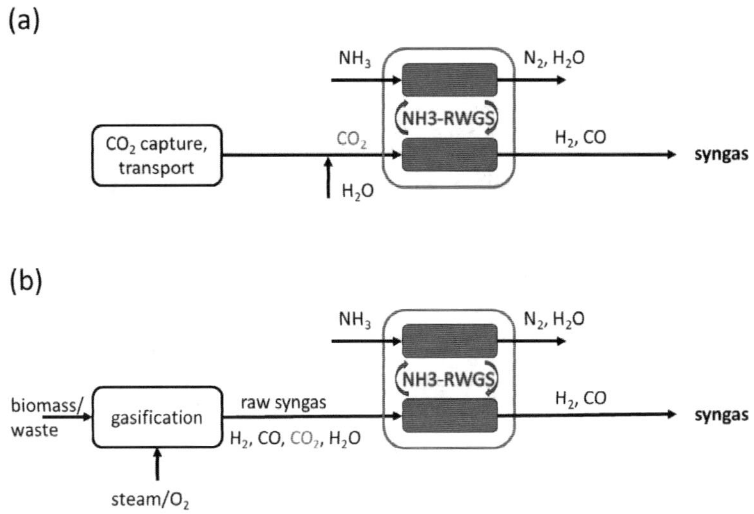

Figure 3. CO_2 sources for the NH3-RWGS process: (a) from CO_2 capture from industrial processes, air or combustion plants, (b) from raw syngas from biomass/waste gasification.

2. Demonstration of basic chemistry

2.1. Experimental

Granular Fe/$MgAl_2O_4$ was produced from fine powders of Fe_2O_3 (50 wt-%, Sigma-Aldrich, U.S.) and $MgAl_2O_4$ (50 wt-%, Baikowski, France) in an Eirich EL1 high-shear mixer. Subsequently, it was calcined at 1150 °C for 3 hours in air to impart mechanical stability. Finally, the calcined granules were sieved to the size range of 250-500 µm. To preclude the effect of initial deactivation from sintering during redox cycling of the material, three cycles of reduction in H_2 until completion and subsequent oxidation in CO_2 until completion were carried out. The obtained material was then used in the experiments with NH_3 either as-is (referred to as "granular") or finely ground with pestle and mortar (referred to as "powdered").

With these materials, experiments in a coupled thermogravimetry / quadrupole mass spectrometry (TG/QMS) setup were carried out to investigate the reduction kinetics (cf. eq. 3), the cracking kinetics (cf. eq. 2) and the oxidation kinetics with CO_2 (cf. eq. 5).

2.2. Results

The results from the TG/QMS experiments are summarized in Figure 4. The reduction and cracking kinetics exhibited a change in activation energy at higher temperature, indicating mass transfer limitations. This was not observed for the oxidation reaction. Nitride formation, which would be detrimental to process performance, was observed at lower temperatures only and could be avoided with higher operating temperatures. The material exhibited no degradation throughout repeated cycling. The basic chemistry of the proposed process was thus successfully demonstrated.

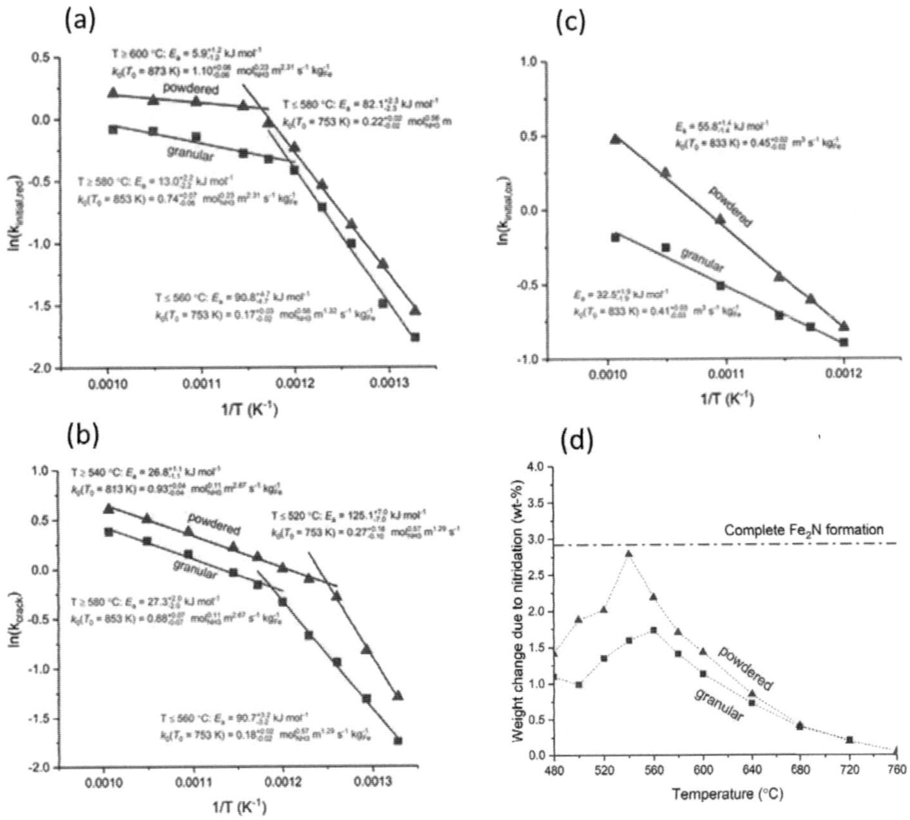

Figure 4. Results from TG/QMS experiments with Fe/$MgAl_2O_4$: (a) Reduction kinetics with 6.95 % NH_3 in He. (b) Cracking kinetics with 6.95 % NH_3 in He. (c) Oxidation kinetics with 5 % CO_2 in He. (d) Nitride formation as a function of temperature.

3. Cold flow model

An acrylic cold flow model was designed to ascertain the functioning of the two-reactor, counter-current moving bed reactor system. The model is shown in Figure 5a. To prevent gas leakage while the granular bed material is moved through the reactor system, a lock hopper system with two pneumatical ball valves is used to remove the bed material from the system. Since the particle volume removed in each cycle of switching the lock hopper valves is precisely known, this system also controls the mass flow rate of bed material through the system. Make-up material is constantly supplied with a vibratory feeder from the top.

To prevent a cross flow of gas between the two reactors, a small seal gas stream is introduced in between the top and the bottom reactor.

Experimentation with the cold model confirmed the functioning of the lock hopper system, as well as the prevention of cross flow by the sealant gas flow. The latter was achieved by using O_2 as a tracer gas and feeding it to the top reactor ("NH3 feed"), and detecting its presence and concentration in the bottom reactor exit ("CO exit") as a function of sealant feed flow rate. A typical pressure balance under conditions where no cross flow is observed is shown in Figure 5b. Based on the learnings from the cold model, a hot lab-scale pilot-plant has been designed and is currently under construction.

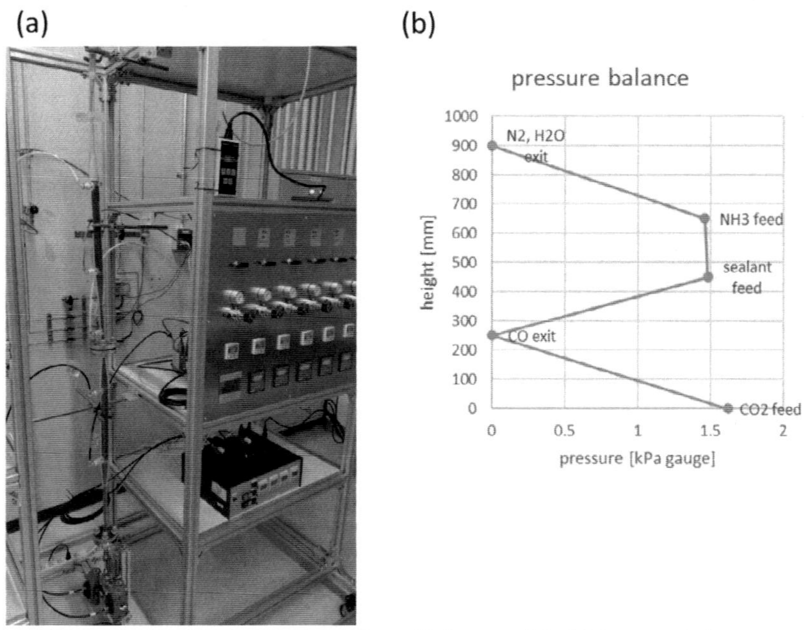

Figure 5. (a)Photograph of acrylic cold flow model. (b) Pressure balance.

4. Conclusions

A novel process has been conceived to produce N2-free syngas from CO_2 and NH_3 The basic chemistry was demonstrated with a $Fe/MgAl_2O_4$ granular material, and a cold flow model of the proposed two-reactor counter-current moving bed reactor was constructed and operated. A hot pilot plant of this process is currently under construction.

References

[1] J. Adanez, A. Abad, F. Garcia-Labiano, P. Gayan, L.F. de Diego, Progress in Chemical-Looping Combustion and Reforming technologies, Prog. Energy Combust. Sci. 38 (2012) 215–282. https://doi.org/10.1016/j.pecs.2011.09.001.

[2] D. Maiti, B.J. Hare, Y.A. Daza, A.E. Ramos, J.N. Kuhn, V.R. Bhethanabotla, Earth abundant perovskite oxides for low temperature CO2conversion, Energy Environ. Sci. 11 (2018) 648–659. https://doi.org/10.1039/c7ee03383d.

[3] M. Wenzel, N.V.R. Aditya Dharanipragada, V. V Galvita, H. Poelman, G.B. Marin, L. Rihko-Struckmann, K. Sundmacher, CO production from CO 2 via reverse water–gas shift reaction performed in a chemical looping mode: Kinetics on modified iron oxide, J. CO2 Util. 17 (2017) 60–68. https://doi.org/10.1016/j.jcou.2016.10.015.

[4] I.S. Metcalfe, B. Ray, C. Dejoie, W. Hu, C. de Leeuwe, C. Dueso, F.R. García-García, C.M. Mak, E.I. Papaioannou, C.R. Thompson, J.S.O. Evans, Overcoming chemical equilibrium limitations using a thermodynamically reversible chemical reactor, Nat. Chem. 11 (2019) 638–643. https://doi.org/10.1038/s41557-019-0273-2.

[5] R. Iwama, H. Kaneko, Integration of Materials and Process Informatics: Metal Oxide and Process Design for CO 2 Reduction, ACS Omega. 7 (2022) 46922–46934. https://doi.org/10.1021/acsomega.2c06008.

[6] I.S. Metcalfe, B. Ray, C. Dejoie, W. Hu, C. de Leeuwe, C. Dueso, F.R. García-García, C.M. Mak, E.I. Papaioannou, C.R. Thompson, J.S.O. Evans, Overcoming chemical equilibrium limitations using a thermodynamically reversible chemical reactor, Nat. Chem. 11 (2019) 638–643. https://doi.org/10.1038/s41557-019-0273-2.

[7] M. Marchese, S. Chesta, M. Santarelli, A. Lanzini, Techno-economic feasibility of a biomass-to-X plant: Fischer-Tropsch wax synthesis from digestate gasification, Energy. 228 (2021) 120581. https://doi.org/10.1016/j.energy.2021.120581.

[8] T. Roshan Kumar, T. Mattisson, M. Rydén, V. Stenberg, Process Analysis of Chemical Looping Gasification of Biomass for Fischer-Tropsch Crude Production with Net-Negative CO2 Emissions: Part 1, Energy and Fuels. 36 (2022) 9687–9705. https://doi.org/10.1021/acs.energyfuels.2c00819.

[9] M.N. Saeed, M. Shahrivar, G.D. Surywanshi, T.R. Kumar, T. Mattisson, A.H. Soleimanisalim, Production of aviation fuel with negative emissions via chemical looping gasification of biogenic residues: Full chain process modelling and techno-economic analysis, Fuel Process. Technol. 241 (2023) 107585. https://doi.org/10.1016/j.fuproc.2022.107585.

IOP Conf. Series: Materials Science and Engineering **1318** (2024) 012008 doi:10.1088/1757-899X/1318/1/012008

Parametric analysis on the sulfone adsorption of using iron impregnated bentonite

M G T Alcaraz[1], A E S Choi[1,*], N P Dugos[1] and M W Wan[2]

[1]Department of Chemical Engineering, De La Salle University, 2401 Taft Ave, Malate, Manila, 0922 Metro Manila, Philippines
[2]Department of Environmental Engineering, Chia Nan University of Pharmacy and Science, 60, Erh-Jen Rd., Sec.1, E707, Jen-Te, Tainan, Taiwan, 71710

angelo.choi@dlsu.edu.ph, peterwan@mail.cnu.edu.tw

Abstract. Sulfur left in liquid fuel is oxidized during combustion and released into the atmosphere, contributing to the worsening of global warming. To address this, oil refineries make use of hydrodesulfurization (HDS) to reduce sulfur content in the fuel. Alternatives to HDS are being studied due to its extreme operating conditions. An alternative for treating sulfur in fuel is oxidative desulfurization, which converts sulfur to sulfones and can be extracted using adsorption methods. This study focuses on the use of bentonite modified with Fe^{3+} to improve its adsorptive performance for removing dibenzothiophene sulfone (DBTO) in model oil. The prepared Fe^{3+}-bentonite adsorbents resulted in lower removal of sulfur compared to raw bentonite. This is due to Fe^{3+} being a hard acid, thus lowering the affinity of bentonite in adsorbing DBTO, which is a soft base. Three parameters were varied in this study, namely adsorption time, adsorbent dosage, and temperature. Results showed that DBTO removal increased with adsorption time as the DBTO molecules occupy more adsorption sites. Increasing adsorbent dosage also increases sulfur removal as it introduces more reaction sites for DBTO removal. The DBTO and bentonite system was also found to be endothermic as desulfurization increased with the temperature rise. With the data collected in this study, it is recommended to explore the possible upscaling of adsorptive desulfurization.

1. Introduction

Industrialization around the world is driven by the accessibility of transportation, and ultimately, liquid fuel. However, this comes at the expense of the worsening of pollution. Petroleum distillates contain different sulfur components. When left in the fuel and oxidized during combustion, sulfur dioxide is released into the atmosphere with the exhaust gases. Sulfur dioxide is harmful to humans and is also considered an indirect greenhouse gas, contributing to the worsening of global warming. Sulfur dioxide in the atmosphere also results in acid rain, which is harmful to humans, the environment, and infrastructure. Thus, it is imperative to treat transportation fuel to lower sulfur content and prevent further damage to the environment and people.

The continuing problem of global air pollution must be acted on by different constituents. Oil refineries bear significant responsibility in mitigating this issue by reducing the sulfur content of their fuels to meet regulatory standards. This is achieved through desulfurization processes integrated into fuel refining. The most prevalent desulfurization process used is hydrodesulfurization (HDS). This method utilizes hydrogen gas to convert sulfur compounds in the fuel into hydrogen sulfide (H_2S). While HDS is widely adopted and effective, it presents challenges due to its high costs associated with extreme operating conditions and limited ability to hydrogenate heterocyclic sulfur compounds [1]. An alternative approach is oxidative desulfurization (ODS) where sulfur compounds in the fuel are treated with the use of oxidant systems [2]. This process has the advantage over HDS as it can be performed at ambient pressure and mild temperatures [3]. Making use of oxidating agents, like H_2O_2, can result in a two-phase system, specifically the aqueous phase and oil phase. This limits the contact of reactants with one another, and so, phase transfer agents are used to overcome mass transfer limitations [4]. Catalysts are also often used in ODS, such as phosphotungstic acid dissolved in the aqueous phase, or supported catalysts that can be recovered through filtration [5]. Through the conversion of sulfur to sulfones, it is made more polar and is easier to extract through adsorption, which is be the focus of this study.

Key characteristics of an effective adsorbent include its adsorption capacity, sulfur selectivity, regenerability, and durability [6]. A common type of adsorbent used is a carbon-based adsorbent, specifically activated carbon (AC). AC finds application in a wide range of industries: water purification, food and beverage processing, odor removal, and industrial pollution control. It is an attractive option for an adsorbent as it has a high surface area, high porosity, and comes at a low cost [7]. High surface area and porosity are good indicators of the potential of a material as a good adsorbent as these pores become the sites for adsorption. Modification of raw AC, such as treatment with an acid or metal impregnation, also helps in increasing its sulfur yield by modifying the functional groups on its surface and contributing to a better interaction with the sulfur in the fuel [8–10]. Another commonly used adsorbent is zeolite, which has high selectivity for compounds because of its structure of pores and chambers that result in the breaking down or exclusion of other molecules [6]. It is also able to perform cation exchange because of its metal-based active site, making it a commendable adsorbent for sulfur removal [7]. High sulfur removal rates were obtained using zeolite that was ion-exchanged with metal ions, specifically copper and silver, and studies also show good regeneration capability of the adsorbent [11,12]. Metal organic frameworks (MOF) are also a good choice for adsorbents as they have high surface area, porosity, and number of active sites. They are composed of an inorganic center, which could be comprised of metal ions, and are linked together by di- or poly-dentate chelating organic bridges called linkers [13]. MOFs as effective adsorbents can be attributed to the high affinity of the sulfur to the immobilized metal ion [14]. Its high surface area and porosity also contribute to the adsorption of sulfur [15].

A relatively novel research focus for adsorbents is on clay minerals. These adsorbents have gained attention for their abundance, affordability, stability, and environmentally-benign nature [16]. Examples of clay adsorbents include kaolinite, bentonite, and palygorskite. Several studies have compared different types of clay minerals as adsorbents. A study examined the performance of kaolinite, montmorillonite, palygorskite, and vermiculite in removing sulfur compounds from petroleum products [17]. The results revealed variations among the clay minerals, with kaolinite exhibiting better performance compared to the other adsorbents, despite vermiculite having a slightly higher surface area. These findings underscore the effect of the mineralogical composition of clay adsorbents on their performance and interaction with sulfur compounds. Studies have also been done to modify the clay adsorbents through impregnation to improve desulfurization capacity. Past studies have shown the formation of sigma bonds between the sulfur and the metal impregnated on the adsorbent [18]. In a related study, the strong attraction of dimethyl sulfide to the impregnated Ag^+ on the surface of bentonite caused the sulfur molecules to barely reach the deeper layers of the modified bentonite [19]. The lower electronegativity of the sulfur molecule allows it to donate lone pair electrons to the metal ion, forming a sigma bond with the adsorbent [18].

The goal of this study is to determine the ability of Fe^{3+}-impregnated bentonite in increasing the adsorption of oxidized sulfur, specifically DBTO, from model oil. The effect of adsorption time, adsorbent dosage, and temperature on DBTO removal was tested through batch experiments.

IOP Conf. Series: Materials Science and Engineering **1318** (2024) 012008 doi:10.1088/1757-899X/1318/1/012008

2. Experimental

2.1. Materials
Bentonite obtained from Seimao Chemical Material Co., Ltd was used. The reagent used for metal impregnation was ferric nitrate nonahydrate ($Fe(NO_3)_3 \cdot 9(H_2O)$) (100%) from Sigma-Aldrich. Materials used for model oil preparation were DBTO (98 wt%) from Tokyo Chemical Industry and toluene (99.8%) from Sigma-Aldrich.

2.2. Metal impregnation
The procedure for the impregnation was patterned from a related study [20]. A 0.1 M $Fe(NO_3)_3$ solution was prepared and mixed with bentonite in a beaker at a 1 g: 10 mL ratio. The mixture was constantly stirred for twelve hours using a magnetic stirrer. The mixture was then centrifuged for 10 min at 40,000 rpm and the adsorbent was transferred to a petri dish to be left to dry in an oven for 12 h at 120 °C. The dried adsorbent was then crushed, ground, and passed through a 100-mesh sieve and stored in a glass bottle.

2.3. Batch adsorption
A 500 ppm S solution was prepared by dissolving DBTO in toluene. Batch adsorption experiments were carried out in a temperature-controlled water bath shaker at 110 rpm by separate usage of raw bentonite (BR) and Fe^{3+}-impregnated bentonite (BF3).

The effect of adsorption time was investigated by placing 1 g of adsorbent in 30 mL of DBTO solution at room temperature. Samples were taken from 1 h to 24 h.

Experiments for testing adsorbent dosage were conducted by mixing 30 mL of DBTO solution with the adsorbent at room temperature for 24 h. Samples with 0.3 g to 1 g of adsorbent were collected.

Investigation of the effect of temperature was carried out by placing 1 g of the adsorbent in 30 mL DBTO solution for 24 h. Temperatures from 30 °C to 50 °C were tested.

3. Results and discussion

3.1. Effect of adsorption time
A study on the adsorption time of the DBTO and bentonite system is important to determine the point at which equilibrium is achieved and aid future process design. The residual sulfur in the model oil after adsorption with BR and BF3 is depicted in the graph in figure 1. Fast adsorption during the first hour was observed, which can indicate that the adsorption sites were quickly filled during this time. The gradual decrease in sulfur concentration with increasing adsorption time can be attributed to the slow access of unoccupied adsorption sites [8]. This also displays the resistance that the DBTO faces as saturation of the vacant sites is reached [21]. At 6 h, the sulfur concentration showed little change until the end of the experiment, hence, can be determined as the time necessary to reach equilibrium.

The experiment also showed that BR was able to remove a higher amount of sulfur as compared to BF3. The lowering of desulfurization performance using the impregnated adsorbent can be attributed to the HSAB principle. The HSAB principle implies that soft acids prefer binding with soft bases, and hard acids prefer binding with hard bases [22]. Since Fe^{3+} is classified as a hard acid while DBTO is a soft base, the impregnation of Fe^{3+} on bentonite resulted in a lower affinity for DBTO [23].

IOP Conf. Series: Materials Science and Engineering **1318** (2024) 012008 doi:10.1088/1757-899X/1318/1/012008

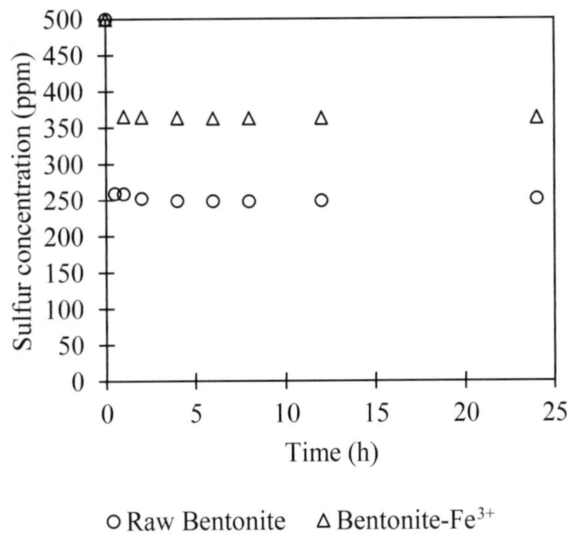

Figure 1. Residual sulfur at different adsorption times after adsorption with BR and BF3.

3.2. *Effect of adsorption dosage*

Studying adsorbent dosage is vital for adsorption processes to be able to achieve maximum sulfur removal at an economic amount of adsorbent. The effect of varying adsorbent dosage on the residual sulfur concentration can be seen in figure 2. The results show that increasing the adsorbent dosage leads to a lower residual sulfur concentration. When a low adsorbent dosage is used, the adsorption sites are quickly filled and become saturated with adsorbate molecules [24]. Therefore, more residual DBTO is left in the solution. An increase in the adsorbent amount increases the available active sites for adsorption. As adsorbent dosage is increased, the surface area also increases and presents more sites for DBTO to adhere to [25].

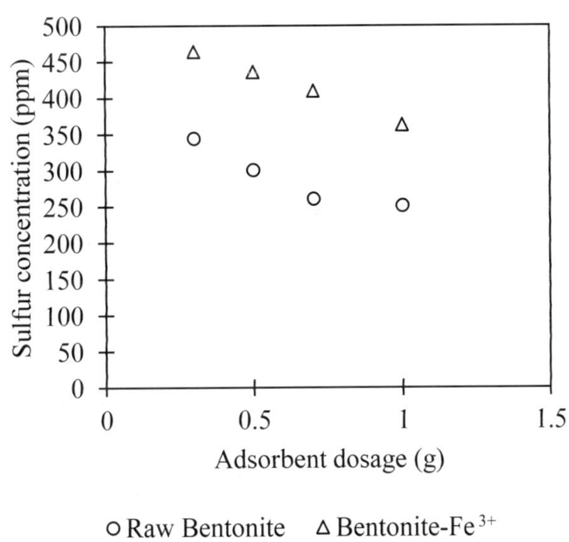

Figure 2. Residual sulfur at different adsorbent dosage after adsorption with BR and BF3.

IOP Conf. Series: Materials Science and Engineering **1318** (2024) 012008 doi:10.1088/1757-899X/1318/1/012008

3.3. Effect of temperature

Temperature is an important parameter to control for unit operations as it affects the reaction rate of the process and the quality of the reactants. The performance of BR and BF3 for sulfur adsorption was tested at varying temperatures, and the results are shown in figure 3. Increasing the system temperature resulted in a decreased residual sulfur due to the increase of kinetic energy of the DBTO particles [26]. The elevated temperature also caused the excitation of molecules and increased mobility of the adsorbate particles due to the reduction of solution viscosity [21]. With higher kinetic energy, higher chances of effective collisions are made available and the more the reaction between the adsorbents and DBTO can progress. The results imply the endothermic nature of the DBTO and BR and BF3 system, wherein an increase in temperature is accompanied by an increase in desulfurization performance.

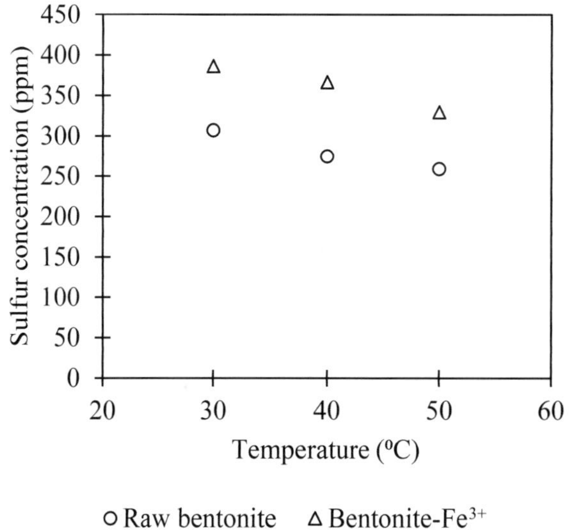

Figure 3. Residual sulfur at different temperatures after adsorption with BR and BF3.

4. Conclusions

ADS can be considered a viable alternative to HDS as it can achieve high desulfurization rates at moderate operating conditions. With its low cost, bentonite is an attractive option as an adsorbent for desulfurization because of its low cost, stability, and environmentally-friendly characteristics. Metal impregnation is widely used to modify adsorbents as it can improve the interactions between sulfur and adsorbent. In this study, bentonite was modified with Fe^{3+} to test if it has a positive effect in increasing DBTO removal from model oil. The experiments showed higher residual sulfur content after adsorption with BF3, as opposed to BR. This is due to the opposition in characteristics of Fe^{3+} and DBTO, wherein Fe^{3+} is classified as a hard acid, while DBTO is considered a soft base. Since soft acids tend to react with soft bases, the impregnation of bentonite with Fe^{3+} reduced its affinity for DBTO.

A parametric analysis of the adsorption of DBTO by BR and BF3 was performed. An increase in adsorption time showed a decrease in residual sulfur content as it allows the DBTO molecules to occupy active sites on the adsorbent. An equilibrium time of 6 h was observed for both BR and BF3. Adsorbent dosage and residual sulfur content showed an inverse relationship as increasing the amount of adsorbent results in more sites for adsorption. With more active sites available for the bonding of DBTO with the adsorbent, a lower concentration of the solution was observed. Lastly, the effect of temperature was investigated and showed that an increase in temperature increased the removal of DBTO. Due to the increase in kinetic energy of the system, the DBTO molecules can have more effective collisions with the adsorbent particles. The DBTO and BR and BF3 system was observed to be endothermic, as the increase in temperature resulted in the progression of the adsorbent-adsorbate reaction. With these insights, it is vital to arrive at the most economical set of parameters partnered with an efficient adsorbent for the effective desulfurization of fuel oil.

5. References

[1] Srivastava V C 2012 An evaluation of desulfurization technologies for sulfur removal from liquid fuels *RSC Adv.* **2** 759–83

[2] Rajendran A, Cui T Y, Fan H X, Yang Z F, Feng J and Li W Y 2020 A comprehensive review on oxidative desulfurization catalysts targeting clean energy and environment *J. Mater. Chem. A* **8** 2246–85

[3] Ettekali N, Allahyari S, Rahemi N and Abedini F 2021 One-pot oxidative-adsorptive desulfurization of model and real fuel using micro-mesoporous SiO2 aerogel supported MoO3 *Microporous Mesoporous Mater.* **326** 111376

[4] De Luna M D G, Wan M W, Golosinda L R, Futalan C M and Lu M C 2017 Kinetics of Mixing-Assisted Oxidative Desulfurization of Dibenzothiophene in Toluene Using a Phosphotungstic Acid/Hydrogen Peroxide System: Effects of Operating Conditions *Energy and Fuels* **31** 9923–9

[5] Flores R, Rodas A and Gasperin R 2019 Oxidative desulfurization of diesel fuel oil using supported Fenton catalysts and assisted with ultrasonic energy *Pet. Sci.* **16** 1176–84

[6] Rezakazemi M and Zhang Z 2018 Desulfurization Materials *Comprehensive Energy Systems* vol 2 (Woodhead Publishing) pp 944–79

[7] Saha B, Vedachalam S and Dalai A K 2021 Review on recent advances in adsorptive desulfurization *Fuel Process. Technol.* **214** 106685

[8] Yaseen M, Ullah S, Ahmad W, Subhan S and Subhan F 2021 Fabrication of Zn and Mn loaded activated carbon derived from corn cobs for the adsorptive desulfurization of model and real fuel oils *Fuel* **284** 119102

[9] Jung B K and Jhung S H 2015 Adsorptive removal of benzothiophene from model fuel, using modified activated carbons, in presence of diethylether *Fuel* **145** 249–55

[10] Shah S S, Ahmad I, Ahmad W, Ishaq M, Gul K, Khan R and Khan H 2018 Study on adsorptive capability of acid activated charcoal for desulphurization of model and commercial fuel oil samples *J. Environ. Chem. Eng.* **6** 4037–43

[11] Cerutti M L M N, Hackbarth F V., Maass D, Chiaro S S X, Pinto R R C, Cardoso M J B, Arroyo P A, Ulson de Souza A A and de Souza S M A G U 2019 Copper-exchanged Y zeolites for gasoline deep-desulfurization *Adsorption* **25** 1595–609

[12] Yang R T, Hernández-Maldonado A J and Yang F H 2003 Desulfurization of transportation fuels with zeolites under ambient conditions *Science (80-.).* **301** 79–81

[13] Kampouraki Z C, Giannakoudakis D A, Nair V, Hosseini-Bandegharaei A, Colmenares J C and Deliyanni E A 2019 Metal organic frameworks as desulfurization adsorbents of DBT and 4,6-DMDBT from fuels *Molecules* **24** 1–23

[14] Huang M, Chang G, Su Y, Xing H, Zhang Z, Yang Y, Ren Q, Bao Z and Chen B 2015 A metal-organic framework with immobilized Ag(i) for highly efficient desulfurization of liquid fuels *Chem. Commun.* **51** 12205–7

[15] Chen M, Ding Y, Liu Y, Wang N, Yang B and Ma L 2018 Adsorptive desulfurization of thiophene from the model fuels onto graphite oxide/metal-organic framework composites *Pet. Sci. Technol.* **36** 141–7

[16] Tian G, Wang W, Zong L and Wang A 2017 MgO/palygorskite adsorbent derived from natural Mg-rich brine and palygorskite for high-efficient removal of Cd(II) and Zn(II) ions *J. Environ. Chem. Eng.* **5** 1027–36

[17] Shakirullah M, Ahmad W, Ahmad I, Ishaq M and Khan M I 2012 Desulphurization of liquid fuels by selective adsorption through mineral clays as adsorbents *J. Chil. Chem. Soc.* **57** 1375–80

[18] Yi D, Huang H and Shi L 2013 Desulfurization of model oil via adsorption by copper(II) modified bentonite *Bull. Korean Chem. Soc.* **34** 777–82

[19] Huang H, Yi D, Lu Y, Wu X, Bai Y, Meng X and Shi L 2013 Study on the adsorption behavior and mechanism of dimethyl sulfide on silver modified bentonite by in situ FTIR and temperature-programmed desorption *Chem. Eng. J.* **225** 447–55

[20] Chen T C, Agripa M L, Lu M C and Dalida M L P 2016 Adsorption of Sulfur Compounds

from Diesel with Ion-Impregnated Activated Carbons *Energy and Fuels* **30** 3870–8
[21] Ullah S, Hussain S, Ahmad W, Khan H, Imran K, Khan S and Khan S 2020 Desulfurization of Model Oil through Adsorption over Activated Charcoal and Bentonite Clay Composites *Chem. Eng. Technol.* **43** 564–73
[22] Pearson R G 1963 Hard and Soft Acids and Bases *J. Am. Chem. Soc.* **85** 3533–9
[23] Xiao J, Li Z, Liu B, Xia Q and Yu M 2008 Adsorption of benzothiophene and dibenzothiophene on ion-impregnated activated carbons and ion-exchanged Y zeolites *Energy and Fuels* **22** 3858–63
[24] Ali F D 2018 Adsorptive Desulfurization of Liquid Fuels Using Na-Bentonite Adsorbents *Al-Nahrain J. Eng. Sci.* **21** 248–52
[25] Chen T C, Sapitan J F F, Ballesteros F C and Lu M C 2016 Using activated clay for adsorption of sulfone compounds in diesel *J. Clean. Prod.* **124** 378–82
[26] Shah S S, Ahmad I and Ahmad W 2016 Adsorptive desulphurization study of liquid fuels using Tin (Sn) impregnated activated charcoal *J. Hazard. Mater.* **304** 205–13

Acknowledgments

Acknowledgment is given to the Department of Science and Technology (DOST) of the Philippines for the provision of its Engineering Research and Development for Technology (ERDT) program and the Ministry of Science and Technology (MOST 111-2221-E-041-002-MY3) for providing financial support.

IOP Conf. Series: Materials Science and Engineering **1318** (2024) 012009 doi:10.1088/1757-899X/1318/1/012009

Fast biodiesel production via potassium ferrate catalysis at room temperature using used cooking oil and refined canola oil

R B Demafelis, M K A Cases, J C Sombilla, V P Migo, A E D Matanguihan and R J B Landoy
Department of Chemical Engineering, College of Engineering and Agro-Industrial Technology, University of the Philippines Los Baños, College, Laguna, Philippines, 4031

rbdemafelis@up.edu.ph, macases1@up.edu.ph, jcsombilla@up.edu.ph, vpmigo@up.edu.ph, admatanguihan@up.edu.ph, and rblandoy@up.edu.ph

Abstract. As biodiesel is continuously aimed to be a better alternative to petroleum-based diesel fuel, the University of the Philippines Los Baños – Interdisciplinary Biofuels Research Studies Center (UPLB-IBRSC) continues to explore process conversion techniques to improve efficiencies in biodiesel production. In this study, the catalytic activity of potassium ferrate in the transesterification of canola oil and used cooking oil for biodiesel production at room temperature was examined for potential reduction in energy consumption. Parameters such as methanol-to-oil molar ratio, catalyst loading, and reaction time were varied to determine their effects on biodiesel yield and purity. Thin layer chromatography was applied to qualitatively analyze the biodiesel, while ImageJ software was used for purity determination. The characterization analysis confirmed that the refined canola oil used in the study has an acceptable FFA content of 0.06%. On the other hand, the used cooking oil showed an FFA content of 3.71%, which is beyond the desired limit of 2%. Hence, neutralization through the addition of potassium hydroxide was performed, resulting in a lowered FFA content of 0.86%. Experimental results of the parametric and optimization study on used cooking oil showed that an optimum conversion of 86.55% was achieved at a 2:1 methanol-to-oil molar ratio, 5 wt% catalyst loading, and 90 minutes reaction time. Consequently, a yield of 83.21% was attained using refined canola oil at the optimum conditions of 6:1 methanol-to-oil molar ratio, 3 wt% catalyst loading, and 69 minutes reaction time. Generally, findings show that catalyst loading and reaction time significantly affect the biodiesel yield and purity only up to a certain optimum value. The methanol-to-oil ratio, on the other hand, showed to have a significant inverse relationship on yield and purity. Compared to the typical biodiesel production process at 60 °C with up to two hours of reaction time, this study which describes a catalytic process at room temperature and a relatively shorter reaction time, showed the potential to reduce significantly the high energy consumption in biodiesel production.

1. Introduction

Due to its inexpensive cost and capacity to meet the demand of sustaining everyday life, fossil fuel is widely used, making its reserves overly exploited throughout the years. However, the production and combustion of fossil fuels, such as coal, petroleum oil, and natural gas, became a major driver of climate change and global warming due to the greenhouse gas emissions and air pollution associated with its use [1]. The presence of greenhouse gases in large quantities in the atmosphere threatens the environment as it further contributes to the Earth's surface temperature rise. With the adverse threats to the environment, the use of biodiesel is seen as one of the possible solutions for a greener and safer alternative source of energy.

Biodiesel is a renewable energy source that can be obtained from renewable and biodegradable materials, making it sustainable as well. Common feedstock includes edible oil and non-edible oil, algae, and waste oil. In the Philippines, the most common feedstock used in biodiesel production is coconut oil, an edible or first-generation feedstock. As reported by the Department of Energy [DOE] [2], the thirteen accredited biodiesel producers in the country are using coconut oil as feedstock and are estimated to produce a total of 708 million liters annually. This amount of production signifies that the Philippines is one of the countries that recognize the capability of biodiesel as a diesel or petroleum fuel alternative [3]. Due to the economic growth and projected aim of increasing the amount of biodiesel mixed in petroleum products from 2% to 5%, an increase in the biodiesel requirement is expected in the coming years [4].

The current biodiesel industry in the Philippines involves the utilization of coconut oil as its feedstock and methanol for transesterification catalyzed by a strong alkali catalyst such as sodium hydroxide (NaOH) and potassium hydroxide (KOH) [5]. This catalyst usually allows the transesterification process to proceed at 60 °C for 180 minutes [6]. However, it raises concerns regarding its economics since the reaction takes significant time, aside from requiring a relatively high temperature. Another problem the industry currently faces is the continuously increasing copra price which contributes to the high selling price of biodiesel.

With this, the study aimed to assess the rapid production of biodiesel at room temperature conditions from canola oil and used cooking oil, categorized under first-generation and second-generation feedstocks, respectively, via transesterification catalyzed by potassium ferrate.

2. Materials and Methods

2.1. Titrimetric Method for Free Fatty Acid Analysis

The titration method was used to analyze the free fatty acid (FFA) content of canola oil and used cooking oil. Determination of the FFA content of oils to be used in transesterification is an essential part of biodiesel production since FFAs can increase the degradation of biodiesel [8]. Hence, it is desired that the oil sample has an FFA content of less than 2% for a better transesterification process [9]. Oil neutralization was performed by adding NaOH to ensure the FFA content was within the recommended range.

To determine the FFA content of the oil samples, a standardized NaOH solution was used as a titrant. Approximately two grams of each oil sample was used, dissolved in 50 mL of ethanol, as well as three drops of phenolphthalein indicator in ethanol. Equation 1 was used to calculate the FFA content.

$$\%FFA = \frac{N_{NaOH}(V_{sample} - V_{blank})(56.1)}{m_{oil}} \tag{1}$$

where V_{sample} is the volume of NaOH solution used to titrate the sample, V_{blank} is the volume of NaOH solution used in blank titration, N_{NaOH} is the standardized concentration of NaOH solution (N), and m_{oil} is the mass of the oil sample (g).

2.2. Optimization of Biodiesel Yield. For process optimization, a Box-Behnken Design (BBD) for response surface methodology was used to analyze the effects of the three reaction parameters, as shown in Table 1. Using the Design Expert 13 software (trial version), 15 experimental runs were generated for each oil sample.

Table 1. Parameters with their corresponding levels for Box-Behnken design of experiment

FACTORS	CANOLA OIL			USED COOKING OIL		
	Low Level	Medium Level	High Level	Low Level	Medium Level	High Level
Catalyst Loading (wt%)	3.0	5.5	8.0	3.0	5.0	7.0
Methanol:Oil (molar ratio)	6:1	12:1	18:1	2:1	4:1	6:1
Reaction Time (min)	30	60	90	30	60	90

The production of biodiesel through transesterification was facilitated at room temperature conditions. An oil amount of 25g was transferred into a 125mL Erlenmeyer flask. The methanol and potassium ferrate mixture was then added to the flask, and the reaction was allowed to proceed. After the corresponding reaction time, the reaction mixture was filtered using filter paper and was transferred into a separatory funnel to remove the spent catalyst and to separate the glycerol and biodiesel. After a two-layer mixture (crude biodiesel and glycerol) was formed, the glycerol part was drained from the separatory funnel to separate it from the crude biodiesel. For about two to three times, warm distilled water was then added to the crude biodiesel and then drained to remove water-soluble impurities and unreacted methanol. Then, the washed biodiesel was heated for 20 to 30 minutes to obtain the final biodiesel product. Meanwhile, the produced biodiesel was subjected to a purity test using Thin Layer Chromatography (TLC), and the corresponding biodiesel yield was computed as follows:

$$\%Biodiesel\ Yield = \frac{(Mass\ of\ biodiesel\ recovered)(\%\ Purity)}{Theoretical\ FAME\ yield} \tag{2}$$

2.3. Biodiesel Purity Test. The purity of the biodiesel obtained from the transesterification with potassium ferrate was tested using Thin Layer Chromatography (TLC). The TLC plates used in this study were Merck TLC Silica Gel 60 F254 20cm by 20cm Aluminum Sheets. Furthermore, the mobile phase was a mixture of hexane, diethyl ether, and formic acid in a volumetric ratio of 80:20:2.

To initiate the purity test, 2µL of the biodiesel sample from each run was spotted on the TLC plates (approximately one inch from the bottom) and was allowed to fully dry. Then, the plates were transferred into the solvent chamber, in which the solvent was allowed to equilibrate. For about an hour, the solvent was allowed to travel up the plate until the solvent frow reached one inch below the top of the plate. The plate was then removed from the chamber and transferred into another chamber that contained iodine pellets for the chromatogram to develop fully. The spots identified from the chromatogram were then analyzed using ImageJ software by determining the peak areas of the spots.

3. Results and Discussion

3.1. Free Fatty Acid Analysis

The FFA content of the oils was determined to assess whether pre-treatment of the feedstock is necessary prior to transesterification. Using the titrimetric method, it was found that the FFA content of the refined and unused canola oil was 0.06%. This result agrees with the typical 0.04% to 0.06% FFA content of unfried canola oil [10]. Furthermore, its FFA content is lower than the 2.0% FFA content limit to be considered a good quality biodiesel feedstock [11]. This result indicates that no prior treatment is necessary for the canola oil. Meanwhile, the FFA content obtained for used cooking oil was found to be 3.71% which is higher than the recommended range of FFA content. Thus, pre-treatment for this oil sample was done. After neutralization, it was found that the FFA content of used cooking oil was reduced from 3.71% to 0.86%, which is now in the range of acceptable FFA content for biodiesel production.

3.2. Biodiesel Purity. The acquired biodiesel product from each run was analyzed via TLC to determine its purity. The analysis revealed that intermediates like monoglycerides, diglycerides, and FFAs are also present in addition to the methyl esters generated. The topmost spot on the chromatogram represents fatty acid methyl esters (FAME) [12]. The spots on the second layer of the TLC plate are the FFAs, and the latter three spots represent 1,3-diglycerides, 1,2-diglycerides, and monoglycerides, respectively. Since silica is polar, a component that is more polar tends to stay deeper into the silica gel or at the stationary phase [13]. Therefore, the component's ability to move along the plate will decrease as it becomes more polar. Diglycerides are less polar than monoglycerides, which explains why they migrated more quickly up the TLC plate, whereas monoglycerides took longer [14]. This outcome can be linked to both components' structural similarities, which include a polar portion—glycerol—and a non-polar portion—a fatty acid chain. The diglyceride is less polar than the monoglyceride because it has two fatty acid chains in its structure as opposed to the monoglyceride's single chain of non-polar region [15]. On the other hand, FAME and monoglycerides are both polar molecules. However, FAME occupies the uppermost position in the chromatogram, given the molecular weight difference between the two. The molecular weights of the constituents have an impact on how the components separate in TLC [16]. It was verified that a component with a greater molecular weight often has a lesser rate of flow in the TLC plate. This indicates that a lower molecular weight component has a greater likelihood of migrating up the plate, whereas a higher molecular weight component is likely to stick at the lower area of the plate.

For purity analysis, after identifying the spots in the chromatograph, the relative area of the peak corresponding to the methyl ester was determined by subjecting the photo of the chromatogram to ImageJ software. As a basis, Figure 1 shows the example plot area generated from one of the runs. The area of the peaks considered in the analysis is done by drawing a line across from where the peak starts to where it ends significantly. Furthermore, it was identified that the most prominent peak on the leftmost part of the graph corresponds to the peak of fatty acid methyl ester. Other peaks are labelled in the figure.

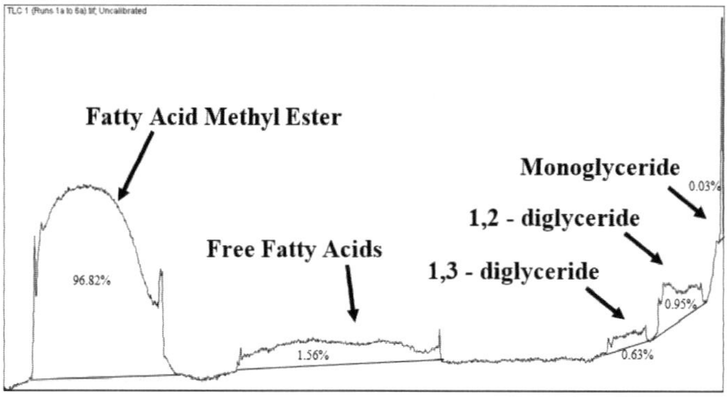

Figure 1. Peak areas generated by ImageJ.

From the analysis of the developed chromatographs, the yield with the lowest average purity (95.24%) from canola oil feedstock was obtained for a reaction time of 60 minutes, the highest catalyst loading of 8 wt%, and the least amount of methanol (6:1 molar ratio). On the other hand, the run that produced the highest purity of 98.96% involved a catalyst loading of 5.5 wt% and a methanol-to-oil molar ratio of 6:1. Conversely, for the used cooking oil feedstock, the yield with the highest recorded average purity was for runs 5 (2:1, 5.0 wt%, 30 min), 7 (2:1, 5.0 wt%, 90 min), 13 (4:1, 5.0 wt%, 60 min), 14 (4:1, 5.0 wt%, 60 min), and 15 (4:1, 5.0 wt%, 60 min) with a purity of 100% while the lowest recorded average was for run 2 (6:1, 5.0 wt%, 60 min) with a purity of 82.67%.

3.3. Yield optimization using Box-Behnken design of experiment. From the results of the optimization study, the highest biodiesel yield obtained from canola oil was 83.19% at a reaction time of 60 minutes, 6:1 methanol-to-oil molar ratio, and catalyst loading of 3 wt%. Whereas, for used cooking oil, the highest yield was recorded at a methanol-to-oil molar ratio of 2:1, catalyst loading of 5.0 wt%, and reaction time of 90 minutes. On the other hand, the lowest biodiesel yield recorded from using canola oil is 53.71%. This yield was obtained when the reaction time is 60 minutes, and the catalyst loading and methanol-to-oil molar ratio are at the highest, 8 wt% and 18:1, respectively. For used cooking oil, on the other hand, the lowest yield of 52.49% was recorded in the run with a catalyst loading of 3 wt%, methanol-to-oil molar ratio of 6:1, and a reaction time of 30 minutes. The results of the optimization study revealed the following empirical model equations to obtain biodiesel yield from canola oil (Equation 3) and used cooking oil (Equation 4).

$$Y = 68.69 - 6.99B - 4.13A^2B \tag{3}$$

$$Y = 71 - 11.59B + 2.76C - 6.11A^2 \tag{4}$$

where A is the coded value for catalyst loading, B is the coded value for methanol-to-oil molar ratio, and C is the coded value for reaction time.

For both oil samples, the methanol-to-oil molar ratio affects the biodiesel yield significantly. Among the values of the coefficients, parameter B has the highest magnitude, which implies that this factor had the most significant influence among the parameters [17]. Particularly, the trend suggests an inverse relationship between the amount of methanol used in the reaction and the biodiesel yield. Hence, the amount of biodiesel produced can be increased at a relatively lower amount of methanol used in the reaction. Meanwhile, a quadratic term involving catalyst loading shows that this parameter significantly affects the biodiesel yield only up to a certain optimum value, and further addition of catalyst beyond this limit will negatively affect biodiesel yield. This can be explained due to the high oxidative property of potassium ferrate, which could result in direct conversion of methanol to formaldehyde at excessive concentrations of the catalyst, thereby decreasing the formation of side products in the form of methoxides which are essential for transesterification [18]. Additionally, reaction time shows to be a significant factor for the transesterification of used cooking oil but not in refined canola oil. Observations from the yield obtained from used cooking oil revealed that reaction time positively affects the biodiesel yield.

Using the generated empirical models, it was found that the optimum conditions for biodiesel production using canola oil are at 3.001 wt% catalyst loading, 6.005:1 methanol-to-oil molar ratio, and 68.986 minutes reaction time. On the other hand, the optimum condition for used cooking oil as feedstock for biodiesel production are 2:1 methanol-to-oil molar ratio, 5 wt% catalyst loading, and 90 minutes reaction time. These conditions were validated through confirmatory runs for canola oil and used cooking oil in which an average of 84.70% (98.73% purity) and 84.25% (100% purity) biodiesel yield were obtained, respectively. These values agree with the obtained theoretical yield using the generated empirical models, having only a percent difference of 0.08% and 2.65% for canola and used cooking oil, respectively.

In comparison to other studies involving biodiesel production from canola oil and used cooking oil, Ilgen and Akin optimized the reaction conditions at a nine-hour reaction time, 3 wt% MgO-loaded catalyst concentration, 6:1 molar ratio of methanol to canola oil, and 60°C reaction temperature yielding 95.05% biodiesel [19]. Similarly, a 99.6% biodiesel yield resulted from processing canola oil when 3 wt% KF/Al_2O_3 catalyst was used at 60°C reaction temperature, but the optimum reaction time and methanol-to-canola oil molar ratio were eight hours and 15:1, respectively [20]. However, only 71.9% biodiesel yield was obtained in a nine-hour reaction period at 60°C reaction temperature, 6:1 molar ratio, and 3 wt% catalyst concentration by using Mg-Al hydrotalcite catalyst in the transesterification with canola oil [21]. Furthermore, a 98% biodiesel yield was obtained from canola oil transesterification using the optimum conditions of 18-hour reaction time, 24:1 methanol-to-canola oil molar ratio, 100°C reaction temperature, and $AlCl_3$ catalyst loading of 5 wt% and tetrahydrofuran as co-solvent [22]. On the other hand, a study concerning used cooking oil as feedstock with optimized conditions of 50°C reaction temperature, 8:1 methanol-to-used cooking oil molar ratio, 1 wt% calcium oxide (CaO) catalyst loading, and 90 minutes reaction time yields an estimated conversion of 96% [23].

4. Conclusion

Comparing the biodiesel production from canola oil and used cooking oil to the conventional biodiesel production from coconut oil in the Philippines, this study shows good potential in terms of improving the economic viability of the production process. Studies show that the optimized parameters for coconut oil as feedstock are 65°C reaction temperature, 6:1 methanol-to-oil molar ratio, 1 wt% potassium hydroxide (KOH) catalyst loading, and 40 minutes reaction time, yielding about 79% conversion [24]. Additionally, looking into the recent study on transesterification of coconut oil catalyzed by CaO/Al_2CO_3, it was found that the optimum yield is 93.29% methyl ester at the optimum conditions of 6:1 methanol-to-oil molar ratio, two hours reaction time, 60°C reaction temperature, and 1.0 wt% heterogeneous catalyst [25]. These processes entail high energy demand that significantly contributes to high production costs. This study proves the possibility of biodiesel production at room temperature in a relatively shorter reaction time with comparable yield and purity that may potentially cut production costs. It is recommended for future studies to conduct a financial assessment on the use of potassium ferrate catalysis for commercial-scale biodiesel production. This may eventually aid the country push for higher biofuel blending for diesel sold nationwide.

References

[1] Perera F and Nadeau K 2022 Climate change, fossil-fuel pollution, and children's health *NEJM.* **386** 2304-2314 (doi: 10.1056/NEJMra2117706)
[2] Department of Energy 2023 List of Biodiesel
[3] Rouhany M and Montgomery H 2019 Global biodiesel production: The state of the art and the impact on climate change *Biofuel Bioprod Biorefin.* **8** (doi: 10.1007/978-3-030-00985-4_1)
[4] Voegele E 2022 Biodiesel consumption in the Philippines to increase in 2022 *Biodiesel Magazine*
[5] Acda M N 2022 Production, regulation, and standardization of biofuels: a Philippine perspective *Value-Chain of Biofuels* 553-570 (https://doi.org/10.1016/B978-0-12-824388-6.00001-4)
[6] Talha N S and Sulaiman S 2016 Overview of catalyst in biodiesel production *ARPN J. Eng. Appl. Sci.* **11**1819-6608
[7] IT-Desiree 2023 Commodity price watch
[8] Demirbas A 2007 Thermal Degradation of Fatty Acids in Biodiesel Production by Supercritical Methanol *Energy Explor.* **25** 64-70
[9] Suraj C K Anand K and Sundararajan T 2017 Investigation of biodiesel production methods by altering free fatty acid content in vegetable oils *Biofuels* **11(5)** 587-595 (https://doi.org/10.1080/17597269.2017.1378993)
[10] Adjonu R Zhou Z Prenzler P D Ayton J and Blanchard C L 2019 Different processing practices and the frying life of refined canola oil *Foods* **8(11)** 527 (doi: 10.3390/foods8110527)
[11] Satar I Permadi A and Ahmed W A 2022 Evaluating physicochemical properties of mutated *Jatropha curcas* seed oils and their feasibility as new feedstocks for biodiesel production *Eng. Agric.* **42** (doi: 10.1590/1809-4430-Eng-Agr.v42nepe20220093/2022)
[12] Fedosov S N Brask J and Xu X 2011 Analysis of biodiesel conversion using thin layer chromatography and nonlinear calibration curves *J. Chromatog. A.* **1218** 2785-2792 (doi: 10.1016/j.chroma.2011.01.067)
[13] Garbarino J 2023 Thin layer chromatography (TLC) for the separation of lipids
[14] Hudson B J F 2003 Fatty acids | Properties *Food Sci. Nutr.* **2** 2297-2300 (doi: 10.1016/B0-12-227055-X/00446-6)
[15] Prajapati H N Dalrymple D M and Serajuddin A T M 2011 A comparative evaluation of mono-, di-, and triglyceride of medium chain fatty acids by lipid/surfactant/water phase diagram,

solubility determination and dispersion testing for application in pharmaceutical dosage form development. *Pharm. Res.* **29(1)** 285-305 (doi: 10.1007/s11095-011-0541-3)

[16] Kamiyama F and Inagaki H 1971 The mechanism of polymer fractionation by thin-layer chromatography. *Bulletin of the Institute of for Chemical Research* **49(2)**

[17] Salam K K Agarry S E Arinkoola A O and Shoremekun I O 2015 Optimization of operating conditions affecting microbiologically influenced corrosion of mild steel exposed to crude oil environments using response surface methodology *Br. Biotech. J.* **7(2)** 68 – 78 (doi:10.9734/BBJ/2015/16810)

[18] Xu Z Hu J Yan Z Yang S Zhou J and Lu W 2009 Potassium ferrate (VI) and decomposed K_2FeO_4 assisted methanol electro-oxidation in alkaline media *Electrochim. Acta.* ***54*(13)** 3548-3552 (https://doi.org/10.1016/j.electacta.2009.01.001)

[19] Ilgen O and Akin A N 2008 Transesterification of canola oil to biodiesel using MgO loaded with KOH as a heterogenous catalyst *Energ Fuel.* **23** 1786-1789 (doi: 10.1021/ef800345u)

[20] Boz N and Kara M 2008 Solid base catalyzed transesterification of canola oil *Chem Eng Commun.* **196(1-2)** 80-92 (doi: 10.1080/0098644080230148)

[21] Ilgen O Dincer I Yildiz M Alptekin E Boz N Canacki M and Akin A N 2007 Investigation of biodiesel production from canola oil using Mg-Al hydrotalcite catalyst *Turk. J. Chem.* **31(5)** 509-514

[22] Soriano N U Venditti R and Arygropoulus D S 2009 Biodiesel synthesis via homogenous Lewis acid-catalyzed transesterification *Fuel* **88(3)** 560-565 (doi: 10.1016/j.fuel.2008.10.013)

[23] Degfie T A Mamo T T and Mekonnen Y S 2019 Optimized Biodiesel Production from Waste Cooking Oil (WCO) using Calcium Oxide (CaO) Nano-catalyst (https://doi.org/10.1038/s41598-019-55403-4)

[24] Abubakar H G and Muhammad I 2018 Determination of Optimal Yield of Biodiesel from Coconut (Cocos Nucifera) Seed Oil *IJISET* **5(4)** 127-131

[25] Usman J Abdullahi B H Idris B Tanko U M Usmar B Ali J B and Abubakar K 2020 Process optimization of biodiesel production from crude coconut oil seeds oil using CaO/Al_2CO_3 as heterogeous catalyst *Sci. Pro. Ser.* **2(1)** 92-97

IOP Conf. Series: Materials Science and Engineering **1318** (2024) 012010 doi:10.1088/1757-899X/1318/1/012010

Fast biodiesel production via potassium ferrate catalysis at room temperature using refined coconut oil and vegetable oil

R B Demafelis, K C Cello, J T Ronquillo, M G Borines, M E Bambase Jr, B T Magadia, A E D Matanguihan, R J B Landoy
Department of Chemical Engineering, College of Engineering and Agro-Industrial Technology, University of the Philippines Los Baños, College, Laguna, Philippines, 4031

rbdemafelis@up.edu.ph, kqcello@up.edu.ph, jtronquillo@up.edu.ph, mgborines@up.edu.ph, mebambase@up.edu.ph, btmagadia@up.edu.ph, admatanguihan@up.edu.ph, and rblandoy@up.edu.ph

Abstract. Traditionally, the transesterification reaction for biodiesel production takes place at elevated temperatures, around 60-65°C, for up to two hours of reaction time, entailing a high energy consumption. As the University of the Philippines Los Baños – Interdisciplinary Biofuels Research Studies Center (UPLB-IBRSC) continuously aims to conduct studies on improving process efficiencies in biofuels production, this research investigated the potential of potassium ferrate as a catalyst for fast biodiesel production at room temperature. Particularly, the catalytic activity of potassium ferrate in the transesterification of refined coconut oil and vegetable oil was examined. The effects of the parameters such as methanol-to-oil molar ratio, catalyst loading, and reaction time on biodiesel yield and purity were studied. Thin layer chromatography, with the aid of ImageJ software, was applied to analyze the purity of the biodiesel through the developed spots' area estimation. Characterization experiments in terms of free fatty acid (FFA) content and acid value were performed to confirm the low FFA content of the feedstocks prior to transesterification. Results of the parametric and optimization study revealed that a biodiesel yield of 87.67% using vegetable oil was experimentally achieved at the optimum conditions of 6.75 wt% catalyst loading, 6:1 methanol-to-oil molar ratio, and 30 minutes reaction time. Using coconut oil, on the other hand, resulted in a higher biodiesel yield of 94.90% at a catalyst loading of 6 wt%, a methanol-to-oil molar ratio of 6:1, and a reaction time of 60 minutes. Generally, findings show that catalyst loading has a positive effect on biodiesel yield and purity only up to a certain optimum amount. Among all the parameters analyzed, the methanol-to-oil molar ratio was found to have the most significant effect as an individual factor. Given the results of this study, it can be concluded that the potassium ferrate-catalyzed transesterification reaction at room temperature can potentially reduce the energy consumption in biodiesel production and the production cost in general.

1. Introduction

Biodiesel, also referred to as Fatty Acid Methyl Ester (FAME), is a renewable and biodegradable fuel consisting of long-chain fatty acid alkyl esters that were produced through the transesterification of oil or fats [1]. As green energy, it can be a suitable alternative to petroleum-based diesel or fossil fuel. In the Philippines, the current biodiesel blend as of July 2022 is B2 (2% biodiesel, 98% petroleum) [2]. Even though the country has long been open to supporting the cleaner way of providing energy for

IOP Conf. Series: Materials Science and Engineering **1318** (2024) 012010 doi:10.1088/1757-899X/1318/1/012010

transportation purposes, the mandated plan for increasing the blend still hasn't been met. The Philippine Energy Plan for 2012-2030 made by the Department of Energy (DOE), which includes reaching the target blend of B5 (5% biodiesel, 95% petroleum) by 2015, is already long overdue. According to the former Energy Secretary of DOE, Zenaida Y. Monsada, the two key factors considered in mandating the increase in the blend are sustainability of implementation and impact on pricing. The sustainability of implementation highlights the availability of raw materials in the Philippines to support the carrying out of a higher blend, whereas the impact on pricing is more focused on the social acceptability in terms of the resulting price considering the resources and energy used during manufacturing [3].

At present, typical biodiesel productions are conducted at around 65 °C, which is the boiling point of methanol. The process is operated under this temperature condition to facilitate the creation of a methoxide ion that attacks the triglyceride to create biodiesel [4]. However, due to the high-temperature requirement, this part of the manufacturing process is accountable for the majority of used-up energy—making biofuel production less sustainable. As for the reaction time, transesterification typically takes 2-4 hours to achieve full conversion, depending on the reaction temperature and other operating conditions. The long reaction time is not economical as it induces higher production costs, hence the failure to meet the second key factor considered in mandating the increase in the blend, which has an impact on pricing. In this direction, the objective of this study was to propose a more efficient method of producing biodiesel at room temperature using a readily available feedstock and a heterogeneous catalyst to attain faster reaction, reduced energy consumption, and lower production costs.

2. Materials and Methods

2.1. Free Fatty Acid Analysis

In producing high-quality biodiesel, a high amount of free fatty acid (FFA) during transesterification is undesirable since it would consume the catalyst to form soap, thereby decreasing the methyl ester yield and purity. Hence, the FFA value of the feedstock to be used was first determined prior to transesterification to ensure that it was low enough to be used for the reaction. For the most common alkali-catalyzed transesterification, the target acid value suggested is less than 2 mg KOH/g which is roughly 1% FFA. The FFA analysis was conducted through the titration method adapted from Kwegyir (2014) [5].

2.2. Biodiesel Yield Optimization

The catalyst-alcohol mixture was prepared in a 125mL Erlenmeyer flask by combining the corresponding amount of potassium ferrate and methanol. Following this, 25g of oil was added to the mixture, which was then put on a magnetic stirrer and agitated. The reaction occurred at room temperature with the flask capped by a glass stopper. The reaction mixture was transferred into separatory funnels once the specific reaction time had passed to improve the separation of the biodiesel, glycerol, and used catalyst. To remove any unreacted methanol, the biodiesel product was collected and heated to 65°C. The methanol-free methyl ester subsequently underwent two to three rounds of washing with distilled water. After being washed, the biodiesel was dried on a hotplate while being agitated continuously at 105°C for 25 minutes. The solution was then allowed to stand for a while until the top layer was entirely free of any distilled water that may have remained after washing. The clear upper layer was collected and weighed before being subjected to thin layer chromatography (TLC).

For process optimization of biodiesel production from vegetable oil, a Box Behnken design of experiment was applied to determine the effects of the three factors (catalyst loading, methanol-to-oil molar ratio, and reaction time) on biodiesel yield. Meanwhile, for coconut oil, optimum conditions of catalyst loading and methanol-to-oil molar ratio were first determined, followed by reaction time optimization. Table 1 and Equation 1 show the parameters with their corresponding levels and the equation used for the calculation of biodiesel yield, respectively.

Table 1. Parameters with their corresponding levels for optimization

FACTORS	REFINED COCONUT OIL			VEGETABLE OIL		
	Low Level	Medium Level	High Level	Low Level	Medium Level	High Level
Catalyst Loading (wt%)	0.15	3.075	6.0	3.0	5.5	8.0
Methanol:Oil (molar ratio)	6:1	11:1	16:1	6:1	12:1	18:1
Reaction Time (min)	5, 10, 15, 20, 25, 30, 45, 60			30	60	90

$$\%Biodiesel\ Yield = \frac{(Mass\ of\ biodiesel\ recovered)(\%\ Purity)}{Theoretical\ FAME\ yield} \quad (1)$$

2.3. Biodiesel Purity Test using Thin Layer Chromatography (TLC)
Ten evenly spaced dots were marked along the TLC plate's bottom margin. On each dot, two microliters of biodiesel from the corresponding run were applied and dried. The TLC plate was then cautiously put into the solvent-filled chamber until the solvent front reached the plate margin at the top. To prevent the biodiesel applied to the marks from washing off, it was ensured that the bottom margin was just over the level of solvent inside the chamber. After approximately an hour or when the solvent front was at the same level as the margin, the plate was removed from the chamber, air-dried for about a minute, and then placed in another chamber containing iodine pellets. The plate was left inside the chamber until the chromatogram was developed, and then it was removed and documented right before the chromatogram faded away. The spots that appeared after colorization were photographed and analyzed through ImageJ, an open-source image analysis software developed by the National Institutes of Health (NIH), which has been used in quantitative analysis coupled with TLC in various publications [6].

3. Results and Discussion

3.1. Analysis of Free Fatty Acid Content
Results of the FFA analysis confirmed that the average computed value of %FFA for refined coconut oil, which is 0.0318%, is less than 0.05% making it a good quality biodiesel feedstock. Also, since the average FFA values obtained from the two trials of vegetable oil is 0.1016%— which is less than 1% as recommended, there is no need to carry out a pre-treatment procedure for the oil sample. Moreover, the obtained data agrees with the findings from the literature, which state that refined vegetable oil contains less than 1% FFA [7]. This confirms that the chosen vegetable oil can indeed be used as a feedstock for the transesterification reaction. Therefore, both of the oils obtained favorable FFA content prior to transesterification.

3.2. Analysis of Biodiesel Purity
The purity of biodiesel was obtained by subjecting the photos of TLC plates to analysis using ImageJ, which functions as a densitometer that does qualitative analysis by getting the relative composition of the sample, as determined by the presence, area, and intensity of other spots observed in the lane of chromatogram per run. This value of purity is used to determine the yield of biodiesel. Since the software is sensitive to color changes, accurate outlining of the spots was performed to increase the readability of the spots.

For both oil samples, it can be seen in Figure 1 that intermediate products, including monoglycerides, diglycerides, and FFAs, were found aside from the desired output which is the methyl esters. The standard used in distinguishing the components observed in the TLC plate is from the study conducted by [8], where the topmost spot corresponds to FAME, right below this spot is the FFA, while the last three detectable spots pertain to 1,3-diglycerides, 1,2-diglycerides, and monoglycerides, respectively. A journal on TLC by [9] was also referred to in identifying the components.

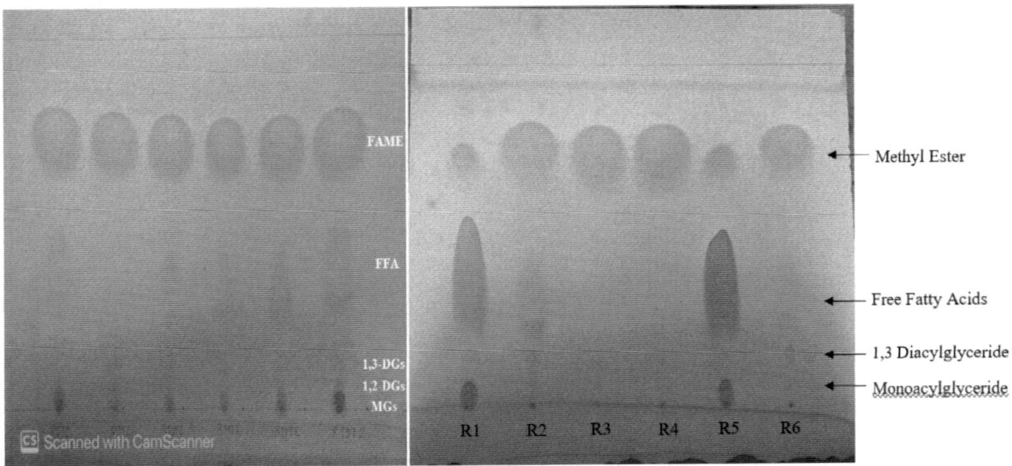

Figure 1. Developed chromatograms of biodiesel from (left) vegetable oil and (right) refined coconut oil

The photo of the TLC plates with the developed chromatogram was then uploaded and processed using the ImageJ software to determine the relative area of the corresponding peak of methyl ester. A sample plot from a run using vegetable oil which was generated from ImageJ is presented in Figure 2. Here, it can be observed that the area of the largest peak, which corresponds to the presence of fatty acid methyl ester (FAME), is way greater than that of monoglycerides'. The data provided by ImageJ was used to calculate the percentage of FAME present in the aforementioned run — which was then found to be 99.58% pure. The presence of monoglycerides accounts for the remaining 0.42%, while no 1,2-DGs, 1,3-DGs, and FFA were found in this particular sample.

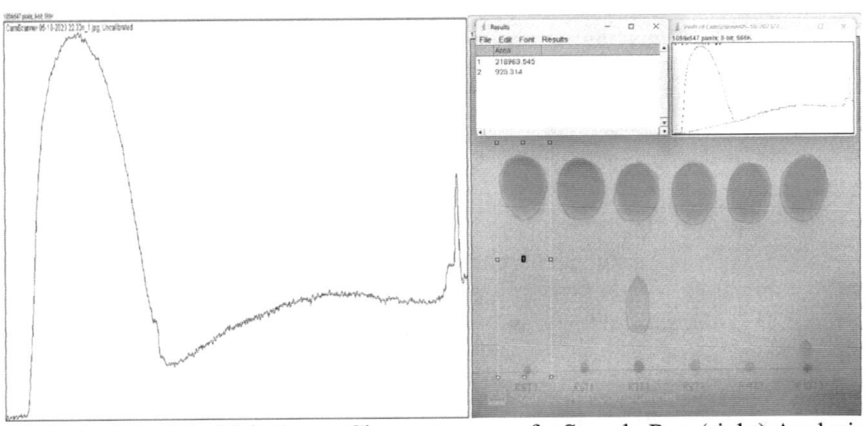

Figure 2. (left) A plot of the Thin Layer Chromatogram of a Sample Run (right) Analysis of Thin Layer Chromatogram of a Sample Run

Results of the purity test revealed that for vegetable oil, the lowest recorded values were 82.44% (3 wt% catalyst loading, 6:1 methanol-to-oil molar ratio, 60 minutes reaction time) and 86.44% (3 wt% catalyst loading, 12:1 methanol-to-oil molar ratio, 30 minutes reaction time). The presence of FFA was apparent for both runs on the TLC plate with a developed chromatogram. By examining the operating conditions, it can be observed that the catalyst loading is at its minimum in these runs. This can be explained as biodiesel purity is decreased at low catalyst loading since the amount of catalyst is insufficient to achieve full conversion and formation of methyl ester [10]. On the other hand, the highest biodiesel purity of 99.40% was achieved at a mid-level catalyst loading of 5.5 wt.%, maximum reaction time of 90 minutes, and methanol-to-oil molar ratio of 6:1 and 18:1. A sufficient catalyst loading and longer reaction time aid in achieving full conversion and formation of methyl ester, thus resulting to a high biodiesel purity.

Meanwhile, for coconut oil, the run with a catalyst loading of 6 wt % and a methanol-to-oil molar ratio of 6:1 obtained the highest average purity of 97.58%. On the other hand, a catalyst loading of 0.15 wt% and methanol to oil molar ratio of 6:1 had the lowest average purity of 22.00%. Generally, it can be observed that increasing the catalyst loading results in higher percent purity. This trend could be due to the increased concentration of active catalytic sites available for the reaction. This enhances the catalyst-substrate interaction, promoting the breaking of ester bonds in the triglycerides present in the oil. The catalyst facilitates the transesterification reaction pathway with lower activation energy allowing for a more efficient conversion of oil to methyl ester. For comparison, biodiesel produced from both vegetable oil and coconut oil showed a significant increase in purity with increasing catalyst loading.

3.3. Optimization of Biodiesel Yield

Based on the results of the optimization study for vegetable oil, it can be observed that the highest biodiesel yield was 87.67% in which the operating conditions are as follows: 5.5 wt% catalyst loading, 6:1 methanol-to-oil molar ratio, and 90 minutes reaction time. Meanwhile, the lowest value was recorded with 58.35% biodiesel yield. For this run, both the catalyst loading and methanol-to-oil molar ratio were at the maximum while the reaction time was 60 minutes. It can also be observed that the run with the second-highest biodiesel yield of 84.26% has the same catalyst loading and methanol-to-oil molar ratio as that with the highest yield. Moreover, for coconut oil, the run with catalyst loading of 0.15 wt% and methanol-to-oil molar ratio of 16:1 obtained the lowest average biodiesel yield of 5.83%. Meanwhile, the run with a catalyst loading of 6 wt % and methanol-to-oil molar ratio of 6:1 had the highest average yield of 53.75%.

Generally, it was found that biodiesel yield increases as the catalyst loading increases and as the methanol-to-oil molar ratio decreases. This trend resulted from the combined effects of catalyst and methanol. Increasing catalyst loading predominantly led to an increased reaction rate, while increasing the methanol-to-oil molar ratio predominantly resulted in excess methanol, increasing the likelihood of esterification reactions and the formation of unwanted by-products reducing the overall yield of biodiesel.

For vegetable oil, the optimum operating conditions are identified to be 6.747 wt.% catalyst loading, 6:1 methanol-to-oil molar ratio, and 30.010-minute reaction time, resulting in a desirability of 1.000 and a biodiesel yield of 87.667%— which is as high as the maximum value attained experimentally. The shorter reaction time for the optimum condition aligns with this study's concept of promoting fast biodiesel production, as the conventional process usually takes three hours or 180 minutes [11]. Among all the possible combinations or solutions provided, these optimum parameters were selected since aside from the low-level value for reaction time, it also consists of the minimum methanol-to-oil molar ratio and a considerable amount of catalyst, hence lessening not just the operating expenses in general but also the raw material costs itself.

Results of the two-factor optimization study for coconut oil showed that the highest biodiesel yield of 52.616 was obtained at a catalyst loading of 6 wt% and 6:1 methanol-to-oil molar ratio. Hence, these parameters were used for the second stage of the experiment, which is reaction time optimization. The transesterification reaction in biodiesel production follows a certain rate equation and kinetics. The reaction time optimization helps determine the optimal duration required to reach equilibrium or near-equilibrium conditions for the reaction. It allows for sufficient time for the reaction to proceed, but not excessively, minimizing the costs associated with longer reaction times. Also, longer reaction times may require higher energy input for heating and mixing the reactants. By optimizing the reaction time, energy consumption can be minimized while still achieving the desired conversion. This leads to cost savings and improved process efficiency. From the results of reaction time optimization, it can be observed that there is no significant difference in biodiesel purity at 30 minutes (95.43%) and 60 minutes (97.72%) reaction time. However, comparing the biodiesel yield, a significant difference from 53.75% to 94.90% was observed with the increase in reaction time. Hence, the considered optimum time for efficient coco methyl ester (CME) production is at 60 minutes.

To verify the biodiesel yield using the obtained optimum parameters for both oil samples, confirmatory runs were performed. For coconut oil, the obtained biodiesel yield at optimum operating conditions (6 wt% catalyst loading, 6:1 methanol-to-oil molar ratio, 60 minutes reaction time) is 94.90% with an average purity of 98.73%. As for the vegetable oil feedstock, a biodiesel yield and purity of 87.67% and 99.37, respectively, were achieved at the optimum values of the parameters (6.75 wt% catalyst loading, 6:1 methanol-to-oil molar ratio and 30 minutes reaction time).

4. Conclusion

In conventional biodiesel production with potassium hydroxide (KOH) as the catalyst, the yield is averaging 94.33% when temperature, reaction time, and alcohol-to-oil ratio were 65°C, 2 hours, and 9:1, respectively [12]. The findings of this study revealed that biodiesel production from coconut oil and vegetable oil through potassium ferrate-catalyzed transesterification could be achieved at room temperature and at a relatively faster reaction time with comparable yield and purity. Considering the energy consumption reduced by conducting the experiment under a room temperature condition rather than the conventional 65°C, and the power costs saved in the short reaction time of 30-60 minutes instead of the conventional two to three hours, the data from this study may be of substantial help to the Philippines' biodiesel industry in terms of improving the economic viability of biodiesel production. For future studies, it is recommended to perform a financial assessment on the application of potassium ferrate catalysis in commercial-scale biodiesel production.

References

[1] Yan Y 2016 Biodiesel - an overview (https://www.sciencedirect.com/topics/agricultural-and-biological-sciences/biodiesel)

[2] Rivera D 2022 Philippines consumption of biodiesel expected to rise further *Philstar.com* (https://www.philstar.com/business/2022/07/25/2197713/philippines-consumption-biodiesel-expected-rise-further)

[3] Lopez S 2016 Biodiesel production from refined coconut oil using NaOH-impregnated CaO catalyst derived from waste chicken eggshells with petroleum ether as co-solvent *Unpublished undergraduate thesis* – Chemical Engineering University of the Philippines, Los Baños, College, Laguna

[4] Gutiérrez-López A N Mena-Cervantes V Y Gonzàlez-Espinosa M A Sosa-Rodriguez F S Vazquez-Arenas J Rodriguez-Ramirez R and Hernàndez-Altamirano R 2022 Green and fast biodiesel production at room temperature using soybean and jatropha curcas L. oils catalyzed by potassium ferrate. *J. Clean. Prod.* (https://www.sciencedirect.com/science/article/pii/S0959652622033169)

[5] Kwegyir E K 2014 Determination of free fatty acid in soybean oil by titration from food and drugs authority, Ghana (https://csirspace.foodresearchgh.site/bitstream/123456789/625/1/Determination_Free_Fatty_Acid_Soybean.pdf)

[6] Sowers M Ambrose R Bethea E Harmon C and Jenkins D 2022 Quantitative thin layer chromatography for the determination of medroxyprogesterone acetate using a smartphone and open-source image analysis. *J. Chromatogr. A.* (https://pubmed.ncbi.nlm.nih.gov/35305455/)

[7] Clifford CB The Reaction of Biodiesel: Transesterification *John A. Dutton Institute for Teaching and Learning Excellence* (https://www.e-education.psu.edu/egee439/node/684)

[8] Lopez I A P 2015 Optimization of sodium hydroxide-catalyzed transesterification of triglycerides

from used cooking oil for ethyl ester production *Unpublished undergraduate thesis* – Chemical Engineering. University of the Philippines, Los Baños, College, Laguna
[9] Kamiyama F and Inagaki H 1971 Core – aggregating the world's Open Access Research Papers (https://core.ac.uk/download/pdf/39208914.pdf)
[10] Erchamo Y S et al 2021 Improved biodiesel production from waste cooking oil with mixed methanol-ethanol using enhanced eggshell-derived CaO Nano-Catalyst *Nature News*
[11] Ramavathu L N Radhika N and Sravani K 2015 Optimized Parameters for Production of Biodiesel from Fried Oil *IJARSET.* (doi: 10.17148/IARJSET.2015.2615)
[12] Ogunkunle O Oniya O O and Adebayo A O 2017 Yield response of biodiesel production from heterogeneous and homogeneous catalysis of Milk Bush Seed (Thevetia Peruviana) oil *Energy and Policy Research* (doi: 10.1080/23815639.2017.1319772)

IOP Conf. Series: Materials Science and Engineering **1318** (2024) 012011 doi:10.1088/1757-899X/1318/1/012011

Carbon Footprint and GHG Reduction Potential of Bioethanol Production in the Philippines

R B. Demafelis, B T. Magadia, AE D. Matanguihan, AA U. Mawili

Department of Chemical Engineering, College of Engineering and Agro-Industrial Technology, University of the Philippines Los Baños, Los Baños, Laguna, Philippines

Email: rbdemafelis@up.edu.ph, btmagadia@up.edu.ph, admatanguihan@up.edu.ph, aumawili@up.edu.ph

Abstract. As the Philippine bioethanol industry concluded its first decade, it is in due time to assess whether the objectives of the Biofuels Act of 2006 (Republic Act No. 9367) are being met. Specifically, the law aims to reduce the country's dependence on oil importation, mitigate toxic and greenhouse gases (GHGs), and promote the use of biofuels from indigenous resources to boost the country's rural economy, by mandating biofuels blending to fossil-based fuels for transport use. This study examined the contribution of seven bioethanol production systems to the country's climate change mitigation efforts by assessing the environmental impacts, particularly the carbon footprint and GHG emission reduction potential of each bioethanol distillery through life cycle assessment (LCA). Results of the environmental impact assessment revealed that the weighted average carbon footprint of the seven bioethanol production systems is at 1,415.66 $gCO_2e\ L^{-1}$, which ranges from 287 to 1,726 $gCO_2e\ L^{-1}$, translating to an average GHG reduction potential of 47.54% or an average avoided GHG of 268,091.15 tons $CO_2e\ yr^{-1}$ for CY 2019-2020. With the environmental hotspots identified, recommendations were made to improve the distilleries' carbon footprints, and a sensitivity analysis was conducted to quantify the benefits from the recommended strategies. By applying all strategies, the average GHG reduction potential of the seven bioethanol production systems can get as high as 85.63%, which can be translated to an average avoided GHG of 482,933.91 tons $CO_2e\ yr^{-1}$.

1. Introduction

As the bioethanol industry in the Philippines concluded its first decade, it is in due time to assess whether the objectives of the Biofuels Act of 2006 (Republic Act No. 9367) are being met, which generally aims to reduce the country's dependence to oil importation, and at the same time provide protection to the public health, the environment, and natural ecosystems by mandating biofuel blending to diesel and gasoline sold in the country. Additionally, biofuel blending aims to mitigate toxic and greenhouse gases (GHGs), to promote the use indigenous resources as biofuel feedstock, to provide alternative and renewable clean energy sources, and to boost the rural economy [1]. The current bioethanol blending mandate in the Philippines is at 10%, implemented since 2011, which was initially set at five percent in 2009 [2]. As of 2020, there are 12 fuel ethanol processing plants operating in the country [3] however, approximately 58% of the bioethanol demand is still being imported to meet the current mandate of E10. Present challenges of the industry include insufficient feedstock supply and high local price of bioethanol [4].

This study examined the contribution of seven bioethanol production systems to the country's climate change mitigation efforts by assessing the environmental impacts, particularly the carbon footprint and GHG emission reduction potentials of each of the seven bioethanol supply chain. Strategies to improve the carbon footprint of bioethanol distilleries were recommended and focused on the identified environmental hotspots within the system. The study also projected the avoided GHG emissions of the bioethanol distilleries under study at different blending scenarios until 2030.

2. Methods

Environmental impacts were assessed through life cycle assessment (LCA) methodology by first setting the functional unit, which is 'grams carbon dioxide equivalent per liter of bioethanol' ($gCO_2e\ L^{-1}$), and the system boundary (Figure 1) of the study. Within the system boundary, life cycle inventory was done by gathering the material and energy inputs and outputs within the system through key informant interviews (KII) with bioethanol distillery personnel. Material and energy balances were conducted to validate the data gathered. Feedstocks, raw materials, and product sources or suppliers of each of the seven bioethanol distilleries were also obtained from KII with the distilleries personnel to determine transportation distances for carbon footprint assessment.

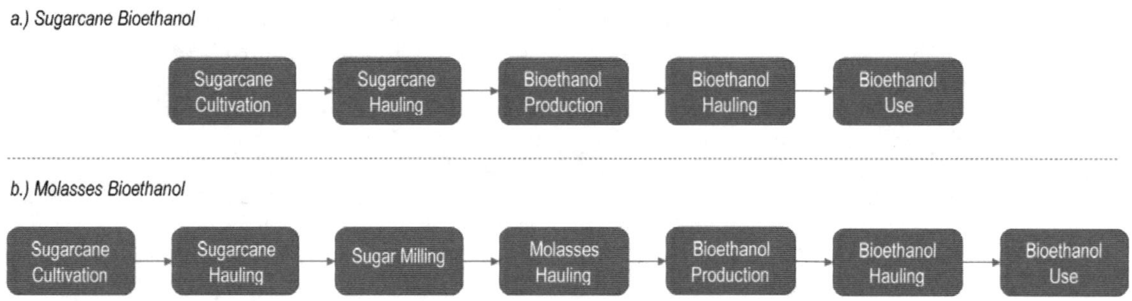

Figure 1. Basic molasses bioethanol production and use pathway from a.) sugarcane and b.) molasses.

For data consistency, all bioethanol production pathways assumed the same sugarcane cultivation practices and yield (Table 1). A hypothetical sugar mill with a capacity parallel to the molasses requirement of the individual bioethanol distilleries were made for material and energy balances. For agricultural inputs and other raw materials (e.g. coal) of known suppliers overseas, it was assumed that the materials are brought by barge or sea transpo1rt, and distances were determined from plant to port and port-to-port (country of origin to nearest port in the Philippines). Meanwhile, for other chemicals of unknown original country of destination, database from Sima Pro was obtained and used the global average market and mode of transportation for each chemical.

In this study, the environmental impact assessment focused on the global warming potential (GWP) by determining the carbon footprint and GHG emission reduction potential of the distilleries. Two carbon accounting calculations were done in this study using the LCA methodology, utilizing (1) Microsoft Excel to manually calculate the carbon footprints from each stage of the life cycle, and (2) SimaPro [5] – a leading LCA software worldwide to analyze and monitor the sustainability performance of a product or service. After the material and energy balance calculations, energy consumptions and materials inputs are inputted in the software, selecting the appropriate entry from the extensive and various database SimaPro offers. For this study, IPCC 2013 GWP 100a V1.03 was employed for impact assessment.

Table 1. Data assumptions for carbon footprint calculations.

Parameters	National Average Values (CY 2019-2020)
Sugarcane Yield	53.04 tons ha^{-1} (CY 2018-2019)
Raw Sugar	1.84 L-kg TC^{-1}
Molasses	4.81 % (ton molasses TC^{-1})
Fuel Economy	1.92 km L^{-1} (for 20 tons truck)
Molasses Price	PhP 11,456.7 ton^{-1} molasses
Raw Sugar Price	PhP 1,484.3 ton^{-1} raw sugar
Bioethanol Price*	PhP 58.7 L^{-1}
Liquefied CO2**	PhP 10,000 ton^{-1}
Dry Ice**	PhP 55,000 ton^{-1}

* *Bioethanol Price Index last September 2020 (SRA)*
** *from key informant interviews*

Average results of the environmental impacts were used to project the avoided GHG emissions of the industry at different blending scenarios until 2030. Strategies to improve the environmental impact results were deduced based from the environmental hotspots identified within the system, and sensitivity analyses were conducted to investigate to what extent can the recommended strategies improve the environmental benefits of bioethanol production in the country.

2.1. *Carbon Life Cycle Inventory*

In principle, life cycle inventory is simply the summation of all the GHG emissions classified under Scopes 1, 2 and 3 as defined by the GHG Protocol Product Life Cycle Accounting and Reporting Standard [6] within the defined system boundary. The general equation to compute for the GWP or carbon footprint of bioethanol is shown below:

$$GWP_{ethanol}\,(kgCO_2e/L) = \sum S1 + \sum(X_{S2})(EF_{X_{S2}}) + \sum(X_{S3})(EF_{X_{S3}}) \quad (1)$$

wherein S1 represents the direct emissions in kgCO2equivalent L-1 within the set system boundary, XS1 and XS2 are energy consumption and material inputs to the system from different supply chain in kilogram per liter ethanol, respectively, and $EF_{X_{S2}}$ and $EF_{X_{S3}}$ are the emission factors of certain energy and material input in gCO2e kg-1.

2.2. *Carbon Allocation*

Since the bioethanol production in the Philippines is a multi-product scenario, economic allocation method was made such that the computed carbon footprint within the system boundary was not solely attributed to bioethanol because there were other co-products and/or by-products that have economic values. The carbon allocation factor was computed based on the following formula:

Carbon Allocation Factor of Product A = Economic Value of Product A/ Total Economic Value of All Products (2)

The first economic allocation was done after sugar milling, wherein the main product is raw sugar, while the co-product is molasses; the latter being the desired product. The second economic allocation was done after bioethanol processing wherein liquid CO2 and/or dry ice are co-products that have economic values.

2.3. Global Warming Potential

GHG emission reduction of bioethanol relative to a fossil fuel reference, expressed in percentage (%), was estimated and was used to estimate the historical GHG avoided by the Philippine bioethanol distilleries over the span of 12 years (2009-2020), as well as the projected potential GHG avoidance of the Philippine bioethanol industry from 2021 to 2030 at varying bioethanol blending scenarios. Equation 3 shows equation for estimating the GHG emission reduction potential of bioethanol.

$$GHG\ Reduction\ (\%) = (GWP_{fossil} - GWP_{bioethanol})/GWP_{fossil}\ X\ 100 \quad (3)$$

3. Results and Discussion

3.1. Carbon Footprint and GHG Reduction Potential of the Seven Bioethanol Production Systems

Based on the actual bioethanol production volume share of each of seven participating bioethanol distilleries and their individual carbon footprint or global warming potential (GWP), the estimated weighted average carbon footprint is at 1,415.66 $gCO_2e\ L^{-1}$, which ranges from 287 to 1,726 $gCO_2e\ L^{-1}$ of bioethanol. This weighted average carbon footprint of the seven bioethanol industry translates to a GHG reduction potential of about 47.54%, with minimum GHG reduction potential at 36% and maximum GHG reduction potential at 89%. Summarized in Table 2 are the individual carbon footprint and global warming potential of seven distilleries under study. The fossil fuel reference carbon footprint used to compare the carbon footprints of seven bioethanol distilleries under study is 83.8 $gCO_2e\ MJ^{-1}$ or 2,698.36 $gCO_2e\ L^{-1}$ (EU Directives 2009/28/EC) [7].

Table 2. Carbon footprint and GHG reduction of the seven bioethanol production system.

Bioethanol Production System (BPS)	Carbon Footprint $gCO2e\ L^{-1}$		GHG Reduction Potential %		Avoided GHG (CY 2019-2020) $tonCO2e\ yr^{-1}$	
	Manual	SimaPro	Manual	SimaPro	Manual	SimaPro
BPS1	1,693.52	1,706.76	37.24	36.75	39,289.41	38,771.40
BPS2	268.39	305.01	90.05	88.70	38,879.51	38,293.65
BPS3	992.63	963.66	63.21	64.29	15,999.73	16,271.50
BPS4	1,088.58	1,063.14	59.66	60.60	45,838.40	46,562.80
BPS5	1,223.30	1,129.83	54.67	58.13	49,488.35	52,624.29
BPS6	1,629.20	1,662.43	39.62	38.39	70,564.66	68,371.47
BPS7	1,739.00	1,713.40	35.55	36.50	15,829.42	16,251.92

Resulting average carbon footprint of the seven bioethanol production systems is higher compared to the sugarcane bioethanol of Brazil reported by Mekonnen *et al.* (2018) at 816.2 $gCO2e\ L^{-1}$ (38.5 $gCO2e\ MJ^{-1}$) and corn-based bioethanol of US at 951.9 $gCO2e\ L^{-1}$ (44.9 $gCO2e\ MJ^{-1}$) [8]. However, the ecoinvent database of SimaPro suggests that the average carbon footprint of the seven bioethanol production systems is comparable to the sugarcane bioethanol of Brazil and corn-based bioethanol of US at 2.34 kgCO2e kg^{-1} bioethanol (1,846.3 gCO2e L^{-1}) and 2.49 kgCO2e kg^{-1} bioethanol (1,964.6 gCO2e L^{-1}), respectively.

3.2 Avoided GHG of the Seven Bioethanol Production Systems for CY 2019-2020 and its Equivalent Monetary Benefit

From the article of Chemnick last March 1, 2021 at Environment and Energy (E&E) News, it was reported that the current social cost of carbon, incorporating the global and varied impacts of greenhouse gases to society, is about USD 51 per ton carbon dioxide (CO_2) [9]. This social cost of carbon was used to translate the avoided GHG of the seven bioethanol distilleries equivalent to 268,091.15 tons CO_2e yr^{-1} for CY 2019-2020 in monetary terms. The avoided GHG of the Philippine bioethanol industry has an equivalent indirect monetary benefit of about 683.63 million PhP yr^{-1} (13.67 million USD yr^{-1}; currency conversion used is 1 USD = 50 PHP) for CY 2019-2020.

3.3 Historical Avoided GHG of the Seven Bioethanol Production System (2009-2020)

Given the weighted average carbon footprint of seven bioethanol production system at 1,415.66 gCO_2e/L and the historical actual local supply of bioethanol from 2009-2020, the annual GHGs avoided were estimated as shown in Figure 2. As the Philippine bioethanol industry grows in volume over the years, it can be observed in
Figure 2 that the annual GHG avoided of the seven bioethanol production system also increases from 2009 to 2020, and has a total cumulative GHG avoidance of about 2,412,417 million tons CO_2e translating to about Php 6.15 billion of equivalent indirect monetary benefits.

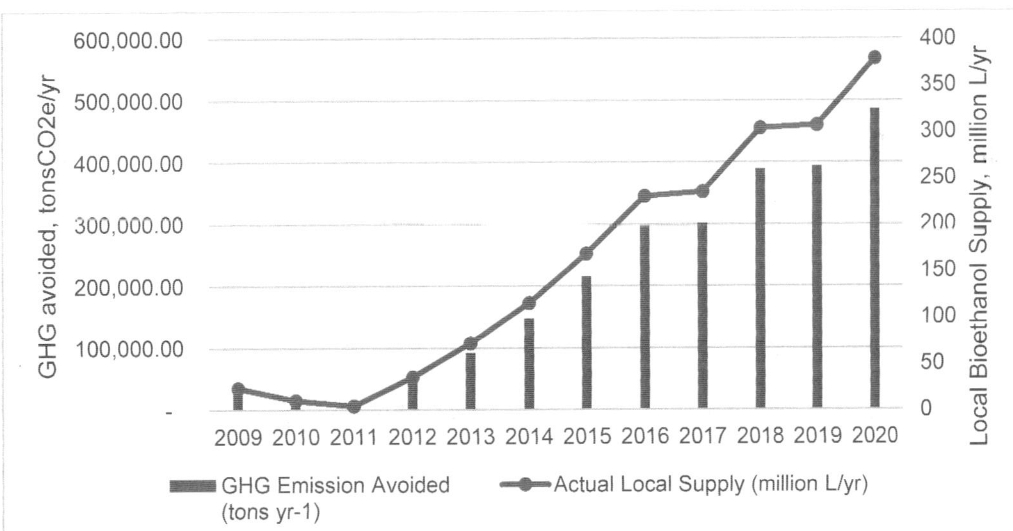

Figure 2. Annual GHG avoided of the seven bioethanol production system based on their actual local supply.

3.4 Avoided GHG Projections of the Seven Bioethanol Production Systems Until 2030 at Varying Bioethanol Blending Scenarios

Again, using the CY 2019-2020 weighted average carbon footprint of the seven bioethanol production systems in the country of about 1,415.66 gCO_2e/L, avoided GHGs were projected until 2030, as shown in Table 3, based on the equivalent ethanol requirement from the projected gasoline demand from 2021 to 2030 at different blending scenarios.

Table 3. Projected avoided GHG of the seven bioethanol production system until 2030 at varying bioethanol blending scenarios.

Year	Gasoline Demand* *million Li*	Bioethanol Blends	GHG Reduction Potential %	Ethanol Supply Requirement *million Li*	Local Supply	GHG Emission Avoided *tons CO_2e/yr*
2021	8,257.80	10%	6.32	825.78	380.50	488,068.14
2025	11,266.27	10%	6.32	1,126.63	1,126.63	1,445,126.78
2025	11,266.27	20%	12.64	2,253.25	2,253.25	2,890,253.56
2030	15,518.58	10%	6.32	1,551.86	1,551.86	1,990,571.46
2030	15,518.58	20%	12.64	3,103.72	3,103.72	3,981,142.93
2030	15,518.58	85%**	53.72	13,190.79	13,190.79	16,919,857.44

**Source: DOE (2021)*
***Optional*

Table 3 shows the potential annual GHG avoidance of the Philippine bioethanol industry if the initial bioethanol blending mandate plans advised by the National Biofuels Board in 2012 will be realized for 2025 to 2030. For 2021, since it will be impossible at this time to produce the required bioethanol local requirement, the local supply assumed is based on the current over-all rated capacities of the bioethanol distilleries in the country.

3.5 Carbon Hotspots and Sensitivity Analyses of the Seven Bioethanol Production Systems

Based on the global warming potential impact assessment of the seven participating bioethanol distilleries, carbon hotspots were identified, and the following general strategies are recommended to lower the over-all carbon footprints of the bioethanol production, thereby improving the GHG reduction potential and avoided GHGs of the seven bioethanol production system:

a. Refrain from using fossil-based fuels such as coal and bunker fuel for steam and/or electric power requirement of the distillery;
b. Invest on a co-generation plant to maximize energy generation of both steam and electric power requirement of the distillery and potential export to the grid;
c. Maximize CO_2 recovery and use from fermentation;
d. Improve process efficiencies, particularly in fermentation;
e. Encourage proper fertilization, and encourage the substantial use of organic fertilizer, together with introduction of high-yielding sugarcane varieties to lessen the carbon footprint of sugarcane production, and thereby, the molasses carbon footprint as well.
f. Find a market for filter cake and wastewater treatment sludge.

By applying all the strategies enumerated above except for items e and f, Table 4 shows the improvements of the seven bioethanol production systems in terms of its environmental impacts.1 Note that items e and f were not considered in the analysis since bioethanol distilleries has no full control over the sugarcane yields and fertilization of sugarcane planters in the farm as well as the market value of filter cake and wastewater treatment sludge.

Table 4. Improvements in base case carbon footprint of the seven bioethanol production system in the country if recommendation strategies are implemented through sensitivity analysis.

Bioethanol Production System (BPS)	Carbon Footprint *gCO2e L⁻¹*	% Diff. from Base Case	GHG Reduction Potential *%*	Avoided GHG (CY 2019-2020) *tonCO2e yr⁻¹*	% Diff. from Base Case
	Improved Case		Improved Case	Improved Case	
BPS1	368.94	78.30	86.33	91,080.32	133.36
BPS2	310.26	-8.22	88.50	38,780.66	0.50
BPS3	319.74	67.31	88.15	22,311.49	38.27
BPS4	370.69	65.55	86.26	66,280.45	43.46
BPS5	280.66	76.15	89.60	81,113.70	58.87
BPS6	740.42	9.87	72.56	129,224.36	4.32
BPS7	235.53	86.36	91.27	40,636.75	153.34

In average, the improved carbon footprint of the seven bioethanol production system reduced to 388.51 gCO2e L^{-1}, which is 72.56% lesser than the average base case scenario (Table 2), and the average GHG reduction potential increased to 85.60%. The equivalent avoided GHG of the improved case is 485,769.63 tons CO2e yr^{-1}, which is 80.08% lesser than the average base case (Table 2). It is important to mention that some distilleries apply the recommended strategies above, which is the main reason why the base case weighted average carbon footprint shows good result. However, there are some bioethanol distilleries that do not employ such practices that could potentially lower their overall carbon footprints. Therefore, sensitivity analyses were done to individual distilleries to quantify the improvement of their bioethanol production carbon footprint.

4. Conclusion

This study examined the environmental impacts of the seven bioethanol production systems in the Philippines and their contribution to the country's climate change mitigation efforts. The average carbon footprint of the seven bioethanol production systems is comparable with the Brazil and US bioethanol carbon footprint. With the recommended strategies identified to improve the environmental impacts, the average carbon footprint of the seven bioethanol production system can be decreased by 72.56%, and the average GHG reduction potential can be increased to 85.60%, which is equivalent to avoided GHG emission of 485,769.63 tons CO2e yr^{-1}.

References

[1] Biofuels Act of 2006, Republic Act No. 9367 § 2, 1 (2007). *Retrieved from: ra_9367.pdf (doe.gov.ph)*

[2] Department of Energy. 2016. Philippine Energy Plan 2016-2030. Taking the Leap towards Sustainable Future. *Retrieved from: https://www.doe.gov.ph/sites/default/files/pdf/pep/2016-2030_pep.pdf*

[3] Department of Energy. 2020. List of Accredited Bioethanol Producers as of 31 March 2020. *Retrieved from: https://doe.gov.ph/sites/default/files/pdf/renewable_energy/list-of-accredited-bioethanol-producers-as-of-2020-03-31.pdf*

[4] US Department of Agriculture. 2021. Philippines: Biofuels Annual. *Retrieved from: https://apps.fas.usda.gov/newgainapi/api/Report/DownloadReportByFileName?fileName=Biofuels%20Annual_Manila_Philippines_10-18-2021.pdf*

[5] PRé Sustainability. (2020). *SimaPro* [Computer software]. https://simapro.com/

[6] Bhatia P, Cummis C, Draucker L, Rich D, Lahd H, & Brown A. 2011. Greenhouse gas protocol product life cycle accounting and reporting standard. *Retrieved from: https://ghgprotocol.org/sites/default/files/standards/Product-Life-Cycle-Accounting-Reporting-Standard_041613.pdf*

[7] Union, E. (2009). Directive 2009/28/EC of the European Parliament and of the Council of 23 April 2009 on the promotion of the use of energy from renewable sources and amending and subsequently repealing Directives 2001/77/EC and 2003/30/EC. *Official Journal of the European Union*, *5*, 2009.Fertilizers Europe (n.d.). Carbon Footprint Reference Values: Energy Efficiency and Greenhouse Gas Emissions in European Mineral Fertilizer Production and Use. Retrieved from: https://www.fertilizerseurope.com/fileadmin/user_upload/publications /agriculture_publications/c arbon_footprint_web_V4.pdf

[8] Mekonnen, MM, Romanelli T L, Ray C, Hoekstra AY, Liska AJ, Neale CM. 2018. Water, energy, and carbon footprints of bioethanol from the US and Brazil. *Environmental science & technology* 52(24), 14508-14518

[9] Chemnick J. 2021. Cost of Carbon Pollution Pegged at $51 a Ton. E&E News - March 1, 2021. *Retrieved from: https://www.scientificamerican.com/article/cost-of-carbon-pollution-pegged-at-51-a-ton/*

Acknowledgment
This research study would not have been possible without the financial support and credible data provided by Sugar Regulatory Administration and Ethanol Producers Association of the Philippines.

Life Cycle Assessment of Bioethanol Produced from Banana Peduncle

K M Daniel[1], J V Escaño[1], J L Palada[1], J M Aberilla[1*]

[1] Sustainable Production and Responsible Consumption (SPaRC) Laboratory, Department of Chemical Engineering

University of the Philippines Diliman, Quezon City 1101

*Email of corresponding author: joaberilla@up.edu.ph

Biofuel production has been promoted in part to reduce dependency on fossil fuels in transportation due to fewer environmental emissions. Feedstocks for biofuels include agricultural waste and non-edible crops, suggesting that large amounts of waste can be used to produce biofuels such as bioethanol, which is currently being blended with gasoline in the Philippines as transportation fuel. However, shifting from non-renewable fuels to bioethanol and other biofuels as the dominant transportation fuel must consider aspects other than fuel usage. The objectives of this study are to perform cradle-to-grave life cycle analysis on bioethanol produced from banana peduncle and determine the contributions of each stage in the life cycle to different environmental impacts and compare the results with those of other transport fuels. The contributions of each stage in the life cycle of bioethanol are modelled in openLCA software with the ReCiPe impact assessment method. It was identified that electricity use in simultaneous saccharification and fermentation, and sulfuric acid consumption are the main contributors to the different environmental impacts. Comparison of the results of banana peduncle bioethanol production stage to those of the first-generation bioethanol production showed that most of the emissions were lower except for those pertaining to freshwater ecotoxicity, human toxicity, marine ecotoxicity, and metal depletion. Comparing fuel use, environmental impacts of conventional gasoline were slightly lower than those of a 10% mix with bioethanol. But impacts of the 85% mix with bioethanol in climate change and marine eutrophication are higher by 29% and 37%, respectively, than those of conventional gasoline. Overall, increasing the ethanol concentration would reduce the emissions during fuel usage but impacts of production will increase. Potential process improvements for environmental performance of bioethanol from banana peduncle are in the pre-treatment methods and fermentation set-up in bioethanol production, use of renewable power, and ethanol blending concentrations with gasoline.

Keywords: environmental impacts, transport fuel, second-generation feedstock, openLCA, blending, ReCiPe

IOP Conf. Series: Materials Science and Engineering **1318** (2024) 012012 doi:10.1088/1757-899X/1318/1/012012

1. Introduction

It has been observed that as the energy demand in the Philippines increases, the main sources of energy in the Philippines, which are coal- or fossil-based sources, are expected to increase production and consequently, increase heavy greenhouse gas emissions [1] during fuel production, not just in the usage of these fuels. Existing solutions in the Philippine context to mitigate these hazardous emissions include blending mandates of biofuels and their gradual increase throughout the years [2]. However, mandated blending percentages have stagnated, instead of continuously increasing, mainly due to insufficient feedstock and lack of refineries for biofuel production. One solution is to find alternative sources for biofuels using crops found in the Philippines including excess crops and agricultural wastes.

One of the Philippines' top produced agricultural crops is the banana fruit [3], and the Philippines was also the top exporter of bananas in Asia, and the third-highest exporter globally in 2019 [4]. During production and harvest of the banana fruit, different types of waste can be generated such as peels, leaves, rachis, and peduncles. Among these, the lignocellulosic rachis and peduncle are seen as potential sources of biofuel production as they can be categorized as sources of second-generation biofuels, wherein the lignocellulosic material is used as the source of sugars to produce biofuels such as bioethanol. Furthermore, there have been studies such as Gupta et al [5] and Hossain et al [6] that have studied the feasibility of banana waste in bioethanol production and have said that such a bioethanol feedstock and plant can be economically and environmentally sustainable. However, contextualization of its use in the Philippines has yet to be studied.

The objective of this study is to conduct a cradle-to-grave analysis of the life cycle of a second-generation bioethanol with banana peduncle as feedstock. Particularly, this study aims (1) to assess the environmental impact of each phase in the life cycle of bioethanol production from banana peduncle, (2) to compare the environmental impacts of a proposed second-generation bioethanol plant with those of a Philippine first-generation plant, and (3) to identify hotspots in the bioethanol life cycle. The study focuses on life cycle assessment (LCA) of the main stages in the bioethanol life cycle (feedstock cultivation, bioethanol production, and fuel usage).

In Section 2, this paper explains its methodology in collecting the necessary data that was entered into the software that was used to evaluate the different environmental impacts. Afterwards, the generated results are shown in Section 3, and these are compared to the results of other research in terms of the production stage and the usage stage in the life cycle of bioethanol produced from banana peduncle. Section 4 provides observations and recommendations for plans for bioethanol production from banana peduncle in the Philippines based on the results regarding the bioethanol's life cycle environmental impacts.

2. Methodology

This study involves the implementation of cradle-to-grave analysis of bioethanol life cycles of two types of bioethanol, first-generation and second-generation. This study will follow the ISO 14040 [7] LCA Framework, which has four main stages of performing LCA, namely, (1) goal and scope definition, (2) inventory analysis, (3) impact assessment, and (4) interpretation.

2.1. Goal and Scope Definition

This section defines the goals, functional unit, product system, and system boundaries of the study. In this study, the functional unit is based on the mileage of an average passenger car, with the goal of assessing the bioethanol life cycle as a transport fuel. The reference passenger car is the Toyota Corolla Altis since it was among the top gasoline cars in the Philippines since 2018 [8]. Thus, the functional unit is determined to be 1,008 km [9].

The general life cycle of bioethanol includes three main stages: feedstock cultivation, bioethanol production, and bioethanol usage. Each stage has corresponding inputs, outputs, and emission streams. Both first-generation and second-generation feedstocks are included in the scope. The common feedstock for the first-generation bioethanol in the Philippines is sugarcane, which is used in two

IOP Conf. Series: Materials Science and Engineering **1318** (2024) 012012 doi:10.1088/1757-899X/1318/1/012012

different sectors: food and biofuel. In evaluating the life cycle of bioethanol produced from sugarcane, the environmental impacts from sugarcane cultivation and harvesting are attributed to the bioethanol produced. The feedstock for the second-generation bioethanol is banana peduncle. While the banana fruit is utilized as food, the banana peduncle is a waste product from the cultivation of the banana fruit. Thus, the environmental impacts from the cultivation of the banana fruit can be associated with the life cycle of the banana fruit itself, since it is the economically desired product. Correspondingly, it is taken that the start of the life cycle of bioethanol from banana peduncle is at the bioethanol production stage.

2.2. *Inventory Analysis*

The Philippines only has first-generation bioethanol facilities on an industrial scale, where the main sources are sugarcane and sugarcane molasses. While current bioethanol production plants and organizations are unable to provide their inventory of a first-generation bioethanol production plant in the Philippines, electrical and water consumption for the production process, provided by Demafelis et al [10], are used to be compared to the LCA results of the second-generation bioethanol plant. The main production process is shown in Figure 1a.

The production process for second-generation bioethanol from banana peduncle is based on three studies of Guerrero and Munoz [11] who provided the general process of bioethanol production from banana waste shown in Figure 1b; Guerrero et al [12] who provided the optimal conditions for acid catalyzed steam explosion pre-treatment based on their laboratory-scale set-up; and Guerrero et al [13] who provided the ethanol yield concentration for simultaneous saccharification & fermentation.

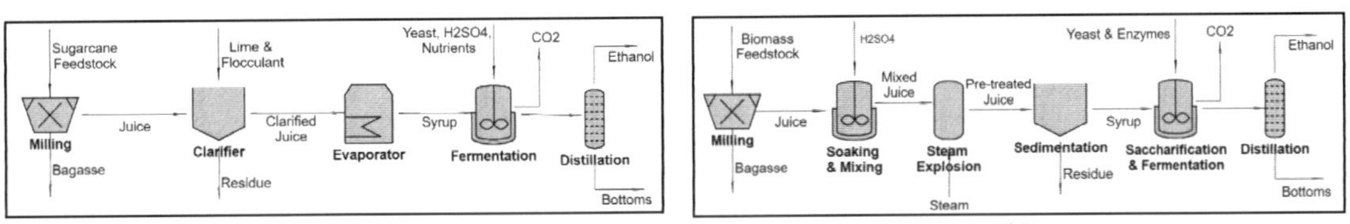

(a) (b)

Figure 1. Process flow diagrams of the production of (a) first-generation from Demafelis et al [10] and (b) second-generation bioethanol from Guerrero and Munoz [11]

The EcoInvent 3.7.1 database [14] is chosen to represent the different inputs and outputs because its database encompasses several sectors and locations. The sugarcane cultivation inventory is based on the available inputs, outputs, and flows in the EcoInvent 3.7.1 database for Brail adopted for the Philippine setting. The tropical regions of Brazil and the Philippines, similar harvest seasons, and combination of manual and machine planting practices are the main similarities of the two countries when it comes to sugarcane cultivation.

Daylan and Ciliz [15] provided the emissions produced during fuel usage with bioethanol concentrations of 0% (E0), 10% (E10), and 85% (E85), which are comprised of carbon dioxide, carbon monoxide, non-methane volatile organic compounds, nitrogen oxides, ammonia, sulfur dioxide, nitrous oxide, and methane. Their study is chosen due to its lignocellulosic biomass feedstock, scope of emissions covered, and availability of E10 and E85 data, which is the current mandated fuel blend of the Philippines and projected increase of bioethanol mix due to the Biofuels Act of the Philippines, respectively [16]. Additionally, Daylan and Ciliz [15] gave the fuel economies for the E0, E10 and E85 fuels tested which are 15.39 km/kg, 14.5 km/kg and 10.9 km/kg, respectively. For this study, the car manufacturing inventory is not included as similar vehicles are assumed for both generations of bioethanols.

2.3. *Impact Assessment*

Life cycle environmental impacts are calculated using OpenLCA v. 1.3 [17]. The impact assessment method is ReCiPe 1.13 [18] which encompasses several environmental categories relevant to the scope (Table 1). The specific emissions that are included in each impact category can be seen in Huijbregts et al [18]. The midpoint impact categories are assessed because they are specific categories wherein the emissions contribute directly to after they are released into the environment. Furthermore, hierarchist characterization factor for ReCiPe 1.13 is chosen as it considers a long-term period and is often the consensus or default model for such assessments [18].

3. Results and discussion

This section shows the results and analysis from the life cycle inventory. Since the life cycle of bioethanol from banana peduncle is comprised of two stages (bioethanol production and bioethanol fuel usage), evaluation of the environmental categories of the second-generation bioethanol production plant ("well-to-tank") and the bioethanol fuel usage ("tank-to-wheel") are discussed first. Lastly, comparisons are made between the bioethanol life cycles ("well-to-wheel") of the E10 and E85 fuels using bioethanol produced from banana peduncle. Additionally, a well-to-tank comparison is made between the first-generation and second-generation bioethanols.

3.1. Well-to-Tank Impacts of Bioethanol from Banana Peduncle

Table 1 shows the environmental impacts of the bioethanol production stage using banana peduncle. The specific definition and contributing emissions of each category can be found in Huijbregts et al [18]. Comparing the production stages for bioethanol in E10 and E85 fuels, it is expected that the impact values would be higher in E85 than in E10 due to increased production for the same functional unit.

Table 1. Impact analysis of bioethanol production stage for bioethanol from banana peduncle for fuel used to drive 1,008 km (functional unit)

Impact Category	Abbreviation	Unit	Impact Value	
			E10	E85
Agricultural Land Occupation	ALOP	m^2a	0.1765	1.4998
Climate Change	GWP100	kg CO_2-eq to air	3.804	32.3341
Fossil Depletion	FDP	kg oil-eq	1.3474	11.4525
Freshwater Ecotoxicity	FETPinf	kg 1,4-DCB-Eq to fresh water	0.8314	7.0672
Freshwater Eutrophication	FEP	kg P-eq to fresh water	0.0032	0.0269
Human Toxicity	HTPinf	kg 1,4-DCB-Eq to urban air	4.4711	38.0042
Ionizing Radiation	IRP_HE	kg U235-Eq to air	0.043	0.365
Marine Ecotoxicity	METPinf	kg 1,4-DCB-Eq to marine water	0.7206	6.1247
Marine Eutrophication	MEP	kg N-eq to marine water	0.0009	0.0076
Metal Depletion	MDP	kg Fe-Eq	1.7362	12.7576
Natural Land Transformation	NLTP	m^2	-0.0005	-0.004
Ozone Depletion	ODPInf	kg CFC-11-Eq to air	$1.8*10^{-7}$	$1.5*10^{-6}$
Particulate Matter Formation	PMFP	kg PM10	0.011	0.0935
Photochemical Oxidant Formation	POFP	kg NMVOC -Eq	0.0124	0.105
Terrestrial Acidification	TAP100	kg SO_2-Eq to air	0.0187	0.1586
Terrestrial Ecotoxicity	TETPInf	kg 1,4-DCB-Eq to industrial soil	0.0003	0.0023
Urban Land Occupation	ULOP	m^2a	0.1423	1.2093
Water Depletion	WDP	m^3 water-Eq	0.009	0.0767

Figure 2 indicates that the soaking and mixing (SM) and simultaneous saccharification and fermentation (SSF) are the stages that contribute most to the different impact categories. The SM stage comprises about 45.65% to 99.11% of impacts in FETPinf, FEP, HTPinf, IRP_HE, METPinf, MDP, NLTP, TETPinf, ULOP, and WDP; while the SSF stage comprises about 48.25% to 91.33% of impacts in ALOP, GWP100, FDP, MEP, ODPinf, PMFP, POFP, TAP100, TETPinf, and ULOP. Impacts from these stages are largely due to electrical power and sulfuric acid consumption. Sulfuric acid consumption is associated with the SM stage where sulfuric acid is consumed as a raw material during production, where it is used in the hydrolysis process, which breaks down the polymer chains to produce sugar molecules [19]. The emissions and impacts of sulfuric acid are based on those available in the EcoInvent 3.7.1 database.

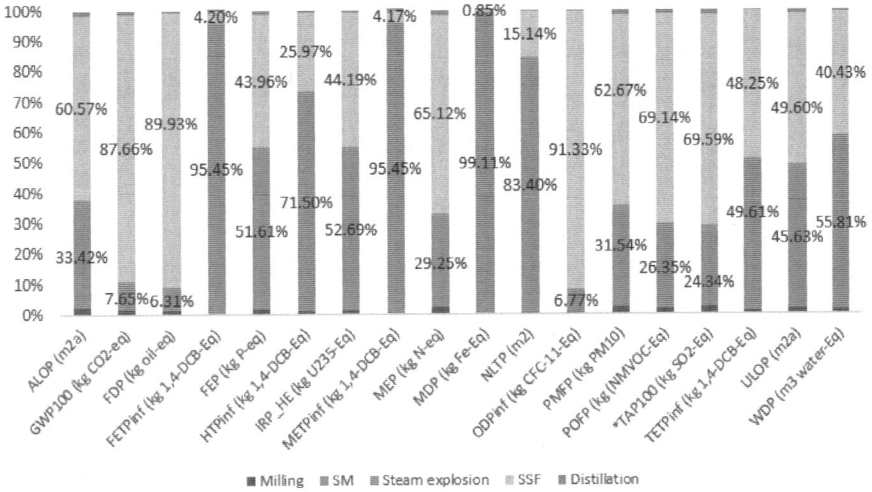

Figure 2. Contributions of each stage in the production process of second-generation bioethanol to each impact category (see Table 1 for abbreviations)

In terms of the different stages in bioethanol production, hotspot analysis reveals that the top two contributors across all categories are the SM and SSF stages. Furthermore, it is revealed that specific to the SM stage, the first ranked contributor to the environmental impacts is sulfuric acid production, while specific to the SSF stage, the two main contributing factors to the environmental impacts are electricity consumption and heat generation.

3.2. Tank-to-Wheel Impacts of Bioethanol

The results show that the impact categories that are affected are GWP100, MEP, PMFP, POFP, and TAP100, as emissions are mainly carbon dioxide, carbon monoxide, non-methane volatile organic compounds (NMVOC), nitrogen oxides, ammonia, sulfur dioxide, dinitrogen oxide, and methane. According to Figure 3, GWP100 and MEP show that these impacts increase with increasing ethanol concentration by as much as 37.2%, while the other three impacts (PMFP, POFP, and TAP100) show an increase in impacts from E0 to E10 by as much as 10.9%, and then a decrease from E10 to E85 by as much as 26.7%. This indicates that increasing the ethanol concentration in fuel tanks shows increased specific emissions that affect GWP100 and MEP and decreased specific emissions that affect PMFP, POFP, and TAP100.

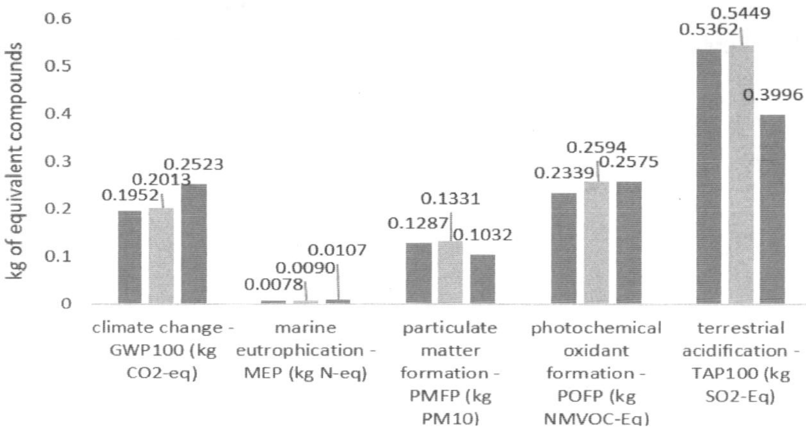

Figure 3. Impact categories and values for fuel usage for 1,008 km travel distance

Climate change and marine eutrophication increase with ethanol concentration due to increased emissions such as carbon monoxide by 20%, dinitrogen oxide by 25.4%, and methane by 262%, which all contribute to the two impact categories. The remaining categories are observed to have higher impacts at E10 than in E85. The higher impact values of E10 than in E85 in categories PMFP, POFP, and TAP100 are because of the higher fuel economy of E10 than E85 [15]. Ethanol-fueled vehicles release more NOx emissions than gasoline-fueled vehicles [20] and the NOx emissions contribute to the impacts PMFP, POFP, and TAP100 [18]. Since E10 has a much higher fuel economy than E85, this explains the higher impacts of the three mentioned categories in E10 than in E85.

3.3. Well-to-Wheel Impacts of Blended Bioethanol

Given that the functional unit of this study is 1,008 km travel distance, a fuel tank with 10% ethanol concentration contains 5 L of bioethanol while 85% ethanol concentration indicates 42.5 L of bioethanol. Thus, these values are used to calculate the balance gasoline in each fuel when comparing the life cycle of the fuel usage of bioethanol from banana peduncle.

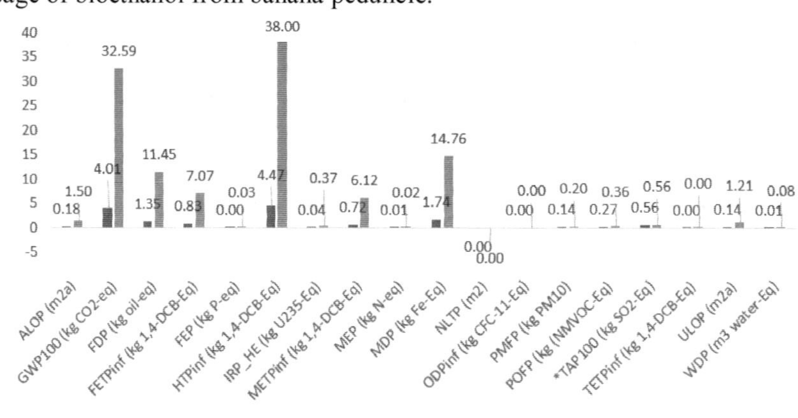

Figure 4. Life Cycle Impacts of E10 and E85 fuel (see Table 1 for abbreviations)

Based on Figure 4, in terms of the impact categories, all impacts of the life cycle of E10 fuel are significantly less than those of E85, except for the terrestrial acidification category, which shows similar impact results. As seen in Figure 3, terrestrial acidification decreases from E10 to E85, indicating that the similar results when evaluating the entire life cycles of E10 and E85 seen in Figure 4 are attained by an increase in emissions that contribute to TAP100 at the production phase. Thus, while increasing the ethanol concentration reduces impacts during fuel usage, the impacts from the production stage increases, such that some increased impacts from the production stage may surpass the reduced impacts during fuel usage. Based on the hotspot analysis conducted, increased electricity consumption from the levels of the E10 fuel to E85 fuel have compensated for the decreased TAP100 impact in the fuel usage phase.

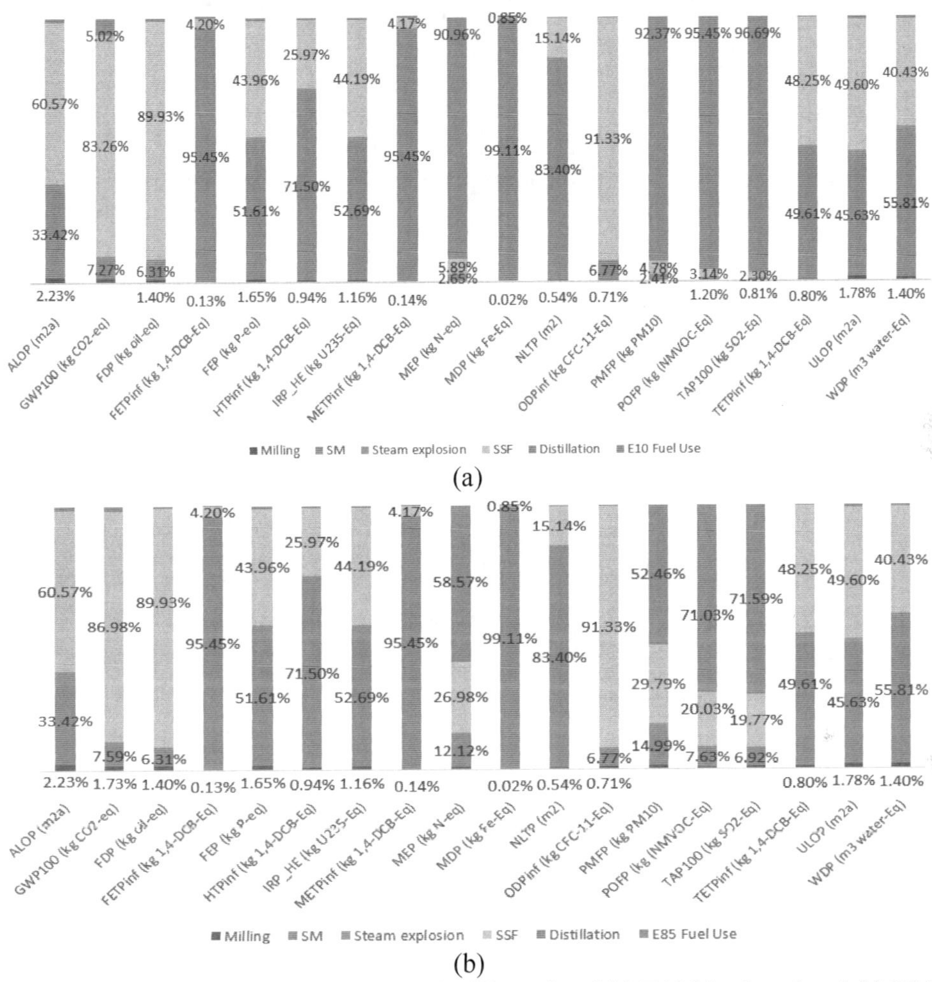

Figure 5. Contribution of each stage in the life cycle of (a) E10 bioethanol and (b) E85 bioethanol (See Table 1 for abbreviations)

Comparison of Figures 5a and 5b show that the top two rankings in hotspot analysis are still SM and SSF for each impact category for both concentrations of fuels except for the impact categories affected by the fuel usage phase, namely, MEP, PMFP, POFP, and TAP100. In these impact categories, for the E10 bioethanol, the contributions of the fuel usage phase to the forementioned categories are at a range from 90% to 97%, while for the E85 bioethanol, the contributions range from 52% to 72%. This indicates that even for the decreased emissions of some of the impact categories due to the use of E85 ethanol, the fuel usage phase still contributes the largest to these impact categories.

However, for these same impact categories, the production stages' contributions increased between E10 to E85. This indicates that while increasing bioethanol concentrations in the fuel mix according to the Biofuels Act can reduce the impacts in the fuel usage stage, the burden of the emissions and impacts shifts to the production phase of bioethanol. This further implies that the main source of these additional

IOP Conf. Series: Materials Science and Engineering **1318** (2024) 012012 doi:10.1088/1757-899X/1318/1/012012

burdens is from chemical production and usage associated with the SM stage in pre-treatment of the banana peduncle feedstock, and from electrical consumption and heat generation in the SSF stage.

In the overall life cycle, lower concentrations of bioethanol in the fuel mix seem to produce lower environmental impacts due to the lessened emissions in the second-generation bioethanol production phase compared to those of higher concentrations of bioethanol in the fuel mix. Thus, to be able to reduce the impacts, the key targets are the SM and SSF stages in the bioethanol production stage.Figure 6 shows a scaled comparison of the overall impacts of both generations of bioethanol according to their indicated units, with impacts from the first-generation bioethanol as the 100% basis. When comparing the impacts of both generations of bioethanol, lower impacts are observed in the second-generation bioethanol production for most impact categories except for FETPinf, HTPinf, METPinf, and MDP. In terms of the process flow, the main difference between the two production processes is that the second-generation production process has a pre-treatment phase for the lignocellulosic banana peduncle. Thus, the greater amount of emissions that are associated with the said categories can originate from the pre-treatment phase, especially when the SM stage, which is part of the pre-treatment phase, contributes 71.5% to 99.11% to FETPinf, HTPinf, METPinf, and MDP impacts, of the second-generation bioethanol plant. It should be noted however, that this comparison is made where only the energy and water consumption of the first-generation bioethanol are included for the first-generation bioethanol production process as these are the only available information regarding the local first-generation bioethanol plant in the Philippines by Demafelis et al [10]. Furthermore, the first-generation bioethanol production lacks emissions from chemical reactions and chemical production. Thus, the magnitude of the emissions from the first-generation bioethanol may still increase and affect the impact values. In the cases of FETPinf, HTPinf, METinf, and MDP, their comparisons may still change, expecting that the first-generation emissions will increase with more complete emissions data.

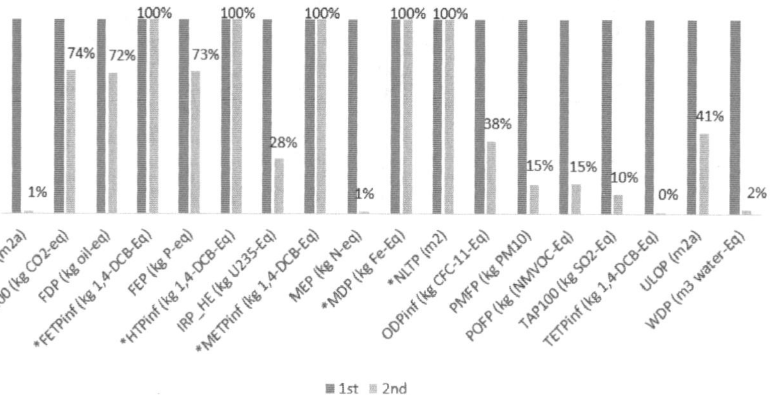

Figure 6. Relative comparison of well-to-tank impacts of sugarcane and banana peduncle bioethanol relative to sugarcane bioethanol (see Table 1 for abbreviations)

With the pre-treatment phase of the second-generation bioethanol production process as the key difference between the bioethanol production processes, this is one of the main contributors to the increased emissions and impacts, especially since FETPinf, HTPinf, METinf, and MDP are mainly from the SM stage according to the hotspot analysis. This suggests that testing of other pre-treatment methods in second generation bioethanol production can be explored to determine the optimal method based on the impact categories. Additionally, the lesser amount of impacts of banana peduncle bioethanol compared to sugarcane bioethanol is significantly due to the feedstock cultivation stage that is included in the first-generation bioethanol life cycle but not for the second-generation bioethanol life cycle. This indicates that the use of agricultural waste does provide an advantage over dedicated cultivation crops for biofuels, especially since the sugarcane cultivation stage contributes heavily to the following categories: ALOP (98.69%), FETPinf (66.67%), IRP_HE (80.75%), MEP (98.62%), MDP (88.17%), ODPinf (84.55%), PMFP (81.27%), POFP (85.03%), TAP100 (86.87%), TETPinf (99.91%), ULOP (56.99%), and WDP (98.12%). This encourages the use and incorporation of agricultural wastes and residues into different production processes of bioethanol. This is also supported by lower impacts of the second-generation bioethanol production, which are 99.8% less impact in TETPinf, 99% less impact in ALOP, 98.9% less impact in MEP, 97.7% less impact in WDP, 90.04% less impact in TAP100, 85.11% less impact in PMFP, 84.71% less impact in POFP, 71.63% less impact in IRP_HE, 62.5% less impact in ODPinf, 58.54% less impact in ULOP, 27.7% less impact in FDP, 26.91% less impact in FEP, and 26.4% less impact in GWP100, relative to the first-generation bioethanol production.

4. Conclusions and Recommendations

Using life cycle assessment, this study estimates the impacts of the life cycle of second-generation bioethanol produced from banana peduncle feedstock. Hotspot analysis shows that in the bioethanol production stage, the soaking and mixing stage (part of the pre-treatment stage) and the simultaneous saccharification and fermentation stage are where most of the impacts originate from due to the use of sulfuric acid and electricity, respectively.

For the cradle-to-gate comparison of bioethanol from sugarcane and bioethanol from banana peduncle, most of the impacts of bioethanol from banana peduncle are lower compared to that from sugarcane, except for freshwater ecotoxicity, human toxicity, marine ecotoxicity, and metal depletion. These higher impacts can be mainly attributed to the soaking and mixing stage of the bioethanol production of second-generation bioethanol, due to the use of sulfuric acid. However, the comparison of both generations of bioethanol regarding these impact categories is still subject to change with a more complete inventory for first-generation bioethanol production process. Overall, the life cycle of bioethanol from banana peduncle shows that the fuel usage, soaking and mixing, and simultaneous saccharification and fermentation stages in the bioethanol production phase are the main contributors of emissions to the different impact categories.

However, when comparing blended fuels E10 and E85, the increased concentration of bioethanol reduces the impacts in the fuel usage stage but increases the impacts in the bioethanol production stage. Thus, if the Biofuels Act in the Philippines is to be followed, then the production stage of bioethanol will be the main source of emissions and impacts if bioethanol from banana peduncle is to be utilized.

This study recommends that to ensure environmental sustainability, deployment of a second-generation bioethanol plant utilizing banana peduncle should be preceded by:

- Increasing the renewable energy mix in the electrical grid in the Philippines;
- Selecting a more environmental-friendly chemical supplier for sulfuric acid;
- Investigating different pre-treatment methods and fermentation set-ups for lignocellulosic biomass feedstock;
- Building an emissions database for the agricultural sector of the Philippines to be used in analysis of feedstock cultivation of first-generation biofuels (for better reference and comparisons);
- Validating the material and energy balance for a bioethanol plant utilizing banana peduncle through process simulation or pilont plant testing; and
- Measuring emissions from engines combusting fuels with different bioethanol concentrations

Acknowledgment
The authors would like to express their gratitude to the Department of Science and Technology – Science Education Institute (DOST-SEI) for providing financial support to the proponents of this study during its implementation.

References

[1] Mondal, M. A., Rosegrant, M., Ringler, C., Pradesha, A., and Valmonte-Santos, R. (2018). The Philippines Energy Future and low-carbon development strategies. *Energy*, ***147***, 142–154. https://doi.org/10.1016/j.energy.2018.01.039

[2] Bedford, R., and Haas, M. (2020). (rep.). *Biofuels Annual*. United States Department of Agriculture Foreign Agricultural Service. Retrieved 2022, from https://apps.fas.usda.gov/newgainapi/api/Report/DownloadReportByFileName?fileName=Biofuels%20Annual_Manila_Philippines_10-26-2020.

[3] Philippine Statistics Authority. (2020). *2020 Selected Statistics on Agriculture.* https://psa.gov.ph/sites/default/files/2 SSA2020 final signed.pdf

[4] Alzate Acevedo S, Díaz Carrillo ÁJ, Flórez-López E, and Grande-Tovar CD. Recovery of Banana Waste-Loss from Production and Processing: A Contribution to a Circular Economy. Molecules. 2021 Aug 31;**26(17)**:5282. doi: 10.3390/molecules26175282. PMID: 34500714; PMCID: PMC8434441

[5] Gupta, G., Baranwal, M., Saxena, S., and Reddy, M. S. (2019). Utilization of Banana Stem Juice as a Feedstock Material for Bioethanol Production. *Clean - Soil, Air, Water*, ***47*(9)**. https://doi.org/10.1002/clen.201900047

[6] Hossain, N., Razali, A. N., Mahlia, T. M. I., Chowdhury, T., Chowdhury, H., Ong, H. C., Shamsuddin, A. H., and Silitonga, A. S. (2019). Experimental Investigation, Techno-Economic Analysis and Environmental Impact of Bioethanol Production from Banana Stem. *Energies*, ***12*(20)**, 3947. https://doi.org/10.3390/en12203947

[7] International Organization for Standardization. (2006). *Environmental Management—Life Cycle Assessment—Principles and Framework* (ISO Standard No. 14040:2006). https://www.iso.org/standard/37456.html

[8] Ghayad, A. (2023). *20 best & worst corolla Years 2023 guide: Data proofs. Engineerine. https://engineerine.com/best-worst-toyota-corolla-years/#:~:text=However%2C%202023%2C%202022%2C%202021,NHTSA%2C%20and%20Kelley%20Blue%20Book*

[9] Toyota Makati, Inc. (2022). *Corolla Altis. Toyota Makati Inc. https://toyotamakati.com.ph/corolla-altis/*

[10] Demafelis, R. B., Alcantara, A. J., Movillon, J. L., Espaldon, M. V.O., Pacardo, E. P., Flavier, M. E., Magadia, B. T., and Matanguihan, A. E. D. (2020). Sugarcane bioethanol processing plant in the Philippines: energetics and water inventory. Journal of Environmental Science and Management, 23(2), 80–88.

[11] Guerrero, A. B., and Muñoz, E. (2018). Life cycle assessment of second generation ethanol derived from banana agricultural waste: Environmental impacts and Energy Balance. Journal of Cleaner Production, 174, 710–717. https://doi.org/10.1016/j.jclepro.2017.10.298

[12] Guerrero, A. B., Ballesteros, I., and Ballesteros, M. (2017). Optimal conditions of acid-catalysed steam explosion pretreatment of banana lignocellulosic biomass for fermentable sugar production. Journal of Chemical Technology & Biotechnology, 92(9), 2351–2359. https://doi.org/10.1002/jctb.5239

[13] Guerrero, A. B., Ballesteros, I., and Ballesteros, M. (2018). The potential of agricultural banana waste for bioethanol production. Fuel, 213, 176–185. https://doi.org/10.1016/j.fuel.2017.10.105

[14] Burhan, S., Srocka, M., Ciroth, A., and Lemberger, P. (2020). (rep.). *ecoinvent v.3.7 in openLCA* (1st ed.). Berlin: GreenDelta GmbH. Retrieved July 15, 2022, from https://nexus.openlca.org/ws/files/19340.

[15] Daylan, B. and Ciliz, N. (2016). Life cycle assessment and environmental life cycle costing analysis of lignocellulosic bioethanol as an alternative transportation fuel. *Renewable Energy*. DOI: http://dx.doi.org/10.1016/j.renene.2015.11.059

[16] Mojica-Sevilla, F. (2021). *Biofuels Annual.*

[17] OpenLCA [Computer software]. (2022). Retrieved from https://www.openlca.org/download/

[18] Huijbregts, M. A. J., Steinmann, Z. J. N., Elshout, P. M. F., Stam, G., Verones, F., Vieira, M. D. M., Hollander, A., Zijp, M., and van Zelm, R. (2017). (rep.). *ReCiPe 2016 v1.1 A*

harmonized life cycle impact assessment method at midpoint and endpoint level Report I: Characterization . Retrieved January 9, 2023, from https://pre-sustainability.com/legacy/download/Report_ReCiPe_2017.pdf.

[19] Aznury, M., Zikri, A., Ningsih, A. S., Chodijah, S., Hanura, F., Aksa, M. A., and Rachmadona, N. (2022). Effect of sulfuric acid and fermentation time on bioethanol production from Empty Fruit Bunch (EFB). *Atlantis Highlights in Engineering*. https://doi.org/10.2991/ahe.k.220205.024

[20] Kim, S. and Dale, B. E. (2006). Ethanol Fuels: E10 or E85 – Life Cycle Perspectives. *The International Journal of Life Cycle Assessment,* 11(2), 117-121. http://dx.doi.org/10.1065/lca2005.02.201.

Chemical activation of biochar derived from coffee husk for the adsorption of phytohormones from waste coconut water

M A M del Barrio[1], V P Migo[1], M C M Detras[1] and C G Alfafara[1]

[1] Department of Chemical Engineering, College of Engineering and Agro-Industrial Technology, University of the Philippines Los Baños, College, Laguna, Philippines 4031

Email: mmdelbarrio@up.edu.ph

Abstract. The potential of NaOH-activated biochar derived from waste coffee husks via pyrolysis for the extraction of phytohormones from waste coconut water was investigated. The raw coffee husk biochar (CHB) and coffee husk activated biochar (CHAB) were characterized and compared in terms of morphology, particle size distribution, elemental composition, surface functional groups, bulk density, phytohormone adsorption efficiency and equilibrium adsorption capacity. The phytohormone adsorption efficiency of the CHB and CHAB produced were 62.36% and 91.33%, respectively. Future work includes flask adsorption experiments and using the phytohormone content extracted from the coconut water for tissue culture studies. Collectively, results from this study present an alternative use of coconut and coffee wastes, and ultimately contribute to the waste management processes of the coconut and coffee industries.

1. Introduction

Coconut is the third most dominant permanent crop in the Philippines, which makes the country the top exporter and second top producer of coconut products worldwide. In 2019, approximately 1.2 million metric tons of coconut oil were produced in the Philippines, which generated huge amounts of waste during production, including coconut husks, shells, and coconut water, which would be the focus of this study [1].

Coconut water has a high BOD (13,000 - 15,000 mg/L) and COD (40,000 mg/L) content, and it is also acidic (pH 4.7 to 5.2), which could pose great harm to the environment, especially bodies of water upon run-off [2-3]. These effects include fish kills, bad odors, and spoiled natural resources, among others. Moreover, the acidic pH of the waste coconut water prevents it from being used as irrigation water in rice paddies. The high BOD and COD levels in waste coconut water would mean that it would still have to undergo costly treatment procedures to reduce these amounts to acceptable levels before they can be discharged to bodies of water [4].

Despite being labeled as "waste", one value-added product that can be obtained from coconut water is plant hormones or phytohormones. These plant growth regulators serve as "multivitamins" that improve plant seedling survival, increase crop yield, and improve crop harvest quality. Industries such as tissue culture nurseries, cut-flower industries, and hydroponic farms use plant growth regulators.

These phytohormones can be extracted from waste coconut water through adsorption onto biochar derived from agricultural wastes. An inexpensive material that can be used as adsorbents for phytohormone extraction from waste coconut water is biochar that can be obtained from waste coffee husks.

In 2019, approximately 6 thousand metric tons of coffee were produced in the country [1]. For the processing of coffee beans, it is estimated that for every ton of coffee fruit, 0.18 tons of coffee husks are generated, and these waste coffee husks have limited uses and are usually dumped out or burnt [5-8]. Several studies on the use of coffee husk as raw materials for adsorbents for the removal of organic and inorganic pollutants have been conducted [9-14]. Biochar technology is one of the recent options to produce adsorbents from agricultural wastes, such as rice, coffee, maize, and corn residues [15-16]. Biochar is defined as a carbon-rich material derived from biomass through pyrolysis [17-18].

Conversion of these agricultural wastes, specifically coffee husks, into biochar is a practical approach to produce a low-cost, value-added, and efficient adsorbent since coffee husk wastes are abundant, available in large quantities, inexpensive, and may have potential as efficient adsorbents due to their unique structures and physicochemical characteristics (i.e., micropore shapes, structural dimensions, and surface functional group composition) [19-21]. Moreover, biochar production using pyrolysis offers economic and environmental advantages [22]. According to [23], renewable energy can be produced through biochar utilization through the conversion of biomass wastes to H_2 and/or CH_4 gases. In addition, secondary pollution that occurs during incineration and landfill of biomass wastes can be avoided.

In this study, biochar was produced using coffee husks through pyrolysis and was chemically activated using sodium hydroxide (NaOH). The purpose of chemical activation was to further improve the adsorption capacity of the raw biochar. The physical and surface properties of both raw and chemically activated coffee husk biochar were assessed. Future studies include batch adsorption experiments to adsorb phytohormones from waste coconut water.

2. Materials and methods

2.1. Feedstock preparation and characterization

Coffee husks (CH) collected from Mendez, Cavite, were washed thoroughly, air-dried for 48 hours, and oven-dried at 80°C for 16 hours. The prepared samples were stored in air-tight containers to prevent the absorption of moisture from the atmosphere. The prepared feedstock was used in pyrolysis for the preparation of coffee husk biochar.

2.2. Coffee husk biochar preparation

For each pyrolysis run, prepared CH were placed inside a stainless-steel pyrolyzer. The pyrolyzer was initially purged with nitrogen at 12.5 psi for 25 minutes to ensure the absence of oxygen. Then, the pyrolyzer was heated until it reached the desired temperature of 500°C [19]. The heating rate (4.65 °C/min) for this set-up was already determined by measuring the temperature levels at certain time intervals until it reached 500°C. Then, the heating was maintained for 1 hour before turning the heater off [19]. The coffee husk biochar (CHB) was collected when the pyrolyzer cooled down to room temperature. The collected CHB was weighed, and the pyrolysis yield was calculated on a dry basis using Equation (1). The prepared CHB was subjected to chemical activation, as described in the next section.

$$\%\ \text{Yield} = \frac{\text{mass of CHB}}{\text{mass of CH}} \times 100\% \qquad (1)$$

2.3. Chemical activation of coffee husk biochar

In a stainless-steel bowl, solid NaOH micropearls (99% purity) were dissolved in distilled water. Then, CHB was gradually added and mixed into the NaOH solution (NaOH:biochar ratio of 1.5 g/g). The mixture was left to stand for 2 hours at room temperature then it was heated on a hotplate to allow the water to evaporate.

For the chemical activation, the mixture was transferred to a stainless-steel lab-scale reactor, where it underwent pyrolysis in a muffle furnace. The reactor was purged with nitrogen to ensure the absence of oxygen. The furnace was heated until it reached 500°C and was maintained at 90 minutes before turning the heater off.

After activation, the coffee husk activated biochar (CHAB) was collected and washed with tap water until the pH became neutral. The neutralized CHAB was washed with distilled water and then oven-dried at 105°C for 12 hours or until it became completely dry. The oven-dried CHAB was collected and weighed, and the chemical activation yield was calculated on a dry basis using Equation (2). The CHAB was stored in air-tight containers at room temperature until further use.

$$\%\ \text{Yield} = \frac{\text{mass of CHAB}}{\text{mass of CHB}} \times 100\% \qquad (2)$$

2.4. Determination of the effect of chemical activation on the coffee husk biochar using single flask adsorption experiments

To determine the effect of NaOH activation of the biochar, single flask adsorption experiments were done. In an Erlenmeyer flask, 1.25 g of activated biochar was added to 25 mL of filtered waste coconut water (biochar loading based on preliminary experiments). The contents were agitated on a shaker for 2 hours at room temperature [19].

The equilibrium concentration of the phytohormones in waste coconut water cannot be directly and cost-effectively measured using the available analytical methods but can be indirectly determined using the absorbance ratio from the UV-Vis spectrophotometer [24]. The absorbance of the waste coconut water before and after adsorption was measured. The % removal and equilibrium adsorption capacity were calculated using Equations 3 and 4, respectively.

$$\%\ \text{Removal} = \left(1 - \frac{A_e}{A_o}\right) \times 100\% \qquad (3)$$

$$q_e' = \frac{q_e}{C_o} = \left(1 - \left(\frac{A_e}{A_o}\right)\right)\left(\frac{V}{W}\right) \qquad (4)$$

where A_e is the absorbance of waste coconut water at equilibrium, A_o is the initial absorbance of waste coconut water, V is the volume of the waste coconut water (mL), and W is the mass of the activated biochar (g). Equation 4 shows that the term q_e can be determined without directly measuring the phytohormone concentration in the waste coconut water using the absorbance ratio obtained via UV-Vis spectrophotometry [24].

2.5. Comparison of coffee husk biochar and coffee husk activated biochar in terms of adsorbent characteristics and adsorption performance

For comparison purposes, the CHB and CHAB produced were subjected to characterization. The morphological and elemental analyses were done using Scanning Electron Microscopy (SEM) with Energy Dispersive X-Ray Analysis (EDX). The functional groups were determined using Fourier Transform Infrared (FTIR) Spectroscopy. In addition, the CHB and CHAB produced were also compared in terms of particle size distribution, bulk density, phytohormone adsorption efficiency, equilibrium adsorption capacity, and yield.

3. Results and discussion

3.1. Characteristics of coffee husk biochar and coffee husk activated biochar

3.1.1. Morphological Analysis. The micrographs obtained using SEM are shown in Figure 1. In Figure 1 (a), irregular edges, holes, and cracks were observed in the coffee husk biochar (CHB). While in Figure 1 (b), sharper and larger cracks and pores were observed after chemical activation. The porous structures of the CHB and coffee husk activated biochar (CHAB) are due to the release of volatile matter present in the CH (thermal cracking) [25]. During pyrolysis, volatile matter is released through the holes, cracks, and the pores [26]. Total thermal degradation of the cellulose and hemicellulose is achieved at 500°C, which makes the porous structure and surface area of the biochar fully developed [27]. The resulting microscopic surface structure of CHB and CHAB after pyrolysis has a greater potential for filtration and adsorption of soil pollutants [28]. In addition, an apparent modification in the surface morphology of the biochar due to the NaOH activation are exhibited by the development of pores and an increase in the pore volume [19], which leads to the increase in surface area available for adsorption.

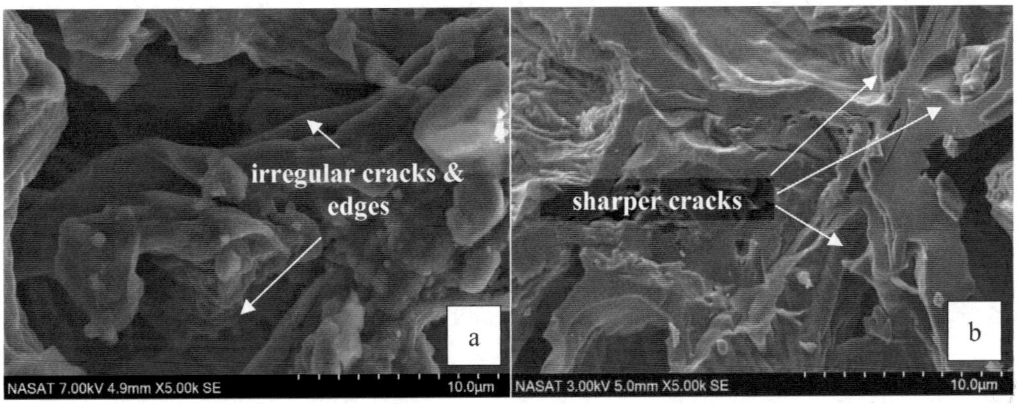

Figure 1. SEM images of (a) coffee husk biochar (CHB) and (b) coffee husk activated biochar (CHAB) at 5000X magnification.

3.1.2. Particle size distribution. Aside from surface morphology analysis, particle size distribution (PSD) of the samples has been determined and summarized in Table 1. The particle size of the samples was observed to decrease from CHB to CHAB, due to the breakdown of biomass that occurs during pyrolysis. The highest mass fraction obtained from CHB was 39.01% at mesh no. 18 (1 mm), and 32.52% of CHAB was obtained at mesh no. 35 (0.5 mm). A smaller particle size would have a larger surface area for adsorption, which was supported by another study, wherein the highest percent removal (96.08%) for crystal violet dye using waste coffee husk was achieved from the smallest particle size (0.15-0.3 mm) as compared to the 89.85% removal using the largest particle size (2.36-4.75 mm) [6]. Moreover, chemical activation using NaOH could increase the surface area of the CHB which is favorable for adsorption [19]. During chemical activation with alkaline (NaOH), solid-solid or solid-liquid reactions occur which generate porosity through hydroxide reduction and carbon oxidation. Evolution of CO, CO_2 and H_2 is observed during these reactions, and additional reactions between the active intermediates produced on the surface and the constituents of the gas phase may occur [29].

Table 1. Particle size distribution of CH, CHB, and CHAB.

Mesh No.	Sieve Opening (μm)	% (wt/wt) CHB	CHAB
10	2000	10.25	6.93
18	1000	<u>39.01</u>	31.71
35	500	31.04	<u>32.52</u>
60	250	14.17	19.42
140	106	5.11	8.09
270	53	0.32	1.13
PAN	0	0.10	0.20

3.1.3. Elemental composition. Energy Dispersive X-Ray Analysis (EDX) was also conducted to determine the elemental composition of the samples. A summary of the results is shown in Table 2. The most abundant element is carbon (C) that comprise 71.90% of CHB and 87.58% of CHAB. Thus, may be attributed to the abundance of C-containing functional groups, followed by oxygen (O) that comprise 26.84% of CHB and 9.82% of CHAB. Presence of inorganic elements in small quantities have also been detected in the samples. Significantly high amounts of alkali metals and other ash-forming elements could cause problems during the pyrolysis process, reduce the burning rate and cause incrustation, reactor corrosion and slag formation, which are undesirable in thermochemical conversion [25].

Table 2. Elemental analysis of CHB and CHAB.

Element	Average (%wt/wt)	
	CHB	CHAB
Carbon (C)	71.90	87.58
Oxygen (O)	26.84	9.82
Calcium (Ca)	0.20	2.03
Magnesium (Mg)	0.12	0.23
Sodium (Na)	0.08	0.18
Potassium (K)	0.82	0.06

During the combustion process, C and H are the important sources of heat [25]. In another study, ultimate analyses were done in coffee husk biochar pyrolyzed at 350°C, 400°C and 450°C and confirmed that during pyrolysis, the increase in temperature causes an increase in the amount of C and a decrease in the amount of N, H, and O in the biochar [25]. Carbon concentration in the solid structure increases as the other components, like volatile matter are released. The decrease in H and O content is due to the breakdown of the weaker bonds that are present in the solid structure [30]. In addition, the increase in Na could be due to chemical activation using NaOH.

3.1.4. FT-IR analysis. The functional groups of CHB, and CHAB analyzed using FT-IR are summarized in Table 3. The reduction in the transmittance of the bands from CH to CHB and CHAB is due to the thermal degradation and carbonization of organic components during pyrolysis and activation. In addition, volatile compounds that are released from the material may either condense or deposit on the surface of the material. The formation of a carbonaceous compound (biochar) and the release of volatile compounds can scatter or absorb light; therefore, decreasing the transmittance values. Qualitative changes in the structure upon pyrolysis and activation could be attributed to the partial removal of extractives and lignin [31].

Table 3. Functional group assignment for the characteristic peaks of CH, CHB, and CHAB.

Wavenumber (cm^{-1})	Functional Group Assignment	Sample	Reference
1600	C—C bond of aromatic groups (stretching); O—H (deformation or bending)	CHB, CHAB	[33]
1400	C=C bond (stretching); C—H bond (alkane) (deformation or bending)	CHB	[32]

3.1.5. Bulk Density. The values of bulk density decreased from CHB (0.67 g/mL) to CHAB (0.59 g/mL), which is similar to the trend obtained in another study, wherein bulk density decreased upon pyrolysis [34]. Carbons with higher density can hold more adsorbate per unit volume [35]. Based on the standards set by the American Water Work Association (AWWA, 1991), the bulk density of adsorbents should be greater than 0.25 g/mL to be of practical use for various environmental applications, including wastewater treatment [36]. Higher bulk density values were obtained from this study; therefore, are suitable adsorbents for wastewater treatment.

3.2. Adsorption characteristics and performance of coffee husk biochar and coffee husk activated biochar

The chemical activation of CHB for 90 minutes at 500°C, with a NaOH:biochar ratio of 1.5, resulted to an average CHAB yield of 65.37% (g CHAB/g CHB). The CHAB samples obtained were compared to CHB samples in terms of adsorption efficiency or % removal, and equilibrium adsorption capacity. For CHB, the % removal obtained was 62.36% with an equivalent equilibrium adsorption capacity of 12.4722 mg/g, while for CHAB, the % removal obtained was 91.33%, with an equivalent equilibrium adsorption capacity of 18.2661 mg/g. Results show that there had been a significant increase in both the % removal and the equilibrium adsorption capacity after chemical activation using NaOH. In addition, the % removal of CHAB was greater than the % removal of the granular activated carbon (GAC) (70.00%), which is commercially available and a commonly used adsorbent.

4. Conclusion

The study evaluated the potential of activated biochar derived from waste coffee husk via pyrolysis for the extraction of phytohormones from waste coconut water. The characteristics (morphology, particle size distribution, elemental composition, FT-IR analysis, and bulk density) of the coffee husk biochar (CHB) and the coffee husk activated biochar (CHAB) were determined and compared.

Morphological analysis showed that the surface of the adsorbent developed sharp cracks and pores upon pyrolysis and chemical activation. In addition, the particles were broken down into smaller pieces, as proven by the particle size distribution analysis. Elemental analysis showed that the samples are comprised mostly by C and O, which can be related to the presence of hydroxyl groups detected in the FT-IR spectra, and minute amounts of other elements. Lastly, the bulk density values obtained indicate that the samples are suitable adsorbents for wastewater treatment.

CHB was produced through pyrolysis of CH at 500°C for 1 hour, with a yield of 38.33%. Then chemical activation using NaOH was done with a yield of 65.37%. The phytohormone adsorption efficiency of CHAB obtained was 91.33% using a NaOH:biochar ratio of 1.5 and a carbonization time of 90 min, as compared to the phytohormone adsorption efficiency using CHB (62.36%) and GAC (70.00%). Therefore, the CHAB produced from this study using coffee husks can be a potential alternative to the commercially available GAC.

Acknowledgments

This research was funded by the Department of Science and Technology - Engineering Research and Development for Technology (DOST-ERDT) and the Department of Science and Technology - Philippine Council for Agriculture, Aquatic and Natural Resources Research and Development (DOST-PCAARRD).

References

[1] Philippine Statistics Authority (PSA) Selected statistics on agriculture [Internet] 2020 Available from: https://psa.gov.ph/sites/default/files/2_SSA2020_final_signed.pdf
[2] Kumar U. 2006Agricultural products and by-products as a low cost adsorbent for heavy metal removal from water and wastewater: A review. Sci. Res. Essay. Nov 1;1(2):33-7.
[3] Santoso U, Kubo K, Ota T, Tadokoro T, Maekawa A. Nutrient composition of kopyor coconuts (Cocos nucifera L.). Food chemistry. 1996 Oct 1;57(2):299-304.
[4] Asian Productivity Organization [APO] Handbook on green productivity [Internet] 2006 Available from: http://www.apotokyo.org/publications/files/gp-hb_gp.pdf
[5] Murthy PS, Naidu MM. Recovery of phenolic antioxidants and functional compounds from coffee industry by-products. Food and Bioprocess Technology. 2012 Apr;5:897-903.
[6] Cheruiyot GK, Wanyonyi WC, Kiplimo JJ, Maina EN. Adsorption of toxic crystal violet dye using coffee husks: Equilibrium, kinetics and thermodynamics study. Scientific African. 2019 Sep 1;5:e00116.
[7] Hien TT. Optimizing the process of transforming coffee husks into biochar by means of

hydrothermal carbonization. Vietnam Journal of Science and Technology. 2016;54(4B):138-.
[8] Guimarães T, de Carvalho Teixeira AP, De Oliveira AF, Lopes RP. Biochars obtained from arabica coffee husks by a pyrolysis process: Characterization and application in Fe (ii) removal in aqueous systems. New Journal of Chemistry. 2020;44(8):3310-22.
[9] Ahmad MA, Rahman NK. Equilibrium, kinetics and thermodynamic of Remazol Brilliant Orange 3R dye adsorption on coffee husk-based activated carbon. Chemical Engineering Journal. 2011 May 15;170(1):154-61.
[10] Boonamnuayvitaya V, Sae-ung S, Tanthapanichakoon W. Preparation of activated carbons from coffee residue for the adsorption of formaldehyde. Separation and purification technology. 2005 Mar 15;42(2):159-68.
[11] Djilani C, Zaghdoudi R, Modarressi A, Rogalski M, Djazi F, Lallam A. Elimination of organic micropollutants by adsorption on activated carbon prepared from agricultural waste. Chemical Engineering Journal. 2012 May 1;189:203-12.
[12] Kante K, Nieto-Delgado C, Rangel-Mendez JR, Bandosz TJ. Spent coffee-based activated carbon: Specific surface features and their importance for H2S separation process. Journal of hazardous materials. 2012 Jan 30;201:141-7.
[13] Khenniche L, Benissad-Aissani F. Adsorptive removal of phenol by coffee residue activated carbon and commercial activated carbon: equilibrium, kinetics, and thermodynamics. Journal of Chemical & Engineering Data. 2010 Nov 11;55(11):4677-86.
[14] Ma X, Ouyang F. Adsorption properties of biomass-based activated carbon prepared with spent coffee grounds and pomelo skin by phosphoric acid activation. Applied Surface Science. 2013 Mar 1;268:566-70.
[15] Rovani S, Rodrigues AG, Medeiros LF, Cataluña R, Lima ÉC, Fernandes AN. Synthesis and characterisation of activated carbon from agroindustrial waste—Preliminary study of 17β-estradiol removal from aqueous solution. Journal of Environmental Chemical Engineering. 2016 Jun 1;4(2):2128-37.
[16] Rodiguez MH, Yperman J, Carleer R, Maggen J, Dadi D, Gryglewicz G, Van der Bruggen B, Hernández JF, Calvis AO. Adsorption of Ni (II) on spent coffee and coffee husk based activated carbon. Journal of Environmental Chemical Engineering. 2018 Feb 1;6(1):1161-70.
[17] Chen S, Zhang B, Xia Y, Chen H, Chen G, Tang S. Influence of mixed alkali on the preparation of edible fungus substrate porous carbon material and its application for the removal of dye. Colloids and Surfaces A: Physicochemical and Engineering Aspects. 2021 Jan 20;609:125675.
[18] Tan X, Liu Y, Zeng G, Wang X, Hu X, Gu Y, Yang Z. Application of biochar for the removal of pollutants from aqueous solutions. Chemosphere. 2015 Apr 1;125:70-85.
[19] thi Quyen V, Pham TH, Kim J, Thanh DM, Thang PQ, Van Le Q, Jung SH, Kim T. Biosorbent derived from coffee husk for efficient removal of toxic heavy metals from wastewater. Chemosphere. 2021 Dec 1;284:131312.
[20] Guilhen SN, Mašek O, Ortiz N, Izidoro JC, Fungaro DA. Pyrolytic temperature evaluation of macauba biochar for uranium adsorption from aqueous solutions. Biomass and Bioenergy. 2019 Mar 1;122:381-90.
[21] Jin J, Li S, Peng X, Liu W, Zhang C, Yang Y, Han L, Du Z, Sun K, Wang X. HNO3 modified biochars for uranium (VI) removal from aqueous solution. Bioresource Technology. 2018 May 1;256:247-53.
[22] Dias JM, Alvim-Ferraz MC, Almeida MF, Rivera-Utrilla J, Sánchez-Polo M. Waste materials for activated carbon preparation and its use in aqueous-phase treatment: a review. Journal of environmental management. 2007 Dec 1;85(4):833-46.
[23] Ren S, Lei H, Wang L, Bu Q, Chen S, Wu J. Hydrocarbon and hydrogen-rich syngas production by biomass catalytic pyrolysis and bio-oil upgrading over biochar catalysts. Rsc Advances. 2014;4(21):10731-7.
[24] Magboo MMC 2015 Pre-scale up studies for the extraction of phytohormones from waste coconut water [Unpublished undergraduate thesis] Department of Chemical Engineering University of

IOP Conf. Series: Materials Science and Engineering **1318** (2024) 012013 doi:10.1088/1757-899X/1318/1/012013

the Philippines Los Baños, College, Laguna.

[25] Setter C, Silva FT, Assis MR, Ataíde CH, Trugilho PF, Oliveira TJ. Slow pyrolysis of coffee husk briquettes: Characterization of the solid and liquid fractions. Fuel. 2020 Feb 1;261:116420.

[26] Biagini E, Narducci P, Tognotti L. Size and structural characterization of lignin-cellulosic fuels after the rapid devolatilization. Fuel. 2008 Feb 1;87(2):177-86.

[27] Lee Y, Ryu C, Park YK, Jung JH, Hyun S. Characteristics of biochar produced from slow pyrolysis of Geodae-Uksae 1. Bioresource Technology. 2013 Feb 1;130:345-50.

[28] Mohan D, Sarswat A, Ok YS, Pittman Jr CU. Organic and inorganic contaminants removal from water with biochar, a renewable, low cost and sustainable adsorbent–a critical review. Bioresource technology. 2014 May 1;160:191-202.

[29] Raymundo-Pinero E, Azaïs P, Cacciaguerra T, Cazorla-Amorós D, Linares-Solano A, Béguin F. KOH and NaOH activation mechanisms of multiwalled carbon nanotubes with different structural organisation. Carbon. 2005 Jan 1;43(4):786-95.

[30] Edreis EM, Yao H. Kinetic thermal behaviour and evaluation of physical structure of sugar cane bagasse char during non-isothermal steam gasification. Journal of Materials Research and Technology. 2016 Oct 1;5(4):317-26.

[31] Silveira Junior EG, Perez VH, de Paula SC, Silveira TD, Olivares FL, Justo OR. Coffee Husks Valorization for Levoglucosan Production and Other Pyrolytic Products through Thermochemical Conversion by Fast Pyrolysis. Energies. 2023 Mar 18;16(6):2835.

[32] Ayalew AA, Aragaw TA. Utilization of treated coffee husk as low-cost bio-sorbent for adsorption of methylene blue. Adsorption Science & Technology. 2020 Jul;38(5-6):205-22.

[33] Couto AM, Protásio TD, Trugilho PF, Neves TA, Sá VA. Multivariate analysis applied to evaluation of Eucalyptus clones for bioenergy production. Cerne. 2013;19:525-33.

[34] Khanmohammadi Z, Afyuni M, Mosaddeghi MR. Effect of pyrolysis temperature on chemical and physical properties of sewage sludge biochar. Waste Management & Research. 2015 Mar;33(3):275-83.

[35] Qureshi K, Bhatti I, Kazi R, Ansari AK. Physical and chemical analysis of activated carbon prepared from sugarcane bagasse and use for sugar decolorisation. International Journal of Chemical and Biomolecular Engineering. 2008;1(3):145-9.

[36] Ravichandran P, Sugumaran P, Seshadri S, Basta AH. Optimizing the route for production of activated carbon from Casuarina equisetifolia fruit waste. Royal Society open science. 2018 Jul 11;5(7):171578.

IOP Conf. Series: Materials Science and Engineering **1318** (2024) 012014 doi:10.1088/1757-899X/1318/1/012014

Preparation of activated biochar from durian shells for the extraction of phytohormones from waste coconut water via adsorption

Julienne Eris L. Paranaque[1]*, Monet Concepcion M. Detras[1], Catalino G. Alfafara[1], and Veronica P. Migo[1]

[1]Department of Chemical Engineering, College of Engineering and Agro-Industrial Technology, University of the Philippines Los Baños, College, Los Baños, Laguna Philippines 4031

*email: jlparanaque@up.edu.ph

Abstract. The demand for activated carbon increases due to its effectiveness in wastewater treatment and other soil applications. However, its growing demand in the market resulted to higher price, thus, this study aimed to develop a cheaper alternative. The use of agricultural waste for biochar production has been the focus of several studies nowadays. Aside from the cheaper cost of the feedstock and the process, the problem on waste generation in the country will be lessened. In this study, activated biochar derived from Durian (*Durio zibethinus*) shells was prepared as adsorbent for the extraction of phytohormones from waste coconut water. Phytohormones or plant hormones are all-natural substitute to synthetic plant growth regulators (PGRs) used for different tissue culture practices in the country. Chemical activation conditions (alkali-to-raw biochar ratio and holding time) were determined to maximize phytohormone adsorption efficiency of the activated biochar. The activation of the durian shell-derived biochar was done using potassium hydroxide as chemical activating agent. Factors such as KOH-biochar ratio (high level: 5:1 (w/w), low level: 1:1 (w/w)) and carbonization holding time (high level : 90 mins, low level: 45 mins) were applied to test its significance on the activation of durian shell biochar with percent phytohormones removal from waste coconut water as a response. The parametric study showed that KOH-biochar ratio as a main effect was significant and its interaction with holding time was also significant. This result will be used in the ongoing study to produce optimized activated biochar and will be used for batch and column adsorption. The prepared durian shell activated biochar will then be used for the extraction of phytohormones from waste coconut water via adsorption.

1. Introduction

Activated carbons have high market value due to its high demand for its various applications. Hence, production of alternatives such as activated biochar became the focus of different studies. Biochar are carbon-rich materials produced in an environment with limited to no supply of oxygen or the process of pyrolysis. This has a porous structure, high pore surface area, high organic carbon content, active functional groups, and high cation-exchange capacity. Biochar has different characteristics depending on the biomass used, residence time, pyrolysis temperature, heating rate, and pressure [1].

Agricultural waste can be utilized to produce biochar. Aside from coconut shells, sugarcane bagasse, corn cobs, and rice husks, the Philippines also generate durian shells as agricultural wastes. Durian, which is known to be "King of fruits", is a tropical fruit found in Southeast Asian countries such as Philippines, Malaysia, Thailand, and Indonesia. It is famous for its outer shell covered with spikes and for its unique pungent odor. Approximately 79 thousand metric tons of durian have been produced in the Philippines at present and Davao Region as the top producer of durian [2]. With these large volumes of production, edible parts such as its flesh and seeds are used for industries leaving the shells and husks as wastes. Paper making projects have been proposed and implemented in several places in the country to utilize durian shell waste [3]. Due to its cellulose, hemicellulose, and lignin content, durian shells are considered by several studies as biomass for biochar production which will then be applied to wastewater treatment.

Aside from the utilization of agricultural wastes into biochar as adsorbent, water treatment has also been the focus of different recent studies. With the Philippines producing tons of coconut and converting them to high value products, wastes are also generated. Waste coconut water is acidic and has high biological oxygen demand that can pose risk to the environment particularly on aquatic species and resources [4] (Kumar et al, 2002). Apart from that, waste coconut water contains nutrients that can still be useful given several processes and treatments.

Phytohormones or plant hormones are organic substances which are naturally occurring and give significant physiological effects to plants even in minute concentrations. This can affect plant development, cell division, elongation, determination, flowering, etc [5]. Several known phytohormones found in coconut water are auxins, gibberellins, cytokinin, abscisic acid, and salicylic acid [6]. During 1970, scientists led by Dr. Juanita Mamaril from the National Institute of Molecular Biology and Biotechnology in the University of the Philippines Los Baños studied the extraction of phytohormones from waste coconut water [7]. Later, phytohormones were produced on a small scale and available only for walk-in customers. Other plant growth regulators are synthetic and imported, hence production of cheaper and organic PGRs are being developed.

In this study, activated biochar derived from durian shells were used which is more natural and cheaper than the use of resins and commercial activated carbon for column and batch adsorption. The effect and significance of the chemical activating agent to biochar ratio and holding time on the % adsorption of the phytohormones were observed.

With the emerging demands on synthetic PGRs and accumulation of agricultural wastes, the study focused on the conversion of durian shells into activated biochar for the extraction of phytohormones from waste coconut water. Utilization of durian shells to produce biochar as adsorbent will also help in minimizing waste generation in the country. Results on this study will be applied on the production of optimized activated biochar that will be used for a separate study on batch and column adsorption of phytohormones in coconut water. The study will be sustainable due to the abundance and availability of waste coconut water and durian shells in the country.

2. Materials and Methods

The activated biochar was produced using dried durian shells and potassium hydroxide as chemical activating agent. These were produced via pyrolysis using furnace. Waste coconut water from VCO plant was collected and used for the extraction of phytohormones.

2.1. Preparation of Durian Shell-derived biochar

Durian shells were collected and pooled from different places in the Philippines. These shells were washed, dried, and sized reduced before subjecting to pyrolysis. Pyrolysis was done to convert the durian shells to biochar. The collected biochar were pooled and prepared for activation to improve the pore surface area and the adsorption capacity for phytohormone extraction.

IOP Conf. Series: Materials Science and Engineering **1318** (2024) 012014 doi:10.1088/1757-899X/1318/1/012014

2.1.1. *Preparation of durian shells*. The collected shells were washed with tap water for the removal of dirt, fungi, and excess residues. The washed durian shells were sun-dried for five days and oven dried at 105 deg C for 12 hours. When the durian shells became brittle enough, these were subjected for size reduction using hammer mill located at the Center for Agri-Fisheries and Biosystems Mechanization (BIOMECH), College of Engineering and Agro-Industrial Technology (CEAT), University of the Philippines Los Baños (UPLB). Fine particles were discarded, and the oversize components were kept in airtight containers.

2.1.2. *Biochar production via pyrolysis*. Dried and milled durian shell samples were placed inside a stainless-steel reactor for pyrolysis. To ensure that the stainless-steel reactors have limited to no oxygen content, nitrogen gas was purged inside for 2 minutes at 12.5 psi. The purged reactors with sample were placed in a furnace until it reached 500 deg C. Upon reaching this desired temperature, the reactors were held for an hour. After an hour, the furnace was turned off and the reactors were allowed to cool down to room temperature. The produced biochar were collected, weighed, and stored in airtight containers. The biochar yield was computed.

2.2. Activation of Durian Shell Biochar
To improve the surface characteristics and adsorption capacity of biochar, chemical activation was done. Using Design Expert ver.13, factors such as KOH-biochar ratio and holding time were tested against percent phytohormones removal as response. The factors were tested if they have significance as main effects as well as their interaction. The appropriate conditions were picked for future use on batch adsorption and column adsorption experiments.

2.2.1. *Chemical Activation using Potassium Hydroxide*. Durian shell biochar (DSB) was activated with potassium hydroxide (KOH). Depending on the ratio for each run, KOH was dissolved in distilled water. Twenty grams of DSB was mixed with the dissolved KOH and allowed to stand for 2 hours at room temperature to ensure the full contact of the biochar and the activating agent. Excess moisture of the mixture was evaporated before transferring to the stainless-steel reactors for pyrolysis at 500 deg C for specified holding time. The sample was then cooled down to room temperature and washed with tap water until neutrality. Distilled water was used for the last washing of the activated biochar. The activated biochar was oven dried at 100 deg C for 12 hours or until bone dried. The percent yield for each run was calculated.

2.2.2. *Experimental Design*. Response Surface Methodology was used as experimental design of the study. The parametric study of the activation of durian shells was done using 2^k factorial with KOH-biochar ratio and holding time as factors and percent phytohormone removal as a response. For the KOH-biochar ratio, the low level used was 1:1 (w/w) and high level was 5:1 (w/w). Meanwhile, for the low and high level of the holding time, 45 minutes and 90 minutes respectively were used. Three replicates were done to reach the 80% minimum power requirement for detecting the effects of the chemical activation. The significance of the factors was tested using Analysis of Variance (ANOVA) at 95% confidence interval. Diagnostic plots were examined and generated using Design Expert® ver13 and the assumptions of ANOVA were verified.

2.2.3. *Single Flask Adsorption Experiment*. The response for the experiment is the percent phytohormone adsorption efficiency of the DSAB from each activation run. Coconut water from virgin coconut oil plant was collected and filtered. The initial absorbance of the coconut water was measured using UV-Vis spectrophotometer. A 0.5 gram of DSAB from each activation run was placed in an Erlenmeyer flask filled with 25 mL of coconut water. The mixture was subjected to mechanical shaker for 2 hours based from the study of Borbo [8]. The mixture was then filtered for absorbance reading of the filtrate. The collected filtrate was subjected to UV-Vis Spectrophotometer to calculate the percent adsorption efficiency.

3. Results and Discussion

3.1. Raw Biochar and Activated Biochar yield

An 18-gram sample of durian shell was prepared in a stainless-steel reactor for pyrolysis. After two hours of holding time of pyrolysis, only 5.5 grams were recovered which corresponds to about 30.5% (w/w) yield. Durian shell contains cellulose, hemicellulose, and lignin by which degrades at increasing temperatures causing the weight loss of the durian shell. The pre-pyrolyze durian shells were found to contain approximately 30.6% alpha cellulose (w/w), 19% hemicellulose (w/w), 11% extractives, and 49.5% holocellulose. In a study in 2021 [9], the mass loss first occurred at a temperature of 110^{0}C which corresponds to the loss of moisture of the durian shell. Volatile matters were the next to release as the temperature increase [10]. Durian shells were found to contain about 70-80% of volatile matter based on proximate analysis done by Chandra [10], Liu [11], and Selvarajoo [9]. Selvarajoo [9] studied the thermogravimetric analysis of durian rind thermal conversion and observed that there were two zones for the durian rind pyrolysis, the active and passive zones. In the active zone which was observed at 300-350 deg C, there was a sharp increase in the weight loss of the samples. The passive zone after which took up to around 600 deg C showed further loss of weight of the samples which gives about 65.55% mass loss [11]. At this temperature, hemicellulose and cellulose of the durian rind simultaneously break down and decompose. Hence, causing the low yield of the biochar.

Chemical activation of durian shell biochar was done to improve the surface characteristics of the biochar. This was done using potassium hydroxide (KOH) as chemical activating agent. After activation of a 20-gram sample durian shell biochar, 15-19 grams of activated biochar were recovered which gave a 75%-95% yield (w/w). Chemical activation can be done using acids, bases, or salts. Based from studies, KOH is the most common activating agent since it produces higher biochar yield and highly developed microporous structures [12]. Mass loss for the biochar upon activation is due to the development of microporous structure of the biochar. In the activation of biochar, KOH reacts with active O-containing species of the durian shell biochar resulting to the removal of O-groups leading to the release of free radials vacating the sites. KOH also etched carbon fragments of biochar which also cause the vacancies of the pores due to the removal of hydrogen protons. OH- groups from KOH fill the vacancies and form new O-groups in the biochar. O-groups and KOH reacts causing the conversion of KOH into K2CO3 at temperature range of 400-700 deg C. This leads to the release of metallic K and gaseous products which results in the increase in oxygen content and biochar porosity [12].

3.2. Parametric Study on the chemical activation of durian shell biochar

The effect of KOH-biochar ratio and pyrolysis holding time, together with their interaction, were tested to see if they have significance on the chemical activation of durian shell biochar. This was determined using 2^k factorial. Factors were tested to see if they are significant using Analysis of Variance (ANOVA) at 95% confidence interval. Diagnostic plots were generated using Design Expert® version13 and ANOVA assumptions were verified. From Figure 1a, it was observed that the residual plots did not show an "S" pattern satisfying the first assumption of ANOVA that the residuals should be normally distributed. The residuals vs. predicted plot from Figure 1b should not exhibit a megaphone pattern by which the plot satisfies, hence, residuals have constant variance. Lastly, Figure 1c showed a slight error in the first points but showed later on varying patterns indicating that these data were independent of each other. Following this observed behavior of the residual plots, the ANOVA assumptions were satisfied indicating that the parametric analysis was significant. The model was significant with a p-value of less than 0.0500. However, only the main effect A, which is the KOH-biochar ratio, and the interaction showed a p-value less than 0.0500 that signify their significance in the model. Main effect B, which is the holding time, showed no effect on the chemical activation of the durian shell biochar. The Analysis of Variance also gave a not significant lack of fit and a difference in R^2 of less than 0.2 which also supports the significance of the parametric analysis.

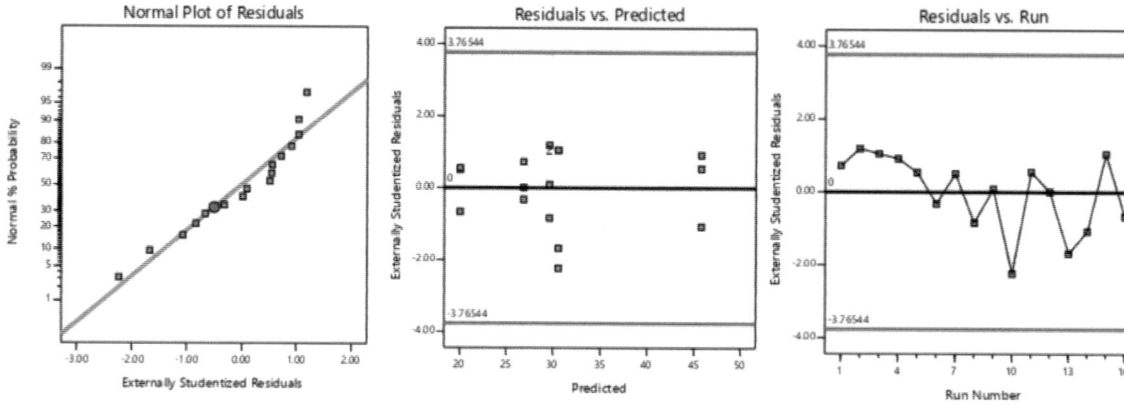

(a) (b) (c)

Figure 1. Diagnostic plots (a) Normal Plots of Residuals (b) Residuals vs.Predicted Plot (c) Residual vs.Run plot

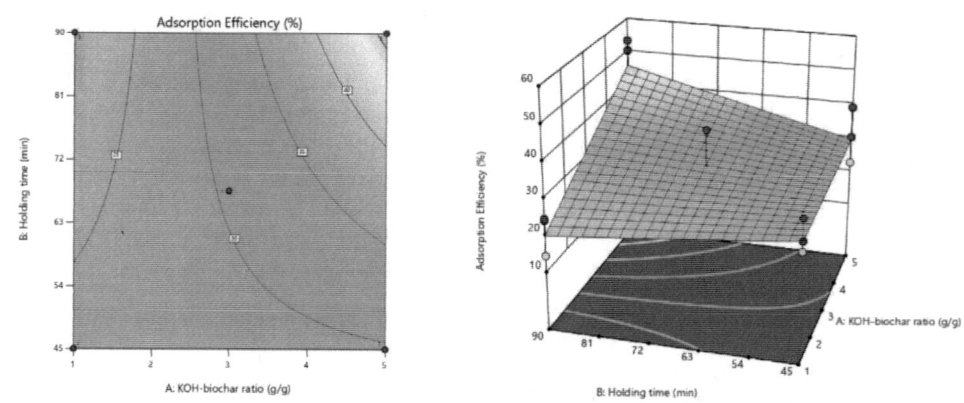

Figure 2. Surface plot generated for the parametric analysis.

In Figure 2, warm colors signify the area where possible optimum is located. In this region, maximum phytohormone removal can also be observed. As observed in the figure, the highest %phytohormone removal was 53% with a KOH-biochar ratio of 5 and a holding time of 90 mins. Meanwhile, the least %phytohormone removal was found to be at 14% designated by the blue region at KOH-biochar ratio of 1 and a holding time of 90 mins. One factor plot and interaction plot were also generated to illustrate the behavior of the factors and their interaction as the levels were increased.

3.2.1. *Effect of KOH-biochar ratio on the % phytohormone removal from waste coconut water.* Based from ANOVA, the model for the parametric analysis was significant with the main effect A, which is the KOH-biochar ratio, to be the only factor that was significant. The surface plot illustrated in Figure 2 and the one factor plot in Figure 3a showed that increasing the KOH-biochar ratio also increase the %phytohormone removal from waste coconut water. This behavior can be attributed to the formation of larger pores and give higher surface area [12]. The high amount of KOH applied for chemical activation

IOP Conf. Series: Materials Science and Engineering **1318** (2024) 012014 doi:10.1088/1757-899X/1318/1/012014

resulted to formation of larger micropores in the surface of the biochar and leaving O-containing and OH- groups by which reactions with phytohormones' functional groups can occur.

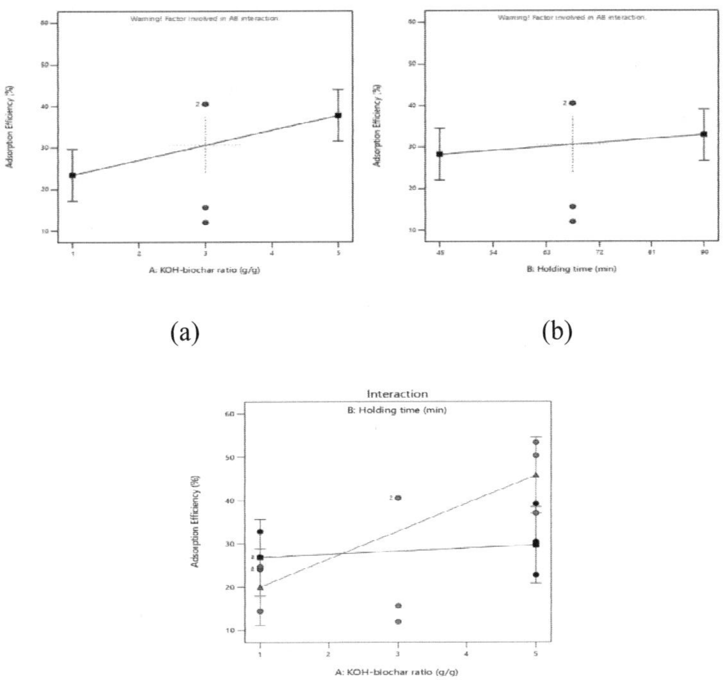

(a) (b)

(c)

Figure 3. One factor plots (a) KOH-biochar ratio (b) holding time (c) interaction

3.2.2. *Effect of holding time of activation on the % phytohormone removal from waste coconut water.* Holding time for pyrolysis shows no significance on the percent phytohormone removal based on the one factor plot in Figure 3b. Increasing the holding time alone did not affect the performance of the activated biochar on the adsorption of phytohormones.

3.2.3. *Effect of the interaction on the % phytohormone removal from waste coconut water.* Increasing the KOH-biochar ratio, together with increasing the pyrolysis holding time, gave significant difference on the percent phytohormone removal. Higher removal was achieved as both factors were increased. The increase in the amount of the activating agent while increasing the holding time cause the improvement of pore structure (due to release of volatile matters [13]) and attachments of relevant functional groups affecting the adhesion of the organic compounds and phytohormones.

4. Summary and Conclusion

The abundance of durian shell as waste was utilized to produce activated biochar for the extraction/removal of phytohormones in coconut water. The activation of the durian shell-derived biochar was done using potassium hydroxide as chemical activating agent. Pyrolysis of the durian shell yielded 30.5% (w/w) while after activation 75%-95% (w/w) yield was achieved.

The effects of the KOH-biochar ratio and holding time were tested against the %phytohormone removal using 2k factorial. With a p-value less than 0.05 and difference in R2 less than 0.2, the model was found to be significant. The KOH-biochar ratio and its interaction with holding time were significant and ANOVA assumptions were satisfied. It was found that increasing the KOH-biochar ratio will also increase the phytohormone removal due to reaction of the surface to the chemical activating agent leading to larger pores. Although holding time was not a significant factor, its interaction with KOH-biochar ratio gave a directly proportional relation to the percent removal.

The results will be used for the continuation of the study particularly for the batch adsorption studies and column adsorption studies.

Acknowledgement
Author acknowledges the support from the Department of Science and Technology – Engineering Research and Development for Technology (DOST-ERDT) and the Department of Science and Technology – Philippine Council for Agriculture, Aquatic and Natural Resources Research and Development (DOST-PCAARRD).

References

[1] Jamaludin N, Rashid S, and Tan T 2019 Synthesis, Technology and Applications of Carbon Nanomaterials (Elsevier Inc) p.109-134
[2] PCAARRD 2002 Durian Industry strategic science and technology program
[3] Phuong B 2012 Money from waste: Woman earn from durian husk
[4] Kumar S Senanayake G Visvanathan C and Basu B 2003 *Energy Convers. Manag.* 44. 10.1016/S0196-8904(02)00235-2.
[5] Zhao B Liu Q Wang B and Yuan F 2021 *J. Agric. Food Chem*, 69(12), 3566–3584. doi:10.1021/acs.jafc.0c07908
[6] Yong JW Ge L Ng YF Tan SN 2009 *Molecules* doi:10.3390/molecules14125144
[7] Magboo M 2015 Pre-scale up studies for the extraction of phytohormones from waste coconut water (unpublished)
[8] Borbo E 2016 Extraction of phytohormones from waste coconut water by adsorption using coconut-shell-derived granular activated carbon (unpublished)
[9] Selvarajoo A 2021 *IOP Conf. Ser.: Mater. Sci. Eng.* **1092** 012042
[10] Chandra TC Mirna MM Sudaryanto Y Ismadji S 2007 *Chem. Eng. J.* 127 pp 121-129
[11] Liu H Liu J Huang H Evrendilek F Wen S and Li W 2020 *Renew. Energ.* doi:10.1016/j.renene.2020.09.044
[12] Chen W Gong M Li K Xia M Chen Z Xiao H Fang Y Chen Y Yang H and Chen H 2020 *J. Appl. Energy* **278** 115730 doi:10.1016/j.apenergy.2020.11573
[13] Osman N B Shamsuddin N and Uemura Y 2016 *Procedia. Eng.* 148, 758–764. doi:10.1016/j.proeng.2016.06.610

IOP Conf. Series: Materials Science and Engineering **1318** (2024) 012015 doi:10.1088/1757-899X/1318/1/012015

Optimization study for chemical activation of biochar derived from sugarcane bagasse for maximum phytohormone adsorption from waste coconut water

Katrina Joy T. Castillo[1], Monet Concepcion M. Detras[1], Catalino G. Alfafara[1] and Veronica P. Migo[1]

[1]Department of Chemical Engineering, College of Engineering and Agro-Industrial Technology, University of the Philippines Los Baños, College, Laguna, Philippines, 4031

Email: ktcastillo1@up.edu.ph

Abstract. Improper treatment and disposal of sugarcane bagasse (SB) in farms and sugar industries can lead to environmental and health concerns. Instead of burning these wastes, bagasse can be utilized for biochar production which can be applied as adsorbents. Additionally, studies reported that the performance of biochar adsorbents can be improved by activation. Hence, this study aimed to optimize the chemical activation conditions of biochar from SB for maximum adsorption of phytohormones from waste coconut water (WCW). Pretreated SB was first converted into biochar though pyrolysis and was chemically activated using potassium hydroxide (KOH). Response Surface Methodology was used to determine the optimum activation conditions to maximize phytohormone adsorption efficiency. Factors considered were KOH-biochar ratio (1:1 to 5:1 g/g) and carbonization holding time (45 to 90 minutes). Analysis of Variance showed that the adsorption efficiency of activated biochar is significantly affected by KOH-biochar ratio. Increasing the ratio created cavities on the surface which serve as adsorption sites; however, further increase in ratio resulted in pore degradation lowering the adsorption performance of activated biochar. Conversely, carbonization holding time did not have significant effect on the response for the range of factors considered. The optimum conditions which were 3.18:1 KOH-biochar ratio and 65.64 minutes holding time resulted to ~82.14% actual phytohormone adsorption efficiency. The characteristics of the optimized activated biochar (SBAB) were then compared with that of the unmodified biochar (SBB) in terms of morphology, surface elemental composition, functional groups, adsorption efficiency, average particle size, and bulk density. The SBAB produced can be applied in further studies focusing on the adsorption of phytohormones from WCW and desorption from the adsorbent to isolate them. Results of this study can also provide a sustainable solution in minimizing the disposal of sugarcane bagasse and waste coconut water while increasing their economic returns and environmental benefits.

1. Introduction

The Philippines is an agricultural country with about 47 % of its total land area allocated to agriculture [1]. Among the agricultural crops in the country, sugarcane is one of the abundant and important crops, having sugar as its main product. Aside from it, sugarcane molasses, a by-product from sugar manufacturing, is being used as feedstock for biofuel production [2]. Unfortunately, sugarcane

IOP Conf. Series: Materials Science and Engineering **1318** (2024) 012015 doi:10.1088/1757-899X/1318/1/012015

processing industries generate approximately 6.4 million MT of sugarcane bagasse, the residual after the juice extraction from the cane stalk [1]. Although bagasse is being applied for power cogeneration in some sugar mills, refineries, and distilleries, they are still in excess and are considered wastes [3]. Instead of burning and causing emissions to the environment, a resource-recovery approach can be employed.

One of the technologies for waste processing includes pyrolysis which is a thermochemical conversion process carried out in the absence of oxygen. Pyrolysis has three main products such as bio-oil, gas, and biochar [4]. Specifically, biochar can be produced from different sources and a sustainable feedstock is agricultural residue including sugarcane bagasse. Bagasse is a lignocellulosic biomass with great potential in terms of its composition, thermochemical behavior, and abundance as feedstock for biochar production. Among other characteristics, bagasse has high contents of volatile combustible matter (VCM) which creates pores on the biochar surface upon VCM release during thermochemical conversion [5]. As the degree of carbonization increases, carbonaceous compounds remain, making biochar a carbon-rich material [6]. The porous structures of biochar also contribute to the increase in its surface area allowing it to be applied as an adsorbent for wastewater treatment through adsorption [7].

Furthermore, the physicochemical properties of sugarcane bagasse-derived biochar can still be improved by employing chemical activation. The activation process aims to enhance the pores, surface area, and surface chemistry of biochar to boost its adsorption performance [8]. During activation, oxygen-containing functional groups can also be formed on the surface of biochar which are necessary for the adsorption of organic substances in wastewaters [9].

Coconut water from matured coconuts is one of the wastes being generated in coconut processing industries and farms. Specifically, it was estimated that about 3.5 billion liters of matured coconut water is generated in the country each year [10]. Since there are no available wastewater treatment facilities particularly in such small-scale coconut producers due to high cost and other limitations, coconut water is typically discharged directly into the environment which can cause negative impacts including ground spoilage and oxygen depletion in water bodies due to its high biological oxygen demand (BOD) and acidic property [11].

However, value-added products can be recovered from coconut water which are phytohormones. These are naturally occurring hormones in plants that can improve plant growth and development, regulate plant's response to environmental stress, and delay aging [12][13]. Aside from these essential roles, phytohormones from coconut water are all-natural making it a better substitute for synthetic plant regulators used in plant propagation industries. Thus, it is beneficial to extract phytohormones from waste coconut water.

A common separation process to extract a desired component (adsorbate) from a gas or liquid mixture onto the surface of a porous solid (adsorbent) is adsorption [14][15]. One of its applications is for the removal of organic and inorganic substances from wastewater using activated carbon and polymeric resin as adsorbents. However, a cheaper and sustainable alternative adsorbent is biochar [9].

In this study, sugarcane bagasse (SB) was used to produce biochar which was then chemically activated to enhance its surface properties. The optimum chemical activation conditions were determined through Response Surface Methodology (RSM) for maximum percentage phytohormone adsorption efficiency. Lastly, the characteristics of the optimized activated biochar from bagasse (SBAB) were then compared with that of the raw biochar (SBB).

2. Materials and Methods

2.1. Preparation of Raw Biochar

Collected SB samples were washed with tap water, air dried, and oven dried at 105°C for 12 hours. The dried samples were size reduced using a hammer mill and were screened using a sieve with mesh number 20 to remove smaller particles. The prepared SB samples were then pyrolyzed at 500°C for 1 hour. SBB produced was then stored in an airtight container.

2.2. Chemical Activation

2.2.1. Activation experiment. Potassium hydroxide (KOH) 90 % flakes was used as the activating agent. A desired amount of KOH, depending on the KOH-raw biochar ratio (g/g) for each run, was dissolved in a fixed volume (124 mL) of distilled water (based on the solubility of KOH in water). Then, 30 g of SBB was gradually added to the mixture. The mixture was allowed to stand for 2 hours at room temperature with periodic mixing to ensure contact. The mixture was then heated on a hot plate until it was dried enough to be transferred to stainless-steel reactors.

The reactors were placed inside the furnace. Purging the system with nitrogen was done for 2 minutes to ensure absence of oxygen. The furnace was then heated to 500°C and was maintained until the desired holding time for each run was reached. After carbonization, the activated biochar was allowed to cool down to room temperature. It was then washed with tap water until neutral, followed by distilled water. Lastly, the washed activated biochar was oven dried at 105°C for 12 hours and was placed in a properly labelled airtight container.

2.2.2. Batch adsorption experiment. To determine the percentage phytohormone adsorption efficiency of the activated biochar for each run, batch adsorption experiment was performed. Initially, waste coconut water sample was collected and filtered using two layers of cheesecloth to remove coarse particles. Then, it was chilled and reclarified using coarse filter paper followed by Whatman #5 filter paper to further remove fat globules.

The initial absorbance of the filtered coconut water was read using UV-Vis spectrophotometer at 267 nm wavelength. An adsorbent loading of 25 g/L was used wherein 0.625 g activated biochar was mixed with 25 mL coconut water. The mixture was agitated using a mechanical shaker for 2 hours at room temperature. After the contact period, activated biochar samples were filtered using Whatman # 5 filter paper and the absorbance of the filtrate from each sample was read. The percentage adsorption efficiency for each activated biochar run can be calculated in terms of the absorbance readings as shown in equation (1).

$$\%\,Adsorption\,Efficiency = \left(1 - \frac{A_e}{A_o}\right)(100) \tag{1}$$

where:
A_e is the equilibrium absorbance of waste coconut water
A_o is the initial absorbance of waste coconut water

2.2.3. Experimental design and statistical analysis. For the experimental design in the determination of the optimum chemical activation conditions for maximum percentage adsorption efficiency, Response Surface Methodology-Central Composite Design (RSM-CCD) was employed using Design Expert® software. CCD was implemented in two blocks which were the factorial and augmented block wherein both blocks had center points added. The factors included KOH-biochar ratio (1:1 to 5:1 g/g) and carbonization holding time (45 to 90 minutes). Analysis of Variance (ANOVA) at 95 % confidence interval was used for the determination of model significance. After the optimization runs were performed, numerical optimization and experimental verification (5 trial runs) were conducted.

2.3. Characterization of SB, SBB, and SBAB

The composition and thermochemical behavior of SB were determined through compositional and proximate analysis (following the American Society for Testing and Materials Standard Test Method for Chemical Analysis of Wood Charcoal), respectively. The characteristics of SB, SBB, and SBAB were then compared in terms of morphology and elemental composition, functional groups, and percentage adsorption efficiency via Scanning Electron Microscopy – Energy-Dispersive X-ray Spectroscopy (SEM-EDX), Fourier Transform Infrared (FTIR) spectroscopy, and batch adsorption experiments, respectively. The average particle size of SB, SBB, and SBAB were also determined in terms of the volume surface mean diameter via standard sieve analysis using sieves of mesh numbers 10, 18, 35, 60, 140, and 270. Water displacement method was used for the bulk density determination.

IOP Conf. Series: Materials Science and Engineering **1318** (2024) 012015 doi:10.1088/1757-899X/1318/1/012015

3. Results and Discussion

3.1. Optimum Conditions of Chemical Activation

After performing the CCD runs in two blocks, Design Expert® suggested that the best-fitted model for the response is a quadratic model. Based on ANOVA at 95 % confidence interval, results of the optimization study for chemical activation of raw biochar showed that the model is significant since the resulting p-value equal to 0.0015 is less than 0.05. In terms of the factors, only the square of factor A (KOH-biochar ratio) had a significant effect on the percentage adsorption efficiency, whereas holding time and its square as well as its interaction between KOH-biochar ratio were still insignificant for the given range. Lack-of-fit is also not significant indicating that the model fits well.

The linearity coefficient, R^2 of 0.6362 suggests that about 63.62 % of the variation can be explained by the model. The residual percentage implies that there can be other factors not considered in this study that might have significant effects on the percentage adsorption efficiency. Additionally, the adjusted and predicted R^2 values are in reasonable agreement with each other having less than 0.2 difference in values. The assumptions of ANOVA were also satisfied since no "S-shaped" pattern was observed in the normal plot of residuals, no megaphone pattern for the residuals vs predicted plot, and the data points in residuals vs run plot were scattered.

The 3D surface plot is presented in figure 1. It can be observed that the highest % adsorption efficiency can be obtained at KOH-biochar ratio and holding time values of approximately 3 (g/g) and 67 minutes, respectively as indicated by the peak in the plot.

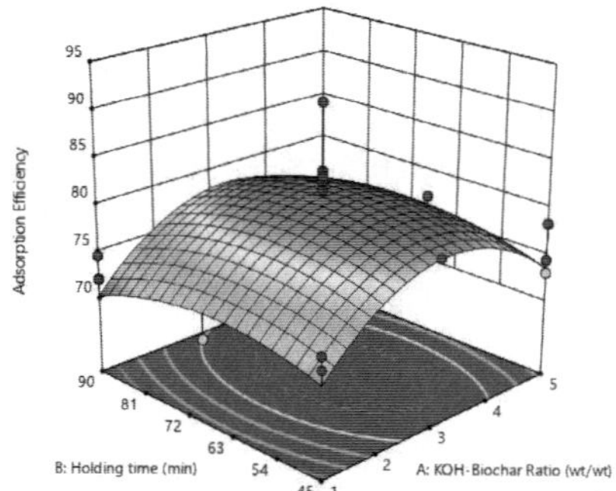

Figure 1. 3D surface plot.

The 3D surface plot shows that lower KOH-biochar ratio resulted to lower % adsorption efficiency of activated biochar, but the addition of KOH improves the efficiency of adsorption up to a certain limit or optimum. At a very low ratio, insufficient KOH was provided; hence, the activation reaction between the alkali and biochar might not have completely proceeded and few pores were developed. As KOH-biochar ratio is increased, more cavities were formed on the activated biochar due to enough etching of the alkali with the carbon network [16]. These cavities and voids serve as the adsorption sites promoting higher adsorption capacity of the adsorbent [17][18]. However, further increase in the KOH-biochar ratio might destroy the porous structure of the activated biochar reducing its adsorption performance. This explains the decrease in the % adsorption efficiency of the activated biochar produced at ratio equal to 5. This also agrees with another study [18] wherein the highest % removal of nitrate is achieved when the activated biochar from sugarcane bagasse is produced at KOH-biochar ratio equal to 3; although the activation process was carried out via microwave pyrolysis. Similarly, further addition of the alkali beyond 3 resulted in a decrease in % nitrate removal. The high amount of KOH elevated its reaction with carbon in biochar which then damaged the walls of the previously developed pores, hence creating larger pores and decreasing the surface area [18].

In terms of the holding time, results of the ANOVA showed that there was no significant effect on the adsorption efficiency for the range (45 to 90 min) considered. Although, other studies reported that increasing the activation time ensures enough contact and reaction between the alkali and biochar for the improvement of pores [18][19]. However, longer time beyond an optimum can reduce the capacity and % removal of the adsorbents due to pore degradation caused by prolonged activation [18][19].

After performing the RSM runs, numerical optimization was done. The goals set were to maximize the % adsorption efficiency while the factors KOH-biochar ratio and holding time were in range. Although holding time is not significant, it was not minimized since a higher % adsorption efficiency was predicted by Design Expert® when holding time is in range. Design Expert® generated optimum values for KOH-biochar ratio and carbonization holding time which were 3.18:1 g/g and 65.64 minutes, respectively with a predicted % adsorption efficiency of 81.11 %. This was experimentally verified with a resulting data mean of 82.14 % which is within the 95 % prediction interval. Moreover, the percentage difference between the actual and predicted mean was 1.27 % which is less than 10 % implying that the value of the actual percentage adsorption efficiency is close to that of the theoretical or predicted value. Hence, the model is valid and the obtained and chosen optimum conditions from the numerical optimization is acceptable in maximizing the percentage adsorption efficiency.

3.2. Characteristics of SB, SBB, and SBAB

The composition and thermochemical behavior of SB were determined for its suitability as feedstock for biochar production. SBB and SBAB were also characterized in terms of their structures and composition to determine their potential as adsorbents for phytohormone from waste coconut water.

3.2.1. Compositional and proximate analyses. SB samples obtained where found to have cellulose (30.81 %), hemicellulose (23.74 %), acid insoluble lignin (17.25 %), acid soluble lignin (2.61 %), and extractives. High contents of cellulose and hemicellulose are responsible for the production of gases during SB pyrolysis. On the other hand, lignin content contributes to biochar yield since lignin requires higher degree of carbonization to completely degrade compared to the other components [5].

Proximate analysis determines the thermochemical behavior of a biomass. Specifically, SB samples contained moisture content (8.26 %), VCM (79.56 %), ash (3.56 %), and fixed carbon (8.62 %). It is desirable for biochar production that the prepared SB samples had low moisture of less than 10 % for a more efficient pyrolysis. The high VCM of the sample can lead to more volatiles being released which leave gaps and cavities onto the biochar surface. In terms of ash, a lower value is desired for better quality of biochar. Furthermore, biochar yield can also be attributed to the fixed carbon (FC) content [5]. Lower FC and lignin of SB can explain the low SBB yield of about 33.41 %.

3.2.2. Morphology. Scanning electron microscopy (SEM) was used to obtain micrographs of the materials at 1000x magnification as shown in figure 2. The SEM image of SBB in figure 2b illustrates an elongated shape and fibrous plant-like structure of bagasse. Due to thermal degradation during pyrolysis, the surface seemed to crack and form crumbs or tiny fragments that were situated on the biochar surface. The cracks can be attributed to the volatile combustible matter (VCM) that were released during biochar production [20]. On the other hand, cavities and a more uneven surface can be observed in figure 2c. These are caused by the etching of KOH with the carbon network of biochar during activation [16]. The voids formed serve as the adsorption sites for the target adsorbate, thus activation of biochar can result to an enhanced surface structure of adsorbents based on SEM analysis.

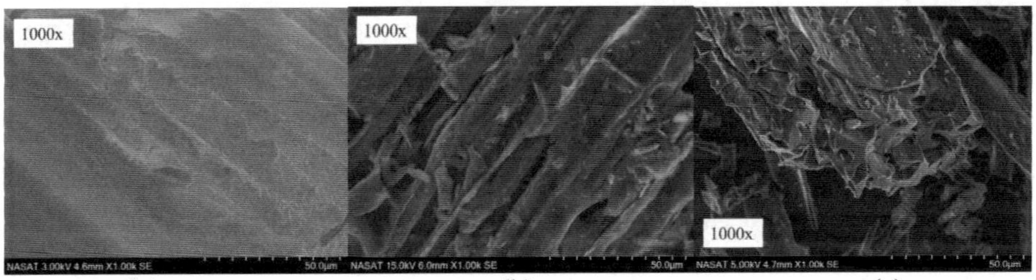

(a) (b) (c)

Figure 2. SEM images of (a) SB, (b) SBB, (c) SBAB.

3.2.3. Elemental composition. The elemental composition of the surfaces of SB, SBB, and SBAB was determined using Energy Dispersive X-ray (EDX) Spectroscopy. Based on table 1, the dominant species for all materials were carbon (C) and oxygen (O). The thermochemical conversion of SB into SBB resulted in increased C content due to carbonization of biomass and decreased O content due to the conversion of organic content into volatile matter which escaped during pyrolysis. From these, the O/C ratio of SBB was determined to be about 0.09 which is less than that of the SB (0.47). Low O/C implies high hydrophobicity of SBB which is characteristic of a good adsorbent used for adsorption of organic substances [21].

After KOH activation, C decreased from 91.06 to 81.00 % while O increased from 8.12 to 16.68 %. Similarly, a study of KOH activation of bamboo biochar resulted in a decrease in C from 90.00 to 80.49 % and increase in O from 7.16 to 17.20 % [22]. As activation proceeds, KOH reacts with the carbon network causing the consumption of carbon, thus lowering its content after the process [22]. Moreover, the OH^- (from KOH-carbon reaction) that was further converted into a more stable O-containing groups explains the increase in oxygen content of the activated biochar [22][23]. Because of the increase in O and reduction of C for SBAB, its O/C increased lowering its hydrophobicity; however, the pores formed in SBAB surface can still enhance adsorption of organic substances.

Table 1. Surface elemental composition of SB, SBB, and SBAB.

Sample	Composition (% w/w)								
	C	O	Al	Mg	Ca	K	Si	P	Na
SB	63.74	29.70	5.22	0.30	0.95	0.90	0.18	-	-
SBB	91.06	8.12	-	0.10	0.15	0.42	0.10	0.18	-
SBAB	81.00	16.68	-	0.54	1.50	0.18	-	0.10	0.15

3.2.4. Functional groups. Based on the FTIR results shown in table 2, SB has the strongest peak at around 3,300 cm^{-1} associated with the hydroxyl groups since cellulose and hemicellulose were still present in the biomass before pyrolysis. For SBB and SBAB, it was observed that OH stretching was more evident after activation possibly due to the increase in O-containing groups caused by the conversion of OH^- from the activation reaction between KOH and carbon [22][23]. The weak peak at 2900 cm^{-1} was still present for SBB implying that alkane groups were not fully removed after pyrolysis. Whereas alkane group was completely removed for SBAB implied by the absence of that peak. Alkane groups might be converted into carbon having sp^2 hybridization which is desirable for activated biochar [24]. Furthermore, the presence of carboxylic acids at around 1690 cm^{-1} was removed after activation. The peak (1200 cm^{-1}) for C-O-C was present in SBAB which indicated that activation initiated the decarboxylation of raw biochar [25]. The peak around 1500–1600 cm^{-1} corresponding to the C=C functional group was most evident in the spectrum of SBAB which is also a desirable characteristic of activated carbons. With this, activation promoted aromatization and carbon matrix formation on the activated biochar. The O-containing groups of SBAB are also favorable to induce interaction between the functional groups of phytohormones for better adsorption.

Table 2. FTIR peaks and corresponding band assignments for SB, SBB, and SBAB.

Peak (cm^{-1})	Band assignment	SB	SBB	SBAB
3,300	OH stretching, hydroxyl	present (strong)	present (weak)	present
2,900	C-H stretching, alkane	present (strong)	present (weak)	-
2,300	C≡N, nitrile	-	present (weak)	present
2,100	C≡C, alkyne	present	present	present
1,690	C=O, carbonyl compounds	present	present	-
1,520-1575	C=C stretching	present	present	present
1,200	C-O-C	-	-	present

3.2.5. Adsorption performance. Adsorption efficiencies of SBB and SBAB were determined through batch adsorption experiment. Result showed that the mean percentage adsorption efficiency for SBB and SBAB were 52.75 % and 79.72 %, respectively. In addition, the equilibrium adsorption capacity also increased from 21.10 mL/g to 31.89 mL/g after KOH activation. Thus, in this study, surface modification through chemical activation using KOH increased the adsorption performance of raw biochar by approximately 51.13 %. As discussed earlier, activation allowed the development of porous structure and adsorption sites of the biochar. Moreover, functional groups were introduced onto the surface promoting interaction between phytohormones and the activated biochar during adsorption.

3.2.6. Average particle size and bulk density. The average particle size of SB, SBB, and SBAB were 2.29 mm 1.74 mm, and 1.15 mm, respectively. Because of the escape of volatiles and material degradation during pyrolysis and activation, SBB and SBAB decreased in size. Extreme temperatures and longer holding time causes particle size reduction due to a decrease in its tensile strength as thermochemical conversion process proceeds [7]. Furthermore, SBAB appeared to have decreased in size caused by the etching of KOH in carbon resulting in cracking and formation of tiny fragments.

Bulk density of a material depends on the nature of feedstock and conversion process involved. The bulk density of SB, SBB, and SBAB were 0.4783 g/cm3, 0.2642 g/cm3, and 0.4801 g/cm3, respectively. In this study, activation increased the bulk density of SBAB probably due to its potential microporous surface compared to raw biochar having larger pores [7].

4. Conclusion

To address environmental and health hazards posed by the burning and disposal of sugarcane bagasse and waste coconut water from farms and agro-industries, a resource recovery approach was explored. In this study, raw biochar from the pyrolysis of sugarcane bagasse was activated and the optimum conditions for activation were determined to maximize phytohormone adsorption efficiency. Based on the statistical analysis, KOH-biochar ratio can significantly affect phytohormone adsorption efficiency. Increasing the ratio up to a certain limit enhances the adsorption performance of activated biochar. The optimum chemical activation conditions using RSM were found to be 3.18:1 (g/g) KOH-biochar ratio and 65.64 minutes holding time to obtain an actual maximum percentage adsorption efficiency of 82.14 %.

Comparisons were made between SBB and SBAB in terms of their characteristics and phytohormone adsorption. It was found in this study that chemical activation improved the performance of SBAB wherein it had a higher % adsorption efficiency (79.72 %) compared to SBB (52.75 %). The SEM micrograph of SBAB also showed a more porous structure than SBB aiding in phytohormone adsorption. The presence of O-containing functional groups was higher for SBAB which also facilitated and improved the interaction and adsorption of phytohormones.

The results of this study can provide a sustainable solution in minimizing the disposal of bagasse and waste coconut water while increasing their economic returns and environmental benefits. Moreover, this study could be used as basis for further activation and adsorption studies.

Acknowledgements

Authors acknowledges the financial support of Engineering Research and Development for Technology (ERDT) and Philippine Council for Agriculture, Aquatic and Natural Resources Research and Development (PCAARRD) both under the Department of Science and Technology (DOST).

References

[1] Zafar S 2021 Agricultural wastes in the Philippines *Bioenergy Consult*
[2] United States Department of Agriculture 2021 Sugar annual
[3] Sugar Regulatory Administration 2015 Sugarcane roadmap 2020
[4] Lehmann J and Joseph S 2009 *Biochar for Environmental Management* (London: Earthscan)
[5] Miranda N T, Motta I L, Filho R M and Maciel M R W 2021 Sugarcane bagasse pyrolysis: a

review of operating conditions and products properties *Renew. Sust. Energ. Rev.* **149**

[6] Lee J, Sarmah A K and Kwon E E 2019 Production and formation of biochar *Biochar from Biomass and Waste* eds Y S Ok, D C W Tsang, N Bolan and J M Novak (Elsevier Inc.) chapter 1 pp 3-18

[7] Downie A, Crosly A and Munroe P 2009 Physical properties of biochar *Biochar for Environmental Management* eds J Lehmann and S Joseph (London: Earthscan) chapter 2 pp 13-34

[8] Sizmur T, Fresno T, Akgul G, Frost H and Jimenez E M 2017 Biochar modification to enhance sorption of inorganics from water *J. Bior. Tech.* **246** pp 34-47

[9] Tareq R, Akter N, Azam M S 2019 Biochars and biochar composites: low-cost adsorbents for environmental remediation *Biochar from Biomass and Waste* eds Y S Ok, D C W Tsang, N Bolan and J M Novak (Elsevier Inc.) chapter 10 pp 169-209

[10] Ison L 2019 Coco farmers to undergo natural vinegar production training *Philippine News Agency*

[11] Coconut Handbook 2021 Chapter 3-composition

[12] Checker V G, Kushwaha H R, Kumari P and Yadav S 2018 Roles of phytohormones in plant defense: signaling and cross talk *Molecular Aspects of Plant-Pathogen Interaction* eds A Singh and I Singh (Singapore: Springer Nature) chapter 7 pp 159-84

[13] Santner A, Calderon-Villalobos L I A and Estelle M 2009 Plant hormones are versatile chemical regulators of plant growth *Nat. Chem. Biol.* **5** pp 301-7

[14] Foust A S, Wenzel L A, Clump C W, Maus L and Andersen L B 1980 *Principles of Unit Operations Singapore* 2nd ed (Singapore: John Wiley & Sons)

[15] Seader J D and Henley E J 2006 *Separation Process Principles* 2nd ed (USA: John Wiley & Sons)

[16] Abechi S E, Gimba C E, Uzairu A and Dallatu Y A 2013 Preparation and characterization of activated carbon from palm kernel shell by chemical activation *Res. J. Chem. Sci.* **3** pp 54-61

[17] Foo K Y and Hameed B H 2012 Microwave-assisted preparation and adsorption performance of activated carbon from biodiesel industry solid residue: influence of operational parameters *J. Bior. Tech.* **103** pp 398-404

[18] Mehrabinia P, Ghanbari-Adivi E, Fattahi R, Samimi H A and Kermanezhad J 2021 Nitrate removal from agricultural effluent using sugarcane bagasse active nanosorbent *J. App. Water Engg.* **10** pp 238-49

[19] Han X, Chu L, Liu S, Chen T, Ding C, Yan J, Cui L and Quan G 2015 Removal of methylene blue from aqueous solution using porous biochar obtained by KOH activation of peanut shell biochar *BioRes* **10** pp 2836-49

[20] Veiga P A D S, Cerqueira M H, Goncalves M G, Matos T T D S, Pantano G, Schultz J, De Andrade J B and Mangrich A S 2021 Upgrading from batch to continuous flow process for the pyrolysis of sugarcane bagasse: structural characterization of biochars produced *J. Environ. Manage.* **285**

[21] Barquilha C E R and Braga M C B 2021 Adsorption of organic and inorganic pollutants onto biochars: challenges, operating conditions, and mechanisms *Bioresour. Technol. Rep.* **15**

[22] Chen Q, Gong M, Li K, Xia M, Chen Z, Xiao H, Fang Y, Chen Y, Yang H and Chen H 2020 Insight into KOH activation mechanism during biomass pyrolysis: chemical reactions between O-containing groups and KOH *Appl. Energy* **278**

[23] El-Hendawy A 2009 An insight into the KOH activation mechanism through the production of microporous activated carbon for the removal of Pb2+ cations *Appl. Surf Sci* **255** pp 3723-30

[24] Daud W M A W and Houshamnd A H 2010 Textural characteristics, surface chemistry and oxidation of activated carbon *J. Nat. Gas Chem.* **19** pp 267-79

[25] Zhao C, Ma J, Li Z, Xia H, Liu H and Yang Y 2020 Highly enhanced adsorption performance of tetracycline antibiotics on KOH-activated biochar derived from reed plants *RSC Adv.* **10** pp 5066-76

Greenhouse Gas Mitigation Potential of the Enhanced Rice Straw Biogas System in the Philippines

C C Gomez[1,4], R B Demafelis[1,5], B T Magadia[1], A E D Matanguihan[1], E P V Estante[1], C J E Nuñez[1], J V Sumague[1], K Chong[2], M Roeder[2] and C Jamieson[3]

[1] Department of Chemical Engineering, College of Engineering and Agro-Industrial Technology, University of the Philippines Los Baños, College, Laguna, 4031 Philippines

[2] Energy and Bioproducts Research Institute, Aston University, Aston Triangle, Birmingham B4 7ET, United Kingdom

[3] Straw Innovations, Lawes Open Innovation Hub, Rothamsted Research, West Common, AL5 2JQ, United Kingdom; with operations in Laguna, Philippines

E-mail: [4]ccgomez2@up.edu.ph; [5]rbdemafelis@up.edu.ph

Abstract. Rice straw is an agricultural waste produced abundantly every rice cropping cycle. Its disposal or removal from the field is a problem to rice farmers every start of the next cropping cycle due to its labor-intensive collection from the field. One traditional practice is soil incorporation; however, rice straw does not degrade easily, and this management practice releases significant amount of greenhouse gas emissions to the environment. An attempt to solve this issue is to utilize rice straw with cattle manure as feedstock for biogas production and use the energy generated for rice post-harvest processing such as grain drying and milling. An Enhanced Rice Straw Biogas facility was constructed in Victoria, Laguna, Philippines and a trial lagoon run was conducted from June to October 2021. The data from the initial run was used as input to the modified Aston-developed Mass Energy Balance Model to determine the biogas production potential of rice straw and cattle manure as bioenergy feedstocks. The proposed system can produce 8,803,567.99 L (296.19 L/kg VS) of biogas which can be converted to 19,502.97 MW electricity and 84,512.86 MJ heat. From the energy generated per batch, 18,417.21 kg rice grain (dry season) or 15,752.79 kg rice grain (wet season) can be dried and milled, with an excess of equivalent kerosene (696.93 L for dry season or 248.46 L for wet season) and electricity (3,745.12 kWh for dry season or 6,182.23 kWh for wet season). Due to biogas utilization, an estimated 3,351 L of fossil-based kerosene and 12,185.91 kWh of electricity equivalent can be avoided. This low GHG emission rice production system (Aston Model) has a theoretical annual carbon footprint of 153,134 kg CO_2e and a 28.21% reduction of greenhouse gas emissions compared to the conventional rice production system.

Keywords: Biogas, Greenhouse gases, Carbon footprint, Life cycle assessment, Rice straw, Cattle manure

1. Introduction

The climatic changes in the environment are intensifying through time as the global greenhouse gas (GHG) emissions continuously reach record high levels. Extreme weather conditions such as super typhoons, massive wildfires, and strong tornadoes encountered are expected to occur at a more frequent and more destructive magnitudes. Glacial retreats and shrinking ice sheets caused by increase in global temperatures result to sea level rise while high carbon dioxide amounts in the atmosphere leads to ocean acidification. These events threaten both flora and fauna into a possible extinction. Impacts of these events greatly cost numerous lives, scarcity in food, water insecurity, as well as disruption on worldwide peace and security due to increased competition on resources. According to scientists, future environmental disasters may still be prevented if the global GHG emissions, in which more than half comes from anthropogenic sources (55.46% of total global GHG emission in year 2016 [1]), will be reduced at an acceptable level.

World Resources Institute Climate Analysis Indicators Tool reported that Philippines emitted 157.6 metric tons of carbon dioxide equivalent ($MtCO_2e$) in the year 2012 [2]. This accounts to 0.33% of the world's total GHG emissions (World emission 47,599 $MtCO_2e$ [2]). From the emissions from the Philippines, 54% originates from energy sector while 33% comes from agriculture sector [2]. These emissions were high due to the dependency on burning fossil fuels, and use of synthetic fertilizers, substrates, and processes with high carbon footprint. In addition, the anthropogenic GHG emissions are projected to increase to provide for the demand of growing population.

Focusing on the Philippine agriculture sector, the main GHG emitter is the rice cultivation. This accounts to 60% of the total emission from agriculture or around 20% of the total GHG emissions of the whole country based on the 2012 statistics [2]. In 2020, Philippines ranked 8th largest rice-producing country in the world with an annual production of 19.44 million metric tons [3]. The country is expected to increase its rice production with the Department of Agriculture targeting a higher yield to cope with the high demand in rice. Unfortunately, high rice production entails not only large amount of rice grain but also massive quantities of by-products such as rice straw, rice husk and rice bran. The ratio of rice straw to rice grain is from 0.7 to 1.4, depending on the variety and cultivation conditions [4]. Hence, an estimated 13.61 to 27.22 million tons of rice straw was produced from the rice production in 2020. Unlike rice husk and rice bran, there is no current economic value for rice straw. The most common practice of farmers to dispose this waste is by incorporating it on soil which releases around 3,500-4,500 kg CO_2e/ha [5]. Adding this to the other emissions during rice cultivation, harvest and post processing worsens the GHG emission from this agricultural sector.

As an initiative to solve this growing concern on GHG emission and rice straw management, Straw Innovations Ltd. constructed an enhanced biogas system utilizing rice straw and cow manure to produce methane and convert it to energy. This industrial research project (March 2020-March 2022) funded by Innovate UK entitled "Enhanced Rice Straw Biogas" was in collaboration with Aston University for the social analyses and with University of the Philippines Los Baños for the environmental analyses. The aim of this paper is to extend the results of the trial lagoon digester run and determine the GHG mitigation potential of the system.

2. Methodology

The methodological framework for the determination of GHG emission mitigation potential of the enhanced rice straw biogas system is shown in Figure 1. Data gathered includes farm data collected from key informant interviews (KIIs) and surveys with rice farmers, biogas production data from the trial research run conducted by Straw Innovations in 2021, and other necessary values obtained from literature.

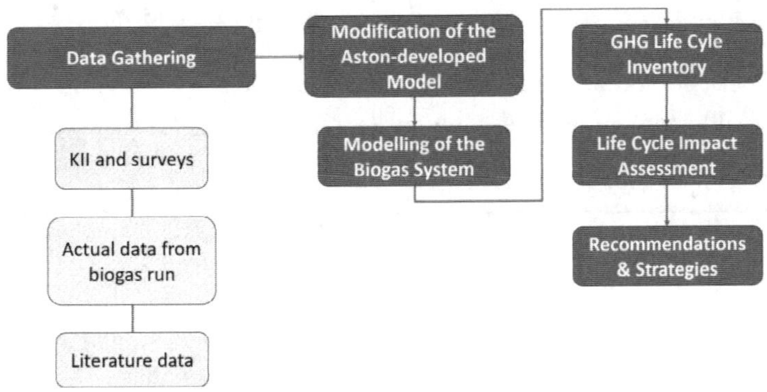

Figure 1. Methodological framework.

2.1. Bioenergy Mass-Energy Balance Model developed in Aston University

A scoping tool designed by a team in Energy and Bioproducts Research Institute (EBRI), Aston University, UK to evaluate the viability of various biomass as feedstocks to the commercially available bioenergy technologies was used in this study This tool allows modelling of typical plant configurations to fit selected feedstock and target energy output range of 10 kW_e to 5 MW_e [6]. This is a part of a project entitled "*Bioenergy for Sustainable Local Energy Services and Energy Access in Africa Phase 2 (BSE-AA2)*", funded by the UK foreign, Commonwealth and Development Office (FCDO) under the Transforming Energy Access (TEA) program.

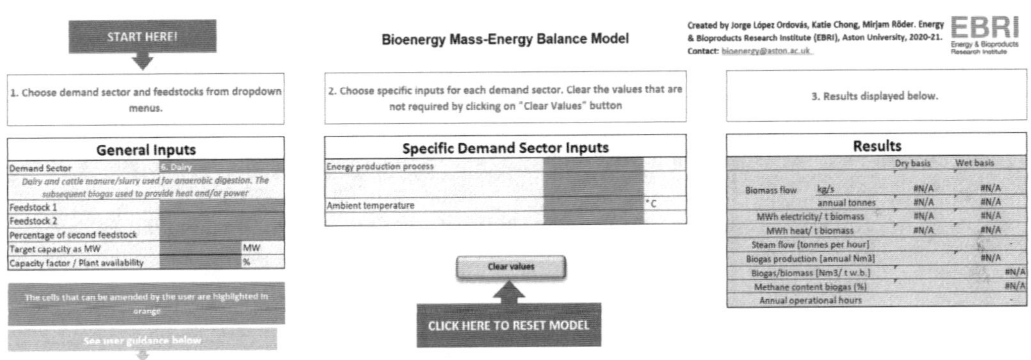

Figure 2. Bioenergy Mass-Energy Balance Model developed in Aston University, UK.

This bioenergy model uses the 1977 modification of Boyle [7] on the Buswell and Mueller equation [8] (shown in Equation 1) to back-calculate the amount of biomass needed to produce a target methane output. Since Equation 1 stoichiometrically provides the chemical conversion of biomass to biogas based on the feedstock elemental composition, the amounts of products can be determined.

$$C_cH_hO_oN_nS_s + \frac{1}{4}(4c - h - 2o + 3n + 2s)H_2O \rightarrow \frac{1}{8}(4c - h + 2o + 3n + 2s)CO_2 + \frac{1}{8}(4c + h - 2o - 3n - 2s)CH_4 + 2NH_3 + sH_2S \quad (1)$$

The Buswell equation is widely used in the calculation of theoretical maximum methane potential, however, the calculated amount cannot be achieved since a complete breakdown of feedstock cannot be achieved [9]. With this, the reaction efficiency was adjusted to 40% [10] to simulate the conversion of rice straw and cattle manure into biogas. This is based on the degree of degradation in seven farm-scale biogas plants utilizing cattle manure with organic substates [10]. Feedstock properties such as elemental composition of cattle manure and rice straw, shown in Table 1, were based on the studies of Font-Palma [11] and Sumagang *et al* [12], respectively. Adjustments on feedstock type and properties, conversion conditions, and processing units were made on the model to simulate the constructed rice straw biogas facility in Victoria, Laguna, Philippines.

Table 1. Feedstock properties of rice straw and cattle manure.

	Moisture content (wt%)	Ash content (wt%)	Volatile matter (wt%)	Fixed carbon (wt%)	Carbon (wt%)	Hydrogen (wt%)	Nitrogen (wt%)	Oxygen (wt%)	Sulfur (wt%)
Cattle manure[11]	70.70	37.20	52.00	10.80	31.01	4.06	2.09	24.98	0.66
Rice straw[12]	5.73	21.00	61.40	17.60	35.50	4.62	0.99	37.83	0.06

2.2. GHG Life Cycle Inventory and Life Cycle Assessment

Life Cycle Assessments were simulated using SIMAPRO 9.4.0.2 version. The functional unit used in the study is 1 kg milled rice grain while the impact method used is the Intergovernmental Panel on Climate Change 2021 Global Warming Potential climate factors for 100 years timeframe that includes CO_2 uptake and biogenic CO_2 emissions (IPCC 2021 GWP100-incl. CO_2 uptake). The values obtained from the data gathering conducted and from the modified Bioenergy Mass-Energy Balance Model were used in the life cycle assessment of the low GHG emission rice production system. The conventional rice production system was also evaluated for comparison. System boundaries of the two rice production systems are shown in Figures 2 and 3.

2.2.1. Conventional Rice Production System. The conventional rice production system in Victoria, Laguna was considered in the study. This system includes rice production from land preparation until milling of rice grain. Hand tractors and hydrotiller powered by diesel were used for the plowing and harrowing of rice fields while levelling was done using a wooden plank pulled by a carabao. Seeds were initially grown in wet beds before transplanting to the field 14 days after germination. Fertilizers were manually scattered by farmers on the field at a specific growth stage of rice plant. For the pesticides and herbicides, farmers spray it on the field if need arises. Water supplied to rice paddies may either be from the National Irrigation Administration (NIA) sourced from river or from deep well. Harvesting was done using combine harvester. Rice grains were collected in sacks and were transported to drying and milling facilities for post-harvest processing while rice straws that were left on the field will be incorporated on soil upon land preparation for the next cropping cycle Amounts of seeds, fertilizer, pesticides, insecticides, and fuel used for irrigation pump and machineries for land preparation and harvesting were accounted from the KII and surveys with seven rice farmers in the area.

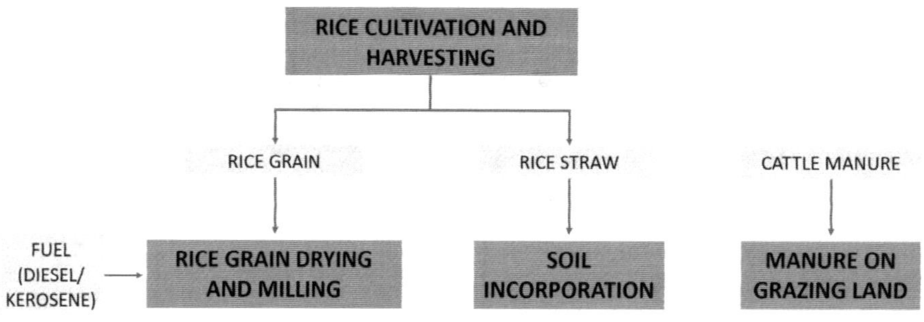

Figure 3. Conventional Rice Production System Boundary.

Calculations for the GHG emissions of rice cultivation is similar with the study of Estante *et al* [13] which was adapted from the 2019 Refinement to the 2006 IPCC Guidelines for National Greenhouse

Gas Inventories was used in the calculations of rice field GHG emissions for methane (CH_4), nitrous oxide (N_2O), and carbon dioxide (CO_2). Energy usage (fuel and electricity) for the drying and milling of rice were obtained from KII with drying and milling facility owners while other values were based on literature.

To balance the inputs of conventional system with the low GHG emission system, the equivalent amount of manure used for biogas production were accounted in the conventional system as cattle manure applied on grazing land. GHG emissions of cattle manure was estimated based on the 2019 Refinement to the 2006 IPCC Guidelines for National Greenhouse Gas Inventories.

2.2.2. Low GHG Emission Rice Production System. Rice cultivation from land preparation until harvesting for the low GHG emission system is similar with the conventional system. Rice straws were collected and co-digested with cattle manure for biogas production. Batch anaerobic digestion trial run by Straw Innovations was conducted in a lagoon digester for 120 days. Fuel usage such as diesel used during the collection of rice straw from nearby rice fields and collection of cattle manure from nearby dairy farm were accounted. Biogas produced was used to generate heat and electricity via Combined Heat and Power (CHP) engine operating at 15% electricity efficiency and 65% heat efficiency conversion. The heat and electricity produced from the biogas substituted the fossil fuel-based energy used to operate rice grain dryer and mill. Wet palay were dried at 13% moisture content at 70% dryer efficiency and while dried rice grains were milled at 70% milling efficiency. Basis for the calculations for drying and milling of rice grain is shown in Table 2. Digestates produced were utilized as soil conditioner and fertilizer.

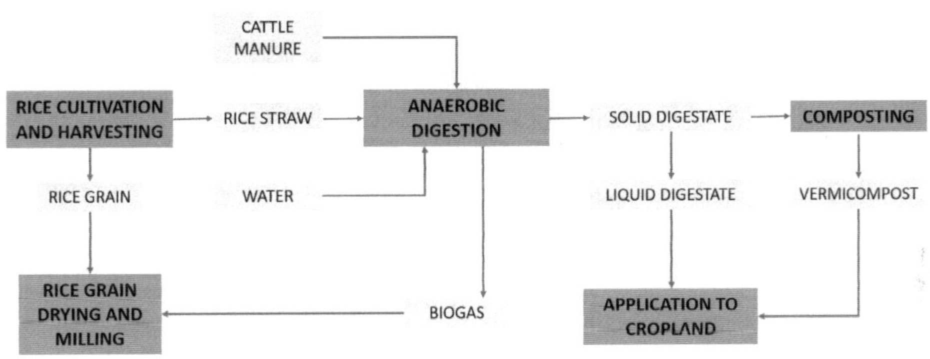

Figure 4. Low GHG Emission Rice Production System Boundary.

Table 2. Rice grain moisture content and energy requirements of recirculating vertical bin dryer and grain milling unit.

	Recirculating vertical bin grain dryer		Rice grain milling unit	
	Dry season	Wet season		
Rice Grain			Unit	6LN-15/155D
Initial moisture (%)	17.00	29.00	Rated output, kg/h	600-800
Target moisture (%)	13.00	13.00	Power requirement, kW	19.87-24.62
Inputs			Efficiency, %	68-72
Wet palay, kg/batch	7,000.00	7,000.00		
Fuel (kerosene), L/batch-unit	120.00	175.00		
Electricity for pumps, kWh/batch-unit	17.90	26.86		
Drying time, h	8.00	12.00		

IOP Conf. Series: Materials Science and Engineering **1318** (2024) 012016 doi:10.1088/1757-899X/1318/1/012016

GHG emissions in CO_2 equivalent were accounted and the environmental hotspots in the rice production systems were identified based from the values obtained per process stage. To determine the GHG reduction potential of the low GHG emission rice production system from the conventional system, Equation 2 was used.

$$GHG\ Reduction\ (\%) = \frac{GWP_{conventional} - GWP_{low\ GHG}}{GWP_{conventional}} x100\% \quad (2)$$

3. Results and discussion

Straw Innovations conducted a trial lagoon digester run in 2021 to test the viability of rice straw and cattle manure as feedstocks for biogas production. Since the study was in experimental stage, the digestion process was not optimized. To determine the potential of the rice straw biogas system as bioenergy source and for mitigating GHG emissions, the same system inputs were used as feedstock amount basis in the Aston Bioenergy Mass-Energy Balance Model (Aston Model). Amounts of feedstocks and the projected biogas and digestates are shown in Table 3.

Table 3. Biogas production inputs and outputs of the actual trial run and the simulated model using Aston Bioenergy Mass-Energy Balance Model.

Models	Input				Output		
	Rice Straw, kg	Cattle Manure, kg	Water, kg	Total, kg	Biogas, L	Solid Digestate, kg	Liquid Digestate, kg
Straw Innovations Run	39,950.00	10,000.00	52,033.43	101,983.43	1,409,744.48	56,841.80	53,970.02
Aston Model	39,950.00	10,000.00	52,033.43	101,983.43	8,803,567.99	47,956.63	53,956.99

The calculated biogas produced, shown in Table 4, was upgraded to meet the CHP biogas composition requirement. With the 8,803,567.99 L biogas produced using the Aston Model, an equivalent of 19,502.97 MW electricity and 84,512.86 MJ heat can be generated. This energy can be used to process the harvested rice grain in the rice fields considered in the system boundary. Since moisture content of harvested palay is higher during wet season, the amount of dried and milled rice grain in this season is lower compared to dry season. Around 18,417.21 kg (dry season) and 15,752.79 kg (wet season) dried and milled rice grain can be processed using the same biogas yield per batch. The generated heat and electricity from the biogas conversion in the Aston Model has excess amounts which can be used to process more rice grain or for other purposes.

Table 4. Biogas composition based on modified Buswell equation (initial) and after upgrading and its equivalent energy conversion using combined heat and power (CHP) engine.

Component	Biogas Composition				Biogas Conversion Using CHP		
	Initial composition		After upgrading		Component	Energy generated, MW	%
	Volume, L	%	Volume, L	%			
CH_4	4,203,050.27	47.74	4,203,050.27	60.00	Electricity	19,502.97	15.00
CO_2	4,329,440.27	49.18	2,802,033.51	40.00	Heat	84,512.86	65.00
NH_3	255,800.03	2.91			Losses	26,003.96	20.00
H_2S	15,277.43	0.17					
Total	8,803,567.99		7,005,083.79		Total	130,019.79	

IOP Conf. Series: Materials Science and Engineering **1318** (2024) 012016 doi:10.1088/1757-899X/1318/1/012016

Table 5. Amount of rice grain processed and excess energy equivalent of biogas utilization.

Component	Conventional System		Aston Model	
	Dry Season	Wet Season	Dry Season	Wet Season
Dried and milled rice grain, kg	18,417.21	15,752.79	18,417.21	15,752.79
Excess kerosene equivalent, L	-	-	696.93	248.46
Excess electricity equivalent, kWh	-	-	3,745.12	6,182.23

Life cycle assessment of the Conventional Rice Production System and Low GHG Emission Rice Production Systems, namely, Aston Model and Straw Innovations System, revealed the carbon footprint of each product stages. The stage with the highest carbon footprint identified was during rice cultivation and harvesting. This is due to high fuel usage during land preparation, irrigation and harvesting, as well as high inputs of synthetic fertilizer. Farm emissions in the conventional system is higher than the low GHG emissions systems by around 25-39% due to emissions of incorporation of rice straw on soil. Although biogas production has an equivalent GHG emission due to fuel used in the transport of rice straw and cattle manure from farm to facility, the calculated amount is still less than the emissions when rice is incorporated in the soil. For the grain drying and milling, there is GHG emission reduction for the Aston Model due to the excess energy produced. GHG emission reduction was also achieved for the use of solid and liquid digestate as organic fertilizers. Shown in Figure 5 are the GHG emissions and GHG reductions per product stage.

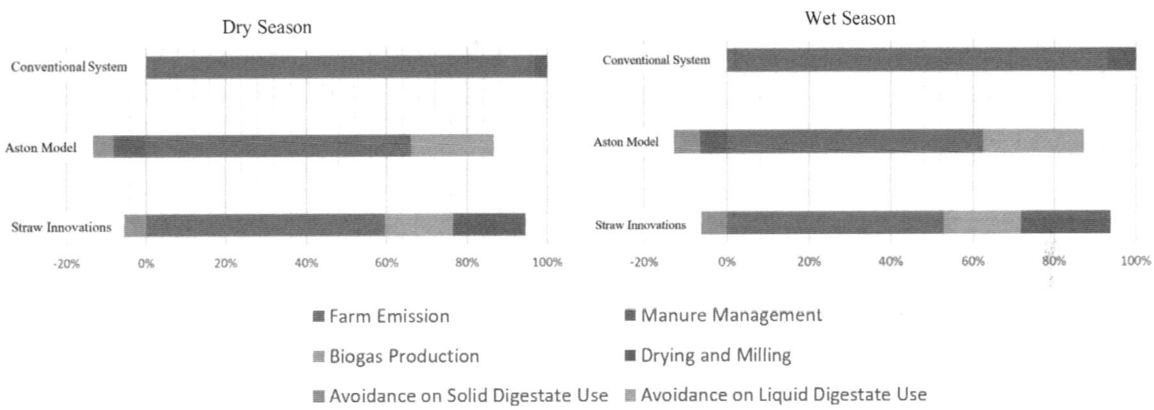

Figure 5. GHG emissions and savings of the three rice production systems.

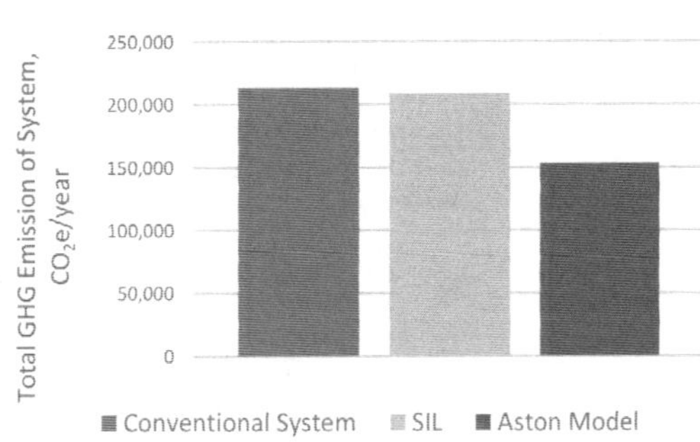

Figure 6. Annual GHG Emission of the rice production systems.

IOP Conf. Series: Materials Science and Engineering **1318** (2024) 012016 doi:10.1088/1757-899X/1318/1/012016

Combining the GHG emissions and GHG savings of the rice production systems for dry and wet seasons, an annual 28.21% GHG mitigation potential for Aston Model can be achieved. This is based on the GHG reduction from the conventional rice production systems. This GHG reduction potential is the theoretical maximum amount that can be achieved if the Straw Innovations rice straw biogas system will be optimized.

4. Conclusion

Findings in the life cycle assessment of the system utilizing rice straw with cattle manure to produce biogas proved the viability of these biomass in reducing GHG emissions in rice production system. With the abundant supply of rice straw generated as waste every cropping cycle in the country, its use for bioenergy production can sustain the energy requirement for rice post-harvest processing as well as supply the excess energy to the grid. However, this can only be achieved upon optimization of the anaerobic digestion and CHP to drying and milling units.

5. Acknowledgement

This paper and the research project it originated would not be possible without the exceptional support of United Kingdom Research and Innovation (UKRI) and our project collaborators in Straw Innovations and Energy and Bioproducts Research Institute (EBRI) in Aston University.

References

[1] Yue X L and Gao Q X 2018 *Adv. Clim. Chang. Res.* **9** pp 243-252
[2] World Resources Institute Climate Analysis Indicators Tool 2.0 2016
[3] International Rice Research Institute 2020 Philippine records its highest rice production in 2020 *Rice Today*
[4] International Rice Research Institute n.d. The value of sustainable rice straw management
[5] Nguyen V H, Topno S, Balingbing C, Nguyen V C N, Roeder M, Quilty J, Jamieson C, Thornley P and Gummert M 2016 *Enegy Rep.* **2** pp 117-122
[6] Chong K *et al* 2021 Bioenergy mass-energy balance model. Energy and Bioproducts Research Institute, Aston University, UK
[7] Boyle M 1977 *Microbial Energy Conversion* **(1)** pp 119-138
[8] Buswell A and Mueller H 1952 *Industrial & Engineering Chemistry* **44(3)** pp 550-552
[9] Zhang Y, Kusch-Brandt S, Salter A M and Heaven S 2021 *Processes* **9**, 1565
[10] Ahlberg-Eliasson K, Nadeau E, Leven L and Schnurer A 2017 *Biomass and Bioenergy* **(97)** pp 27-37
[11] Font-Palma C 2019 *Carbon* **5**, 27
[12] Sumagang M V P, Nguyen V H, Detras M C M, Alfafara C G, Borines M G, Capunitan J A and Gummert M 2020 *Renewable Energy* **148** pp 953-963
[13] Estante E P V, Demafelis R B, Matanguihan A E D, Magadia B T, Gomez C C, Nuñez C J E, Sumague J V and Jamieson C 2023 *Philipp. J. Crop Sci.* **48(1)** pp 41-53

Valorization and investigation of volcanic ash as an aggregate building brick component

Erish T. Daranciang[1,2,†], Kenneth B. Tria[1], Elizabeth O. Santos[1] and Ner C. Rodriguez[1,‡]

[1] *Standards and Testing Division, Industrial Technology Development Institute*, DOST Compound, Bicutan, Taguig City, Metro Manila 1631 Philippines

[2] *National Graduate School of Engineering, University of the Philippines Diliman*, Quezon City, Metro Manila 1101 Philippines

E-mail: [†]`etdaranciang@itdi.dost.gov.ph`, [‡]`ncrodriguez@itdi.dost.gov.ph`

Abstract. The devastating series of Taal volcanic eruptions from 2020 to 2022 covered several locations in the Philippines under a thick layer of ashes, destroying properties and halting business operations. In response, the Biñan City Government in Laguna, Philippines, proposed the development of eco-friendly bricks by leveraging the pervasive amounts of spewed volcanic ash as part of its solid waste management program. The materials recovery facility of the city produced several brick blends of combined Taal volcanic ash, cement mortar, and plastic waste materials. This study investigates the potential of Taal volcanic ash as an aggregate building brick component by measuring the compressive strength of the formulated bricks. With different blends of aggregate components, the bricks were tested for their durability based on the ASTM C67/C67M-20 standard after curing between 15 to 30 days. Per the ASTM C62-17 standard, all brick blends that were cured for 30 days including those with volcanic ash and plastic waste as aggregate components fall under the negligible weathering grade. This indicates that these bricks will be durable in building and structural applications where the average compressive strength requirement for bricks is at least 10.3 MPa. A regression model was also fitted with an adjusted r^2 value of 0.9413 and a p-value of $< 2.2 \times 10^{-16}$ using the experimental data.

1. Introduction

With more than four decades of repose since its phreatic activity in October 1977, the Taal volcano erupted on 12 January 2020 and displayed a state of unrest and phreatomagmatic bursts from 01 July 2021 to 27 March 2022 [1, 2, 3, 4]. The volcanic ash plume and ashfall from this series of hydrovolcanic eruptions and phreatomagmatic activities covered several locations in the Philippines under a thick layer of ashes, which destroyed properties and disrupted normal daily activities and business operations. In this work, a novel effort to abate the effect of such a devastating phenomenon is highlighted.

1.1. Solid Waste Management

As a response to the pervasive amounts of volcanic ash spewed during the Taal volcanic eruption in 2020, the Biñan city government in Laguna, Philippines involved the reuse of dry volcanic ash as valorized raw materials into its existing solid waste management program to develop eco-friendly bricks [5, 6, 7, 8, 9]. This waste management initiative started in 2016 by leveraging

IOP Conf. Series: Materials Science and Engineering **1318** (2024) 012017 doi:10.1088/1757-899X/1318/1/012017

shredded plastic waste to produce eco-friendly bricks through the city government's materials recovery facility. The aftermath of the Taal volcanic eruption immediately led to daily clearing operations in Biñan, which in turn, provided a route to utilizing the collected volcanic ash to produce bricks in the local materials recovery facility. As the Philippines faces an ongoing waste crisis [10, 11, 12], this inventive response combines the collected ashes with cement mortar (i.e., a mixture of cement, sand, and water) and discarded plastic to develop bricks for infrastructure projects in the city. The durability of such bricks made from valorized materials needed to be verified through laboratory testing [5, 6, 7, 8, 9].

1.2. Related Works
The valorization of volcanic ash to produce bricks for structural and engineering applications is a well-researched subject. Its use in cement and concrete has been extensively reviewed and recently published [13, 14, 15]. These reviews also highlighted the pozzolanic activity and properties of volcanic ash; its industrial applications in refractories, geopolymers, and ceramics; and how its physicochemical, mechanical, and structural properties and composition affect these engineering applications. Similarly, several studies indicating that volcanic ash is suitable and effective as an aggregate brick component for cement mortar and concrete are widely published [16, 17, 18, 19, 20, 21]. In these works, one of the commonly studied properties includes compressive strength and how it is affected by certain parameters including blend formulation and curing conditions among others. Various standard methods and specifications were also used per study locale. This signifies the novelty of conducting a study on the valorization of Taal volcanic ash together with plastic waste materials as aggregate components for bricks given the waste pollution in the country [10, 11, 12].

2. Materials and Methods
This study investigates the potential of Taal volcanic ash as an aggregate building brick component by conducting compressive strength studies on brick preparations. The materials and methods used in this study are described in the following sections.

2.1. Aggregate Components
As facilitated by the local government unit in Biñan, Laguna and its *City Environment and Natural Resources Office* (CENRO), clearing operations to collect and abate the pervasive amounts of Taal volcanic ash were conducted in the city's main and secondary roads as well as in the barangays or local neighborhoods of the city. The ash collected from these locales was used in the formulation of bricks.

Preliminary characterization of Taal volcanic ash was conducted in the chemistry testing laboratory of the *Industrial Technology Development Institute* in 2020 [22]. The results of the chemical analyses conducted as shown in Table 2 indicate that the Taal volcanic ash samples satisfy the requirements of the *Philippine National Standard PNS 07:2018* for Type II Portland cement for silica (SiO_2) and magnesium oxide (MgO). This standard guides the mandatory product certification of Portland cement and blended hydraulic cement with pozzolan in the Philippines as per the *Department of Trade and Industry Administrative Order No. 17-06, series of 2017*. The satisfactory results for SiO_2 and MgO confirm the potential of Taal volcanic ash samples as a cement aggregate. However, content specifications for alumina (Al_2O_3), ferric oxide (Fe_2O_3), total alkali (Na_2O + 0.658 K_2O), and loss on ignition are not met. Mixing other additives or aggregates in the brick preparations may help in lowering the concentration values of such compounds found in Taal volcanic ash. Shredded plastic waste materials were also collected by CENRO as part of its solid waste management program. These waste materials were mixed with cement mortar to produce bricks.

From this, an investigation of the effect of the Taal volcanic ash addition in these eco-friendly bricks ensued. The brick production process of CENRO is briefly discussed in this work.

2.2. Brick Production

The CENRO materials recovery facility in Biñan produced several brick formulations of volcanic ash, sand, cement, and plastic waste materials. The CENRO facility can mold and produce 5,000 bricks a day with its hydraulic press-powered brick-making machine. Upon collection of plastic waste materials and volcanic ash and preparation of Portland cement and sands (i.e., white and brown varieties), the production of bricks commenced. The manufacture of these eco-friendly bricks started with the crushing and shredding of plastic waste materials. These plastic materials were passed through a conveyor belt to undergo size reduction in the crushing/shredding machine. Weighing and measurement of the raw materials followed to ensure proper dispensing of the aggregate components of the bricks.

Next, the blending of the aggregate components with water according to the dispensed formulation happened in the mixing bin. While this process maintained the required consistency, this also ensured that the combined mass will be homogeneous and uniform in color. After mixing, the bricks were molded and formed in the hydraulic press-powered brick-making machine. Finally, the molded bricks were exposed to a curing process at ambient conditions. This final process maintained proper hydration of the combined mass to ensure that the brick mass would attain its maximum strength while it sets and hardens.

In this work, formulations of brick samples were produced with different blends of white and brown sands, pure Portland cement, water, plastic waste materials, and volcanic ash as well as curing periods of 15, 25, and 30 days. The formulation summary of the bricks is given in Table 1. The produced bricks were 200 mm long, 100 mm wide, and 75 mm thick, as reported in their nominal values.

Table 1. Formulation summary of the bricks produced by CENRO

Brick sample code	A	B	C	D	E	F	G	H
White sand content, w/w	0.6	0.2	0.6	0.2	0.6	0.6	0.2	0.6
Brown sand content, w/w	0.0	0.0	0.3	0.0	0.0	0.3	0.0	0.0
Volcanic ash content, w/w	0.0	0.4	0.0	0.4	0.0	0.0	0.4	0.0
Plastic waste content, w/w	0.3	0.3	0.0	0.3	0.3	0.0	0.3	0.3
Cement and water content, w/w	0.1	0.1	0.1	0.1	0.1	0.1	0.1	0.1
Curing period in days	25	15	15	15	15	30	30	30

2.3. Experimental Design

The experimental design described in this work was also developed by CENRO with prior consultation with the authors. Table 1 shows the formulation summary of the produced bricks. There are a total of eight (8) brick formulations in the experiment. Five (5) trials of the compressive strength test were conducted for each sample code. For this work, each trial was treated as a replicate. This design has 40 experimental runs corresponding to 40 brick specimens.

The independent variables or experimental factors include the blend composition of sand, volcanic ash, plastic waste, cement, and water content and the curing period of the brick formulations. The compressive strength of bricks is the sole dependent variable.

2.4. Laboratory Testing

The durability of the bricks was tested by measuring the compressive strength using the *MATEST Cyber-Plus Evolution C109N* testing machine. This equipment was calibrated per the *ISO 7500-1:2018* standard. The calibration results indicate that the equipment has been categorized under *Class 3* for having a relative accuracy error between ±3.0% and a relative repeatable error below 3.0%. This establishes the suitability of this machine for the experiment.

The compressive strength test followed the *ASTM C67/C67M-20* standard method. The uniform rate of compressive load was 1.80 kN/s. With this speed, the brick specimens were tested flatwise. This means that the cross-sectional area, where the compressive force was applied, was calculated from the length and width dimensions of the bricks. The exact dimensions and mass of the produced bricks were measured at 0.01-mm and 0.1 g accuracy, respectively. The statistical analysis was done using the *R Statistical Software version 4.3.1* in an *RStudio 2023.06.1 Build 524* environment [23]. Lastly, compliance with the durability requirements in the *ASTM C62-17* standard for building bricks was checked.

3. Results and Discussion

3.1. Statistical Analysis

Together with the mass and cross-sectional area, the compressive load and strength values were measured in the laboratory test performed on the 40 brick specimens. The boxplot in Figure 1 captures the spread and locality of data of the compressive strength.

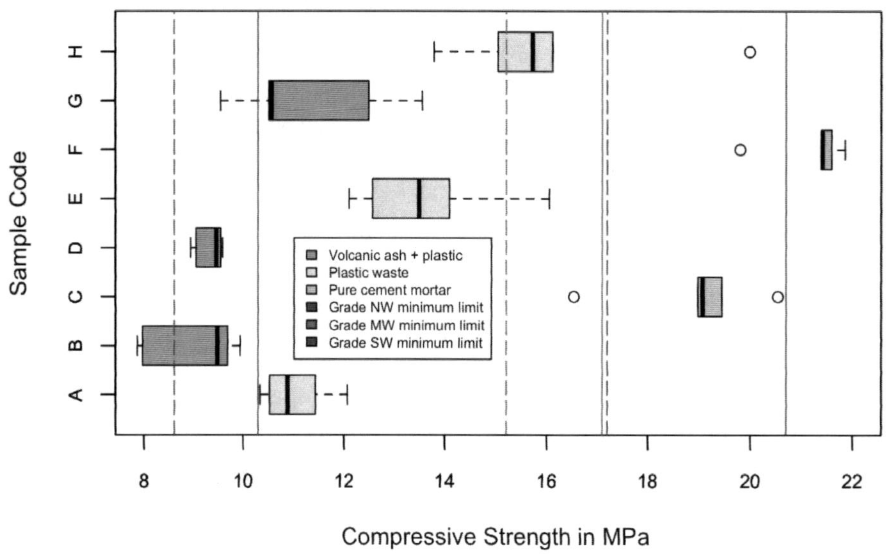

Figure 1. Boxplot of the compressive strength values per brick sample formulation with overlaying minimum specification limits of ASTM C62-17. Generated using R.

Brick specimens with sample codes C and F yielded higher compressive strength values relative to the other six sample formulations. This is depicted in the boxplot in Figure 1. Specimens under these codes essentially corresponded to pure cement mortar bricks and served as the *control group*. *Outlying* data points and *skewness* of data, however, were detected in the boxplot. Excluding such outliers, the data points in the compressive strength distribution for these bricks are closely dispersed around the mean.

Specimens under sample codes A, E, and H replaced brown sand with plastic waste aggregate. These bricks resulted in a relatively lower compressive strength than the pure cement mortar bricks, as also visualized in Figure 1. The boxplot displays a wider dispersion of compressive strength data for these brick specimens compared to the pure cement mortar bricks. On the other hand, those under sample codes B, D, and G produced the lowest set of compressive strength values as shown in Figure 1. A wide dispersion of data points is observed for sample codes B and G. Nonetheless, brick specimens under sample code G yielded compressive strength values comparable to those bricks blended with plastic aggregate only.

In all these blend groups, increasing the curing time up to 30 days seemed to have improved the compressive strength. The compressive strength values of these brick specimens, together with the identified independent variables in Table 1, can also be analyzed using a regression model. With this, the data in Table 1 and the measured mass, cross-sectional area, and compressive load and strength of the bricks are fitted using an *ordinary least squares regression model* in the R statistical software. The mass and cross-sectional area of the brick specimens were also included in the model as independent variables. The regression modeling summary is captured in Figures 2 and 3.

As shown in Figure 2, the model coefficients for the volcanic ash, plastic waste, and cement paste contents were not defined because of the *perfect multicollinearity* or singularities detected

RStudio: Notebook Output

```
Call:
lm(formula = strength ~ mass + area + whitesand + brownsand +
    volash + plastic + cementwater + curing, data = viva)

Residuals:
    Min      1Q  Median      3Q     Max
-2.0058 -0.5771 -0.1149  0.6742  2.1446

Coefficients: (3 not defined because of singularities)
              Estimate Std. Error t value Pr(>|t|)
(Intercept) 115.244798  50.889682   2.265  0.03003 *
mass          0.014318   0.001763   8.123 1.79e-09 ***
area         -0.007211   0.002559  -2.817  0.00801 **
whitesand     1.808611   1.396334   1.295  0.20396
brownsand    12.231440   1.961379   6.236 4.26e-07 ***
volash              NA         NA      NA       NA
plastic             NA         NA      NA       NA
cementwater         NA         NA      NA       NA
curing        0.158500   0.025031   6.332 3.20e-07 ***
---
Signif. codes:  0 '***' 0.001 '**' 0.01 '*' 0.05 '.' 0.1 ' ' 1

Residual standard error: 1.081 on 34 degrees of freedom
Multiple R-squared:  0.9488,    Adjusted R-squared:  0.9413
F-statistic:   126 on 5 and 34 DF,  p-value: < 2.2e-16

                   mass       area       load   strength  whitesand   brownsand     volash     plastic cementwater      curing
mass         1.00000000  0.1027822  0.8774493  0.8750119  0.7496333  0.64763134 -0.7496333 -0.64763134          NA  0.08782541
area         0.10278218  1.0000000 -0.1137805 -0.1249418  0.2256702 -0.29015940 -0.2256702  0.29015940          NA  0.12193749
load         0.87744932 -0.1137805  1.0000000  0.9999319  0.6967826  0.81510184 -0.6967826 -0.81510184          NA  0.32391081
strength     0.87501192 -0.1249418  0.9999319  1.0000000  0.6934864  0.81721699 -0.6934864 -0.81721699          NA  0.32289505
whitesand    0.74963328  0.2256702  0.6967826  0.6934864  1.0000000  0.44721360 -1.0000000 -0.44721360          NA  0.20620302
brownsand    0.64763134 -0.2901594  0.8151018  0.8172170  0.4472136  1.00000000 -0.4472136 -1.00000000          NA  0.05123155
volash      -0.74963328 -0.2256702 -0.6967826 -0.6934864 -1.0000000 -0.44721360  1.0000000  0.44721360          NA -0.20620302
plastic     -0.64763134  0.2901594 -0.8151018 -0.8172170 -0.4472136 -1.00000000  0.4472136  1.00000000          NA -0.05123155
cementwater          NA         NA         NA         NA         NA          NA         NA          NA           1          NA
curing       0.08782541  0.1219375  0.3239108  0.3228951  0.2062030  0.05123155 -0.2062030 -0.05123155          NA  1.00000000
Analysis of Variance Table

Response: strength
          Df Sum Sq Mean Sq  F value    Pr(>F)
mass       1 594.73  594.73 508.4894 < 2.2e-16 ***
area       1  36.25   36.25  30.9919 3.147e-06 ***
whitesand  1   9.11    9.11   7.7887   0.00856 **
brownsand  1  50.02   50.02  42.7652 1.731e-07 ***
curing     1  46.90   46.90  40.0961 3.201e-07 ***
Residuals 34  39.77    1.17
---
Signif. codes:  0 '***' 0.001 '**' 0.01 '*' 0.05 '.' 0.1 ' ' 1
```

Figure 2. Results from the linear regression model fitting, correlation analysis, and analysis of variance performed using R.

from the data. In effect, these factors were excluded from the regression model generated with an adequately high adjusted r^2 value of 0.9413. Only the mass, cross-sectional area, sand content, and curing period of the brick specimens are significant to the compressive strength response variable as confirmed in the analysis of variance. The multicollinearity detected can be attributed to the limited factor levels of the independent variables. This means that further experiments involving appropriate variations in the mixture of the aggregate components of the bricks should be conducted. The residuals of the regression model generated were also analyzed using the residuals plots for *normality* and *variance equality* checking and detection of *outliers* and *influence points* as shown in Figure 3.

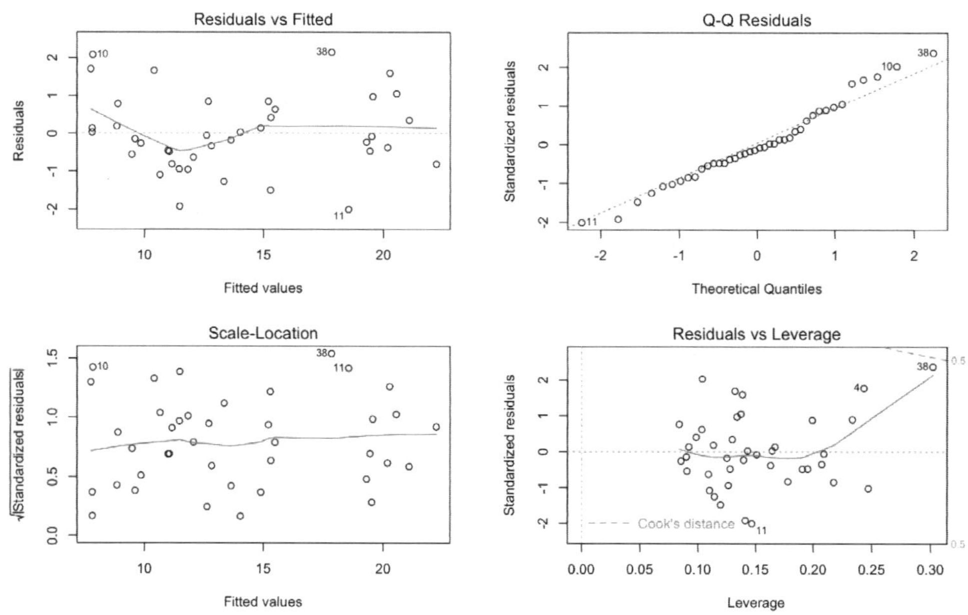

Figure 3. Residual plots generated using R.

In the subplot of the residuals versus the fitted values, the residuals seem to randomly bounce around the zero line, suggesting that the assumption of linearity is reasonable. However, the curved band (shown as a red line) formed by the residuals around the zero line indicates that the equality of the variances of the error terms may not be supported by the data. The scale-location subplot indicates otherwise as the red line is approximately horizontal and the standardized residuals form a random pattern. This suggests *homoskedasticity* and satisfies the *variance equality assumption*.

A comparison of the standardized residuals and the theoretical quantiles is also shown in the Q-Q normal subplot. The seemingly straight line formed by the residuals suffices to pass the fat-pencil test. This satisfies the *normality assumption*.

Labeled residuals in the subplots indicate that these are points of interest in checking for outliers and influence points. Since all standardized residuals are within the *Cook's distance boundary* as seen in the residuals versus leverage subplot, the residuals detected as potential outliers and influence points in Figure 3 may be overruled.

All these considered, the regression model may be regarded as well-defined. This is also supported by the F-statistic ($F = 126$) and its p-value ($p < 2.2 \times 10^{-16}$), which indicate the significance of the model. Nonetheless, sufficient experimental runs and replication are needed to properly model the distribution of the response variable.

3.2. Compliance with the Standard

With a well-defined regression model of the data, compliance with the durability requirements of the ASTM C62-17 standard for bricks used in buildings or masonry structures was also checked. Figure 1 shows the boxplot with overlaying minimum specification limits for compressive strength as required in ASTM C62-17. Solid lines in the figure correspond to the minimum average requirement for five (5) bricks, while the dotted lines represent the individual strength requirement per brick.

In the ASTM C62-17 standard, grades including SW (severe weathering), MW (moderate weathering), and NW (negligible weathering) classify bricks according to their resistance to damage. Despite that only the pure cement mortar bricks satisfy the Grade SW requirement for compressive strength as shown in Figure 1, brick blends with volcanic ash and plastic waste still demonstrated compliance with Grade NW. This indicates that the brick blends will perform satisfactorily and will be durable in building and structural applications where the average compressive strength requirement for bricks is at least 10.3 MPa (1,500 psi).

With the curing period as a significant regression model variable, 30-day curing for brick specimens with sample codes F, G, and H may be attributed to the boost in the compressive strength making it possible for these brick specimens to reach the minimum limits for the various grades. Bricks blended with plastic waste (without volcanic ash) and cured for 30 days, satisfy the individual compressive strength requirement for Grade MW. This denotes that such bricks are durable for buildings and structures where the individual compressive strength requirement per brick is at least 15.2 MPa (2,200 psi).

4. Conclusion and Future Works

This study explored the use of Taal volcanic ash as an aggregate component for building bricks. With various blends of pure Portland cement paste, white and brown sand, plastic waste materials, and volcanic ash, the bricks were tested for their durability by measuring the compressive strength based on the ASTM C67/C67M-20 standard after curing between 15 to 30 days. After testing, a regression model of the compressive strength was fitted from the data with an adjusted r^2 value of 0.9413 and a p-value of $< 2.2 \times 10^{-16}$. The model satisfied the normality and variance equality assumptions with its significant regressors including the mass, cross-sectional area, sand content, and curing period of the brick specimens.

In addition, all brick blends that were cured for 30 days including those with volcanic ash and plastic waste as aggregate components fall under the negligible weathering (NW) grade according to the ASTM C62-17 standard specifications for building bricks. This indicates that the developed bricks will be durable in building and structural applications where the average compressive strength requirement for bricks is at least 10.3 MPa.

For future works, the study can be extended to explore the pozzolanic activity and applications in refractories and geopolymers and to further improve the structural and mechanical properties of concrete and cement preparations. Moreover, improvement in the experimental design is recommended. Proper implementation of factor screening may be included to determine the most influential variables in the measured response variable. Randomization, blocking, and sufficient replication of experimental runs should be explored. The design and analysis may be further extended using a response surface methodology for mixture experiments to focus on the numerical optimization of the mechanical properties with respect to the brick formulations and physicochemical or mechanical treatments of the aggregate components.

Lastly, other physical properties including weathering index, water absorption through 5-hour boiling, saturation coefficient, and initial rate of absorption should be considered as well to further ensure compliance with the ASTM C62-17 standard.

Acknowledgments
The authors would like to acknowledge the assistance provided by the CENRO materials recovery facility for producing the eco-friendly bricks and the city government of Biñan, Laguna for championing the utilization of local waste materials for structural and engineering applications. Continued encouragement and support from former colleagues Dr. Rosalinda Torres and Engr. Adelaida Senica are immensely appreciated.

References

[1] Delos Reyes P J, Bornas M A V, Dominey-Howes D, Pidlaoan A C, Magill C R and Solidum Jr R U 2018 *Earth-Science Reviews* **177** 565–588 ISSN 0012-8252
[2] Lagmay A M F, Balangue-Tarriela M I R, Aurelio M, Ybanez R, Bonus-Ybanez A, Sulapas J, Baldago C, Sarmiento D M, Cabria H, Rodolfo R, Rafael D J, Trinidad J R, Obille E and Rosell N 2021 *Scientific Reports* **11** 15703 ISSN 2045-2322
[3] Martinez-Villegas M M, Reniva P D, Sanico L R D, Loza A R, Seda R G, Doloiras D F and Pidlaoan A C 2022 *Frontiers in Earth Science* **10** ISSN 2296-6463
[4] DOST-PHIVOLCS 2023 Phivolcs local active volcanoes archive, eruption history URL `https://wovodat.phivolcs.dost.gov.ph/volcano/erupt-history`
[5] Adel R 2020 *Biñan government turns volcanic ash into 'Taal eco bricks'* (Philstar) URL `https://bit.ly/PhilStarTaalBricks`
[6] Agence France-Presse 2020 *Filipinos turn Taal Volcano's ash, plastic trash into bricks* (GMA News Online) URL `https://bit.ly/GMANewTaalBricks`
[7] Borbon C 2020 *Look! Philippine city turn Taal volcano's ash, plastic into bricks* (Gulf News) URL `https://bit.ly/GulfNewsTaalBricks`
[8] Magsambol B 2020 *Biñan City is turning Taal Volcano's ash into bricks* (Rappler) URL `https://bit.ly/RapplerTaalBricks`
[9] Mateo J 2020 *From Volcanic Ash To Taal Bricks: Biñan To Help Batangas Rebuild* (OneNews.PH) URL `https://bit.ly/OneNewsTaalBricks`
[10] SEA Circular 2020 *Country Profile for the Philippines* (Bangkok, Thailand: SEA circular Project, UN Environment Programme) URL `https://www.sea-circular.org/country/philippines/`
[11] Liamzon C, Benosa S, Aliño M and Baconguis B 2021 *Sachet Economy: Big Problems in Small Packets* (Quezon City, Philippines: Global Alliance for Incinerator Alternatives (GAIA)) URL `https://www.no-burn.org/sachet-economy/`
[12] World Bank 2023 *Combating the Plastic Waste Crisis in the Philippines: Implementing Extended Producer Responsibility with Lessons Learned from Korea* East Asia and Pacific Region: Marine Plastics Series (Washington DC: World Bank Group) URL `https://bit.ly/PHWaste2023WorldBank`
[13] Siddique R 2012 *Resources, Conservation and Recycling* **66** 40–44 ISSN 0921-3449
[14] Djobo J N Y, Elimbi A, Tchakouté H K and Kumar S 2017 *Environmental Science and Pollution Research* **24** 4433–4446 ISSN 1614-7499
[15] Lemougna P N, Wang K t, Tang Q, Nzeukou A N, Billong N, Melo U C and Cui X m 2018 *Resources, Conservation and Recycling* **137** 177–190 ISSN 0921-3449
[16] Hossain K 2006 *Cement and Concrete Composites* **28** 535–545 ISSN 0958-9465
[17] Hossain K and Lachemi M 2007 *Cement and Concrete Research* **37** 759–766 ISSN 0008-8846
[18] Olawuyi B and Olusola K 2010 *Civil Engineering Dimension* **12** 23–28 ISSN 1979-570X
[19] Lemougna P N, MacKenzie K J D and Melo U F C 2011 *Ceramics International* **37** 3011–3018 ISSN 0272-8842
[20] Karolina R, Syahrizal, Putra M A and Prasetyo T A 2015 *Procedia Engineering* **125** 669–674 ISSN 1877-7058
[21] Kupwade-Patil K, Chin S H, Johnston M L, Maragh J, Masic A and Büyüköztürk O 2018 *Journal of Materials in Civil Engineering* **30** 04018190 ISSN 1943-5533
[22] Dablio A R C, Damian R L, Encarnacion E K P, Piquero L M, Tayag E D, Kalacas N A P, Lagmay M S, Sergote J J L and Jane J B 2021 Preliminary characterization of taal volcanic ash as potential raw material for the construction industry and safety evaluation of its leachates DOST-ITDI Terminal Report, DOST-PCIEERD Completed Project URL `https://projects.pcieerd.dost.gov.ph/project/8813`
[23] R Core Team 2023 *R: A Language and Environment for Statistical Computing* R Foundation for Statistical Computing Vienna, Austria URL `https://www.R-project.org/`

IOP Conf. Series: Materials Science and Engineering **1318** (2024) 012018 doi:10.1088/1757-899X/1318/1/012018

Cellulose acetate production using cacao pod husks from local sources for potential industrial applications

Juniper V. Magallanes- Nava[1,2], Ramuel John I. Tamargo[1,*]

[1]Department of Chemical Engineering, College of Engineering, University of the Philippines Diliman, Quezon City, Philippines, 1101
[2]School of Technology, University of the Philippines Visayas, Miagao, Iloilo, Philippines, 5023
[*]ritamargo@up.edu.ph

Abstract. Cacao (*Theobroma cacao*) is one of the most important high-value crops worldwide. Cacao fruit is mainly composed the cacao beans, pulp, and husks. The cacao pod husk (CPH) accounts for almost 60-80% of the weight of the whole fruit. However, oftentimes, these CPH are discarded as waste. Amongst the potential use of CPH is the isolation of its cellulose. The lack of available studies exploring the cellulose content of CPH of different varieties of cacao in the Philippines, and the comparison of heating methods for its extraction, highlights a significant research gap. This knowledge will provide a more comprehensive understanding of its composition for potential applications. In this study, different sources and varieties of cacao pod husks were used to produce cellulose through alkali treatment and bleaching with conventional heating and microwave synthesis methods. The result showed that the microwave synthesis approach produces a comparable yield to the conventional method but at a lower treatment time. The cellulose produced was successfully converted into cellulose acetate as confirmed via FTIR analysis. The cellulose acetate was dissolved in different solvents as preparation for its use as one of the components for the development of bioplastic. Characterization of the film produced showed the potential of the cellulose acetate from cacao pod husk for bioplastic application.

1. Introduction

Cacao (*Theobroma cacao*) trees are usually grown in the tropical forest zones. They are mainly grown for its beans, which are converted to chocolates, cocoa powder, cocoa butter, and other cocoa products. In 2021, the global chocolate confectionary market size value was $77.6 billion and is predicted to have a compound annual growth rate of 4% from 2022 to 2031. With this, the global market is expected to reach $118.6 billion by 2031 [1]. Côte d'Ivoire (Ivory Coast) and Ghana are the top producers of cacao. Mostly these cacao trees are grown by small-scale farmers in the tropics [2].

The main components of the cacao fruit are its beans, pulp, and husks. The largest part of the cacao fruit is the cacao pod husks, which is about 60-80% of the weight of the whole fruit [3,4,5,6]. Traditionally, these CPHs are viewed as waste material and are often disposed of or left to concerns and lost economic opportunities.

In recent years, there has been a growing recognition of the value of CPH, encouraging researchers and innovators to develop strategies, methodologies, and approaches to transforming this into valuable products. CPH has been found to be a potential bio-resource of a variety of value-added products from fuel to agriculture and food [7]. Cellulose, bioactive compounds, and ethanol has been extracted and synthesized from CPH [8,9,10]. It has been used to develop fuel briquette, cellulose triacetate, cellulose acetate, bioplastic film, dietary fiber, and biocomposite films [11,12,13].

One of the strategies in valorization is cellulose extraction. An emerging technology in synthesis and extraction is microwave-assisted strategies. It offers uniform heating, shorter time, and less solvent use [7]. Microwave-assisted processes help achieved higher pressure and temperature rapidly, thus increasing the reaction rates [14]. Some studies included microwave-assisted technologies in the extraction of various materials from cacao pod husk such as cellulose [15], pectin [16], and phenolic compounds [17].

This study presented an environment-friendly approach to access cellulose from locally available CPH and converting it to cellulose acetate for potential industrial application.

2. Materials and Methods

2.1. Materials

The CPH was collected from two sites: Bago City, Negros Occidental, and Lantapan, Bukidnon. For the extraction of cellulose, sodium hydroxide was from RCl Labscan Limited; hydrogen peroxide from DKL Laboratory; microcrystalline cellulose from Sigma-Aldrich.

2.2. Extraction of CPH Powder

CPH was sun-dried for 72 hours, then oven-dried in an oven for 24 hours at 80°C prior to grinding. CPHs were milled and sieved. The cellulose of the CPH powder was extracted using the procedure by Azmin et al. (2020) with some modifications. Ten (10) grams of the powder was subjected to alkaline treatment by soaking in 4% NaOH for three hours at 100°C with constant stirring. The ratio of biomass to solvent is 1:20. After this, the CPH was filtered and washed with distilled water until the pH becomes neutral. Then, it was bleached using 3% hydrogen peroxide for 180 minutes at 70°C on a hot plate to reach the desired colour of the cellulose. The bleached sample was then washed with distilled water to remove the bleach residue until neutral. It was then dried in an oven at 105°C until constant weight. For the microwave (MW) heating method, same conditions as above except for the alkaline treatment settings which was at 50°C and 40 minutes. For the bleaching, microwave setting was at 50°C and 60 minutes. The product was weighed for computation of percent yield.

2.3. Characterization of CPH Cellulose

The product was analyzed using instrumental and chemical analysis. For the instrumental analysis, FTIR spectroscopy and Scanning Electron Microscopy were used. The FTIR graph was compared to the existing literature for the identification of the functional groups present. SEM images were also examined. The pH, bulk and tapped density, and solubility was also determined.

2.4. Production of Cellulose Acetate

The conversion of CPH cellulose to cellulose acetate followed the procedure of Asriza et al. (2021) with modifications. Cellulose was activated using 25 ml of glacial acetic acid for 1 hour at room temperature with constant stirring. For the acetylation, acetylating agent (15 ml of acetic anhydride) is mixed with a strong Lewis acid catalyst (6 drops of concentrated sulfuric acid). This mixture is mixed to the activated cellulose for 3 hours at 40 – 45°C. It was then continued for 20 hours at room temperature, then 25 ml of 60% glacial acetic acid was slowly added. The resulting mixture was filtered. Demineralized water was added dropwise to the filtrate until white precipitate was formed. The white precipitate was filtered, neutralized with distilled water, and dried in an oven at 90°C until constant weight.

2.5. Production of Bioplastic Film

One (1) gram of cellulose acetate and 10 ml of solvents (Cyrene™ or acetone) were mixed at room temperature. Once the solids fully dissolved, 10 ml of glycerol was added. The solution was casted on a clear glass container until solidified. It was washed with distilled water, and oven-dried for an hour at 40°C.

3. Results and Discussions

3.1. CPH Cellulose

3.1.1. Cellulose Extraction Yield

Table 1 shows the cellulose yield from CPH using different heating methods and varieties of CPH. 1st and 2nd batches of cacao pod husk are from Lantapan, Bukidnon. The variety of cacao from these farms is Forastero. The 3rd batch of the cacao pod husk is from Bago City, Negros Occidental. It is of the Criollo variety. T-test was done to determine significant difference showed that the yield significantly varies per variety (Criollo > Forastero), but there was no significant difference between heating methods. MW heating offered an advantage of having comparable yield to conventional heating but at shorter alkaline and bleaching treatment time and lower temperature.

Table 1. Cellulose yield per method and variety.

Method	Code[a]	Average
Conventional	C-B1	27.89%
	C-B2	27.64%
	C-B3	30.32%
Microwave Synthesis	MS-B1	28.30%
	MS-B2	28.57%
	MS-B3	31.55%

[a] B1 & B2 – 1st and 2nd batch from Lantanan, Bukidnon; B3- 3rd batch from Bago City, Negros Occidental

3.1.2. FTIR Analysis

The chemical bonds present in raw cacao pod husk (CPH), alkaline-treated CPH powder (CPHA), cacao pod husk cellulose from conventional method (CPHCC), and cacao pod husk cellulose from microwave-assisted extraction (CPHCM) are examined using Fourier Transform Infrared Spectroscopy (FTIR) analysis. The FTIR spectra were shown in Figure 1. In all samples, a broad band was seen at 3,305–3,330 cm^{-1}. This is due to the O-H vibration of the adsorbed water molecule [20]. This band has a broad appearance. As its pretreatment process progressed, this peak appeared sharper and had lower intensity. It can be correlated to the destruction of intra-and intermolecular hydrogen bonds [21]. The band at 2,918–2,880 cm^{-1} for all samples is due to C-H stretching which indicated the presence of organic molecules (alkanes).

There was a band present at 1,741 cm^{-1} for CPH indicating the presence of ester linkages in lignin and/or hemicellulose [20]. The lignin peaks were highlighted in Figure 1 (with dash vertical lines). The lignin peaks appeared in the raw CPH powder (CPH)and the alkaline-treated CPH (CPHA) at 1,598.72 cm^{-1} (aromatic ring) and 1,641.15 cm^{-1} (C-O stretching of ether linkage) wavenumbers consistent with the study of Ndruru et al. (2019). It had a higher intensity in the CPH as compared to CPHA. This was expected as the purpose of the alkaline treatment is to disrupt the waxes, hemicellulose, and lignin protective layer that encloses the cellulose. Once it is bleached, this peak disappeared as shown in the FTIR spectra of CPHCC, CPHCM, and the commercial MCC. This proves that the bleaching process further removes the lignin present in the biomass. The presence of the C- O stretching band at 1,028 – 1,022 cm^{-1} for all samples confirmed the presence of cellulose.

Comparing the FTIR spectra of CPHCC and CPHCM with the commercial MCC proved that the isolated products from both conventional and microwave-assisted methods are cellulose.

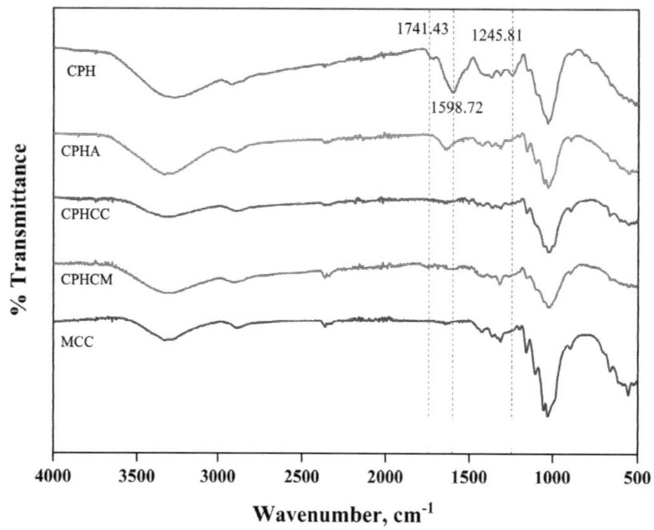

Figure 1. FTIR spectra of raw cacao pod husk (CPH), alkaline-treated CPH Powder (CPHA), cacao pod husk cellulose-conventional (CPHCC), cacao pod husk cellulose-microwave synthesis (CPHCM) & commercial microcrystalline cellulose (MCC)

3.1.3. SEM Imaging

To visualize the surface morphology of the cacao pod husk and the cellulose, SEM imaging was done. The surface of the CPH powder is lamella-shaped and contains waxes and other encrusting substances including hemicellulose, pectin, and lignin. These serve as the protective layer of cellulose [20, 22].

Figure 2 showed the changes in the structure of the CPH before and after undergoing alkaline treatment and bleaching. The untreated cacao pod husks (*a and b)* have surfaces that contain waxes, pectin, hemicellulose, and lignin. It is irregular in shape, compact, and rigid.

For both conventional and microwave-assisted extraction (*c and d*), there were notable changes in the surface morphology, specifically, an increase in porosity. The waxy layer has been removed by the treatment to expose the cellulose component.

In comparison with the commercial MCC, cellulose from CPH showed the presence of pores (*e and f*). The commercial MCC appeared denser than the isolated cellulose from CPH. These pores indicate an increase in the amorph region which results from a decrease in molecular weight as compared to commercial MCC [15]. Other than the pores present in the CPH cellulose, it is observed that it has a similar surface to the commercial MCC.

3.1.4. Physical Properties of the CPH Cellulose

The pH of the cellulose is 6.21-6.41. It is within the standard of pH 5.0 -7.5 set by EU Food Improvement Agents.

Bulk and tapped density are important properties in relation to the flowability and compressibility of the powder. Bulk density is the ratio of the mass to the volume occupied (including voids).Once tapped for a certain amount of time, the resulting mass to volume ratio is called tapped density.

For the solubility test, CPH cellulose was insoluble in the following solvents: deionized water, chloroform, ethanol, methanol, and polar aprotic solvents (DMSO, DMF, Cyrene^{TM}). It was soluble in concentrated sulfuric acid and concentrated hydrochloric acid and slightly soluble in 1 M NaOH. This is consistent with the results of the solubility test in the study of Adeleye et al. (2022). Medronho et al (2012) also describe cellulose as insoluble in water. This insolubility is generally attributed to its strong intermolecular hydrogen bonding, but Medronho et al. (2012) showed evidence that cellulose is amphiphilic and its low solubility in water is due to its hydrophobic part. The solubility profile is an important reference for future application studies.

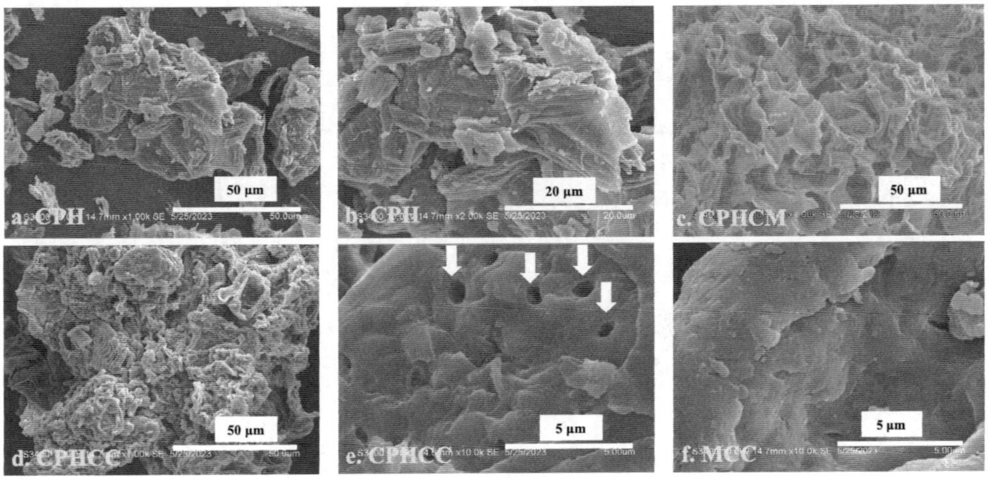

Figure 2. SEM images of (a) cacao pod husk (1000x), (b) cacao pod husk powder (2000x), (c) cellulose-microwave synthesis, (d) cellulose-conventional heating(1000x), (e)cellulose-conventional heating(10000x) & (f) commercial microcrystalline cellulose(10000x).

Table 2. Physical properties of the CPH cellulose.

Property	Unit	Value	Ref. (Adeleye et al., 2022)
Appearance	Very light yellow to off-white powder		Off white
Odor		Odorless	Odorless
pH		6.3±0.0938	6.7±0.37
Bulk Density	g/ml	0.4141±0.0445	0.273±0.02
Tapped Density	g/ml	0.4883±0.0109	0.337±0.001

Table 3. Solubility of the CPH cellulose.

Solvent	Solubility[a]	Ref. (Adeleye et al., 2022)
Deionized water	Insoluble	Insoluble
Chloroform	Insoluble	Insoluble
Acetone	Partially Soluble	
Ethanol	Insoluble	Insoluble
Methanol	Insoluble	
Dimethyl Sulfoxide (DMSO)	Insoluble	
Dimethylformamide (DMF)	Insoluble	
Cyrene^{TM}	Insoluble	
Concentrated Sulfuric Acid	Soluble	Soluble
Concentrated Hydrochloric Acid	Soluble	Soluble
1.0 M Sodium Hydroxide	Partially Soluble	

[a] Insoluble – no change in appearance; Partially Soluble – cloudy appearance; Soluble – total dissolution of cellulose.

3.2. Cacao Pod Husk Cellulose Acetate (CPH CA).

The acetylation of the CPH cellulose were shown at wavenumbers (cm^{-1}) 1736, 1365, and 1215 in Fig.3. These ester bonds correspond to the C=O stretching, C-H stretching, and C-O stretching. This is consistent with the study by Tuates et al. (2020). These peaks also appeared in the commercial CA (CCA). These indicate successful synthesis of cellulose acetate from CPH.

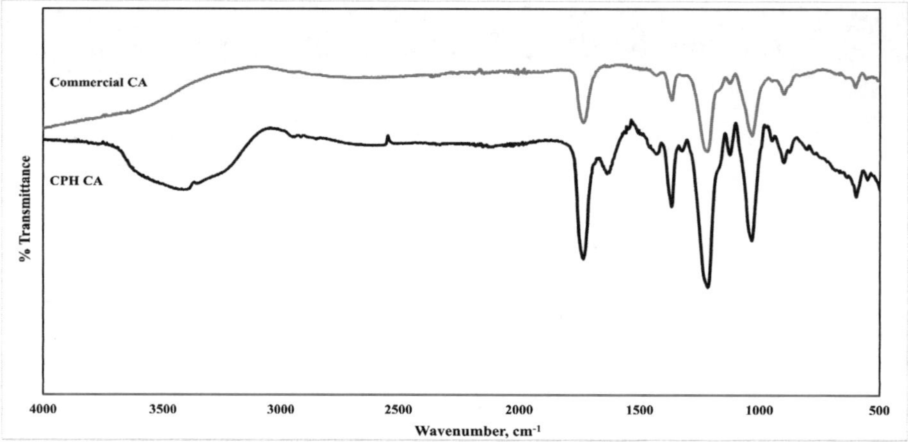

Figure 3. FTIR spectra of CCA and CPH CA.

3.3. CPH CA Bioplastic Film

Using Cyrene™ as solvent, there was a formation of bioplastic film when commercial cellulose acetate was used *(See Fig. 4a)*. It was clear to white in color, rough surface, and flexible when touched. However, when using CPH-derived cellulose acetate, there was no film developed. There is a possibility that the CPH CA did not completely dissolve in the solvent, thus there is insufficient amount of filler for the film to develop.

Table 4. Characteristic of the film produced.

Cellulose Acetate Source	Solvent	Formation of Film	Physical Characteristic of the Film Formed
Commercial	Cyrene™	Yes	Clear to white color with rough surface; Do not break when touched
CPH	Cyrene™	No	---
CPH	Acetone	Yes	White color with rough surface; breaks when held

When the solvent was changed to acetone in CPH CA, there was a formation of a film *(See Fig.4b)*. It has white in color with rough surface, and paper-like texture and softness. It was brittle. This might be due to its high hydrophilicity; thus, the product is easily torn [27]. There might be a need to increase the amount of plasticizer or add another material as filler and plasticizer. Further investigation must be conducted. However, the result showed the potential of CPH to be converted to cellulose acetate for bioplastic film development.

4. Conclusion

In this study, cellulose was successfully isolated from cacao (*Theobroma cacao)* pod husks through conventional heating and microwave-assisted heating methods with comparable yields. However, microwave-assisted method showed had shorter reaction time. Analysis of the physicochemical properties of the cellulose, FTIR and SEM imaging confirmed the production of cellulose. This study showed a greener alternative pathway for the extraction of cellulose.

Moreover, the cellulose was successfully converted to cellulose acetate as confirmed by FTIR analysis. This cellulose acetate has the potential to be developed into bioplastic film.

(a) (b)

Figure 4. Bioplastic formed from (a) commercial cellulose acetate and (b) CPH cellulose acetate.

Acknowledgement
This work was funded by the UP System Balik PhD Program (OVPAA-BPhD-2020-01). Furthermore, the authors acknowledge the Office of the Chancellor of the University of the Philippines Diliman, through the Office of the Vice Chancellor for Research and Development, for funding support through the Ph.D. Incentive Award. The researchers also express their gratitude to the Department of Science of Technology Science Education Institute (DOST-SEI) through the Engineering Research and Development for Technology (ERDT), and the University of the Philippines Visayas for the additional funding support.

References

[1] Indranil, C., Vidit, G., & Roshan, D. (2022, Oct). *Chocolate Confectionery Market Research.* Retrieved from Allied Market Research: https://www.alliedmarketresearch.com/chocolate-confectionery-market-A27762

[2] Laliberte, B. (n.d.). A Global Strategy for the Conservation and Use of Cacao Genetic Resources, as the Foundation for a Sustainable Cocoa Economy. CacaoNet . Retrieved from Global Strategy.

[3] Campos-Vega, R., Nieto-Figueroa, K., & Oomah, B. (2018). Cocoa (Theobroma cacao L.) pod husk: Renewable source of bioactive compounds. *Trends in Food Sci. & Tech. 81*, 172-184.

[4] Lu, F., Rodriguez-Garcia, J., Van Damme, I., Westwood, N. J., Shaw, L., Robinson, J. S., . . . Charalampopoulos, D. (2018). Valorization strategies for cocoa pod husk and its fraction. *Cur. Opin. Green Sustain. Chem.*

[5] Daud, Z. K., Aripin, A., Awang, H., & Hatta, M. (2013). Chemical Composition and Morphological of Cocoa Pod Husks and Cassava Peels for Pulp and Paper Production. *Austr.J. Basic Appl. Sci. 7(9)*, 406-411.

[6] Cruz, G., Pirila, M., Huuhtanen, M., Carrion, L., Alvarenga, E., & Keiski, R. (2012). Production of Activated Carbon from Cocoa (Theobroma cacao) Pod Husk . *Civ.Environ.Eng. 2(2).*

[7] Baidoo, M., Asiedu, N., Darkwah, L., Arhin-Dodoo, D., Zhao, J., Jerome, F., & Amaniampong, P. (2022). Conventional and Unconventional Transformation of Cocoa Pod Husks into Value-Added Products. In M. Samer, *Biomass, Biorefineries, and Bioeconomy.* Cairo, Egypt.

[8] Dimawarnita, F., Indriyantini, P., & Perwitasari, U. (2022). Isolation and Characterization of α-Cellulose from Cocoa Pod Husk using Peracetic Acid. *International Conference on Biomass and Bioenergy.* IOP Conf.Ser.Earth Environ.Sci., Volume 1187.

[9] Belwal, T., Cravotto, C., Ramola, S., Thakur, M., Chemat, F., & Cravotto, G. (2022). Bioactive Compounds from Cocoa Husk:Extraction,Analysis and Applications in Food Production Chain. *Foods 11(6).*

[10] Yogaswara, R., Billah, M., Saputro, E., & Erliyanti, N. (2021). A Kinetic Study in Fermentation of Cocoa Pod Husk using Zymomonas Mobilis. *IOP Conf. Ser. Mater. Sci. Eng.* IOP

Publishing.

[11] Lubis, M., Gana, A., Maysarah, S., Ginting, M., & Harahap, M. (2018). Production of bioplastic from jackfruit seed starch (Artocarpus heterophyllus) reinforced with microcrystalline cellulose from cocoa pod husk (Theobroma cacao L.) using glycerol as plasticizer. *IOP Conf. Ser. Mater. Sci. Eng. 309.*

[12] Ramirez, T., Sapin, A., Asaytono, A., & Alfafara, C. (2017). Biotechnological Approach for Converting Cacao Pod Husk into High Dietary Fiber and Antioxidant-rich Bakery Ingredient.

[13] Sanyang, M., Sapuan, S., & Haron, M. (2017). Effect of Cocoa Pod Husk Filler Loading on Tensile Properties of Cocoa Pod Husk/Polyactic Acid Green Biocomposite Films. *AIP Conf. Proc. 1891*. AIP Publishing.

[14] Gawande, M., Shelke, S., Zboril, R., & Varma, R. (2014). Microwave-Assisted Chemistry: Synthetic Applications for Rapid Assembly of Nanomaterials and Organics. *Acc.Chem.Res 47(4)*, 1338-1348.

[15] Ndruru, S., Wahyuningrum, D., Bundjali, B., & Arcana, I. (2019). Green simple microwave-assisted extraction (MAE) of cellulose from Theobroma cacao L. (TCL) husk. *IOP Conf. Ser. Mater. Sci. Eng. 541(1).*

[16] Sarah, M., Hasibuan, I., Misran, E., & Maulina, S. (2022). Optimization of Microwave-Assisted Pectin Extraction from Cocoa Pod Husk. *Molecules.*

[17] Dewi, S., Stevens, L., Pearson, A., Ferrari, R., Irvine, D., & Binner, E. (2022). Investigating the Role of Solvent Type and Microwave Selective Heating on the Extraction of Phenolic Compounds from Cacao (*Theobroma cacao L.*) Pod Husk. *Food Bioprod. Process., 134*, 210-222.

[18] Azmin, S., Hayat, N., & Nor, M. (2020). Development and Characterization of Food Packaging Bioplastic Film from Cocoa Pod Husk Cellulose Incorporated with Sugarcane Bagasse Fibre*.J.Bioresour.Bioprod.*, 259-266.

[19] Asriza, R.O., Ropalia, Humaira, D., Ryaldi, G.O., & Zomi(2021). Characterization of Cellulose Acetate Functional Groups Synthesized from Corn Husk (*Zea mays*). *IOP Conf. Ser.:Earth Environ.Sci.926.*

[20] Akinjokun, A.I.;Petrik,L.F.;Ogunfowokah,A.O.;Ajao,J.;Ojumu,T.V. (2021). Isolation and characterization of nanocrystalline cellulose from cocoa pod husk (CPH) biomass wastes*.Heliyon 7(4).*

[21] Ciolacu, D., Ciolacu, F., & Popa, V. (2010). Amorphous Cellulose - Structure and Characterization. *Cellul. Chem. and Technol..*

[22] Daud, Z., Awang, H., Kassim, A., Hatta, M., & Aripin, A. (2014). Cocoa Pod Husk and Corn Stalk: Alternative Paper Fibers Study on Chemical Characterization and Morphological Structures. *Adv. Mater. Res. Vol.911*, 331-335.

[23] National Center for Biotechnology Information. (2023, June 1). *Cellulose*. Retrieved from PubChem Compound for CID 16211032: https://pubchem.ncbi.nlm.nih.gov/compound/Cellulose.

[24] Adeleye, O., Bamiro, O., Albalawi, D. A., Iqbal, H., Sanyaolu, S., Femi-Oyewo, M., . . . Menaa, F. (2022). Characterizations of Alpha-Cellulose and Microcystalline Cellulose Isolated from Cocoa Pod Husk as a Pharmaceutical Excipient. *Materials (15).*

[25] Medronho, B., Romano, A., Miguel, M., & Stigsson, L. (2012). Rationalizing Cellulose (in)solubility: Reviewing Basic Physicochemical Aspects and Role of Hydrophobic Interactions. *Cellulose 19(3)*, 581-587.

[26] Tuates Jr, A.M., Carriedo, A.G., Caparino,O.A., & Monserate, J. (2020). Cellulose Acetate from Cocoa Pod Husk – Chitosan Blend (CACPH-CS) as a Biosorbent. *Asian Journal of Postharvest and Mechanization 3(1),* 11-25.

[27] Lusiana, S.W., Putri, D., Nurarizah, Ida, Z., & Bahruddin (2021). Bioplastic Properties of Sago – PVA Starch with Glycerol and Sorbitol Plasticizers. *J.Physics.: Conf. Ser. 1351.*

Production of Bio-plasticized Polyvinyl Chloride (PVC) Films Synthesized from Microalgae (*Chlorella vulgaris*)

M L Ampong[1], D D Catapat[1], D C Panaligan[1], G M Pantaleon[1], J R Clemeña[2], R V Rubi[3], F D Lavilles[2], P J Gildo[2], D Pangayao[2,4].

[1]College of Engineering, University of the City of Manila
[2]Chemical Engineering Department, College of Engineering, University of the City of Manila
[3]Chemical Engineering Department, College of Engineering, Adamson University
[4]Environment and Biotechnology Division, Industrial Technology Development Institute, Department of Science and Technology, Bicutan, Taguig City, Philippines

*Email: itsmarkampong@gmail.com

Abstract. Bio-based plasticizers for PVC films are being explored to replace conventional phthalate-based plasticizers due to the threats they pose to human health and the environment. In the present study, fatty acids from *Chlorella vulgaris* were extracted via Microwave-assisted Extraction and were epoxidized using esterification and transesterification reactions in a magnetically stirred reactor to produce EFAME plasticizer. FTIR Analysis determined the presence of an epoxy group as the peak C=O stretch (1739.79 cm^{-1}) was detected. PVC films were synthesized at different concentrations of EFAME (30%, 40%, and 60%). At 40% of EFAME, the PVC exhibited high displacement in the tensile strength and elongation at break test, which was used as a basis to produce PVC-A, PVC-B, and PVC-C, and was compared to phthalate-plasticized PVC film, PVC-D. SEM-EDX Analysis verified the presence of PVC and EFAME in the plasticized PVC films as Carbon, Oxygen, and Chlorine elements were detected. PVC A had the lowest weight loss percentage at 40.96% indicating it has developed high migration resistance among the extraction mediums used. DSC analysis confirmed the plasticizing effect of EFAME as the glass transition temperature of bioplasticized PVC films decreased. Using One-way ANOVA, the physicochemical tests of PVC films plasticized with EFAME showed significantly better results than PVC films plasticized by DOP. Thus, the present study indicates that the bio-based plasticizer derived from *C. vulgaris* is a better and a potential alternative for phthalates in plasticizing PVC films.

Keywords: Microalgae, *Chlorella vulgaris*, Plasticizer, Polyvinyl Chloride, MAE, FTIR, SEM-EDX

1. Introduction

Polyvinyl chloride, also referred to as PVC, is one of the most frequently produced synthetic polymers of plastic in the world. Plasticizers are common plastic additive compounds used to improve the flexibility of materials, accounting for roughly 90% of all PVC applications [1]. The plasticizer's small molecular size enables it to penetrate through the amorphous sections of the polymer structure and alter the forces acting between the molecules, making the polymer chains move and bend more easily. This study chose DOP and its plasticizing characteristics to be potentially replaced by the bio-plasticizer that was synthesized.

Over the years, the number of traditional polymers has increased drastically. Thus, algae-based biopolymers have generated significant recent research interest worldwide as one of the possible solutions for creating sustainable economic development [2]. Since *C. vulgaris* is a type of microalgae that contains a variety of fatty acids, it is possible to epoxidize some of the unsaturated fatty acids present in *C. vulgaris*.

Organic compounds such as fatty acid methyl esters (FAMEs) have been discovered as viable replacements for producing epoxy compounds that can be utilized as eco-friendly plasticizers. [3]. Lipid epoxidation has been accomplished with plant-based oils derived from oilseeds like soybean, sunflower, canola, and fiber. However, few research have addressed the potential of microalgae which contain a considerable amount of lipids and double bonds for epoxidation. Microalgae possess large quantities of lipids, proteins, and carbohydrates, providing them with multiple advantages as a non-food raw material for producing valuable compounds [4].
In the present study, an algae-based plasticizer for PVC production was utilized via Microwave-assisted Extraction (MAE) which is an optimal method for extracting oils from microalgae [5]. The lipid content of *C. vulgaris* underwent esterification and its unsaturated fatty acids were epoxidized to produce a bio-based plasticizer. The bio-based plasticizer was then utilized in the production of PVC film at three different compositions (50%, 70%, and 100%), and their properties were analyzed with the existing plasticizer.

2. Materials and Methods

2.1. Acquisition of Raw Materials

The oven-dried *C. vulgaris* was purchased from the University of the Philippines – Los Baños (UPLB) under the department of the Institute of Biological Sciences. The chemicals and reagents used in this study were purchased from DKL Laboratory Supplies.

2.2. Extraction of Fatty Acids from C. vulgaris

2.2.1. Microwave-Assisted Extraction (MAE). Samples were placed in a 3-inch covered, glass petri dish and were heated in an HM020M (Hanabishi) microwave which had a power rating of up to 700W and operated at 2450MHz. 80g of NaCl was dissolved in 800 mL of distilled water. 40g of biomass was added into the solution to get 5% (w/v) solid loading [6]. Hexane was added as a non-polar solvent to effectively assist in the extraction of lipids from the samples. The mixture was then centrifuged for 5 minutes at 3500 rpm.

2.2.2. Post-treatment of Samples. A small volume of hexane was added to the treated samples which were centrifuged at 3500 rpm for 5 minutes. The upper layer was transferred in a beaker and was heated at 68.7°C to recover the hexane and the fatty acid.

2.3. Conversion of Fatty Acids to EFAME

Fourteen grams of fatty acid and 22.5g methanol were mixed in a 125ml Erlenmeyer flask. After mixing, a catalyst, 12.5g sulfuric acid was added. The esterification reaction was carried out for 6 hours at 70°C [7] and was transferred in a separating funnel. Afterward, the upper solution was neutralized using 5% (w/v) sodium bicarbonate and washed 5 times with deionized water to obtain FAME weighing 15.34g. The FAME was combined with 1.53g of strong acid cation exchange resin. 9.9g of 50% hydrogen peroxide (V/V) and 2.3g of formic acid were added to the solution. It was stirred magnetically for 6 hours at 60°C [7] and was transferred in the separating funnel then washed 5 times with deionized water to obtain the EFAME bioplasticizer.

2.4. Characterization of EFAME using FTIR-ATR

The FTIR-ATR technique was used to produce the infrared spectra for plasticized films and identify the functional groups present in the plasticizer. The data collected was in the 4000 to 500 cm^{-1} scan range.

2.5. Preparation of Plasticized PVC Samples

PVC Powder was mixed with the EFAME and DOP Plasticizer, and Tetrahydrofuran was later added to the beaker. The composition of PVC films plasticized with EFAME at varying

IOP Conf. Series: Materials Science and Engineering **1318** (2024) 012019 doi:10.1088/1757-899X/1318/1/012019

compositions of plasticizer is shown in Table 1. The solution was magnetically stirred until transparent and was transferred onto a glass petri dish which was then vacuum-dried for 3 days.

Table 1. Composition of PVC Films Plasticized with EFAME

PVC Film	EFAME (%)	PVC Powder (g)	THF (g)
*PVC-0	0	2	12.53
PVC-30	30	2	12.53
PVC-40	40	2	12.53
PVC-60	60	2	12.53

*Control variable

2.6. Preliminary Evaluation using Tensile Strength and Elongation at Break Test

The tensile properties of each PVC film measuring 15.0 by 15.0 mm, were assessed using 500N Shimadzu Autograph (Tensile Tester) AGS-J series in accordance with ASTM D882. The test speed and test length (grip separation) were adjusted to 50 mm/min and 50 mm, respectively. Three samples of each PVC film were examined at storage and test temperature of 23°C and 50±5 %RH.

2.7. Characterization and Physicochemical Testing of Pure and PVC 40 Films

Since PVC films containing 40% EFAME (PVC-40) demonstrated better tensile performance compared to other PVC samples, they were chosen for additional analysis and comparison with phthalate-plasticized PVC films (PVC D)*.

The PVC films were synthesized and prepared in various compositions of EFAME and DOP as illustrated in Table 2. These samples underwent characterization based on their morphology and were tested for their physicochemical properties.

Table 2. Composition of PVC Films Plasticized with EFAME and DOP

PVC Film	EFAME (%)	DOP (%)	PVC Powder (g)	THF (g)
PVC A	100	0	2	12.53
PVC B	70	30	2	12.53
PVC C	50	50	2	12.53
*PVC D	0	100	2	12.53

*Control variable

2.7.1. Morphological and Elemental Analysis. The PVC film samples were subjected to morphology analysis using a Scanning Electron Microscopy with Energy Dispersive X-ray Analysis (SEM-EDX) method, with a dimension of 10mm per sample in accordance with ASTM F1375-92(2020).

2.7.2. Differential Scanning Calorimetry (DSC) Test. The determination of the glass transition temperature (T_g) of PVC films was conducted using DSC method in accordance with ASTM. The PVC films were cut into uniform dimensions of 10 mm for each of the PVC samples. Heating was conducted at a rate of 10 °C/min from 25-500 °C.

2.7.3. Migration Stability Test. The migration stability of the plasticizer was tested according to ASTMD1239-98 at 25°C and 50% relative humidity. The samples were weighed and have been submerged in various solvents, including petroleum ether, ethanol, distilled water, acetic acid, and n-hexane which was done in duplicates. The plasticized PVC films remained immersed in 30 mL of solvent for 24 hours before they were reweighed.

2.7.4. Volatility Test. PVC Samples were immersed in duplicates in a container containing activated carbon for 24 hours, and the temperature was maintained at an ambient temperature, in accordance with ISO 176:2005. The samples were dried for 1 hour and washed with deionized water. Weight variations were measured both before and after the storage process.

3. Results

3.1. Conversion of Fatty Acid to EFAME

The extracted fatty acid, characterized by a pale-yellow appearance, was weighed, and recorded at 14 grams, resulting in a yield of 35%. Through esterification reaction, a total of 15.34g of FAME was obtained. The synthesis of EFAME underwent an epoxidation reaction resulting to 13.948g of EFAME, giving a yield of 99.63%.

3.2. Characterization of EFAME using FTIR Spectra

There were 14 peaks present in the FTIR Spectroscopy shown in Figure 1. The spectra showed the presence of the C=O stretches of acids and esters, the =C–H stretches of alkenes, and the -OH stretches of acids and hydroxyls. The main peaks found were at 842.89 cm^{-1}, 1170.79 cm^{-1}, 1361.74 cm^{-1}, 1463.97 cm^{-1}, 1739.79 cm^{-1}, 2924.09 cm^{-1}, and 3429.43 cm^{-1}. There are 3 functional groups present which are Carbonyl, Alkane, and Alcohol Groups. On the right side of the graph, which is referred to as the Fingerprint region from 500-1500 cm^{-1}, shows 5 groups present which are Epoxy, Carbonyl, Methyl, and Alkane Groups.

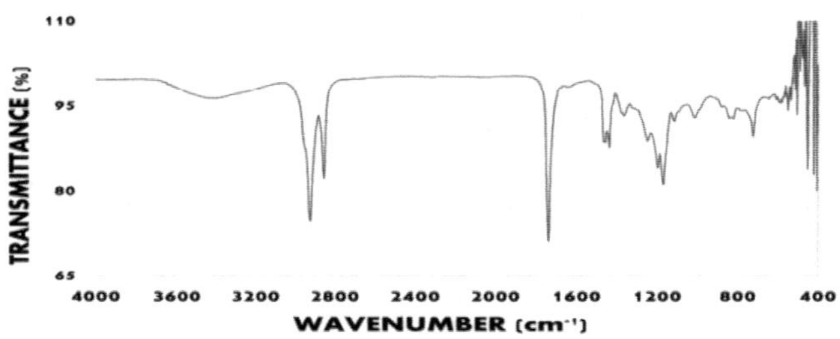

Figure 1. FTIR Spectra of EFAME

3.3. Preliminary Evaluation using Tensile Strength and Elongation at Break Test

The introduction of EFAME to the PVC films results in a lower value in tensile strength and an improvement in flexibility, specifically at a 40% concentration of EFAME. The increase in the concentration of plasticizer in PVC films decreased the tensile strength from 9.6 N/15mm down to 4.3 N/15mm at 60% concentration. Plasticized PVC-40 exhibited the highest displacement at 86.3mm.

Table 3. Tensile Properties of PVC Films

	Maximum Tensile Strength (N/15mm)				**Maximum Displacement (mm)**			
% Plasticizer	**0**	**30**	**40**	**60**	**0**	**30**	**40**	**60**
Sample	***PVC-0**	**PVC-30**	**PVC-40**	**PVC-60**	***PVC-0**	**PVC-30**	**PVC-40**	**PVC-60**
1	11.5	8.2	6.9	3.1	88.7	66.9	103.7	52.2
2	9.4	6.6	7.1	4.6	64.7	48.8	89.1	63.0
3	7.8	7.4	5.2	5.2	68.3	45.8	66.3	84.6
Average	**9.6**	**7.4**	**6.4**	**4.3**	**73.9**	**58.8**	**86.3**	**66.6**

*Control variable

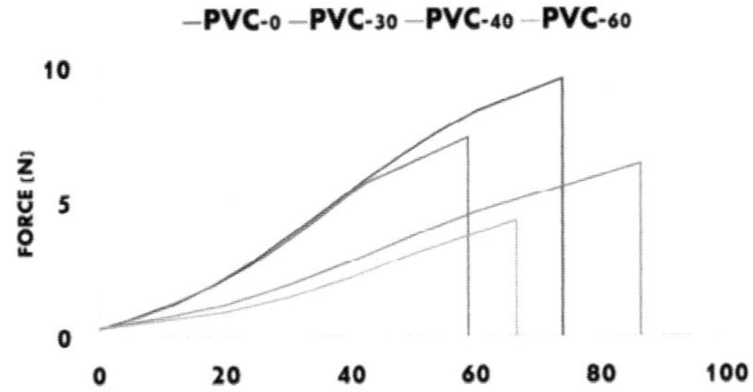

Figure 2. Tensile Stress-strain Curves of Different PVC Films

3.4. Morphological and elemental analysis using SEM-EDX

The analysis showed that the films exhibited a smooth and uniform surface, indicative of good film formation for bioplasticized PVC films. From the given data, it can be seen that all PVC Films has peak elemental composition of Carbon (C), Oxygen (O), and Chlorine (Cl), with traces of Calcium (Ca), Silicon (Si), and Aluminum (Al), as can be seen in Figure 4.

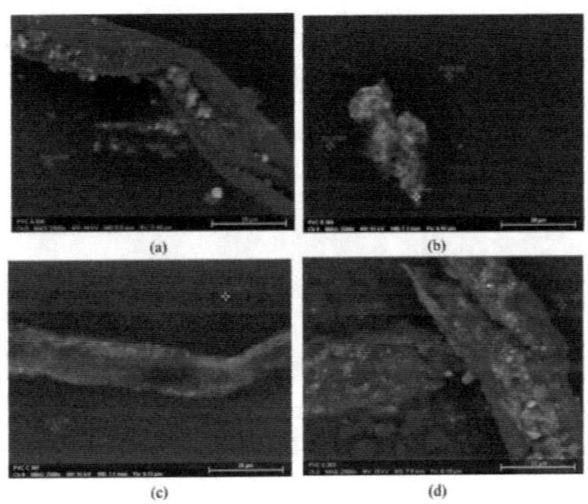

Figure 3. SEM-EDX spectrometric image of (a) PVC A, (b) PVC B, (c) PVC C, (d) PVC D.

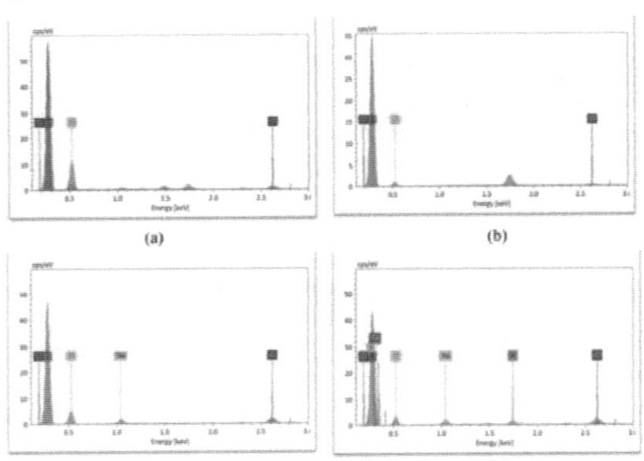

Figure 4. Elemental Distribution of PVC A (a), PVC B (b), PVC C (c), and PVC D (d).

3.5. Physicochemical Tests of Pure and PVC 40 Films

3.5.1. Differential scanning calorimetry (DSC) Test. The corresponding DSC curves are illustrated in Figure 5. PVC C exhibits a notably low T_g of 30.26°C. The values of T_g for PVC A, PVC B and *PVC

D are recorded as 32.74°C, 32.84°C and 31.96°C, respectively. Previous study has found the T_g of pure PVC film or unplasticized PVC is 81.0°C [9]. However, the obtained results for EFAME-Plasticized PVC films were lower when compared to the T_g value of pure PVC films.

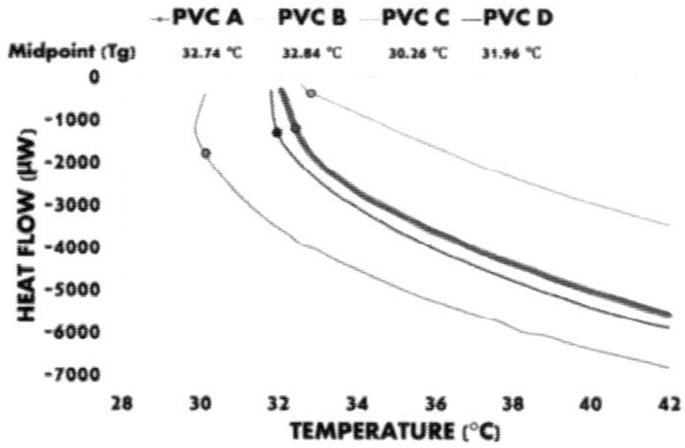

Figure 5. DSC Curves of PVC A (a), PVC B (b), PVC C (c), and *PVC D (d).

3.5.2. Migration Stability Test. To assess the migration stability of the samples, the bio-plasticized PVC samples were compared against phthalate-based PVC sample using their calculated weight loss percentages from different extraction mediums. From Figure 6, an increasing trend was observed for non-polar solvents, n-hexane and petroleum ether, while a decreasing trend was observed for the remaining polar solvents.

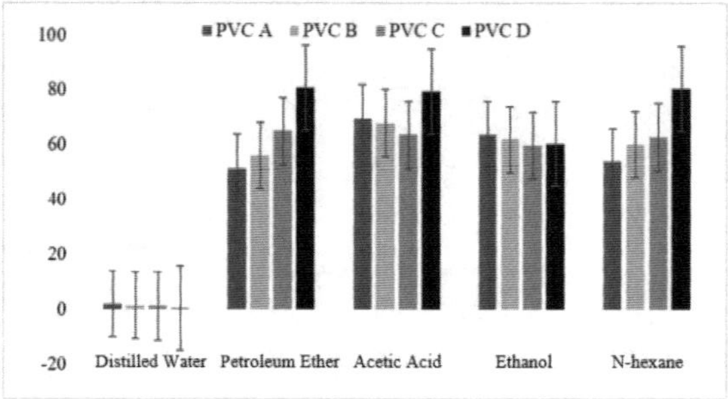

Figure 6. Migration Resistance Test using different Extraction Mediums

3.5.3. Volatility Test. A decreasing trend was observed among the bio-plasticized PVC films, indicating that a decrease in the plasticizer content decreases the plasticizer's migration. Activated carbon successfully absorbed the plasticizer molecules within its pores when the plasticizer migrated from the PVC film. The One-Way ANOVA was employed to conduct statistical analysis on the treatment groups with a p-value <0.00, indicating a significant difference in terms of migration resistance and volatility.

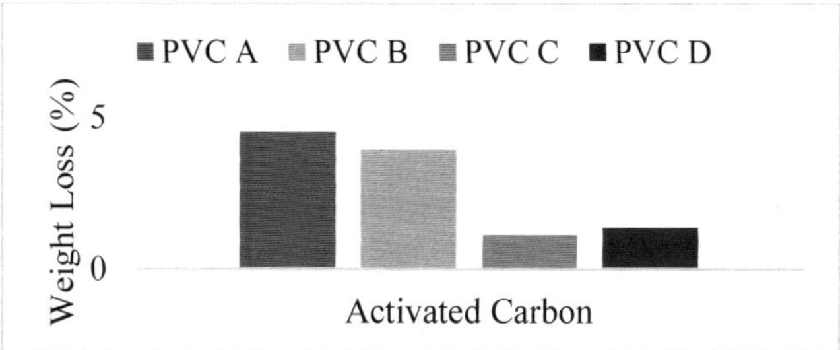

Figure 7. Volatility Test using Activated Carbon

IOP Conf. Series: Materials Science and Engineering **1318** (2024) 012019 doi:10.1088/1757-899X/1318/1/012019

4. Discussion

4.1. Conversion of Fatty Acid to EFAME

During the synthesis of FAME, the darkening of the mixture over time can be attributed to the degradation or oxidation of the reactants or products [10]. During the initial minutes of the synthesis of EFAME, the color of the mixture reverts back to yellow. The reaction between the oxidizing agent and the double bond results in the formation of an epoxy compound. Epoxy compounds typically have a yellow color. Therefore, the yellow color of EFAME reaction indicates the formation of epoxidized compounds, such as oxirane groups [11].

4.2. Characterization of EFAME using FTIR

The strong broad band of stretching vibrations corresponding to -OH group found at 3429.43 cm^{-1} indicates the presence of alcohol from the esterification of the fatty acid chains through the addition of methanol. The carbonyl group found in peak 11 at 1739.79 cm^{-1} can be characterized by the stretching of the C=O of the ester carbonyl groups [11], indicating that the fatty acid has been successfully converted to FAME. Transesterification was confirmed by an increase and shift in carbonyl region at 1739.79 cm^{-1}, whereas the -OH region decreased concurrently, indicating an interaction between the methyl ester and the hydroxyl group. The CH_3 Bending Vibrations found at 161.74cm^{-1} suggests the presence of methyl groups attached to the oxygen atom on the ester functional group to form EFAME. The absorption band at 842.89 cm^{-1} indicates the formation of epoxide. This implies the successful conversion of the double bond within the fatty acid into an epoxy bond.

4.3. Preliminary Evaluation using Tensile Strength and Elongation at Break Test

The tensile strength testing was done at ambient temperature, considering this condition, the plasticized PVC compound exhibited lower tensile strength and elastic modulus with an increase of elongation tensile failure. [12]. The current findings are consistent with previous studies [13][14] that found that adding plasticizer reduces tensile strength. This is because plasticizers disrupt the regularity and packing of polymer chains, making them less tightly bonded and more prone to sliding past one another [15]. PVC films with higher displacement capabilities can stretch and deform more before reaching their breaking point, indicating greater resistance to fracture [1]. Therefore, PVC-40 with 40% plasticizer was used as the basis of producing samples with varying EFAME and DOP concentrations.

4.4. Morphological and Elemental Analysis

The presence of the bioplasticizer resulted in slight changes in surface texture compared to DOP-plasticized PVC films (PVC-D) as seen in the SEM. However, it is evident that the PVC films contain surface contaminants which was further examined in the EDX. The contaminants appeared as foreign particles on the film's surface, distinct from the film matrix. Carbon, which makes up majority of the elemental composition for all PVC Films, is an essential component of both PVC and EFAME. Oxygen is a fundamental component of both PVC and EFAME. PVC contains oxygen atoms in the form of carbonyl (C=O) groups in its structure. EFAME, being an ester compound, contains oxygen atoms in its ester functional group (-COO-). During the synthesis of PVC, these chlorine atoms become part of the polymer structure, resulting in the presence of chlorine throughout the material. When PVC is plasticized with EFAME, the chlorine atoms from the PVC backbone remain in the plasticized material. The EDX analysis of plasticized PVC film detected the chlorine which confirms the presence of plasticized PVC in the film, however it is evident that it is present in smaller amounts than Carbon and Oxygen, since the plasticizer added and reacted with the PVC Film contains more carbon and oxygen molecules.

4.5. Differential scanning calorimetry (DSC) Analysis
The determination of the glass transition temperature (T_g) establishes the boundary region separating predominantly glassy from highly rubbery states. The T_g of the polymer will decrease when a plasticizer is added. The properties of the polymer shift from those of a hard, brittle solid to those associated with a soft, flexible rubber if the T_g of the polymer is sufficiently decreased [16]. EFAME possesses a significant amount of unsaturation and can interact with PVC to produce crosslinks, which can lower the T_g of the PVC. Due to the fact that DOP lacks this crosslinking capability, EFAME is a more efficient plasticizer than DOP [17]. PVC C has the lowest T_g at 30.26°C and considering the application's operating temperature range; it may not be suitable for high-temperature resistance due to softening and loss of dimensional stability, similar to other PVC film samples. PVC C with 50% EFAME and 50% DOP blend has a better T_g and produces the best PVC films among the other samples [15]. As a result, we conclude that the reduction in T_g was larger when EFAME and DOP were combined.

4.6. Migration Stability and Volatility Tests
PVC A exhibited the lowest combined average weight loss percentage, indicating that it has developed high migration resistance. Bioplasticizers like EFAME have higher molecular weights, lower vapor pressure, and lower volatility compared to DOP, which reduces the tendency of the plasticizer to migrates out of the PVC film [19]. Since EFAME plasticizer contains an epoxy group, it is more polar than DOP, exhibiting increased affinity with the PVC matrix. This stronger interaction restricts the mobility of bioplasticizer molecules, minimizing the chances of migration from the film. Thus, bioplasticizers generally demonstrate better compatibility with PVC compared to DOP, resulting in enhanced plasticizer and PVC matrix interaction, reducing the likeliness of migration.

5. Conclusion
In the present study, EFAME was successfully synthesized from *C. vulgaris* via esterification and transesterification reactions. FTIR Analysis of EFAME showed the transformation of double bonds of the unsaturated fatty acids into epoxy group. PVC-40 exhibited high tensile properties which was used as a basis to produce the bio-plasticized PVC samples and compared with pure plasticized PVC film. Further characterization and testing revealed the bio-based plasticizer offers improved properties in terms of functionality and performance. Thus, the bio-based plasticizer demonstrated promising potential as a viable substitute for phthalate-plasticized PVC.

Acknowledgment
Our sincerest gratitude to Dr. Rugi Vicente Rubi and Engr. Jarlie R. Clemeña in terms of this undergraduate study in producing bio-plasticized PVC films synthesized from *C. vulgaris*. We also greatly appreciate Engr. Michael T. Tavas of University of the City of Manila – College of Engineering, Engr. Albert D.C. Evangelista of Adamson University, John Rafael N. Granada and Michael S. Lagmay of University of the City of Manila – College of Science, for their guidance and assistance in conducting the experiment. And lastly, Dr. Milagrosa R. Martinez-Goss of University of the Philippines Los Baños – Institute of Biological Sciences, for extending her knowledge and expertise about the collection, handling, and harvesting of the microalgae.

References
[1] Czogała J, Pankalla E, and Turczyn R 2021 Recent Attempts in the Design of Efficient PVC Plasticizers with Reduced Migration. Materials **14** (4) 844 [DOI:10.3390/ma14040844]
[2] Devadas V V, Khoo K , Chia W Y, Chew K W, Munawaroh S H, Lam M K, Lim J W, Ho Y C, Lee K T, and Show P L 2021 Algae biopolymer towards sustainable circular economy. Bioresource Technology **325** 124702 [DOI:10.1016/j.biortech.2021.124702]

[3] Sustaita-Rodríguez A, Ramos-Sánchez V H , Camacho-Dávila A A, Zaragoza-Galán G, Espinoza-Hicks J C, and Chávez-Flores D 2018 Lipase catalyzed epoxidation of fatty acid methyl esters derived from unsaturated vegetable oils in absence of carboxylic acid. Chemistry Central Journal **12** (1). [DOI:10.1186/s13065-018-0409-2]
[4] Hidalgo P, Navia R, Hunter R, Gonzalez M, and Echeverría A 2019 Development of novel bio-based epoxides from microalgae (*Nannochloropsis gaditana)* lipids. Composites Part B: Engineering **166** 653–662 [DOI:10.1016/j.compositesb.2019.02.049]
[5] Pradana Y S, Sadewo B R, Haryanto S A, and Sudibyo H 2021 Selection of oil extraction process from Chlorella species of microalgae by using multi-criteria decision analysis technique for biodiesel production. Open Chemistry **19** (1) 1029–1042 [DOI:10.1515/chem-2021-0092]
[6] Zghaibi N, Omar R, Kamal S M M, Biak D R A, and Harun R 2019 Microwave-Assisted Brine Extraction for Enhancement of the Quantity and Quality of Lipid Production from Microalgae *Nannochloropsis sp*. Molecules **24** (19) 3581 [DOI:10.3390/molecules24193581]
[7] Liu D, Jiang P, Nie Z, Wang H, Dai Z, Deng J, and Cao Z 2020 Synthesis of an efficient bio-based plasticizer derived from waste cooking oil and its performance testing in PVC. Polymer Testing **90** 106625 [DOI:10.1016/j.polymertesting.2020.106625]
[8] Wilkes C E, Summers J W, Daniels C A, Berard M T, and Al E 2005 PVC handbook. Hanser.
[9] Bouchoul B, Benaniba M T, and Massardier 2017 [DOI:10.1590/0104-1428.14216]
[10] Blanchfield J 2016 Characterization and Degradation of Fatty Acid Methyl Esters and Biodiesel in Artificial Seawater Under UV Exposure Using Raman Spectroscopy [digital.lib.washington.edu:443/researchworks/handle/1773/36524]
[11] Hidalgo P, Álvarez S, Hunter R., and Sánchez A 2020 Epoxidation of Fatty Acid Methyl Esters Derived from Algae Biomass to Develop Sustainable Bio-Based Epoxy Resins. Polymers, **12** (10) 2313 [DOI:10.3390/polym12102313]
[12] Stolp L J 2019 Development Of Biobased Plasticizers from Vegetable Oils For Poly(Vinyl Chloride), PVC [hdl.handle.net/11299/213052]
[13] Kumar S 2019 Recent Developments of Biobased Plasticizers and Their Effect on Mechanical and Thermal Properties of Poly(vinyl chloride): A Review. Industrial & Engineering Chemistry Research, **58** (27), 11659–11672. [DOI:10.1021/acs.iecr.9b02080]
[14] Patil S S, and Jena H M 2021 Performance assessment of polyvinyl chloride films plasticized with Citrullus lanatus seed oil based novel plasticizer. Polymer Testing 101 107271 [DOI:10.1016/j.polymertesting.2021.107271]
[15] Lim S, Andriyana A, Show P L, Lim S, Pang Y L, and Ngoh G C 2022 A Comprehensive Review on the Emerging Roles of Nanofillers and Plasticizers towards Sustainable Starch-Based Bioplastic Fabrication. Polymers **14** (4) 664 [DOI:10.3390/polym14040664]
[16] Sichina W n.d. Measurement of Tg by DSC [thermalsupport.com/wpcontent/uploads/2018/05/PETech-09]
[17] Ma Y, Liao S, Li Q, Guan Q, Jia P and Zhou Y 2020 Physical and chemical modifications of poly(vinyl chloride) materials to prevent plasticizer migration – Still on the run. Reactive and Functional Polymers, **147** 104458 [DOI:10.1016/j.reactfunctpolym.2019.104458]
[18] Chen J, Wang Y, Huang J, Li K, and Nie X 2017 Synthesis of Tung-Oil-Based Triglycidyl Ester Plasticizer and Its Effects on Poly(vinyl chloride) Soft Films. ACS Sustainable Chemistry & Engineering, **6** (1) 642–651. [DOI:10.1021/acssuschemeng.7b02989]
[19] Lenzi L, Esposti M D, Braccini S, Siracusa C, Quartinello F, Guebitz G M, Puppi D, Davide M, and Fabbri P 2023 Further Step in the Transition from Conventional Plasticizers to Versatile Bioplasticizers Obtained by the Valorization of Levulinic Acid and Glycerol. ACS Sustain Chem Eng **11** (25) 9455-9469 [DOI: 10.1021/acssuschemeng.3c01536]

Adsorption of phytohormones from waste coconut water using surface-activated biochar derived from cacao pod husk (*Theobroma cacao* L.)

Jude Martin A Banzuela[1*], Monet Concepcion M Detras[1], Mart Merwin C Magboo[1], Catalino G Alfafara[1] and Veronica P Migo[1]

[1]Department of Chemical Engineering, College of Engineering and Agro-Industrial Technology, University of the Philippines Los Baños, College, Laguna

*email: jabanzuela@up.edu.ph

Abstract. The study investigated the potential of cacao (*Theobroma cacao* L.) pod husk as feedstock for the production of activated biochar. The activated cacao pod husk biochar is to be used for the extraction of phytohormones present in coconut water. Cacao pod husk was dried, milled, sieved, and pyrolyzed at 500°C for 1 hour. The resulting raw biochar was then activated using $ZnCl_2$. The chemical activation process was optimized by employing Central Composite Design with $ZnCl_2$-biochar ratio and holding time as factors. The response was the phytohormone removal/adsorption efficiency from the coconut water. The goal of the optimization was to maximize phytohormone extraction performance. The low and high levels for $ZnCl_2$-biochar ratio were 1:1 (w/w) and 5:1 (w/w), respectively. For the holding time, these were set at 30 minutes and 90 minutes, respectively. The optimum chemical activation conditions for phytohormone extraction performance were at 3.1:1 zinc chloride-biochar ratio and 30 minutes holding time. The predicted phytohormone adsorption (extraction efficiency) was 96.0%. Experimental verification of optimum conditions is ongoing. The results demonstrated that surface-activated biochar derived from cacao pod husk is an effective adsorbent for the extraction of phytohormones from waste coconut water. This approach not only helps mitigate the negative environmental impacts associated with waste disposal in the chocolate and coconut oil industries but also provides an opportunity to create value-added products from these resources, contributing to sustainability efforts.

1. Introduction

Coconut is one of the main crops produced in the Philippines. In 2019, the Philippines has produced a total of 14.765 million metric tons of coconut [1], making it the second top producer of coconut worldwide, next to Indonesia with 17.13 million metric tons of coconuts [2]. One of the Philippines' top agricultural export commodities is coconut oil. In 2020, the country has exported coconut oil which accounted for 52.48% of the global market [3]. Processing large quantities of coconut generates huge volumes of waste coconut water [4].

Waste coconut water poses threat to the environment when disposed untreated. It has high biological oxygen demand (BOD), up to 10,500 mg/L, which can rapidly deplete the oxygen level in water bodies resulting to the death of aquatic organisms [5], [4]. Its acidic nature affects the soil acidity, which has a deleterious impact on surrounding crops [6], [7]. Thus, treatment is needed before its disposal. However, most of the treatments are expensive, discouraging many coconut industries to treat their wastes before disposal. There have been several value-adding processes done to waste coconut water as an alternative to waste coconut water treatment. These include production of vinegar, beverage, *lambanog,* and *nata de coco*. However, most of these products generate smaller revenues since they are already available at an excess market supply. Thus, other approaches are needed to reduce the pollution loads from coconut oil industry, as well as boost the economic returns of waste coconut water.

Value-added products, like phytohormones, can be extracted from coconut water. Phytohormones, also known as plant growth regulators (PGRs), are chemical compounds that regulate all aspects of plant growth, development, and even death [8]. The quality and quantity of harvest have been known to improve as a result of the phytohormone's growth-promoting ability [9], [7]. Coconut water contains phytohormones such as cytokinin, auxins, gibberellin, and abscisic acid [10]. However, they are only present in minute quantities and need a series of concentration processes. One way to extract phytohormones in coconut water is by adsorption process.

Adsorption is a process where chemical species adhere onto the surface of particles [11]. Activated carbon is the most used adsorbent for removing pollutants from water and gas-phase streams. It can be produced from almost any feedstock containing carbon. Currently, the potential of agricultural residues as feedstocks for biochar production is being investigated.

Cacao pod husk (CPH) is a waste generated by cocoa industry after the removal of cocoa beans from the fruit. It constitutes to 70-75% of the total weight of cacao fruit [12]. Processing large amounts of cacao fruits would generate large amounts of waste since only the beans are utilized by cocoa industries. This would lead to serious solid waste problems. The cellulosic and hemi-cellulosic content of CPH makes it a potential precursor for activated carbon production [13], [12]. Different studies show that it has low moisture content, which is favourable for pyrolysis, and a high volatile combustible matter content, which is good for the formation of porous structures [14], [15], [16], [17]. In this study, activated carbon derived from CPH was used as adsorbent for the extraction of phytohormones in waste coconut water.

2. Materials and methods

2.1. Materials

2.1.1. Feedstock preparation. Dried cacao pod husks were obtained from a local cacao farm in Cavite. These were washed with water to remove dirt and other impurities, air-dried, and then oven-dried at 105°C for at least 8 hours. The cleaned and dried feedstocks were then milled using a hammer mill, and then screened using a sieve with Mesh No. 16. Only the oversize components were used in the study. The screened feedstocks were then stored in airlock containers to prevent it from absorbing moisture from the atmosphere.

2.1.2. Waste coconut water preparation. The study used coconut water as a source of phytohormones for adsorption experiments. The coconut water used in the study was obtained from a stall in Los Baños Public Market. The coconut water was carefully collected after splitting open mature coconuts. This was then clarified using cloth filter, and filtered using coarse filter paper to remove visible particles or dirt that were accidentally included during collection. The filtered volumes of coconut water were transferred into small plastic bottles, and stored in the refrigerator at 0°C to 4°C, to preserve its quality.

2.1.3. Chemical activator. Zinc chloride ($ZnCl_2$) was used to chemically activate the raw cacao pod husk-derived biochar. Technical-grade zinc chloride powder used in the study was purchased from Dalkem Corporation.

2.1.4. Carbonization and activation equipment. The feedstock was pyrolyzed and activated using a muffle furnace. The muffle furnace is not connected to a nitrogen supply. Hence, to achieve

IOP Conf. Series: Materials Science and Engineering **1318** (2024) 012020 doi:10.1088/1757-899X/1318/1/012020

an inert atmosphere for pyrolysis, the reactor containing the sample, as well as the inside of the furnace, were purged manually with nitrogen gas. Figure 1 shows the muffle furnace used in the study.

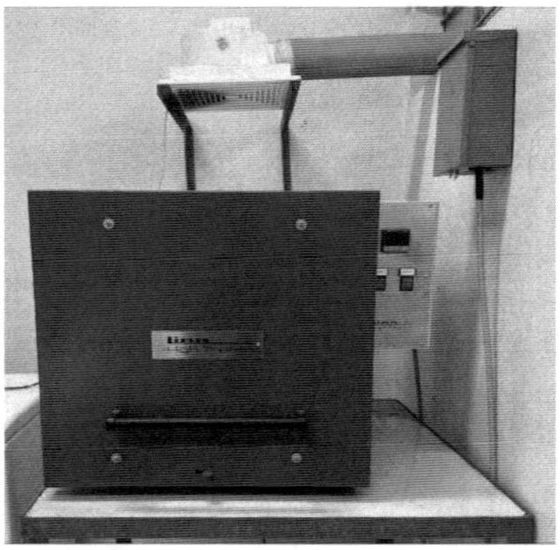

Figure 1. Muffle furnace used in the production and chemical activation of raw cacao pod husk biochar (CPH-B).

2.2. Biochar production

A muffle furnace was used to produce cacao pod husk-derived biochar (CPH-B). Dried feedstocks were placed in a reactor, and then purged with nitrogen gas for 2 minutes at 12.5 psi to ensure the absence of oxygen. The reactor was then placed inside the furnace and the furnace was manually purged with nitrogen gas for 4 minutes. It was then heated to 500°C at the rate of 19.33°C/min and was held for 1 hour. The system was then allowed to cool to room temperature before collecting the biochar. The yield of biochar was computed using the Equation (1).

$$Y = \frac{mass\ of\ biochar\ produced}{mass\ of\ feedstock} x100\% \quad (1)$$

2.3. Chemical activation procedure for production of activated biochar

Raw cacao pod husk biochar (CPH-B) obtained from previous pyrolysis experiment was activated using zinc chloride powder at holding times specified by the experimental design. The experimental design is described in Section 2.5. Target amounts of zinc chloride to achieve the desired zinc chloride-biochar ratios in the experimental design were dissolved first in 50 mL distilled water before adding 20 g of raw CPH-B. The resulting mixture was allowed to stand for 2 hours at room temperature before drying (i.e., evaporation of water) on a hot plate. The dried mixture was then placed in a 155 mL stainless steel cylinder reactor and activated in a furnace at 500°C for 30, 60, and 90 minutes. After activation, the cacao pod husk-derived activated biochar (CPH-AB) was immersed in 0.5M HCl solution to leach out zinc compounds and was heated at 90°C for 30 minutes. The activated biochar was then washed with hot water followed by distilled water until the pH of the washing became neutral. The washed CPH-AB was oven-dried for 12 hours at 105°C before storing in airtight containers. The yield of activated biochar was computed using Equation (2).

$$Y = \frac{mass\ of\ activated\ biochar}{mass\ of\ biochar} x100\% \quad (2)$$

2.4. Single flask adsorption for determination of phytohormone extraction (adsorption) performance as the response

Single flask adsorption experiment was done for each activation run to determine the effects of chemical activation conditions on phytohormone adsorption performance. The ratio of the absorbance of the filtrate at 267nm and initial absorbance of the coconut water (multiplied by 100), was used to quickly and indirectly determine the percentage of phytohormones adsorbed onto the surface of CPH-AB. A specific amount of CPH-AB, based on the effective adsorbent loading from preliminary experiment, was added to a flask containing 25 mL coconut water. Then, adsorption was allowed to take place, by agitating the mixture for 2 hours using a tabletop shaker. After adsorption, the solid CPH-AB was removed by filtration, and the filtrate was collected for absorbance reading. The initial and final absorbances were used to compute the percentage of phytohormones adsorbed on CPH-AB using Equation (3).

$$\%Removal = \left(1 - \frac{equilibrium\ absorbance}{initial\ absorbance}\right) x100\% \tag{3}$$

2.5. *Experimental design for the chemical activation of biochar*

Response Surface Methodology (RSM) using face-centered Central Composite Design (CCD) was employed in the study. Zinc chloride-biochar ratio and holding time were the two main factors considered with the percent phytohormone removal from coconut water as the response. The low level for zinc chloride-biochar ratio was set to 1:1 (w/w) while the high level was set to 5:1 (w/w). On the other hand, the low level for holding time was set to 30 minutes while the high level was set to 90 minutes. For the factorial part of the design, three replicates were made and four center points were added. The design was augmented after detecting the presence of curvature. The design was made and analyzed using Design Expert 13 (trial version). Optimum chemical activation conditions were determined using desirability approach.

3. Results and discussion

3.1. *Yields and phytohormones removal*

The production of activated biochar in the study involved two steps: pyrolysis and chemical activation. Cacao pod husk was subjected to pyrolysis at 500°C for 1 hour with heating rate of 19.33°C/min yields 40.13% biochar. The biochar produced are chemically activated by varying the chemical activator ($ZnCl_2$)-biochar ratio and holding time. The cacao pod husk used and activated biochar produced are shown in Figure 2a and 2b, respectively. The activation yield and percent phytohormone removal efficiency for each activation condition are shown in Table 1.

Figure 2. (a) Cacao pod husk and (b) activated cacao pod husk biochar.

Table 1. Yield and phytohormones removal at different activation conditions.

Activation Condition			
Zinc chloride-biochar ratio (wt/wt)	Holding time (min)	Activation Yield (%)	Phytohormones Removal (%)
1:1	30	83.3995	93.3099
	90	79.8705	93.2596
3:1	60	81.0675	95.2842
5:1	30	83.3185	93.9638
	90	79.6578	92.7565

IOP Conf. Series: Materials Science and Engineering **1318** (2024) 012020 doi:10.1088/1757-899X/1318/1/012020

The yield of activated biochar ranged from 79.87% to 83.40% for the activation conditions considered. Based on the results, at constant zinc chloride-biochar ratio, the activated biochar yield decreased as the holding time increased. On the other hand, at constant holding time, the activated biochar yield decreased minimally when the zinc chloride-biochar ratio was increased. From the results of the single flask adsorption experiments, the percent phytohormone removal efficiency ranged from 92.76% to 95.28%. The highest phytohormone removal was obtained at the center point, corresponding to 3:1 zinc chloride-biochar ratio and 60 minutes holding time. The phytohormone removal for the low and high levels of each factor are close to each other by 0.05% to 1.21%. At constant zinc chloride-biochar ratio, the phytohormone removal decreased as the holding time increased. At lower holding time, increasing the amount of zinc chloride increases the phytohormones removal. However, at higher holding time, increasing the zinc chloride decreases the phytohormones removal. These trends are discussed further in Section 3.3.

3.2. Model equation and statistical validation

The coded equation showing the relationship between phytohormone removal and the two factors considered was a quadratic model as shown in Equation 4.

$$Phytohormones\ Removal = 94.53 + 0.0286A - 0.2115B - 0.2892AB - 3.18A^2 + 1.26B^2 \quad (4)$$

where A is the coded variable for zinc chloride-biochar ratio, B is the coded variable for holding time, and AB is the interaction between A and B. This equation can be used to predict the response (phytohormones removal) of a specific chemical activation condition (zinc chloride-biochar ratio, and holding time) applied to CPH-B. Coefficients of each factor tell the impact of that factor to the response. Disregarding their signs, the higher the coefficient of a factor, the higher is its effect on the response. From Equation 4, A^2 has the highest coefficient followed by B^2, AB, B, and A. Although the coefficient of B is higher than A, the squared form of A is higher than the squared form of B. Thus, overall, zinc chloride-biochar ratio has the most impact on phytohormones removal.

Analysis of variance (ANOVA) was applied to evaluate the effects of the factors on the phytohormones removal. The results of ANOVA for the chemical activation of CPH-B are shown in Table 2.

Table 2. Analysis of variance (ANOVA) for the chemical activation of CPH-B.

Source	Sum of Squares	df	Mean Square	F-value	p-value	
Block	0.0574	1	0.0574			
Model	26.19	5	5.24	3.59	0.0212	significant
A (Zinc chloride-biochar ratio)	0.0102	1	0.0102	0.007	0.9345	
B (Holding time)	0.6262	1	0.6262	0.4298	0.5209	
AB	1	1	1	0.6889	0.418	
A^2	21.41	1	21.41	14.69	0.0013	
B^2	3.94	1	3.94	2.7	0.1185	
Residual	24.77	17	1.46			
Lack of Fit	0.9219	2	0.46095	0.2899	0.7524	not significant
Pure Error	23.85	15	1.59			
Cor Total	51.01	23				
R^2	0.5139		Predicted R^2		0.1187	
Adjusted R^2	0.3709		Adeq Precision		7.1818	

At 5% level of significance, the model was shown significant, while the lack of fit was desirably insignificant [18], [19], [20]. Among the model terms, only the quadratic effect of zinc chloride-biochar ratio (A^2) is significant.

3.3. Contour plot and response surface graphs

The quadratic model was used to generate the 3D surface graph in Figure 3. The graph illustrates the changes in the phytohormones removal as the zinc chloride-biochar ratio and holding time are varied at constant activation temperature of 500°C and an impregnation time of 2 hours. From the graph, increasing the zinc chloride-biochar ratio increases the phytohormones removal to a certain level and then decreases as more zinc chloride is used in the chemical activation process. The increase in phytohormones removal with increasing zinc chloride-biochar ratio (up to a value of about 3:1(w/w)) can be attributed to the increase in pores and channels that were developed in the matrix of activated carbon due to the dehydrating effects of $ZnCl_2$, leading to the release of volatiles during activation [21]. However, the newly formed pores can collapse and become larger at excessive amounts of $ZnCl_2$, (observed at zinc chloride-biochar ratio over 3:1(w/w)), resulting to a decrease in the surface area of the activated biochar [21]. Thus, lower phytohormone removal is observed at higher zinc chloride-biochar ratio. Holding time has minimal effects on the phytohormone removal. However, at extended holding time, the phytohormone removal started to decrease. This could be attributed to the collapse and constriction of porous structures, along with the sealing-off of some pores due to sintering at longer holding time [22], [23], [21]. Based on these results, cacao pod husk-derived activated biochar with high phytohormone removal performance can be produced at higher zinc chloride-biochar ratio of about 3:1 (w/w) and shorter holding time (about 30 minutes). The determination of optimum values based on the statistical model are discussed in Section 3.4.

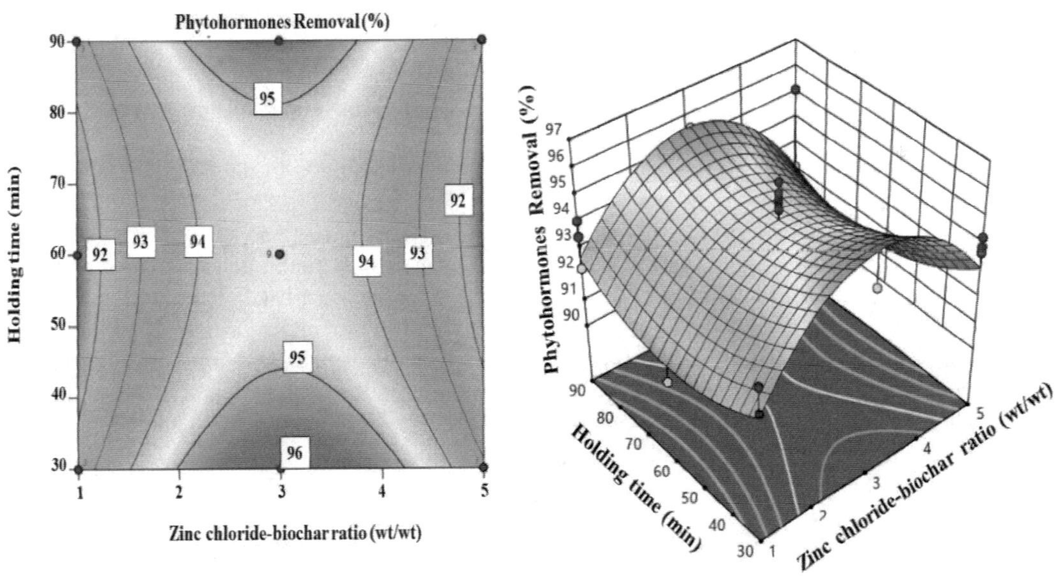

Figure 3. Contour plot and 3D surface plot for the percent phytohormone removal efficiency as a function of zinc chloride-biochar ratio and holding time.

3.4. Numerical Optimization

The goal of performing optimization was to maximize the phytohormone removal. The factors were set to be within the range of its low and high levels in the experiment, and optimum conditions were selected based on overall desirability score [18]. Optimization was performed using Design Expert 13 (trial version), and the results are shown in Table 3.

Table 3. Optimization results given by Design Expert 13.

Number	Zinc chloride-biochar ratio	Holding time	Phytohormones Removal	Desirability	
1	3.1	30	96.004	0.984	Selected
2	3.115	30	96.004	0.984	
3	3.058	30	96.003	0.984	
4	3.146	30	96.002	0.984	
5	2.775	30	95.92	0.969	
6	2.918	90	95.579	0.905	
7	2.941	90	95.578	0.905	
8	2.321	30	95.521	0.895	

There were 8 solutions given by Design Expert with overall desirability values ranging from 0.895 to 0.984. These solutions are all desirable since their desirability are close to unity. Solution 1 to 4 have the same and highest desirability, and comparable values for each factor and response. Solution 1 was selected since it has the highest expected phytohormone removal, and has the second lowest zinc chloride-biochar ratio. Thus, the optimum conditions for the chemical activation of CPH-B are chosen to be 3.1:1 zinc chloride-biochar ratio and 30 minutes holding time. This condition is expected to have a phytohormone removal of 96.0%. Experimental verification of the optimum conditions is still ongoing.

4. Conclusion

The study investigated the potential of using cacao pod husk (*Theobroma cacao* L.) as feedstock for producing activated biochar, which can then be utilized to extract the phytohormones in waste coconut water. The cacao pod husk was pyrolyzed at 500°C for 1 hour and the resulting biochar was activated using $ZnCl_2$. The chemical activation process was optimized by employing face-centered Central Composite Design (CCD) with zinc chloride-biochar ratio (w/w) and holding time (minutes) as factors and phytohormones removal (%) as response. The experimental design and analyses were done with the help of Design Expert 13 (trial version) software.

A quadratic model describing the phytohormone removal as a function of the factors was generated. ANOVA was used to analyze the model and the results showed that the model was significant. The statistically validated quadratic model was used to generate the contour and 3D surface graphs. The model shows that increasing the zinc chloride-biochar ratio increases the phytohormones removal to a certain level and then decreases as more zinc chloride is used in the chemical activation process. On the other hand, changing the holding time has minimal effects on the phytohormones removal. Numerical optimization was done to maximize the phytohormones removal and the optimum conditions were chosen based on the overall desirability score. The optimum conditions were found to be at 3.1:1 zinc chloride-biochar ratio and 30 minutes holding time with predicted phytohormone adsorption efficiency of 96.0%. Experimental verification of the optimum conditions is ongoing.

References

[1] Philippine Statistics Authority (PSA) 2020 *2020 Selected Statistics on Agriculture.* Retrieved August 2, 2021 from https://psa.gov.ph/sites/default/files/2_SSA2020_final_signed.pdf

[2] Burton J 2021 The world leaders in coconut production. Economics. Retrieved August 5, 2021 from https://www.worldatlas.com/articles/the-world-leaders-in-coconut-production.html

[3] Daily Guardian 2022 Ph exports continue to recover boosted by surge in coconut oil sales. Daily Guardian. Retrieved December 4, 2022, from https://www.dailyguardian.com.ph/ph-exports-continue-to-recover-boosted-by-surge-in-coconut-oil-sales/

[4] Magboo M M C 2015 Pre-scale up studies for the extraction of phytohormones from waste coconut water. Unpublished undergraduate thesis – Chemical Engineering. University of the Philippines Los Banos, College, Laguna

[5] Asian Productivity Organization [APO] 2006 Handbook on green productivity. Retrieved May 1, 2022 from http://www.apo-tokyo.org/publications/files/gp-hb_gp.pdf
[6] Grimwood B E, Ashman F, Jarman C G and Dendy D A V 1971 Coconut palm products: their processing in developing countries. Food & Agriculture Org
[7] Borbo E E 2016 Extraction of phytohormones from waste coconut water by adsorption using coconut shell-derived granular activated carbon. Unpublished undergraduate thesis – Chemical Engineering. University of the Philippines Los Banos, College, Laguna
[8] Dilworth L L, Riley C K, and Stennett D K 2017 Plant Constituents. *Pharmacognosy* 61–80
[9] Fu J H, Sun X H, Wang J D, et al. 2011 Progress in quantitative analysis of plant hormones. *Chinese Sci Bull*, 56: 355–366
[10] Yong J W H, Ge L, Ng Y F and Tan S N 2009 The chemical composition and biological properties of coconut (Cocos nucifera L.) water. *Molecules*, 14, 5144-5164
[11] Helmenstine A M 2019 What Adsorption Means in Chemistry. Retrieved June 6, 2022 from https://www.thoughtco.com/definition-of-adsorption-605820
[12] Cruz G, Pirilä M, Huuhtanen M, Carrión L, Alvarenga E, et al. 2012 Production of activated carbon from cocoa (Theobroma cacao) pod husk *J Civil Environment Engg* 2:109
[13] Oladayo, A 2010 Proximate composition of some agricultural wastes in Nigeria and their potential use in activated carbon production *J Appl Sci Environ Manag* 14: 55-58
[14] Syamsiro M, Saptoadi H, Tambunan B H and Pambudi N A 2012 A preliminary study on use of cocoa pod husk as a renewable source of energy in Indonesia. *Energy Sustain Dev* 16:74–77
[15] Titiloye J O, Abubaker M S, and Odetoye T 2013 Thermochemical characterisation of agricultural wastes from West Africa. *Ind Crop Prod* 47:199–203
[16] Martinez-Angel J D, Villamizar-Gallardo R A, and Ortiz-Rodriguez O O 2015 Characterization and evaluation of cocoa (Theobroma cacao L.) pod husk as a renewable energy source. *Agociencia* 49:329–345
[17] Kilama G, Lating P O, Byaruhanga J, and Biira S 2019 Quantification and characterization of cocoa pod husks for electricity generation in Uganda. *Energy, Sustainability and Society*, 9(1)
[18] Vinayagamoorthy R, Manoj I V, Narendra Kumar G, Sai Chand I, Sai Charan Kumar G V and Suneel Kumar K 2018 A central composite design based fuzzy logic for optimization of drilling parameters on natural fiber reinforced composite. *Journal of Mechanical Science and Technology*, 32(5), 2011–2020
[19] Soltani R D C, Rezaee A, Khataee A R, and Safari M 2014 Photocatalytic process by immobilized carbon black/ZnO nanocomposite for dye removal from aqueous medium: Optimization by response surface methodology. *Journal of Industrial and Engineering Chemistry*, 20(4), 1861–1868
[20] Hassani A, Kiranşan M, Darvishi Cheshmeh Soltani R, Khataee A, and Karaca S 2015 Optimization of the adsorption of a textile dye onto nanoclay using a central composite design. *Turkish Journal Of Chemistry*, 39, 734–749
[21] Hock P and Ahmad Zaini M A 2018 Activated carbons by zinc chloride activation for dye removal – a commentary. *Acta Chimica Slovaca*. 11. 99-106
[22] Mohanty K, Das D and Biswas M N 2006 Preparation and characterization of activated carbons from Sterculia alata nutshell by chemical activation with zinc chloride to remove phenol from wastewater. *Adsorption*, 12(2), 119–132
[23] Saygili H, Guzel F 2015 High surface area mesoporous activated carbon from tomato processing solid waste by zinc chloride activation: Process optimization, characterization and dyes adsorption *Journal of Cleaner Production* 113, 995—1004

IOP Conf. Series: Materials Science and Engineering **1318** (2024) 012021 doi:10.1088/1757-899X/1318/1/012021

Effect of Reaction Time and Solvent Ratio on the Acetylation of Isolated Cellulose Fiber from Tropical Water Hyacinth

AC Tilendo[a], JQ Filipinas[b], KMA Marohomsalic[c], SA Edres[d], LV Ginco[e], AB Villame Jr.[f]

Chemical Engineering Department, Mindanao State University – Main, Marawi City, 9700, Lanao del Sur, Philippines

[a]amierson.tilendo@msumain.edu.ph, [b]judilyn.filipinas@msumain.edu.ph, [c]kristinemarfe.marohomsalic@msumain.edu.ph, [d]sahaf.edres@msumain.edu.ph, [e]ginco.ll18@s.msumain.edu.ph, [f]villame.ab78@s.msumain.edu.ph

Abstract. Isolated cellulose fibers from water hyacinth were used to create cellulose acetate (CA) polymers. The cellulose was acetylated using an acetic acid and acetic anhydride reaction, with sulfuric acid acting as the catalyst. The degree of substitution (DS) of synthetic CA was examined, along with the effects of acetylation time and solvent ratio. The study indicated that the appropriate DS for CA was found to be 2.47 and the matching ideal operating parameters included a reaction duration of 16 hours, and a ratio of 4:1 between acetic acid and acetic anhydride at temperature of 50°C. The results of the study showed that the cellulose isolated from water hyacinth was successfully acetylated, which was supported by the existence of distinct peaks in the Fourier transform infrared (FTIR) spectra. This suggests that the process of acetylating the isolated cellulose fibers is a straight-forward process that can be employed to create a cost-efficient and environmentally friendly CA. This CA has the potential to cost-effectively and sustainably replace expensive raw materials derived from fossil fuels in the polymer industry.

1. Introduction

Cellulose acetate (CA) is being used in several applications such as film, fiber, and coating manufacture. It is also used in the production of numerous items, including cigarette filters and eyeglass frames. Calcium alginate is a favorable material for membranes because of its benefits including low flux, good salt rejection qualities, cost-effectiveness, simplicity of fabrication, and non-toxic nature. It is derived by the alteration of cellulose. Cellulose consists of units known as β-1,4-anhydro-D-glucopyranose, which have very reactive hydroxyl groups. Hydroxyl groups may be substituted with acetyl groups, leading to the creation of CA. CA membranes are used in water desalination [1].

Providing clean water to all is a fundamental human right, and it remains a significant challenge in many parts of the world. Innovations have been made, but due to the limitations of the projects, they have not been sustained. Hence, a more sustainable way for water purification utilizes energy-efficient membrane technologies. Cellulose acetate, derived from biomass such as water hyacinth, is an excellent example of biopolymers that have many important properties and can be blended with other compounds for water purification purposes.

IOP Conf. Series: Materials Science and Engineering **1318** (2024) 012021 doi:10.1088/1757-899X/1318/1/012021

Traditionally, wood and cotton have been the main material resources used for the industrial acetylation of cellulose. However, due to their limited availability, agricultural wastes and invasive plant species have become attractive renewable resources for synthesizing cellulose acetate on a large scale. These resources include the water hyacinth, which is widely cultivated globally and produces large amounts of waste material-al that can be used for cellulose acetate synthesis [2].

Current studies have successfully synthesized CA from various agricultural waste sources and a wide range of plants. The biodegradability of CA is contingent upon the degree of substitution (DS). DS is determined by the number of hydroxyl groups in the repeating unit that have been chemically changed to be replaced by a certain substituent. The maximum DS is 3 which results in very low biodegradability.

The production of CA involves a two-step acetylation process, where cellulose reacts with acetic acid, excess acetic anhydride, and sulfuric acid as a catalyst. This is followed by a hydrolysis reaction to achieve a desirable DS of around 2.45-2.5, resulting in CA with high solubility and favorable melt properties for various consumer product applications [3].

In this study, preparation of CA from isolated cellulose of water hyacinth plant was investigated with varying acetylation time and acetic acid/acetic anhydride ratio. CA was synthesized by a homogenous procedure and characterized utilizing its DS, proportion of acetyl groups, and infrared spectroscopy (FTIR).

2. Materials and Method

2.1 Materials

The raw materials used for this study are isolated cellulose acetate (CA) (10 g, from water hyacinth), glacial acetic acid, sulfuric acid (0.2 mL), acetic anhydride, dis-tilled water (10 mL), sodium hydroxide (0.25 M, 5 mL), ethanol (5 mL), and hydrochloric acid 0.25 M (10 mL). Moreover, the glass wares and laboratory equipment used for the experiment are beaker (100 mL), Erlenmeyer flask (200 mL), stirring rod, pipette (25 mL), aspirator, and drying oven.

2.2 Cellulose Acetate Synthesis

The next step after separating cellulose is to generate cellulose acetate (CA) by dissolving cellulose in a solution of glacial acetic acid and stirring at 40°C for 1 hour. Next, 0.1 mL of sulfuric acid, acting as the catalyst, will be added together with a combination of glacial acetic acid. The solution will be stirred for an extra 45 minutes at the same temperature. When the mixture reaches a temperature of 18 °C, acetic anhydride will be added. The next step involves adding sulfuric acid and glacial acetic acid to the mixture, initiating the acetylation process which will continue for a duration of 8 to 24 hours. Cellulose acetate will become a solid after undergoing acetylation and hydrolysis by gradually adding distilled water and stirring until the powder is completely submerged. The precipitate will be cleansed, filtered, and dried in an oven at 50 °C [4].

2.3 Determination of Degree of Substitution (DS) of CA

A mixture containing 0.1 grams of CA, 0.25 moles of NaOH, and 5 milliliters of ethanol was prepared. The combination was left undisturbed for 24 hours. Next, the mixture was treated with 10 mL of 0.25 M HCl and left undisturbed for 30 minutes. A 0.25 M NaOH solution was used to titrate the combination with phenolphthalein as an indicator. Three iterations of this method were completed. Equation 1 was used to calculate the fraction of acetyl groups [4]:

$$\%GA = \{[(Vb_i + Vb_t) * \mu_b - (V_a * \mu_a)] * M * 100\}/m_{ac} \qquad \text{(Equation 1)}$$

Where: Vb_i *= volume of NaOH added to the system (L)*
Vb_t *= volume of NaOH spent in titration (L)*
μ_b *= NaOH concentration (M)*
V_a *= volume of HCl added to the system (L)*
μ_a *= HCl concentration (M)*

M = *molar weight of acetyl group (43 g mol^{-1})*
m_{ac} = *weight of cellulose acetate sample (g)*

2.4 Experimental Design
In this study, researchers will imply varying acetylation time and acetic acid/anhydride ratio to produce CA with a DS of about 2.45-2.5. The design of the experiment of the study uses response surface methodology (RSM) to explore the relationship between the response variables. Three different acetylation time will be implemented that is 8, 16, and 24 hours and different concentration ratio of glacial acetic acid and acetic anhydride that is 3:1, 4:1, and 5:1 with constant temperature. The data gathered in the experiment is subjected to characterization using Fourier transform infrared spectroscopy (FTIR) that determines the presence of functional group in the synthesized CA. The DS determines the biodegradability of the CA that should be within the aim of the study.

2.5 Data Treatment and Statistical Analysis
An experimental study was carried out to analyze differences in the occurrence of functional groups using FTIR and DS by a one-way analysis of variance (ANOVA). ANOVA is a reliable statistical technique used to show variations among two or more means or components using significance tests. It also illustrates a technique for doing multiple comparisons of different population means. The ANOVA test contrasts the variance between sample means with the variation within each sample. The data will undergo analysis and the hypothesis will be evaluated using appropriate statistical methods.

3. Results

3.1 Characterization of cellulose acetate (acetylated cellulose)

3.1.1. Fourier transform infrared spectroscopy (FTIR)
Functional group modifications resulting from chemical treatment of cellulose extracted from water hyacinth were examined using FTIR spectroscopy. The observed peaks in this study align with the assigned IR bands for various functional groups as reported in [8].

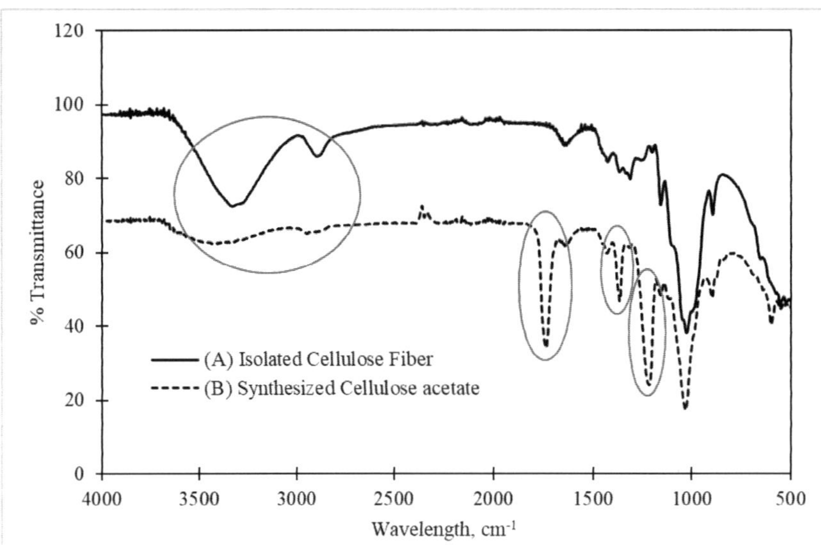

Figure 1. ATR-FTIR Spectra of (A) Isolated Cellulose Fiber and (B) Synthesized Cellulose acetate

Figure 1 show the FTIR spectroscopy spectra for both isolated cellulose fibers and cellulose acetate (CA) generated. The absorption peak at 3300 cm-1 in spectrum (A) is caused by the stretching vibrations of hydroxyl (-OH) groups in the cellulose sample. The hydroxyl groups consist of molecules such as absorbed water, aliphatic primary alcohols, secondary alcohols, and those found in cellulose and hemicellulose extractives. In Figure 1, spectrum (A) displays a decrease in the intensity of the absorption band associated with the stretching vibrations of hydroxyl (-OH) groups, in contrast to spectrum (B). This indicates a reduction in the hydroxyl group content of the isolated cellulose fibers after the reaction took place. Additionally, a shoulder observed near the -OH stretching vibrations at 2900 cm^{-1} is attributed to the stretching vibrations of C-H bonds, which correspond to the presence of aliphatic groups within polysaccharides such as cellulose. FTIR spectroscopy investigation identified distinct absorption peaks in the produced cellulose acetate. The sample exhibited an ester carbonyl absorption peak at 1740 cm^{-1}, a carbonyl hydrogen (C-H) peak at 1353 cm^{-1} in the acetyl group, and an absorption peak at 1247 cm^{-1} corresponding to the (C-O) bond in the O-C=O group. The increased strength of these peaks suggested the creation of ester linkages in the cellulose acetate. The unique peaks confirmed the acetylation technique used on the cellulose extracted from water hyacinth.

In Figure 1, spectrum (B) demonstrates the absence of any peaks or stretching in the range of 2000-1800 cm^{-1}. The lack of free acetic anhydride in the examined acetylated cellulose fibers is shown by this result. This discovery aligns with the data reported by the authors in reference [9], indicating the lack of free acetic anhydride in the FT-IR spectra of altered rice straw designed for thermo-plastic purposes.

3.2 Design of experiments on acetylation process using developed experimental design using Minitab 17 Software

In this study, researchers implied varying acetylation time and acetic anhydride concentration to produce a cellulose acetate (CA) with a degree of substitution (DS) about 2.4-2.5. A study was conducted to analyze the correlation between the duration of acetylation and the ratio of acetic acid to acetic anhydride on two specific responses: the identification of functional groups through Fourier transform infrared spectroscopy (FTIR) and the assessment of the degree of substitution (DS) using response surface methodology (RSM). Table 1 illustrates the experimental design used to get specific outcomes in the inquiry. The laboratory data from the experiment was analysed using FTIR to determine the functional groups in the cellulose acetate generated.

Table 1: Experimental design matrix and response of DS

FACTORS		RESPONSE
Acetic Acid / Acetic anhydride ratio	Time (hr)	Degree of substitution
3:1	8	2.99
3:1	24	4.25
4:1	4.7	2.80
4:1	16	2.76
4:1	16	2.92
4:1	16	2.47
4:1	16	2.92
4:1	24	3.00
4:1	27	3.78
5:1	24	2.84
6:1	16	2.79
8:3	16	3.19

3.3 Analysis of Variance (ANOVA)

The analysis of variance in Table 2 indicates that the projected model is statistically significant, since the model F-value of 8.68 and P-value of 0.007 are both below the threshold of 0.05. The model may be used to explore the design space. Equation 1 displays the appropriate fitted model for the degree of substitution (DS) in cellulose acetylation.

Table 2. Analysis of Variance (ANOVA).

	Coefficient	R squared (predicted)	R squared	F-value	T-Value	P-Value
Model	2.658	0.5576	0.8699	9.36	23.15	0.005
AcAc / AcAn ratio	-0.2868	-	-	9.18	-3.03	0.019
Time (hr)	0.2439	-	-	7.79	2.79	0.027
Ratio * Ratio	0.2268	-	-	7.26	2.69	0.031
Time * Time	0.2945	-	-	10.58	3.25	0.014
Ratio * Time	-0.3330	-	-	6.29	-2.51	0.041

* Where AcAc is the Acetic acid concentration and AcAn is the Acetic anhydride concentration.

The model's R-squared value of 0.8699 shows good agreement, while the adjusted R-squared value of 0.5576 reflects the model's importance. 86.99% of the variance in the data is explained by the model. The P-value of 0.005 is statistically significant since it is below the threshold of 0.05.

The empirical mathematical model equation of DS is shown in Equation 2:

$$DS = 5.74 - 1.617R + 0.0573T + 0.2561R^2 + 0.00460T^2 - 0.0442RT \quad \text{(Equation 2)}$$

Where R is the acetic acid / acetic anhydride ratio and T is the acetylation time in hours.

3.4. Degree of substitution (DS)

Figure 2a and 2b show a three-dimensional surface response and interaction plot demonstrating the relationship between DS and two factors related to the acetylation of cellulose. The acetylation process was notably influenced by the ratio of acetic acid to acetic anhydride and the reaction time in terms of substituting hydroxyl groups with acetyl groups. The catalyst speeds up the process, while warmth encourages the slow breakdown of crystallites. The optimal circumstances included a reaction time of 16 hours and an acetic anhydride degree of substitution (DS) of 2.47. The standard deviation in this investigation is less than the figure of 2.52 reported by Candido et al. for cellulose acetate derived from sugarcane bagasse [4].

3.5. Mechanism of the reaction

Cellulose acetate production begins by activating cellulose by dissolution in a solution of acetic acid and sulfuric acid. Sulfuric acid functions as a catalyst by protonating the hydroxyl groups on cellulose, enhancing its reactivity. After activating the cellulose, acetic acid and acetic anhydride are introduced into the reaction mixture. Acetic anhydride is often used because to its higher reactivity with the hydroxyl groups present in cellulose. Cellulose hydroxyl groups undergo esterification with acetyl groups from acetic acid/acetic anhydride to create an ester bond.

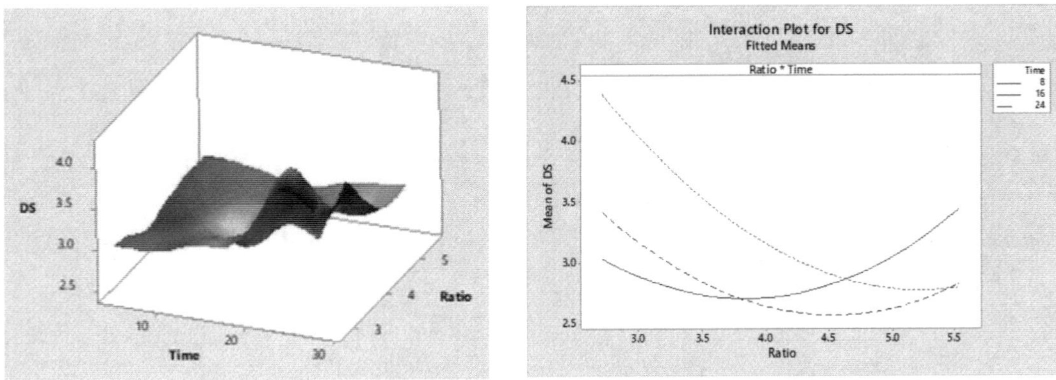

(a) (b)

Figure 2. (a) Response Surface Plot of DS as a function of acetic acid/acetic anhydride ratio and acetylation time. (b) Interaction plot for DS.

This process is known as esterification, and it results in the substitution of the hydroxyl groups with acetyl groups which also constitutes to the degree of substitution of the cellulose acetate and having water as a by-product

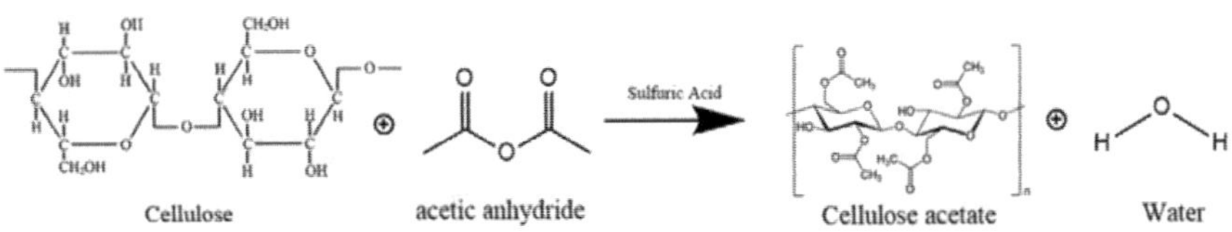

Figure 3a: Main reaction of the process

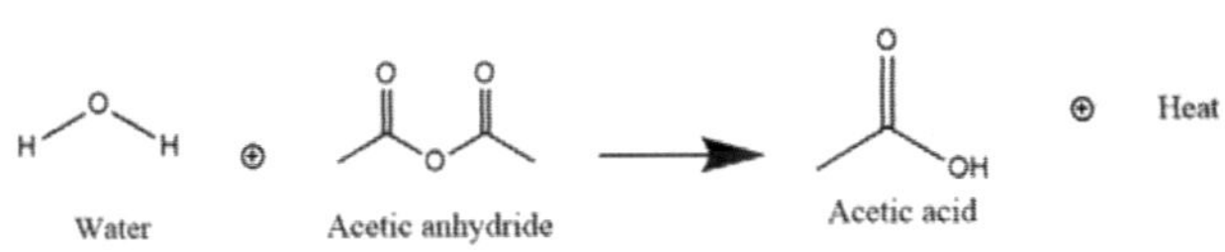

Figure 3b. Side reaction of the process

With the presence of water, acetic anhydride can hydrolyze, described in figure 3b, yielding acetic acid and releasing energy as heat.

3.6. Effect of Acetic acid/acetic anhydride ratio

The micro and macro morphology of the isolated cellulose has a significant im-pact on the rate and desirability of acetylation, and the need for an efficient acetylation of the isolated cellulose is heavily dependent on the accessibility of the cellulose contents of the pretreated isolated cellulose as well as the individual cellulose crystallites towards acetylation. To determine the best processing conditions, the degree of substitution (DS) was determined. Then, the effect of changing the glacial acetic acid and ace-tic anhydride ratio on the isolated cellulose sample was investigated. The primary im-pact of the acetic acid/acetic anhydride ratio on the DS of cellulose acetate (CA) and the surface plot of DS is shown in Figures 4a and 4b.

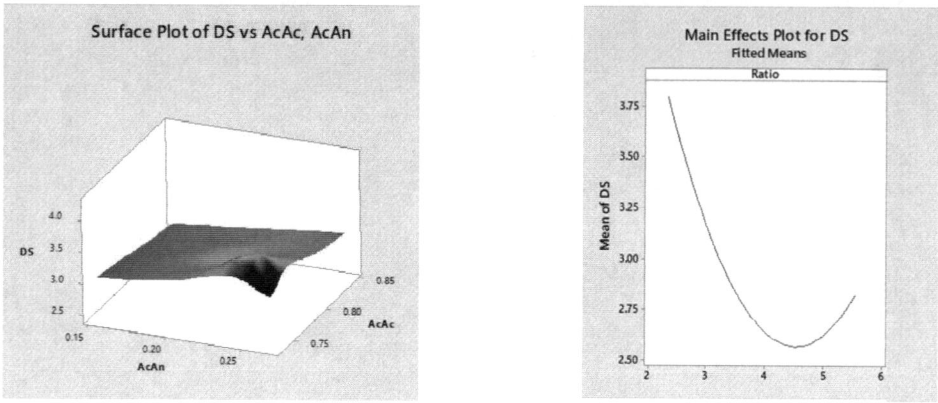

(a) (b)

Figure 4. (a) Response Surface Plot of DS as a function of acetic anhydride and acetic acid concentration. (b) Main effect plot for DS.

Figures 4a and 4b show how changing the proportions of glacial acetic acid and acetic anhydride affects isolated cellulose. As the ratio approaches the optimal value, the degree of substitution decreases due to the prevalence of side reactions with a high amount of acetic anhydride until the desired degree of substitution is reached. Subsequently, the degree of substitution increases indefinitely as the ratio of glacial acetic acid and acetic anhydride rises. Glacial acetic acid functions as a solvent that enhances cellulose swelling by increasing the accessibility of cellulose hydroxyl groups to the acetylating agent, acetic anhydride. Increasing the solvent's ratio significantly impacts acetylation, leading to a DS of 2.79 when using a 6:1 ratio of glacial acetic acid to acetic anhydride with 10 g of cellulose. Increasing the concentration of acetic anhydride or glacial acetic acid may lead to the breakdown of carbohydrates, resulting in a higher DS and lower biodegradability.

3.7. Effect of Acetylation time

The degree of substitution (DS) of cellulose acetate (CA) is affected by the acetylation time. There is a limit to the DS that can be achieved, as excessive acetylation can lead to degradation of the CA. Therefore, the acetylation time must be carefully controlled to achieve the desired DS without causing degradation. DS is the average number of acetyl groups per anhydro glucose unit in CA. Figure 3 shows the main effect plot of acetylation time to the DS of CA.

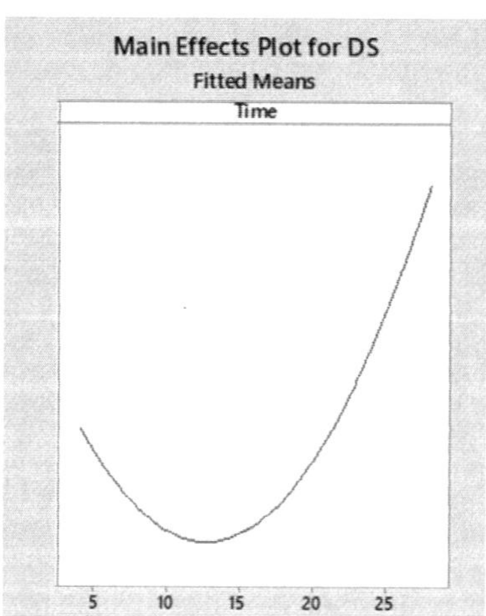

Figure 5. Main effect plot for DS.

Studies suggest that the longer the acetylation reaction occurs, the higher DS will be. This is because longer acetylation times allow for more acetyl groups to be added to the cellulose acetate molecules which is evident from the plot. However, figure 3 shows a different approach. The longer the acetylation time, the lower the DS until it reaches the optimum value. Further acetylation will then result in an indefinite increase of DS. This result shows an effect from the second variable – solvent ratio. The desired value of DS of 2.47 was achieved at about 16 hours of acetylation.

3.8. Optimization and validation of the model

The model in this study was optimized, and the response surface analysis conducted with Design Expert software revealed that the predicted optimal degree of substitution (DS) for cellulose acetate (CA) is 2.56. This is achieved with an acetic acid to acetic anhydride ratio of 4.4558, a reaction time of 15.20 hours, and a temperature of 50°C.

Additional tests were conducted to confirm the optimization outcome derived from the response surface analysis. The discrepancy between the experimental data and the expected value was less than 5%, likely due to inherent experimental mistakes during bench work. Figure 3 displays the contrast between the anticipated and experimental values of the optimization.

Table 3. Validation of the predicted model data

	AcAc/AcAn ratio	Acetylation time (hr)	Predicted DS	Experimental Ds
Model	4.558	15.20	2.56	-
Experimental	4.1	16	-	2.47
Error %			3.64%	

4. Discussion

The cellulose acetate (CA) produced had a DS of 2.47 and a 39.98% acetyl group content. The DS result confirmed the importance of the acetylation duration and acid to solvent ratio as previously suggested. Increasing the amount of acid to solvent enhanced the accessibility of the cellulose chain, hence improving the efficiency of the activation step in the acetylation process. The enhanced activation of cellulose in the acetylation process's second phase was a result of the greater accessibility of acetic anhydride to cellulose's free hydroxyl groups. The activation phase is essential for the production of CA with a DS of 2.0 or above [4].

In Addition, the DS of CA was indeed affected by varying the ratio of acetic acid and acetic anhydride used in the reaction. Increasing the molar fraction of mixed anhydride and the concentration of acetic anhydride can promote esterification and increase the DS [5][6]. The optimum ratio of acetic acid and acetic anhydride for the synthesis of CA was found to be 4:1 under the acetylation time of about 16 hours. However, the DS does not always have a direct correlation with the crystallinity index (Ic) of CA. The re-action time and temperature, as well as the carboxylic content of oxidized cellulose, can also affect the DS of acetyl groups in CA. The presence of hemicelluloses in the cellulose structure can also impact the DS. Additionally, the resistance of cellulose fibers to acid hydrolysis can affect the determination of the DS [7]. The effects of time, temperature, and concentration of acetic anhydride in the reaction have been studied to deter-mine the acetylation of cellulose derivatives [9].

The drop in the band detected at 3400 cm−1 in Figures 4a and 4b is due to the replacement of free hydroxyl (OH) groups by acetyl groups, as shown by the stretching vibrations of OH groups in cellulose. During the acetylation process, the free OH groups present in cellulose undergo a chemical re-action with acetic anhydride, resulting in the substitution of the OH groups with acetyl groups. The acetyl groups introduce new functional groups into the cellulose structure, altering its chemical properties and leading to modifications in its Fourier transform infrared spectroscopy (FTIR) spectrum. Consequently, the decrease in the band at 3400 cm−1 provides evidence of the successful acetylation process and the incorporation of acetyl groups into the cellulose structures [10].

In the same manner, the rise in the band observed at 1750 cm−1 signifies the stretching vibrations of the C=O bond in the ester carbonyl group, which is a characteristic feature of the acetyl groups incorporated into the modified cellulose structure. This rise in intensity indicates an increase in the number of ester carbonyl groups resulting from the acetylation process. The presence of the acetyl groups is confirmed by the absorption of infrared radiation at this specific wavelength. This rise in the band at 1750 cm−1 provides clear evidence of the successful formation and integration of acetyl groups into the cellulose, highlighting the chemical modifications that occurred during the acetylation reaction [10].

Likewise, the decrease in the band observed at 1650 cm−1 corresponds to the absorption of water molecules. This decrease in intensity suggests a reduction in the presence of water in the sample. Water molecules typically exhibit absorption in the infrared region around 1650 cm−1 due to the bending vibrations of the O-H bonds. The decrease of this band indicates a decrease in the amount of water present in the sample, possibly due to the drying or purification processes employed. Consequently, the re-duction in the absorption band at 1650 cm−1 signifies a decrease in the presence of water molecules and suggests a drier or more purified state of the tested material [11].

The band's increase at 1360 cm−1 is due to the vibration of the C-H bond found in the acetyl groups inside the altered structure. This heightened intensity suggests an increase in the quantity of C-H bonds inside the acetyl groups due to the acetylation process. The existence of acetyl groups in the changed material is verified by the absorption of infrared light at a certain frequency [12].

The rise in the band at 1220 cm−1 indicates the vibration of the C-O bond, which connects cellulose with the acetyl group. This increase in intensity suggests an increase in the quantity of C-O bonds due to the creation of ester connections in the acetylation process. The C-O bonds linking the cellulose backbone to the acetyl groups are verified by the absorption of infrared light at a certain wavelength. The increase in the band at 1220 cm−1 indicates the successful formation of C-O bonds, demonstrating the chemical linkage between cellulose and the acetyl group in the altered structure [12-13].

5. Conclusions

The acetylation of isolated cellulose fibers is a complex process where acetic anhydride combines with the material in the presence of sulfuric acid as a catalyst and glacial acetic acid as a solvent. Several factors, such as reaction duration and the proportion of glacial acetic acid to acetic anhydride, significantly impact the acetylation process. Cellulose acetate synthesis was verified by Fourier transform infrared spectroscopy (FTIR) research, showing the chemical conversion of cellulose from water hyacinth into cellulose acetate (CA) due to acetylation.

The optimal degree of substitution (DS) for CA was achieved through experimental conditions involving a glacial acetic acid to acetic anhydride ratio of 4:1, a reaction time of 16 hours, and a reaction temperature of 50°C. The resulting DS value obtained was 2.47, with a deviation of 3.64% from the value predicted by the model. This demonstrates that acetylation of isolated cellulose can be a straightforward and cost-effective process for producing biodegradable cellulose acetate, which serves as an environmentally friendly alternative to fossil fuel-based materials in the polymer industry. CA can be further utilized to create films for packaging applications and water purification purposes.

Acknowledgements

The authors would like to express a heartfelt gratitude to the National Research Council of the Philippines (NRCP) for their generous funding and support throughout the duration of this project. The financial assistance provided by the NRCP has not only enabled the realization of our research goals but has also allowed us to explore new frontiers of knowledge and contribute to the betterment of society.

Their recognition and endorsement of our project have been invaluable, providing us with the necessary resources and opportunities to push the boundaries of our field.

We would like to express our sincere thanks to the NRCP for their trust and belief in our research team. Their support has not only empowered us but has also inspired us to strive for excellence and make meaningful contributions to our respective fields of expertise. We look forward to continuing our partnership with the NRCP and working together to shape a brighter future through research and innovation.

References

[1] Cheng, H., Dowd, M., Selling, G., & Biswas, A. (2010). Synthesis of cellulose acetate from cotton byproducts. Carbohyrate Polymers, 80, 449-452,.

[2] Brodeur, G., Yau, E., Badal, K., & Collier, J. (2011). Chemical and Physicochemical Pretreatment of Lignocellulosic Biomass: A Review. Enzyme Res. Retrieved from https://www.ncbi.nlm.nih.gov/pmc/articles/PMC3112529/

[3] M. Niaounakis, "The Problem of Marine Plastic Debris," Management of Marine Plastic Debris, 2017.

[4] Candido, R., Godoya, G., & Gonc¸alves, A. (2017). Characterization and application of cellulose acetate synthesized from sugarcane bagasse. Elsevier. Retrieved September 29, 2022, from http://www.sciencedirect.com/science/journal/01448617

[5] Huang, Wu, Wang, Wolcott (2019). One-Step Activation and Surface Fatty Acylation of Cellulose Fibers in a Solvent-Free Condition. ACS Sustainable Chem. Eng., 19(7), 15920-15927. https://doi.org/10.1021/acssuschemeng.9b01974

[6] Tang, Wang (2013). Study on Esterification Characteristics of TEMPO-Oxidized Cellulose. AMR, (749), 49-53. https://doi.org/10.4028/www.scientific.net/amr.749.49

[7] Nakai, Demura (1972). Acid Saccharification of Cellulose Acetate. Agricultural and Biological Chemistry, 9(36), 1537-1541. https://doi.org/10.1080/00021369.1972.10860442

[8] Nandiyanto, A., & Ragadhita, R. (2019). How to Read and Interpret FTIR Spectroscope of Organic. Indonesian Journal of Science & Technology, 2-3. doi:http://dx.doi.org/10.17509/ijost.v4i1.15806

[9] Hu, C., Reddy, N., Yan, K., and Yang, Y. Acetylation of chicken feathers for thermoplastic applications. Journal of Agricultural and Food Chemistry, (2011), 59(19), 10517-1052

[10] Rodrigues Filho, G., Monteiro, D. S., Meireles, C. D. S., Assunc¸ ão, R. M. N., Cerqueira, D. A., Barud, H. S., et al. (2008). Synthesis and characterization of cellulose acetate produced from recycled newspaper. Carbohydrate Polymers, 73, 74–82

[11] Barud, H. S., Araújo Junior, A. M., Santos, D. B., Assunc¸ ão, R. M. N., Meireles, C. S., Cerqueira, D. A., et al. (2008). Thermal behavior of cellulose acetate produced from homogeneous acetylation of bacterial cellulose. Thermochimica Acta, 471, 61–69.

[12] Cao, Y., Wu, J., Meng, T., Zhang, J., He, J., & Li, H. (2007). Acetone-soluble cellulose acetates prepared by one-step homogeneous acetylation of cornhusk cellulose in an ionic liquid 1-allyl-3-methylimidazolium chloride (AmimCl). Carbohydrate Polymers, 69, 665–672

[13] Egot, M., & Alguno, A. (2018). Preparation and Characterization of Cellulose Acetate from Pineapple (Ananas comosus) Leaves. Trans Tech Publications. doi:10.4028/www.scientific.net/KEM.772.8

IOP Conf. Series: Materials Science and Engineering **1318** (2024) 012022 doi:10.1088/1757-899X/1318/1/012022

Effect of size and loading of waste single-used plastic (SUP) aggregates on a bio-based high density polyurethane composite

A C Tilendo[1,a], K M M Uy[2,b], J Q Filipinas[1,c], K M A Marohomsalic[1,d], J M Khalid[1,e], and N N Sultan[1,f]

[1] Chemical Engineering Department, Mindanao State University – Main, Marawi City, 9700, Lanao del Sur, Philippines
[2] Lesstics, Inc., Iligan City, 9200, Lanao del Norte, Philippines

[a]amierson.tilendo@msumain.edu.ph, [b]kennomichael.uy@gmail.com, [c]judilyn.filipinas@msumain.edu.ph, [d]kristinemarfe.marohomsalic@msumain.edu.ph, [e]khalid.jm70@s.msumain.edu.ph, [f]sultan.nn59@s.msumain.edu.ph

Abstract. The utilization of waste single-use plastic (SUP) as fillers by containment in a polyurethane matrix was investigated as an alternative solution to the detrimental disposal of these waste materials. High-density bio-based rigid polyurethane composites were prepared by adding polyol, isocyanate, and additives, along with densified SUP, which served as aggregate fillers to enhance the bulk density and flexural strength of rigid polyurethane while increasing the thermal conductivity. The aggregated SUPs were prepared using agglomeration-densification machine and exhibited size variations of 12.44 mm (A), 14.12 mm (B), and 15.56 mm (C). The resulting aggregates were added to the rigid polyurethane matrix at varied loadings of 40%, 50%, and 60% v/v. The green composites were cured in heated hydraulic press at constant temperature and pressure. Based on the obtained results, it was found that the formulation with a SUP size of C and a loading of 60% produced the composite of highest bulk density due to relatively high density of agglomerated SUP as the size enlarges. Additionally, the SUP size of A and a loading of 50% (50A) exhibited the highest flexural strength value. This behaviour was attributed to the increased filler-matrix interfacial area achieved with the smallest size aggregates. However, a further increase in aggregate loading resulted in a decrease in this property as common attribute for composite materials. On the other hand, it was revealed that increasing the loading of SUP aggregates provided a lesser insulating effect due to the increase in thermal conductivity resulting from the presence of filled pores within the polyurethane foam. The findings of this study contributed to understanding the effects of aggregate size and loading on the mechanical and thermal properties of the composite material, providing valuable insights for the development of energy-efficient and environmentally-friendly polyurethane structures.

1. Introduction

Plastic waste has emerged as a formidable environmental challenge due to its non-biodegradable nature. With the steady growth of the global population, the volume of plastic waste generated annually continues to escalate. In response to this crisis, recycling has surfaced as a pivotal solution for mitigating the accumulation of plastic waste. Transforming waste plastics into functional products not only curtails the quantity of discarded plastic but also taps into the potential of waste materials and industrial byproducts, such as recycled concrete aggregate, glass, ceramic, fly ash, and slag [1]. This trend holds the promise of enhancing sustainability while reducing the ecological burden of plastic waste.

However, integrating waste single-used plastics (SUP) into recycled materials and composites presents unique challenges. The conventional utilization of SUP as a filler in polymeric matrices has highlighted certain limitations. Research has demonstrated that the tensile strength of composite materials is directly correlated with the density of the filler material [2-4]. Yet, the incorporation of shredded SUP poses complications. The arrangement of SUP flakes within the polymeric matrix often leads to weak transverse strength and reduced interfacial bonding between the filler and matrix. These challenges are exacerbated by properties such as water permeability, low bulk density, and compromised mechanical integrity. To counter these limitations, densifying the shredded SUP fillers into aggregates is an innovative strategy that aims to reinforce interfacial adhesion and elevate the overall material properties of the resulting composites.

In this context, polyurethane (PU) matrices have garnered attention as versatile materials with a wide range of industrial applications. PU offers exceptional attributes such as robust abrasion resistance, long-term stability, and impressive load-bearing capabilities. These qualities, coupled with the potential for deriving PU from sustainable raw materials like vegetable oils, have spurred extensive research endeavors. This attraction to bio-based PU stems from its potential to revolutionize various industries, from foams to coatings and even structural panels.

The present research seeks to address the overarching objective of enhancing the properties of bio-based high-density PU foam. The central focus is on the incorporation of densified waste SUP aggregates. This study also delves into the influence of filler size and loading on key material characteristics such as bulk density, flexural strength, and thermal conductivity. By exploring the potential of this novel approach, the research strives to contribute to both environmental sustainability and material innovation.

2. Materials and Methods

2.1. Materials

Raw materials needed for this study were waste SUP aggregates with varied sizes (12.44 mm, 14.12 mm, 15.56 mm) previously densified using agglomeration-densification machine have density of 0.919 g/cm^3, 0.921 g/cm^3, and 0.923 g/cm^3, respectively. Synthesized soybean oil phosphate ester polyol (SOPEP) has OH No. 200-250 and acid No.<5. Polymeric methylene diphenyl diisocyanate (PAPI™ 27 from Dow Chemical, has an NCO functionality and content of 2.7 and 31.4 wt%, respectively), and defoamers (INV 690). The 15.4cm x 25.5 cm x 1 cm mold coated with release agent is prepared.

2.2. Procedures

2.2.1 Mixing proportions. The mixing reagents employed in this research exhibit distinct formulations dependent on the sizes and loading (40%, 50%, 60%) of SUP aggregates. A designated 40A (with 40% loading and 12.44 mm aggregate size), precise quantities of SOPEP and defoamer were meticulously measured using an analytical balance. The reagents were then combined within a plastic container and subjected to thorough mixing using a specialized mixer. Subsequently, isocyanate was introduced to the mixture and re-mixed before foam rises with NCO:OH ratio of 1.5:1.

2.2.2 Molding and curing. The molding process was initiated in the research study. The mold and an acetate film were both coated with oil to function as lubricating agents, preventing the composite material from adhering to the surfaces. Before the slurry was poured into the mold, it was carefully lined with a sheet of acetate film, ensuring easy release of the molded product. Another acetate film was placed over the upper surface of the mixture within the mold to complete the encapsulation. A curing period of 7 days was determined to be sufficient for the composite to achieve optimal strength and structural integrity. Following this curing duration, the composite was successfully removed from the mold, marking its readiness for subsequent phases of testing and evaluation.

2.3 Characterization and testing.
To achieve standardized measurements, established protocols were employed. The American Standard for Testing Material (ASTM D-790) was adopted as the benchmark for evaluating the flexural strength of the composite, a procedure conducted using the Universal Testing Machine Shimadzu AGS-X Series from Shimadzu Corp., based in Kyoto, Japan. For the determination of bulk density, the %open and close cell method (ASTM D1622-03) was utilized. The thermal conductivity of each sample was a key metric under investigation, with assessment conducted via a FOX 200 heat flow meter (Laser-Comp, Wakefield, MA). The procedure adhered to ASTM C518-2017 guidelines, with samples measuring 150 × 150 × 20 mm.

3. Results and Discussions

3.1. Bulk density
The bulk density of the prepared composite samples is shown in figure 1. Notably, the graph highlights a distinctive trend, with the composite labeled as 50C (characterized by 50% loading and 15.56 mm aggregate size) showcasing the highest bulk density value among the studied formulations. A comparative analysis across the three distinct sizes—A, B, and C—reveals that size C, possessing the largest dimensions, exhibits superior density and compaction when compared to sizes A and B. This observation can be attributed to the inherent density of individual fillers, which cumulatively contribute to the overall bulk density of the composite [5]. The interplay between size and inherent density evidently has a discernible impact on the bulk density of the composite materials.

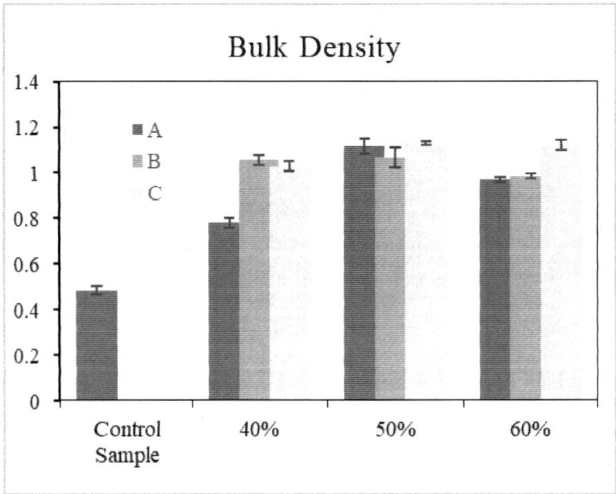

Figure 1. Bulk density of composites at varying sizes and loading of SUP aggregates

Of particular interest was the examination of the loading parameter's influence on bulk density. The results demonstrated a notable correlation between loading percentage and bulk density. Elevating the loading up to 50% (v/v) resulted in a discernible increase in bulk density. This observation suggests that the incorporation of supplementary material contributed to the increased density of the composite matrix. However, surpassing this 50% v/v threshold initiated a distinct trend, as further increases in loading began to minimally affect bulk density. This phenomenon can be attributed to a complex interplay between filler material incorporation and the resulting packing density within the composite

IOP Conf. Series: Materials Science and Engineering **1318** (2024) 012022 doi:10.1088/1757-899X/1318/1/012022

structure. The saturation effect observed beyond the 50% loading point implies that an optimal loading range exists, beyond which the composite's bulk density begins to decline. This delicate balance underscores the need for meticulous consideration of loading percentages to effectively harness the benefits of increased loading without compromising bulk density.

3.2. Flexural strength

The flexural strength of the sample after SUP was added has been determined in figure 2, figure 3, and figure 4. All composite samples have flexural strength higher than the control sample which strongly suggests that the filler has indeed improve the said property. An intriguing trend emerged, where the size of the filler material demonstrated a notable impact on flexural strength. Specifically, an increase in filler size led to a deterioration in flexural strength, a phenomenon evident in both the 50% and 60% loading scenarios. This reduction in flexural strength can be attributed to the diminished interactions between the filler and the matrix, resulting from the larger aggregate size.

Upon analyzing the flexural strength values at different loading levels, distinct patterns emerged. At a loading of 40%, the flexural strength ranged from 3.2 to 8.1 MPa. The 50% loading exhibited a more consistent range of 7.9 to 8.7 MPa, demonstrating excellent flexural strength across all filler sizes. Conversely, at a loading of 60%, the flexural strength spanned from 2.7 to 6.1 MPa, revealing a trend of diminished strength at higher loading proportions. Corroborative studies [6-7] demonstrated a similar trend, where an increased ratio of waste plastic inversely correlated with flexural strength. This concurrence reinforces the established relationship between the incorporation of waste plastic and its impact on the mechanical integrity of composite materials.

A notable observation was made in the case of the 50% loading across all filler sizes, where the composite exhibited exceptional flexural strength. This finding underscores the significance of optimal loading proportions in achieving favorable mechanical properties. Investigation into filler size effects unveiled that the smaller size A (12.44mm) SUP aggregates notably enhanced the flexural strength of the composite at 50% loading, resulting in a stress of 8.5205 N/mm². This outcome reflects the intricate interplay between aggregate size and mechanical performance.

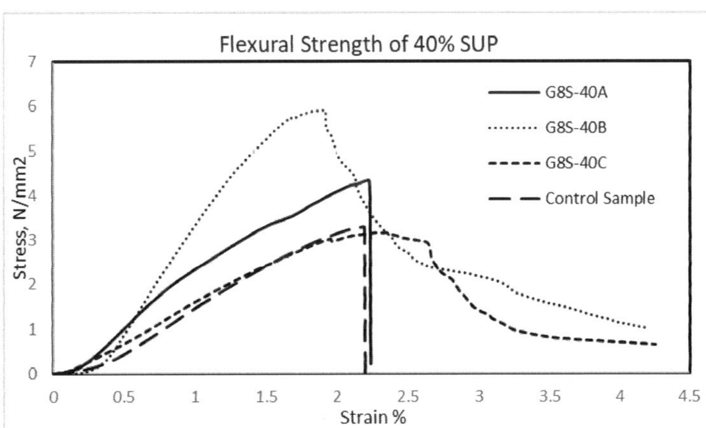

Figure 2. Flexural strength of composites at 40% at varying filler sizes

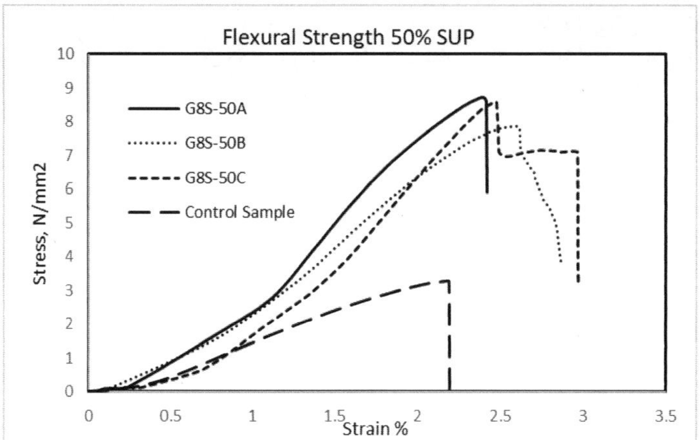

Figure 3. Flexural strength of composites at 50% at varying filler size

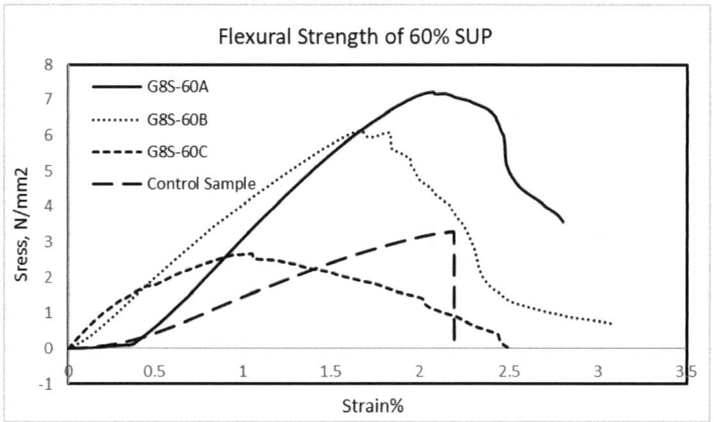

Figure 4. Flexural strength of composites at 60% at varying filler size

3.3. Thermal conductivity

Thermal conductivity serves as a crucial indicator of a material's propensity for heat transfer through its thickness. A higher thermal conductivity value is indicative of reduced insulation capability, as heat can more easily traverse the material. Figure 5 visually encapsulates the complex relationship between filler size, loading, and thermal conductivity. Generally, all composites exhibit an increase in thermal conductivity which explains the replacement of all air pockets and cell structures by the filler, thereby, facilitating heat transfer [8]. Notably, altering the size of the filler material yielded diverse outcomes across distinct filler loading levels. However, a small trend emerged with increasing filler loading: the average thermal conductivity of the composite consistently rose. This phenomenon can be attributed to the effective replacement of void spaces within the foam's cell structures and pores by the densely packed filler material, resulting in diminished voidage as the filler loading escalated.

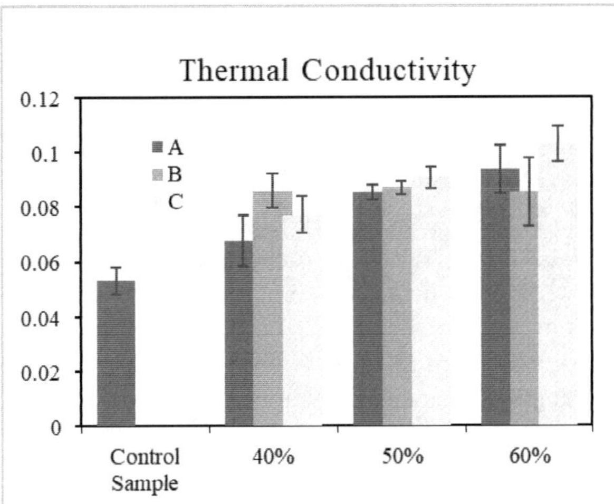

Figure 5. Thermal conductivity of composite at varying sizes and loading

4. Conclusion

In this study, SUP underwent a densification process and was subsequently incorporated as aggregate fillers into the composite. This integration was carried out with variations in both loading and size. The investigation encompassing bulk density, flexural strength, and thermal conductivity of the prepared composite samples shed light on the pivotal roles of size and loading in determining these essential material properties. Through evaluation of composite performance, a noteworthy conclusion emerges: the composite formulation designated as 50A (comprising filler size A and loading of 50%) demonstrates a harmonious blend of properties. This composite formulation boasts several remarkable features. It exhibits one of the highest bulk densities and flexural strengths, indicating its robust mechanical characteristics. Simultaneously, this formulation maintains minimal thermal conductivity at the 50% loading level, highlighting its capacity for efficient thermal insulation. The distinctive combination of high mechanical performance and commendable thermal insulation attributes positions the composite with a filler size and loading of 50A as a promising contender for structural applications. This composite formulation presents an intriguing solution for diverse applications, including ceilings, roof tiles, and wall panels, where stringent mechanical demands must be met while concurrently providing crucial thermal insulation. This dual functionality is of paramount importance in addressing environmental concerns and striving for energy efficiency, all while adhering to cost considerations.

References

[1] A. J. Babafemi, B. Šavija, S. C. Paul, and V. Anggraini, “Engineering properties of concrete with waste recycled plastic: A review,” Sustainability (Switzerland), vol. 10, no. 11. MDPI, Oct. 25, 2018. doi: 10.3390/su10113875.

[2] A. Camann, K. Dragsbaek, S. Krol, J. Sandgren and D. Song, "Properties, Recycling and Alternatives to PE Bags," Worcester Polytechnic Institute, 2010.

[3] Z. Ismail and E. Al-Hashni, "Use of waste plastic in concrete mixture as aggregate replacement," Waste Manag., vol. 28, pp. 2041-2047, 2008.

[4] F. Alqahtani, G. Ghataora, M. Khan and S. Dirar, "Novel lightweight concrete containing manufactured plastic aggregate.," Constr. Build. Mater., vol. 148, p. 386–397, 2017.

[5] S. Yu, X. Li, Y. Zhao, and M. Zou, “A Novel Lightweight Polyurethane Composite for Application on Ultra-High-Voltage Insulator Core Filler”, Polymers (Basel), 12(11): 2737, 2020

[6] A. Topacio, M. Eng, R. C. Arkuino, K. V Romano, and J. C. Abutin, “Development of Hydrophobic Composite Roof Tiles Utilizing Recycled Plastic Materials,” 2018. [Online].

IOP Conf. Series: Materials Science and Engineering **1318** (2024) 012022 doi:10.1088/1757-899X/1318/1/012022

Available: https://www.researchgate.net/publication/328081623
[7] A. Ashraf, M. Hashim, and A. Professor, "Use of waste plastics as fine aggregates substitute in concrete" Int J Sci Eng Res, vol. 7, no. 4, 2016, [Online]. Available: http://www.ijser.org
[8] A. K. Singh, et al. Measurement of thermal conductivity of composites: A review. International Journal of Engineering Research and General Science, 2(3), 246-254, 2018

Effect of rubber surface pretreatment method on the rubberized precast concrete

Danielle Espiritu[1], Dale Tristan O. Garcia[1], Maridil Joy B. Isidro[1], Christine Jim I. Labadan[1], Maria Francesca S. Sumagui[1], Vince Harvey O. Alano[1*], and Erison C. Roque[1]

[1]Chemical Engineering Department, Adamson University, San Marcelino St., Ermita, Manila, Philippines, 1000

*vhoalano@gmail.com

Abstract. Disposal of waste tires is one of the environmental problems that require urgent attention. Their durability and potential for leaching out toxic chemicals have been a problem associated with disposal. This led many researchers to conduct studies on its potential for reuse. Among them is its potential as alternative to natural aggregates in concrete. However, rubber normally negatively affects the mechanical properties of concrete due to its hydrophobic nature resulting to poor adhesion with cement. In this study, the researchers explored the effects of different pre-treatment methods of rubber from waste tires to improve its hydrophilicity, its workability when mixed Type I Portland cement, and the mechanical properties of the resulting "rubberized" concrete. The pretreatment methods considered included: 1) soaking rubber in polyvinyl alcohol at different concentrations (8%, 10% and 12%); 2) soaking rubber in a mixture of sodium hydroxide, potassium permanganate, and saturated sodium bisulfite solution; and 3) exposure to UVA/B/C radiation. The rubber pieces from waste tires were first shredded into 0.3 - 0.5mm pieces prior to pretreatment. Hydrophilicity was measured through the contact angle test using an Optical Theta Lite Tensiometer. Workability of the concrete mix was measured through slump test (ASTM C143). The results show that rubber soaked in PVA 10% yielded the lowest contact angle (better hydrophilicity) and the most similar workability with normal concrete. Several concrete specimens using the pretreated rubber aggregates were then produced and cured for 7 days and 28 days before subjecting to several tests to determine the compressive, tensile, and flexural strengths. Pretreatment by soaking in PVA yielded concrete with mechanical properties comparable and in some instances better than normal concrete.

1. Introduction

The world is facing a global environmental issue concerning the disposal of waste automobile tires (Hooton et al., 2001). Millions of tires are discarded annually all over the world and more than 50% end up in landfills (Thomas et al., 2015). It is forecasted that by the year 2030, this number could increase to as high as 5 billion aggravating the environmental hazards it poses including the rapid depletion of available site for disposal and the toxic chemicals that can leach out of the waste tires upon prolonged exposure to weather and other elements (Xue & Shinozuka, 2013). Several reuse applications have been explored to mitigate this problem. These include reusing waste tires as playground equipment, erosion control, highway crush barriers, guard rail posts, noise barriers and as fuel in cement kilns and pulp and paper plants, but the application largely remains limited, and millions of tons of rubbers are still disposed in landfills (Toutanji, 1996).

Other economically and environmentally desirable methods of recycling waste tires are being explored. One such is the use of waste tires as additive to concrete (Batayneh et al., 2008). This has been the topic of several research. For instance, Khatib & Bayomy (1999) studied the possibility and practicality of employing recycled tire rubbers in Portland cement concrete. The potential of "rubberized" concrete as a structural material to enhance the dynamic performance and lessen seismic response of concrete structure was the focus of the study done by Xue & Shinozuka (2013). However, past research, such as that of Onuaguluchi & Panesar (2014), showed that the integration of rubber aggregates in concrete diminishes its mechanical properties particularly its compressive, tensile, and flexural strengths due to the hydrophobic properties of rubber inhibiting the binding to the cement.

Several research have been conducted to address this gap by pretreating the rubber aggregates before mixing with cement. The studies of Ma & Wang (2013) and Su (2015) showed a slight increase in mechanical strengths of the rubberized concrete using pretreated rubber. Nevertheless, the intended use of these rubberized concrete has still been limited to applications where strength is not a primary concern such as in energy and impact reduction (Khatib & Bayomy, 1999; Twumasi-Boakye, 2014). There are still not many studies that explored the use of rubberized concrete in the construction industry for load-bearing applications.

Hence, this study explored the effects of different pretreatment methods on the hydrophilicity of rubber aggregates and on the resulting concrete mix in terms of workability and mechanical properties i.e., compressive, tensile, and flexural strengths.

2. Materials and Methods

2.1. Reagents and materials

The chemicals used in this study were 98% sodium hydroxide (NaOH), 99% potassium permanganate ($KMnO_4$), 98% saturated sodium bisulfite ($NaHSO_3$), 90% sulfuric acid (H_2SO_4), polyvinyl alcohol (PVA) grade BP-24 (86-89 mol%), 99% deionized water (for preparing the solutions), and tap water.

The materials used for the concrete mix were Type 1 Portland cement, shredded waste rubber tires (0.3-0.5 mm), gravel, and sand.

2.2. Pretreatment of fine rubber aggregates

2.2.1. Pretreatment by soaking in NaOH, $KMnO_4$, and $NaHSO_3$

Fine rubber aggregates (FRA) were first soaked in 5% NaOH solution for 24 hours. The FRA were then rinsed using tap water and soaked in a solution of 5% $KMnO_4$ with constant stirring for 2 hours. The pH level of the solution was kept between 2-3 through the addition of H_2SO_4. The temperature was kept at 60°C. The rubber particles were then washed with tap water and soaked in saturated $NaHSO_3$ solution. The mixture was heated and kept at 60°C for one hour to promote sulphonation reaction. The FRA were then removed from the solution and dried in an oven for 6 hours at 100 °C.

2.2.2. Pretreatment by soaking in PVA

The FRA were first placed in a vacuum desiccator for 6 hours under 920 mbar. The FRA were then soaked in PVA solutions of different concentrations (8, 10 and 12% w/w) for 24 hours at room temperature (23 ± 2 °C). The FRA were then oven-dried at 60 °C for 24 hours.

2.2.3. Pretreatment by exposure to UV light

FRA were exposed to UVA/UVB/UVC radiation for eight minutes.

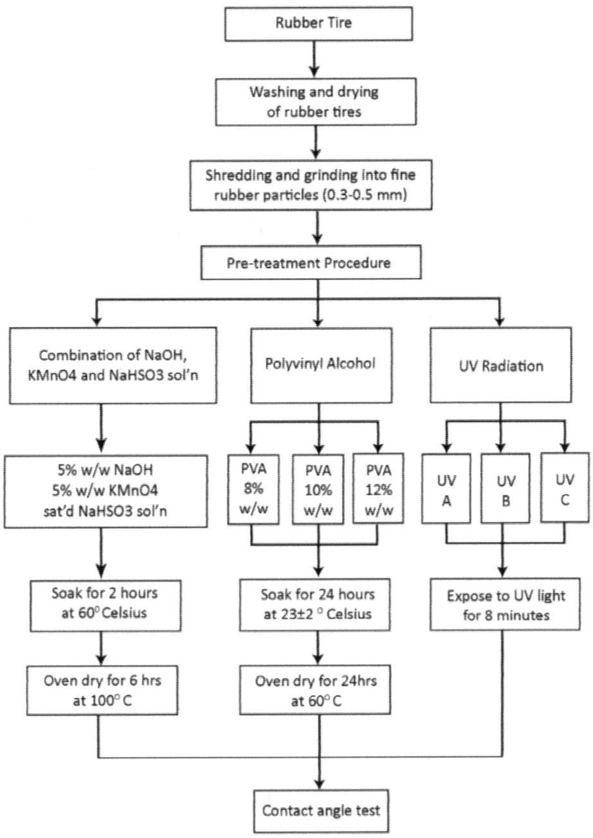

Figure 1. Pretreatment processes for the fine rubber aggregates from waste tire.

2.3. Analysis of pre-treated FRA

Hydrophilicity was measured via the sessile drop method using a tensiometer. Sessile drop method is used to determine the contact angle. Contact angle gives an indication of the hydrophilicity (< 90°) or hydrophobicity (> 90°) of the material. This is executed by placing a drop of liquid on the surface and measuring the angle at which the three phases meet. Images of the droplet of water on top of the rubber particle with the contact angle were taken as well as multiple readings for statistical analysis.

2.4. Preparation of rubberized concrete

The pre-treated FRA were mixed with Type I Portland cement, gravel, sand, and tap water. Three mixtures were prepared according to the ratios shown in Table 1: 1) normal concrete (Control), 2) concrete with untreated FRA (UFRAC), and 3) concrete with pre-treated FRA (PFRAC). For both UFRAC and PFRAC, 5-6% weight of the fine aggregates was replaced with the 0.3-0.5 mm FRA.

Table 1. Proportions of concrete components for different mixes

Mixture	Description	Cement[1]	Water[1]	FRA[1]	Fine aggregate[1]	Coarse aggregate[1] (< 25 mm)
Control	Control	380	190	0.0	858	927
UFRAC	Replacing 5% by weight rubber particle for fine aggregate	380	190	46.4	812	927

Mixture	Description	Cement[1]	Water[1]	FRA[1]	Fine aggregate[1]	Coarse aggregate[1] (< 25 mm)
PFRAC	Replacing 5% by weight rubber particle for fine aggregate	380	190	46.4	812	927

The concrete specimens were casted using conventional rod-and-hammer compaction method. The mold was made of cut PVC pipe (4 inches diameter) with an attached pipe clamp to remove the concrete without causing damage.

The test specimens were stored in a room with at least 90% relative humidity at temperature of 27 ± 2°C. After 24 hours, all the specimens were removed from the mold. The specimens were then cured in a water bath with an average water temperature of 23 ± 2°C. Water was replaced every 7 days of the total 28 days of curing.

A total of 162 specimens were produced: 126 PFRAC specimens, 18 UFRAC specimens, and 18 control specimens were produced.

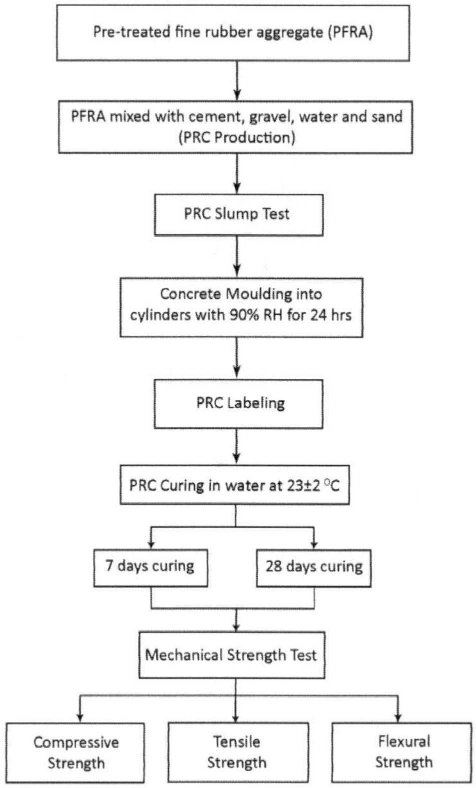

Figure 2. Preparation of concrete specimens.

2.5. Testing of concrete specimens

The concrete specimens were divided into two batches. The first batch was tested after seven days of curing and the second batch was tested after 28 days of curing.

2.6. Slump test of the fresh concrete mixture

The slump test is the most common method for determining the fresh concrete flow properties and its workability. The test was adopted was based on ASTM C143, standard test method for slump of hydraulic-cement concrete. Samples were taken from the fresh concrete mix prepared prior to precasting of the concrete.

The concrete was compacted and molded in a frustum of a cone. The cone had a base diameter of 20 cm, a top diameter of 10 cm and 30 cm height. The tamping rod used has the following specifications: 60 cm length and 16 mm diameter. The test was conducted under free vibration condition. The slump of the concrete was then recorded as the vertical distance between its original position at the top center surface and the displacement.

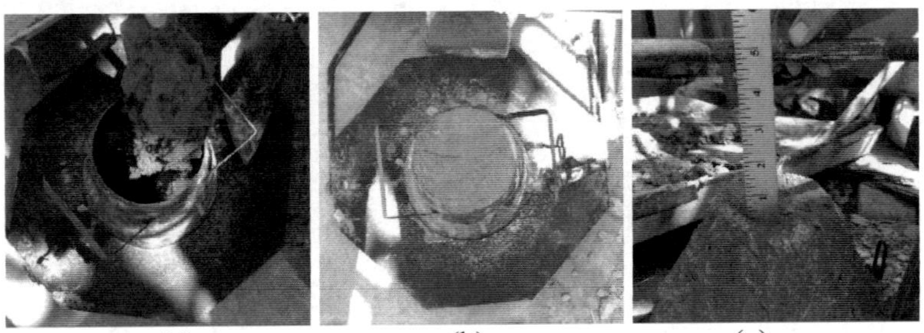

(a) (b) (c)

Figure 3. Slump test: (a) loading of concrete into the cone, (b) molding of concrete, and (c) measurement of displacement.

2.7. Mechanical strength test

A Universal Testing Machine with a maximum load capacity of 1000 KN was used to determine the mechanical strength of the specimens. Compressive, tensile, and flexural strengths were examined based on the standards shown in Table 2.7.

Table 2. Standard test methods

Parameter	Test standards
Compressive strength	ASTM C39
Tensile strength	ASTM C496
Flexural strength	ASTM C78

Each test was carried out in three replicates. The ultimate load value at which the specimen was able to handle before the onset of failure was recorded.

2.7.1. Compressive strength test

ASTM C39 was applied in the testing of the compressive strength as the cylindrical concrete specimens have a unit weight above 800 kg/m^3. The test included applying axial load to the specimen until failure occurs. Compressive strength was determined by dividing the maximum load by the cross-sectional area of the specimen. Rate of loading corresponded to a stress rate on the specimen of 35$\pm$ 7 psi/s (0. 25 $\pm$ 0.005 MPa/s).

2.7.2. Splitting tensile strength test

The test applied a diametral compressive force along the length of a cylindrical-shaped concrete continuously with no shock present at a constant rate of 100-200 psi/min or 0.7-1.4 MPa/min splitting tensile stress until failure had been observed. Load applied along the length of the cylindrical specimen was distributed with the help of plywood bearing strips. Splitting tensile strength was obtained by calculating the maximum load applied divided by the proper geometrical factor.

2.7.3. Flexural strength test

ASTM C78 covers the measurement procedure of flexural strength by means of a beam sample specimen having a third-point loading. Obtained results using this test method are also referred to as the modulus of rupture. Load was applied at a constant rate between 125 – 175 psi/min or 0.86 -1.21

IOP Conf. Series: Materials Science and Engineering **1318** (2024) 012023 doi:10.1088/1757-899X/1318/1/012023

MPa/min while elevating the extreme fiber stress up until rupture occurs. The loading rate is calculated using Equation 1:

$$r = \frac{Sbd^2}{L} \quad \textit{Equation 1}$$

Where:

r loading rate, MN/min
S rate of increase in extreme fiber stress, MPa/ min
b average width of the specimen, mm
d average depth of the specimen, mm
L span length, mm

The calculation of modulus of rupture depended on where the fracture in the specimen occurred, whether within the middle third of the specimen length (Equation 2) or outside the middle third of the specimen length by not more than 5% (Equation 3).

$$R = \frac{PL}{bd^2} \quad \textit{Equation 2} \qquad R = \frac{3Pa}{bd^2} \quad \textit{Equation 3}$$

Where:

R modulus of rupture, psi, or MPa,
P maximum applied load indicated by the testing machine, lbf, or N,
L span length, in., or mm, b = average width of specimen, in., or mm, at the fracture, and
a average distance between line of fracture and the nearest support measured on the tension surface of the beam, in., (or mm).
b average width of the specimen, mm
d average depth of specimen, in., or mm, at the fracture.

3. Results and Discussion

3.1. Contact Angle Test

Of the seven concrete mixes that used pre-treated rubber as fine aggregate, five were noted to have mean contact angles lower than that of the untreated rubberized concrete (URC) as shown in Table 3. This indicates that these pre-treatment methods have indeed increased the rubber's hydrophilicity. Of these 5, PVA 10% showed the lowest mean contact angle at 79.6°, a significant reduction in the mean contact angle compared with that of the URC (at 110.1°). Contact angle less than 90° indicates that the surface of the material had greater capacity to bind with water and other fluid in the mixture. Hydrophilicity of the rubber particles is favorable to the concrete mix since one of the leading causes of the decline in mechanical properties of rubberized concrete is the poor adhesion of rubber to the cement paste (Sweitzer, 2017).

COMBI gave the highest contact angle value among all trials at 127.2°. The mean contact angles of COMBI, PVA 8%, PVA 12%, and UVB are similar with that of the URC suggesting that the pre-treatment did not bring any significant increase in hydrophilicity. UVA and UVC showed considerable reduction in contact angles but remain higher than 90°.

Table 3. Contact angle test results for different concrete specimens

TRIAL	URC	PVA 8%	PVA 10%	PVA 12%	COMBI	UVA	UVB	UVC
1	118.1	124.0	63.0	112.3	127.2	99.9	94.2	99.5
2	105.4	117.5	76.1	117.5	96.8	95.0	119.8	96.2
3	106.9	107.1	99.6	103.0	105.3	94.9	111.2	85.3
MEAN	110.1	116.2	79.6	110.9	109.8	96.6	108.4	93.7

IOP Conf. Series: Materials Science and Engineering **1318** (2024) 012023 doi:10.1088/1757-899X/1318/1/012023

3.2. Slump test

Table 4 presents the slump test values for the different concrete mixes. PVA 10% yielded values similar to normal concrete (NC) at 2.8 in. This means that in terms of workability, this rubberized concrete mix is behaving similarly to normal concrete. Among the other specimens, PVA 8% and UVA gave the largest values at 4.2 in and 4.1 in, respectively. Too high or too low slump measurement is an indicator of the potential undesirable workability of concrete. Other pretreatment methods (PVA 12%, COMBI, UVB, and UVC) also showed higher slump test values than NC and also even that of URC indicating that the pre-treatment had made the workability to be even less than using untreated rubber aggregates.

Table 4. Slump test results for different concrete specimens

TRIAL	NC	URC	PVA 8%	PVA 10%	PVA 12%	COMBI	UVA	UVB	UVC
1	2.5	3.1	4.2	2.6	3.5	4.2	3.7	4.1	4
2	3	2.7	3.8	2.9	4.2	3.8	4.2	3.8	3.4
3	2.8	4	4.7	2.8	4.1	4	4.5	3.4	4.2
MEAN	2.8	3.3	4.2	2.8	3.9	4	4.1	3.8	3.9

3.3. Mechanical Properties

3.3.1. Compressive strength and load

Table 5 summarizes the test results of all three replicates done for all specimens. Figures 4.a and 4.b show the mean values of the compressive load and compressive strength for both 7-day and 28-day curing periods, respectively. As shown in Figure 4.b, the highest compressive strength observed for the 7-day curing period is PVA 12% at 15.9 MPa. All other rubberized concrete specimens, except for UVB, have the same or higher values than NC. At the 28-day period, NC had the highest value at 17.4 MPa followed by URC at 17.2 MPa. PVA 12% and PVA 10% are not far behind at 16.7 and 16.3 MPa, respectively. Of all the samples, UVA, UVC, and COMBI had the lowest values.

Table 5. Compressive strength test results

Specimen	Trial	Compressive Load (kN)		Compressive Strength (MPa)	
		7-days	28 days	7-days	28-days
NC	1	85.6	128.9	9.3	13.6
	2	108.8	179.0	11.2	19.2
	3	95.0	181.2	10.4	19.4
	Mean	96.5	163.0	10.3	17.4
URC	1	110.0	158.2	12.0	16.6
	2	123.2	202.7	13.5	21.7
	3	142.5	123.6	15.2	13.2
	Mean	125.2	161.5	13.6	17.2
PVA 8%	1	143.4	189.9	15.4	20.0
	2	117.0	124.4	12.8	13.3
	3	147.7	134.8	16.1	14.4
	Mean	136.0	149.7	14.8	15.9
PVA 10%	1	124.4	139.3	13.1	16.1
	2	156.9	142.9	16.5	16.5
	3	120.9	145.9	14.0	16.2
	Mean	138.9	142.7	14.5	16.3
PVA 12%	1	141.6	134.8	15.2	15.6
	2	148.3	179.4	16.8	18.9
	3	147.4	147.0	15.8	15.7
	Mean	145.8	153.7	15.9	16.7
COMBI	1	132.5	132.2	14.2	15.2

Specimen	Trial	Compressive Load (kN)		Compressive Strength (MPa)	
		7-days	28 days	7-days	28-days
	2	122.1	153.8	13.1	13.2
	3	137.2	142.7	15.3	13.0
	Mean	130.6	142.9	14.2	13.8
UVA	1	101.3	134.0	10.7	14.1
	2	103.1	121.1	10.8	13.0
	3	85.1	113.3	9.0	11.9
	Mean	96.5	122.8	10.2	13.0
UVB	1	97.9	172.8	10.9	18.5
	2	82.8	125.8	7.1	13.5
	3	95.5	147.0	10.2	15.7
	Mean	92.0	148.5	9.4	15.9
UVC	1	124.4	151.2	13.1	14.1
	2	156.9	124.5	16.5	11.6
	3	120.9	145.8	14.0	13.3
	Mean	134.1	140.5	14.5	13.0

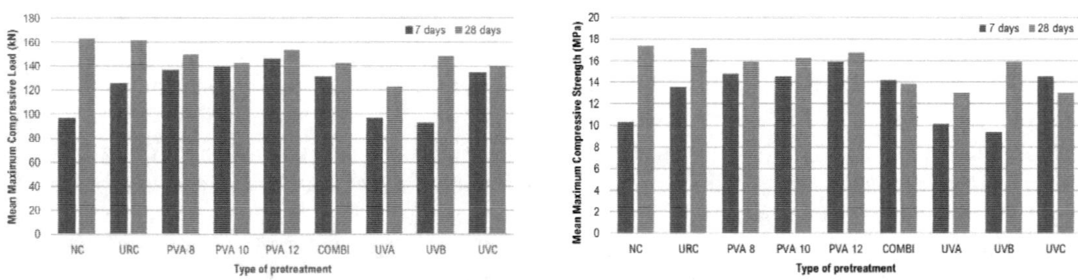

(a) (b)

Figure 4. Compressive strength test results for different concrete specimens: (a) mean maximum compressive load in kN, (b) mean maximum compressive strength in MPa.

3.3.2. Tensile Strength and Load

The results of the test show that the mean maximum tensile strength of both URC and PVA 8% are close to that of NC with difference of only 14% and 6%, respectively. The tensile strength values of all the other specimens are significantly lower than NC. PVA 10% and PVA 12% had significantly lower tensile strength than PVA 8% suggesting a reduction in tensile strength as the PVA concentration is increased. For the UV-treated samples, UVA, UVB, and UVC all exhibited lower tensile strength values than NC and URC. Combi had the lowest 28-day strength among all specimens.

Table 6. Tensile strength test results

Specimen	Trial	Compressive Load (kN)		Compressive Strength (MPa)	
		7-days	28 days	7-days	28-days
NC	1	16.1	79.6	0.5	8.5
	2	30.4	50.0	1.1	5.4
	3	23.1	71.2	0.8	7.6
	Mean	23.2	66.9	0.8	7.2
URC	1	21.0	58.0	0.7	6.3
	2	23.8	57.0	0.8	6.1
	3	19.3	59.3	0.7	6.4
	Mean	21.4	58.1	0.7	6.3
PVA 8%	1	38.8	49.6	4.3	5.2
	2	58.5	77.4	6.3	8.8

Specimen	Trial	Compressive Load (kN)		Compressive Strength (MPa)	
		7-days	28 days	7-days	28-days
	3	37.1	54.7	4.3	6.2
	Mean	44.8	60.6	5.0	6.7
PVA 10%	1	23.3	31.3	0.8	3.1
	2	22.0	49.4	0.7	5.2
	3	22.1	41.7	0.8	4.5
	Mean	22.0	40.8	0.8	4.3
PVA 12%	1	57.2	37.4	6.1	4.2
	2	64.2	34.6	7.3	5.9
	3	35.9	33.8	3.9	3.8
	Mean	52.4	35.2	5.8	4.6
COMBI	1	37.3	42.3	4.0	4.3
	2	38.6	40.5	4.1	4.3
	3	36.4	46.7	4.1	4.0
	Mean	37.5	43.2	4.1	4.2
UVA	1	26.0	58.0	0.9	6.1
	2	23.6	40.3	0.8	4.2
	3	20.1	28.4	0.7	3.0
	Mean	23.2	42.2	0.8	4.4
UVB	1	18.0	45.0	0.6	4.8
	2	10.3	29.9	0.3	3.3
	3	15.7	54.7	0.5	5.9
	Mean	14.7	43.2	0.5	4.7
UVC	1	16.9	48.4	0.6	4.4
	2	18.6	48.7	0.6	5.6
	3	14.1	45.9	0.5	5.0
	Mean	16.5	47.6	0.6	5.0

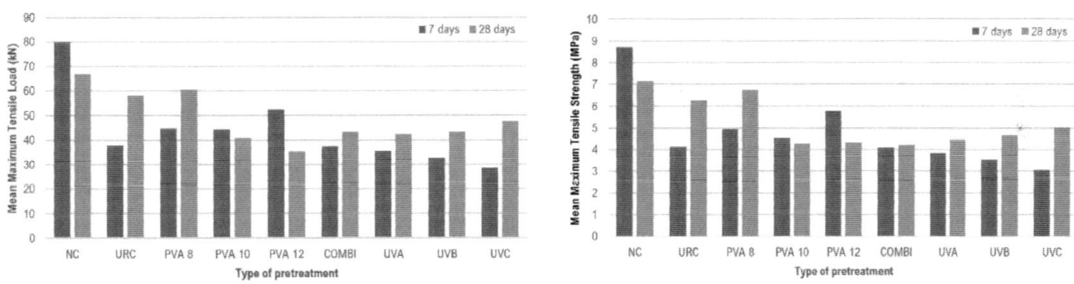

(a) (b)

Figure 5. Tensile strength test results for different concrete specimens: (a) mean maximum tensile load, (b) mean tensile maximum strength.

3.3.3. Flexural Strength and Load

The results show that all specimens have higher 28-day flexural strength values than NC. Among all the samples, PVA 12%, PVA 10%, UVA, and UVB exhibited significantly higher values than NC by 65%, 57%, 54%, and 53%, respectively. PVA 8% and COMBI both have values similar to that of NC at about 10% difference only. Based on the results, the flexural strength of concrete was reduced after the 28 days curing period in the case of NC, URC, PVA 8%. For the rest of the samples, the flexural strength increased. This was more pronounced in the case of UVB and UVC where the flexural strength grew by 73% and 35%, respectively, from the 7-day curing period.

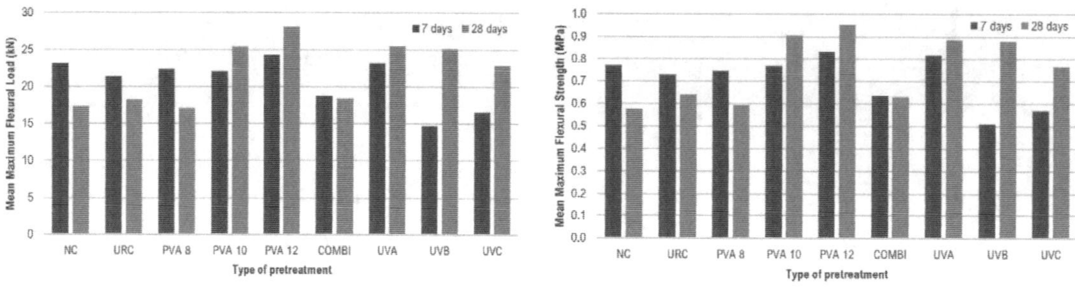

(a) (b)

Figure 6. Flexural strength test for different concrete specimens: (a) mean maximum flexural load, (b) mean maximum flexural strength.

Table 7. Flexural strength test results

Specimen	Trial	Compressive Load (kN)		Compressive Strength (MPa)	
		7-days	28 days	7-days	28-days
NC	1	16.1	16.0	0.5	0.5
	2	30.4	18.3	1.0	0.6
	3	23.1	17.6	0.8	0.6
	Mean	23.2	17.3	0.8	0.6
URC	1	21.0	17.4	0.7	0.6
	2	23.8	18.2	0.8	0.6
	3	19.3	19.0	0.7	0.7
	Mean	21.4	18.2	0.7	0.6
PVA 8%	1	19.2	14.9	0.7	0.5
	2	24.2	16.5	0.8	0.6
	3	23.5	19.8	0.8	0.7
	Mean	22.3	17.1	0.7	0.6
PVA 10%	1	23.3	25.3	0.8	0.9
	2	22.0	21.0	0.7	0.7
	3	22.1	29.9	0.8	1.1
	Mean	22.0	25.4	0.8	0.9
PVA 12%	1	25.0	28.0	0.9	0.9
	2	24.2	31.3	0.8	1.1
	3	23.7	24.8	0.8	0.9
	Mean	24.3	28.1	0.8	1.0
COMBI	1	22.9	18.3	0.8	0.6
	2	16.1	19.3	0.5	0.7
	3	17.5	17.5	0.6	0.6
	Mean	18.8	18.4	0.6	0.6
UVA	1	26.0	26.0	0.9	0.9
	2	23.6	27.0	0.8	0.9
	3	20.1	23.6	0.7	0.8
	Mean	23.2	25.6	0.8	0.9
UVB	1	18.0	18.6	0.6	0.7
	2	10.3	24.5	0.3	0.9
	3	15.7	32.3	0.5	1.1
	Mean	14.7	25.1	0.5	0.9
UVC	1	16.9	27.6	0.6	0.9
	2	18.6	20.7	0.6	0.7
	3	14.1	20.4	0.5	0.7
	Mean	16.5	22.9	0.6	0.8

4. Conclusion

Using pretreated rubber from waste tires as fine aggregate had shown significant effects on the mechanical strength of the concrete. Several pretreatment methods were explored including soaking the rubber particles in polyvinyl alcohol of varying concentrations (8%, 10%, and 12%), soaking the rubber particles in a mixture of sodium hydroxide, potassium permanganate, and sodium bisulfite, and exposing the rubber particles to ultraviolet light (UVA/B/C).

Among these, pretreatment using PVA 10% solution yielded rubber with the least contact angle (mean value of 79.6°) indicating a reduction in the rubber's hydrophobicity (increased hydrophilicity). Increased hydrophilicity is desirable as it allows better binding with the cement. The result of the slum test also showed that the workability of PVA 10% has the same workability as normal concrete.

Based on the mechanical tests conducted on the concrete specimens, PVA pre-treatment yielded 28-day compressive strength, tensile strength, and flexural strength values comparable to normal concrete. In terms of compressive strength, PVA 12% gave the highest value among the pretreated samples being only 4% lower than NC. While URC had the closest value with NC (about 1% difference), the pretreatment using PVA had resulted to concrete with higher compressive strength even at the 7-day period. In terms of tensile strength, PVA 8% had the highest value among the pretreated specimens being 7% lower than NC. Similar to the compressive strength results, URC also had the closer value with NC although PVA 8% had 20% higher tensile strength at 7-day curing period.

For the flexural strength, PVA 10% and 12% outperformed NC significantly in the 28-day curing period by 57% and 65%, respectively. Both samples also outperformed URC which had values close to NC (11% higher than NC).

This study shows that the pretreatment of fine rubber from waste tires with PVA solution may yield concrete that can potentially be used as alternative to normal concrete. Pretreatment had been shown to increase the strength of the specimen at the 7-day curing period better than normal concrete. The use of rubberized concrete in applications that expose it to tension is still limited although this can be addressed with the use of metallic components, i.e., rebars. Flexural strength is one of the properties enhanced by pretreatment.

Further study is recommended to review the applicability of rubberized concrete in applications where thermal conductivity, sound absorption, electrical resistance, and other properties are required.

References

[1] Batayneh M K, Marie I, and Asi I 2008 Promoting the use of crumb rubber concrete in developing countries. *Waste Management* 28(11), 2171–2176

[2] Hooton R, Nehdi M, and Khan A 2001 Cementitious Composites Containing Recycled Tire Rubber: An Overview of Engineering Properties and Potential Applications. *Cement, Concrete and Aggregates* 23(1), 3

[3] Khatib Z K and Bayomy F M 1999 Rubberized Portland Cement Concrete *Journal of Materials in Civil Engineering* 11(3), 206–213.

[4] Ma X and Wang Z 2013 Effect of Ground Waste Concrete Powder on Cement Properties *Advances in Materials Science and Engineering* 2013, 1–5

[5] Onuaguluchi O and Panesar D K 2014. Hardened properties of concrete mixtures containing pre-coated crumb rubber and silica fume *Journal of Cleaner Production*, 82, 125–131

[6] Su H 2015 Properties of concrete with recycled aggrgates as coarse aggregate and as-received/surface-modified rubber particles as fine aggregate

[7] Thomas B S and Gupta R C 2016 A comprehensive review on the applications of waste tire rubber in cement concrete. *Renewable and Sustainable Energy Reviews* 54, 1323–1333

[8] Toutanji H A 1996 The use of rubber tire particles in concrete to replace mineral aggregates. *Cement and Concrete Composites* 18(2), 135–139

[9] Twumasi-Boakye R 2014 Ground Tire Rubber as a Component Material in Concrete Mixtures for Paving Concrete

[10] Xue J and Shinozuka M 2013 Rubberized concrete: A green structural material with enhanced energy-dissipation capability. *Construction and Building Materials* 42, 196–204

IOP Conf. Series: Materials Science and Engineering **1318** (2024) 012024 doi:10.1088/1757-899X/1318/1/012024

Preparation of Jackfruit *Artocarpus Heterophyllus* Seed Starch-Grafted Polyvinyl Acetate by Emulsion Polymerization

K M Almarez[1], B J S Calasanz[1], A M Coritana[1], D I C Jaraula[1], S T Malimban[1], E C Roque[1], J G Olay[1], A V Melendres[1]

[1]Department of Chemical Engineering, College of Engineering, Adamson University, 900 San Marcelino St., Ermita, Manila 1000, Philippines

E-mail:ariel.melendres@adamson.edu.ph

Abstract. Petrochemical-based materials are depleting for the past decades and alternative methods to replace them are being studied over the years. This includes a wide range of probable applications such as adhesive industry. Starch, in its native form, possesses inadequate bond strength and viscosity. The said properties limit the possible applications of starch thus improvement should be employed. Jackfruit seed (JFS) starch is a viable substitute for other starches that is commonly used in the industry. This research developed a starch grafted polyvinyl acetate formula that can be used as an alternative to non-renewable and food consumable resources. Parameters such as monomer/starch ratio, temperature, and amount of initiator were observed and how it affected the graft percent, shear strength, and viscosity of the samples. This process was conducted via emulsion polymerization using PPS as initiator and SDS as emulsifier, Ethanol was also utilized to wash away homopolymers. Each run was done under ambient conditions. The samples were triplicated in the software to ensure the data credibility. From the statistical analysis, it was concluded that all three parameters which are temperature, monomer/starch ratio, and initiator amount were all significant. The appearance of 1735 cm^{-1} band in FTIR spectrum of the grafted starch and the visible roughness in the spherical-shaped granules in SEM photographs indicated that the VAc has been successfully grafted onto jackfruit starch. DSC analysis showed that Tg of starch-g PVAc was shifted to a little higher value as compared to Tg of native jackfruit starch which suggests that material can be used in a wider temperature range.

1. Introduction

Numerous studies focused on finding alternative methods to replace petrochemical-based materials are on-going over the years. This encompasses a wide range of industries from pharmaceuticals to adhesives. One of the renowned procedures in this movement is incorporating starch in chemical-based polymers and vice versa (Li et al., 2014; Wang et al., 2013, 2012; Xu et al., 2016; Zhang et al., 2015; Meimoun et al., 2018). Many types of research delved into the effectiveness of starch-based adhesives in the past years since the adhesive strength of starch alone is not sufficient for commercial use (Sun et al., 2017; Gunorubon, 2012, Melendres et al, 2021). Starch, in its native form, possesses inadequate bond strength and viscosity. The said properties limit the possible applications of starch thus improvement should be employed. In the industry, the commonly used starch is obtained from various cereals as well as tubers or roots such as corn, wheat, rice, potatoes, and cassava/tapioca starch (Yashini et al., 2022).

Jackfruit seed (JFS) starch is a viable substitute for other source of starch. A considerable amount of starch is found in JFS (Kringel et al., 2020: Shetty et al., 2016). In the report of Madruga et al. (2014), the starch content of soft seeds is 92.8% and 94.5% for hard seeds. JFS starch utilization in different areas has been studied lately as a bioplastic, a disintegrant in dissolving tablets, and a paper adhesive (Lubis et al., 2017; Suryadevara, et al., 2017; Chu et al., 2015). Some studies about the extraction and application of jackfruit seed starch had been reported (Nguyen et al., 2022, Luciano et al., 2017). However, from past researches, modification of starch extracted from JFS through grafting with polyvinyl acetate by emulsion polymerization has not yet been conducted. Thus, the researchers in this report studied the modification of starch from *Artocarpus heterophyllus Lam* (jack fruit). In this starch is extracted from jackfruit seeds grafted with polyvinyl acetate by emulsion polymerization using the monomer vinyl acetate in combination with initiators.

2. Materials and Methods

2.1. Reagents and Raw Materials

Distilled water was purchased at a local convenience store. Vinyl acetate monomer (99%) and potassium persulfate (99%), were purchased from Belman Laboratories. Ethanol (98%) and sodium dodecyl sulfate (SDS) (98.5%) were purchased from Alysons' Chemical Enterprises, Inc. Jackfruit seeds were obtained from fruit merchants at the local market of Gahak, Kawit, Cavite.

2.2. Methodologies

2.2.1. Extraction of Jackfruit Seed Starch

After scrubbing and peeling, 200g of JFS and 200 mL of water were mixed in a blender. The slurry was filtered through a 200-mesh sieve, the filter cake was discarded. Combined filtrates added with distilled water were centrifuged for 5 minutes. The supernatant was discarded and the brown sediment on top was scraped off. The collected crude starch was cleaned by washing with distilled water until the water became clear. The starch was dried in an oven at 60 °C until a constant weight was achieved after which the it is ground.

2.2.2. Grafting of Starch with Polyvinyl Acetate

The reaction was conducted in a beaker with an overhead stirrer at about 450 rpm. JFS starch (10 g) and 100 mL of distilled water were mixed prior to the grafting reaction for 30 minutes at 60 °C. The reaction temperature was cooled to room temperature after starch gelatinization. 0.21 g of Sodium dodecyl sulfate (SDS) was added to the mixture, then followed by the initiator dropwise. Vinyl acetate (monomer) was added to the system so that the ml monomer/ml starch ratio of 0.7, 1.2, and 1.7 were obtained. The grafting reaction was held under different initiator quantity: 2g, 5g, and 8g and heated at different reaction temperatures: 80 °C, 90 °C, 100 °C. Once the reaction time is over, the solution was gradually mixed with ethanol until a homogenized mixture was obtained and was stirred for 30 minutes. The precipitate from the solution was filtered using a sintered glass funnel and washed with distilled water thereafter drying the material in a vacuum oven at 60 °C until a constant weight is reached. To determine the graft percent, the following formula is used.

$$G\% = \frac{wt\ of\ grafted\ polymer\ on\ starch}{wt\ of\ starch\ used\ in\ grafting} x100$$

2.2.3. Characterization of modified starch (starch-g PVAc)

The viscosity of the sample was measured using the Rotary Viscometer Model Series VR 300. The instrument was obtained from Adamson University.

2.2.4. Fourier Transform Infrared Analysis
Characterization of the sample with FTIR was conducted with Model Spectrum Two FT-IR Perkin 89392 which was obtained at Adamson University. This machine was used for the confirmation of the grafting process.

2.2.5. Scanning Electron Microscopy
FESEM instrument, Dual Beam Helios Nanolab 600i was used in obtaining the SAM micrograph of the samples. The samples were characterized for identification of changes in the structure and morphology before and after the grafting procedure. The samples were tested at the Advance Device and Materials Laboratory (ADMATEL).

2.2.6. Differential Scanning Calorimetry
Differential Scanning Calorimetry (DSC) analysis of samples was performed using Perkin Elmer DSC 4000. DSC was used in the analysis and determination of thermal stability. The samples were tested at the Advance Device and Materials Laboratory (ADMATEL).

2.2.7. Shear Strength Test
The shear stress of the samples was examined in accordance with the ASTM D906 method. Wood pieces were cut into sizes such as 3.5in x 1in x 1/16in. As a preparation for the analysis, these wood pieces were bonded under 0.5 MPa pressure for 24 hours, stored at 25 °C and 50% humidity for 48 hours. The wood samples were clasped with 2 flat plate grips (screw type) and were pulled at 10 mm/min shear rate. The shear strength of the material, σ_M, is calculated as follow:

$$\sigma_M = F_{max}/A$$

where: F_{max} is maximum failing load and A is the joint area of the two wood pieces.

3. Results and Discussions

3.1. Characterization of starch grafted PVAc.
Figure 1-a. shows the result of the FESEM imaging on the jackfruit seed starch grafted-PVac. The jackfruit seed starch appears to be of spherical shape with an average granule size of 6-13 μm. SEM analysis results of the jackfruit seed starch displays semi-round shape granules. The morphological structure of the jackfruit seed starch didn't have a uniform particle size due to the sieving process carried out at the extraction of seed starch. The effect of the grafting reaction on the adhesive structure was directly exhibited in the SEM photograph in figure 1-b at 1000x magnification. The SEM photograph shows roughness in the spherical-shaped granules. According to Zheng et al. (2017), these changes demonstrate the impact of the graft copolymerization of jackfruit starch with vinyl acetate.

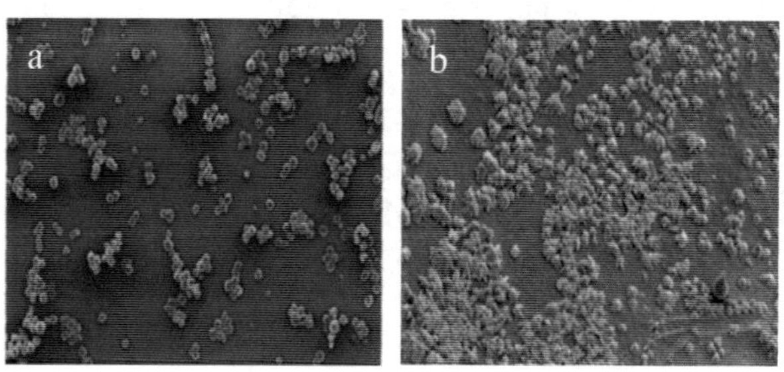

Figure 1. SEM Micrographs in 1000x resolution of a) jackfruit seed starch, b) starch-g PVAc.

The chemical structure of jackfruit seed starch and graft polymer was recorded on a Spectrum Fourier Transform Infrared (FTIR) using KBr pellets as the beamsplitter at the range of 550-4000 cm^{-1}. To demonstrate the grafting reaction's verification, the FTIR spectra for jackfruit seed starch grafted at

80, 90, 100°C with varying parameters were shown in figures 2-B, 2-C, and 2-D, respectively. The FTIR spectrum for raw starch is presented in figure 2-A where a characteristic band is shown between wavenumbers 3300 – 3500 cm^{-1} and is due to the frequency of O-H stretching. Approximately 2750 – 2850 cm^{-1} is designated for the saturated hydrocarbons, meanwhile, the glycosidic linkage (C-O-C bond) is between 1640-1650 cm^{-1} and C=O bond within the range 1700 – 1750 cm^{-1}. This additional peak affirms the existence of the Vinyl Acetate groups in the graft copolymer at all temperatures with varying parameters (Wang et al., 2012).

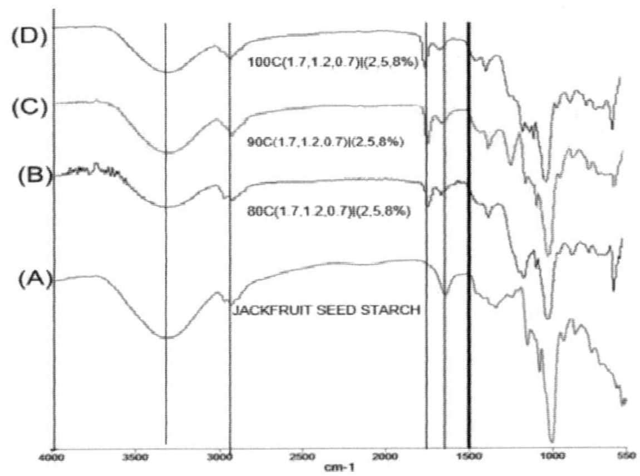

Figure 2. FTIR spectra of raw starch and starch-g PVAc at different treatments.

Figure 3 shows the result of DSC Analysis on the jackfruit seed starch grafted with PVAc compared with native starch in the study of Madruga (2014). The study presented observations on the glass transition temperature being 40°C. No endothermic peaks were observed from the range 30-80 C for JFS grafted PVAc. However, a slight curve can be observed from 70-75°C, which can be a glass transition of the polymer. According to the analysis results, the glass transition temp was found to be at 73.72 C and heat capacity to be at 0.049 J/g C. The comparison of the different endotherms in the graph shows that there is an improvement regarding starch's gelatinization temperature. The higher thermal stability suggests that material can be used in a wider temperature range.

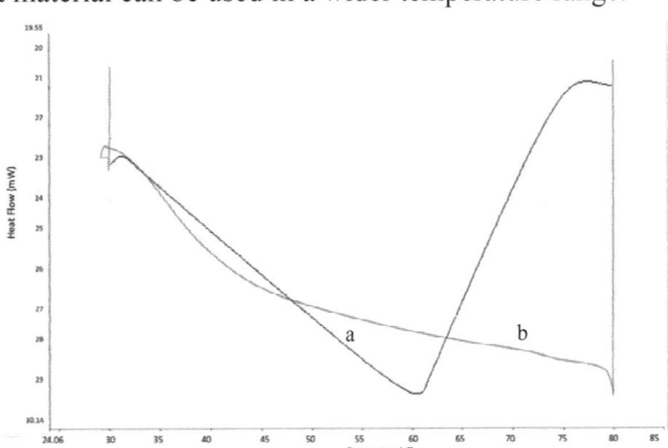

Figure 3. Comparison of (a) Native Starch (Madruga, 2014), (b) JFS grafted PVAc

IOP Conf. Series: Materials Science and Engineering **1318** (2024) 012024 doi:10.1088/1757-899X/1318/1/012024

3.2. Effects Monomer/Starch Ratio with grafting percent, shear strength and viscosity

The trend observed in figure 4-a for the grafting percent when the monomer/starch ratio was increased was increasing until a certain point, then started decreasing. In a similar study by Chu et al., 2015, it was observed that increasing monomer/starch ratio more than the optimum would cause a decrease in the amount of reactant ingredient in the polymerization medium. This causes a decrease in collision probability thereby decreasing the polymer yield percent. The effect of monomer/starch ratio was observed by conducting runs that have the other parameters held constant. The runs were done at temperature of 80°C and 5g initiator. In these experiments, the trends analyzed in a particular parameter were made sure that it is caused only by the change in monomer/starch ratio on the properties of starch-g PVAc. The effect of increasing monomer/starch ratio to the shear strength is shown at figure 4-b where it can be seen that the highest shear strength was achieved at 1.2 monomer/starch ratio and then started to decrease until 1.7. According to Zhang et al. (2015), shear strength decreases when the monomer/starch ratio was increased after reaching the optimum value due to the higher degree of homo-polymerization than the graft copolymerization. The effect of the monomer/starch ratio on the viscosity is shown in in figure 4-c. In this graph, the viscosity increases with increasing monomer/starch ratio. This result is similar to the result of Wang et al, 2017 and Wang et al., 2012. The continued increased in the solidification process is responsible for the increase in viscosity.

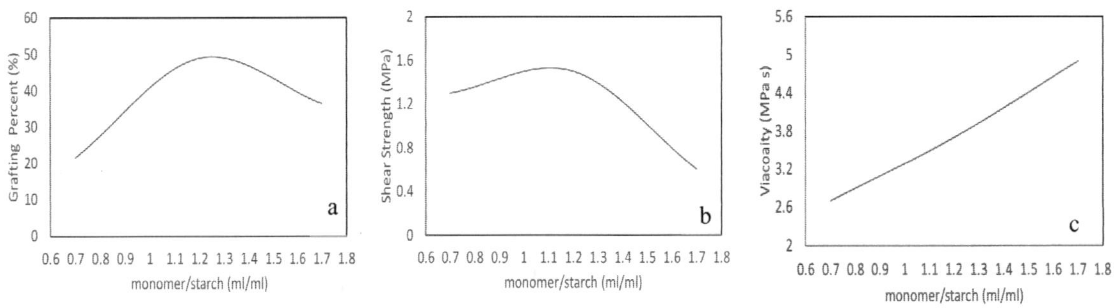

Figure 4. Effect of increasing monomer/starch ratio with a) grafting percent, b) sheart strength, and c) viscosity

3.3. Effects of Temperature with grafting percent, shear strength and viscosity

The effects of temperature on the properties of starch-g PVAc is shown in the following graphs. It can be seen in figure 5-a that increasing the temperature causes a slight increase in the grafting percent. According to some studies, temperature increases with increase in grafting percent (Mahto et al., 2014 and Nikolic, 2012). In these reports, the increase in temperature causes in increase in radical propagation rate thus increasing the grafting percent. In figue 5-b, the effect of temperature in the shear strength is shown where the result is seen as a decrease in the shear strength with increasing temperature. In a study conducted by Xu et al. (2016), the effects of temperature on the shear stress at the peak state were attributed to an increase in the magnitude of negative shear-induced pressure at the peak state with temperature. Increasing the temperature will cause an increase in the viscosity, as shown in figure 4-c. Furthermore, increasing the temperature will cause an increase in the mixture's attractive forces, which then increases the viscosity of the mixture. As the temperature is increased, this could cause increased grafted monomers, thus increasing the viscosity of the product formed (Zhang et al., 2015).

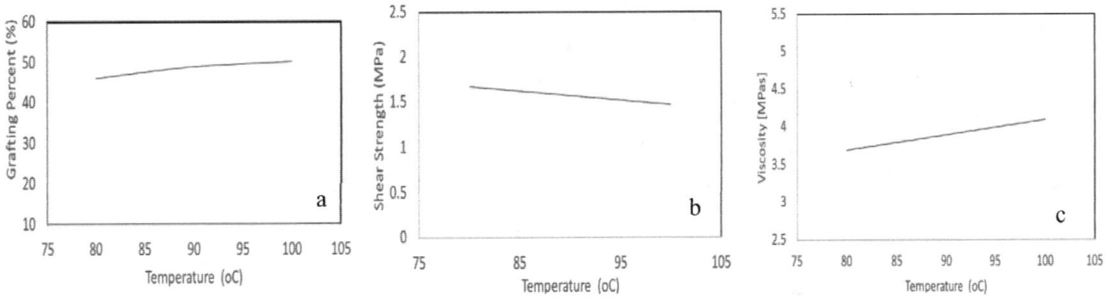

Figure 5. Effect of temperature with a) grafting percent, b) sheart strength, and c) viscosity

3.4. Effects of the amount of initiator with grafting percent, shear strength and viscosity

Table 1 describes the effect of the amount of initiator on the properties starch-g PVAc. The runs were done at constant monomer/starch ratio and temperature to make sure that the trend analyzed in the responses were caused only by the change in the amount of initiator.

Table 1. Effect of Initiator on grafting percent, shear strength and viscosity.

Amount of Initiator, g	Monomer/ Starch Ratio (ml/ml)	Temperature, °C	Graft Percent	Shear Strength	Viscosity
2	0.7	90	24.6	1.15	3.3
8			28.6	1.64	2.91
5	1.2	80	48.8	1.67	3.53
8			49.8	1.78	3.07
2	1.2	100	49.7	1.3	4.3
8			50.3	1.68	4.03
2	1.7	90	38.1	0.61	5.1
8			41.7	1.07	4.7

units: Graft Percent(%), Shear Strenght (Mpa), Viscosity (MPas)

As shown in Table 1, as the amount of initiator increases, grafting percent and shear strength increases at constant monomer/starch ratio and temperature. Also, the viscosity is observed to decrease when the amount of initiator is increased. Base on the study of Worzakowska (2016), the increase in the grafting percent is associated with the increase in the number of radical species on the starch and the polymerization medium, which acts in favor of grafting and thereby enhancing the polymer yield. As shown in the table. as the amount of initiator increases, shear strength increases which according to Mahto et al. (2014), the grafting percent increase has a great(2014), the grafting percent increase has a great effect on the product's shear strength. Higher grafting percent indicates that more PVAc is grafted onto the starch structure, which causes an increase in the shear strength. The higher the number of active sites created leads to more linkage points between the macromolecules to the adhering surface. The increase in the number of active sites caused by increasing the initiator amount resulted in shorter polymer chains causing deterioration in the product's viscosity due to the failure to entangle with the neighboring chains or adhesive surface (Wang et al, 2017). High viscosity is undesirable since it cannot spread easily. Low viscosity is preferred since its ability to flow easily will lead to a possible elimination, if not lessening of wastes (Wang et al., 2012).

4. Conclusion

The appearance of 1735 cm^{-1} band in the FTIR spectrum of the grafted starch indicates that the VAc has been successfully grafted onto jackfruit starch. Also, the visible roughness in the spherical-shaped granules in SEM photographs has demonstrated the impact of graft copolymerization of jackfruit starch with the vinyl acetate monomer. The value of Tg from DSC showed that the product can be used in a wider temperature range. It was observed that the highest graft percent was achieved at a monomer/starch ratio of 1.2. For the shear strength, it was determined that 1.2 monomer/starch ratio gave the highest value which implies that this level acquired higher graft copolymerization than homo-polymerization. By increasing monomer/starch ratio, the graft percent and shear strength increases up to the optimum level of 1.2 and then decreases abruptly. The increasing monomer/starch ratio increases the viscosity which indicates the increasing graft copolymerization and homo-polymerization Increasing the temperature of the sample increased the graft percent and viscosity while the shear strength decreased. It was observed that increasing the initiator amount, increases the graft percent and shear strength while viscosity decreases.

5. References

[1] Chu HJ, Wei HL, Zhu J. Ultrasound enhanced radical graft polymerization of starch and butyl acrylate. Chemical Engineering and processing: Process intensification. 2015 Apr 1;90:1-5.

[2] Gunorubon AJ. Production of cassava starch-based adhesive. *Research Journal in Engineering and Applied Sciences*. 2012;1(4):219-4.

[3] Kringel DH, Dias AR, Zavareze ED, Gandra EA. Fruit wastes as promising sources of starch: Extraction, properties, and applications. *Starch-Stärke.* 2020 Mar;72(3-4):1900200.

[4] Li, Z., Wang, J., Cheng, L., Gu, Z., Hong, Y., & Kowalczyk, A. (2014). Improving the performance of starch-based wood adhesive by using sodium dodecyl sulfate. *Carbohydrate polymers*, *99*, 579-583.

[5] Luciano CG, Landi Franco CM, Ayala Valencia G, do Amaral Sobral PJ, Freitas Moraes IC. Evaluation of extraction method on the structure and physicochemical properties of starch from seeds of two jackfruit varieties. *Starch-Stärke.* 2017 Nov;69(11-12):1700078

[6] Madruga MS, de Albuquerque FS, Silva IR, do Amaral DS, Magnani M, Neto VQ. Chemical, morphological and functional properties of Brazilian jackfruit (Artocarpus heterophyllus L.) seeds starch. *Food chemistry. 2014* Jan 15;143:440-5.

[7] Mahto D, Rani P, Mishra S, Sen G. Microwave assisted synthesis of polyacrylamide grafted soya peptone and its application as water soluble adhesive. *Industrial Crops and Products.* 2014 Jul 1;58:251-8.

[8] Meimoun J, Wiatz V, Saint-Loup R, Parcq J, Favrelle A, Bonnet F, Zinck P. Modification of starch by graft copolymerization. *Starch-Stärke.* 2018 Jan;70(1-2):1600351.

[9] Melendres, A., Cruz, R. V., Espenilla, M. B., & Monsada, A. (2021). Absorbency and Corrosion Inhibition Property of Polyamide Epoxy and Superabsorbent Polymer Composite Material. *Chemical Engineering Transactions*, *86*, 1297-1302.

[10] Nguyen TK, That NT, Nguyen NT, Nguyen HT. Development of starch-based bioplastic from jackfruit seed. *Advances in Polymer Technology.* 2022 Mar 26;2022:1-9

[11] Shetty, M., Fernandes, C. M., Suraksha, P., Karkera, N., Hegde, K., Shet, V. B., & Rao, C. V. (2016). Production of eco-friendly adhesive from jackfruit seed, 4, 102–106.

[12] Sun J, Li L, Cheng H, Huang W. Preparation, characterization and properties of an organic siloxane-modified cassava starch-based wood adhesive. *The Journal of Adhesion.* 2018 Mar 21;94(4):278-93.

[13] Suryadevara V, Lankapalli SR, Danda LH, Pendyala V, Katta V. Studies on jackfruit seed starch as a novel natural superdisintegrant for the design and evaluation of irbesartan fast dissolving tablets. *Integrative medicine research.* 2017 Sep 1;6(3):280-91.

[14] Wang Y, Xiong H, Wang Z, Chen L. Effects of different durations of acid hydrolysis on the properties of starch-based wood adhesive. *International journal of biological macromolecules.* 2017 Oct 1;103:819-28.

[15] Wang Z, Li Z, Gu Z, Hong Y, Cheng L. Preparation, characterization and properties of starch-based wood adhesive. *Carbohydrate Polymers.* 2012 Apr 2;88(2):699-706.

[16] Worzakowska, M. (2016). Starch-g-poly (benzyl methacrylate) copolymers: characterization and thermal properties. *Journal of Thermal Analysis and Calorimetry*, *124*, 1309-1318.

[17] Xu, Q., Wen, J., & Wang, Z. (2016). Preparation and properties of cassava starch-based wood adhesives. *BioResources*, *11*(3), 6756-6767.

[18] Yashini M, Khushbu S, Madhurima N, Sunil CK, Mahendran R, Venkatachalapathy N. Thermal properties of different types of starch: *A review. Critical Reviews in Food Science and Nutrition.* 2022 Oct 28:1-24.

[19] Zhang, Y., Ding, L., Gu, J., Tan, H., & Zhu, L. (2015). Preparation and properties of a starch-based wood adhesive with high bonding strength and water resistance. *Carbohydrate Polymers*, *115*, 32-37.

[20] Zheng, X., Cheng, L., Gu, Z., Hong, Y., Li, Z., & Li, C. (2017). Effects of heat pretreatment of starch on graft copolymerization reaction and performance of resulting starch-based wood adhesive. *International journal of biological macromolecules*, *96*, 11-18.

Life Cycle Assessment of Blended Cement Product: A Case Study

B Martinez[1], R Demafelis[2], J Alcantara[2], A E Matanguihan[2], B Magadia[2], F J Echavez[3], S A Frogoso[3]

[1]Department of Chemical Engineering, Technological Institute of the Philippines, 363 P. Casal St. Quiapo, 1001, Manila, Philippines
[2]Department of Chemical Engineering, University of the Philippines – Los Baños, Batong Malake, Los Baños, 4031, Laguna, Philippines
[3]Holcim Philippines Inc., 7/F Two World Square McKinley Hill, Fort Bonifacio, Taguig City, 1634. Philippines

Abstract. Large-scale cement production has led to serious environmental issues since the process is energy and resource intensive. Moreover, the cement industry is one of the sources of anthropogenic CO_2 emissions. As a matter of fact, in the Philippines, the CO_2 emission increased by 88.23% in over a decade. Hence, this research conducted a life cycle assessment of blended cement products in a cement manufacturing facility in Bulacan, Philippines from cradle to gate. The research utilized functional units of one metric ton of clinker, and one metric ton of blended cement. For the life cycle assessment (LCA), SimaPro 9.3.0.3 software was employed, and the impact assessment was conducted using the IPCC 2021 Global Warming Potential (GWP) at a 100-year timeframe and the Cumulative Energy Demand Method. Results of the research showed that the baseline carbon footprint and cumulative energy demand for the clinker are 995 kg CO_2-eqv and 4.74 x 10^3 MJ, respectively. Also, the baseline carbon footprint and cumulative energy demand for the blended cement are 760 kg CO_2-eqv and 4.13 x 10^3 MJ, respectively. Further, the evaluation of the carbon footprint of clinker, and blended cement in this research is comparable with the results from similar studies, with slight variation on certain instances. The energy utilization in the production of clinker, and blended cement in this research is significantly lower than the result of other studies

1. Introduction

Cement is an essential construction material known to be emission and energy-intensive. In 2020, initial estimates suggested that the global cement production will reach 4.3 Gt [1]. The global production of cement is projected to surge in response to the escalating demands driven by population growth, urbanization, and expanding infrastructure development. However, it is vital to acknowledge that large-scale cement production, primarily due to its substantial energy consumption, generation of solid waste, and emission of harmful flue gases, poses a significant threat to our environment, potentially causing severe environmental deterioration. Hence, understanding the industry's environmental impact is crucial for achieving sustainable development goals [2].

In the production of cement, the majority of carbon dioxide (CO_2) emissions result from two primary sources: the calcination of limestone, and the burning of fossil fuels [3]. Half of these emissions arise

IOP Conf. Series: Materials Science and Engineering **1318** (2024) 012025 doi:10.1088/1757-899X/1318/1/012025

directly from the calcination process, 40% stem from fuel consumption, and the remaining 10% are indirect emissions originating from electricity generation. [4]. The CO_2 emission from the cement industry in the Philippines increased by 88.23% in over a decade; it recorded 9.6 megatons of CO_2 in 2019 [5]. Worldwide, the direct carbon dioxide (CO_2) emissions per unit of cement production increased by an annual rate of 1.8% between 2015 and 2020. Nevertheless, there has been a promising trend in the cement industry where the intensity of thermal energy and electricity consumption has shown a consistent decline. This positive shift can be attributed to the adoption of more efficient grinding equipment and the replacement of traditional wet-process kilns with dry-process kilns, often incorporating staged preheaters and pre-calciners. In particular, the thermal energy intensity associated with the production of clinker, the primary component of cement, has witnessed an annual reduction of 0.8%, ultimately reaching a global average of 3.2 GJ per metric ton. Furthermore, the electricity intensity in the cement production process has displayed a notable yearly decline of 1.9%, resulting in a consumption rate of 84 kWh per metric ton [1 However, it is crucial to recognize that the cement industry is not only a significant emitter of carbon dioxide (CO_2) but also releases various other pollutants into the environment [6]. These pollutants not only contribute to environmental hazards but also pose health risks, particularly in terms of respiratory health. Also, Al-Dadi et al. [7] determined that the quality of the land is affected by the mining operation, material storage, waste disposal, and atmospheric deposition of cement production.

Today, many companies are taking significant steps to cut their carbon emissions after realizing the substantial economic and social costs of these emissions. In 2020, the Global Cement and Concrete Association (GCCA), which comprises forty member companies representing 40% of the global cement industry, committed to achieving carbon-neutral concrete production by 2050 [8]. This historic move marked the cement industry as the first subsector with a global association setting a net-zero goal [1].

Researchers are actively investigating methods and technologies to make cement production more sustainable. A vital tool in this endeavor is a comprehensive approach that evaluates the environmental impact of cementitious products. Life cycle assessment (LCA) is a thorough evaluation approach that can gauge and measure the complete environmental influence of cement from resource extraction to its end-of-life phase.

In Southeast Asia, there is a growing effort to evaluate the sustainability of the cement industry. For example, researchers used life cycle assessment to assess the environmental impacts of cement production in Myanmar [9][10], Indonesia [11], and Hong Kong [12]. Few sustainability studies found in the recent literature assessed the environmental impact of cement in the Philippines. However, the country is committed to reducing GHG emissions by pursuing development using a low-carbon approach under the Paris Agreement. In their Nationally Determined Contribution (NDC) submitted to the UNFCCC in April 2021, the Philippines made a commitment to reduce and avoid projected greenhouse gas (GHG) emissions by 75% between 2020 and 2030. This pledge comprises 2.71% unconditional reductions and 72.29% conditional reductions in the agriculture, waste, industry, transport, and energy sectors. The commitment is rooted in the expected cumulative economy-wide emissions of 3,340.3 megatons of CO2-eqvuivalents for the same period [13].

This research will determine the environmental performance of a blended cement product manufactured in a cement plant in Bulacan in terms of carbon footprint and cumulative energy demand using the Life Cycle Assessment. Additionally, this method will allow for comparing different upstream scenarios in manufacturing blended cement. Furthermore, the carbon footprints and cumulative energy demand of blended cement production will be differentiated with the global cement production data. The result of this research may serve as a basis to produce an Environmental Product Declaration for the said blended cement product, to evaluate the environmental sustainability of concrete products and construction projects made from this material, and to identify process inefficiencies that may eventually translate to internal cost reductions, environment- friendly operation, and a better image for the brand.

2. Materials and Methods

Life cycle assessment has been used globally in various sustainability studies in the cement industry. This method determines the aspects of the manufacturing process that damage the environment. Furthermore, it allows the comparison of different strategies to minimize the negative impacts of the operation. The results of LCA studies provide information that is helpful for the decision and policy-makers in addressing the growing challenges of the cement industry. This research used the standardized LCA methodology described in the ISO 14040[14] and ISO 14044 series [15].

2.1. Goal and Scope Definition

The goal is to assess the environmental impacts of blended cement production based on primary data from a process plant in Norzagaray, Bulacan, Philippines in terms of carbon footprint (CF) and cumulative energy demand (CED). The results from this assessment are intended to be the basis of an environmental product declaration for the blended cement.

The research's system boundary includes the entire cradle-to-gate phase, covering the extraction of raw materials and their conveyance to the manufacturing facility, material processing and upgrading, and cement production. This assessment considers emissions falling under Scope Emission 1 to 3. However, life cycle stages beyond the factory gate have been omitted due to the various product applications. The functional unit being analyzed comprises one metric ton of clinker, and one metric ton of blended cement.

Figure 1 illustrates the system boundary for the cement production examined in this research. It encompasses all inputs and outputs related to raw material preparation, clinker manufacturing, and cement production [16]. The boundary also encompasses the movement of raw materials and fuels from their sources to the plant. Nonetheless, infrastructure and facilities are excluded from this system, as their influence is comparatively minor compared to the impact of the various process stages.

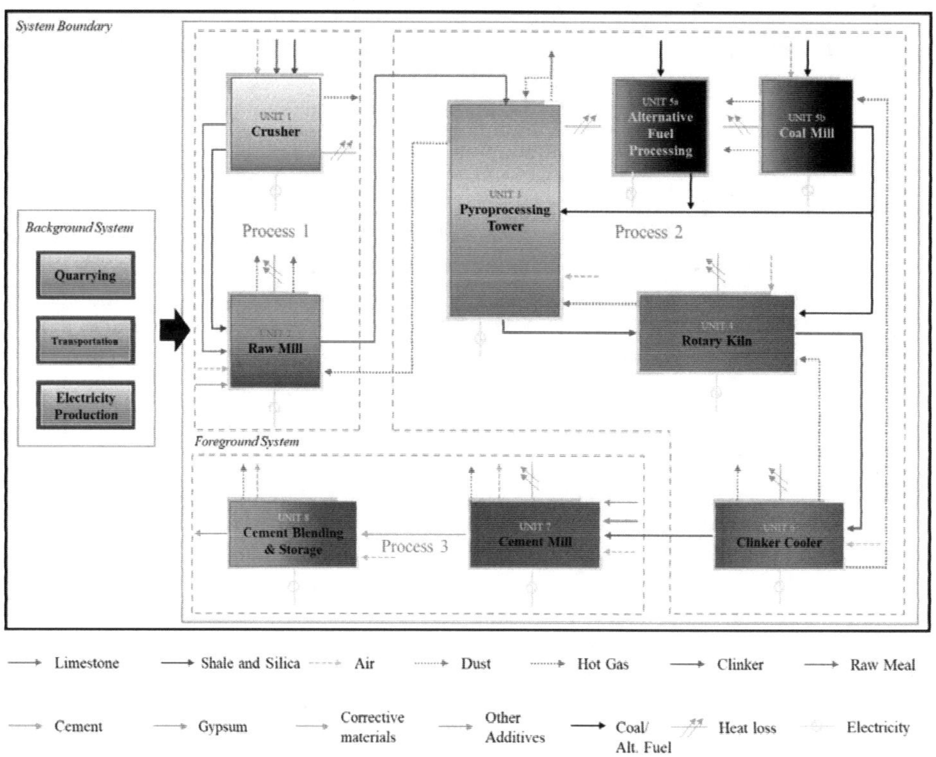

Figure 1. Schematic of the cement manufacturing plant

IOP Conf. Series: Materials Science and Engineering **1318** (2024) 012025 doi:10.1088/1757-899X/1318/1/012025

2.2. Life Cycle Inventory and Technical Assumptions

The inventory data source for the foreground system is the 2021 production data of the cement manufacturing plant located in Norzagaray, Bulacan. Nonetheless, certain data remains undisclosed due to an author's non-disclosure agreement with the cement company. The research obtained inventory data for the background system from the Ecoinvent 3.8 database, encompassing the estimated effects of raw material and fuel production and transportation, accounting for the actual distances between supplier locations and the cement plant.

The Ecoinvent 3.8 database comprises life cycle inventory (LCI) data from various sectors, including energy generation, logistics, construction materials, chemical manufacturing, metal manufacturing, and agriculture, among others. This comprehensive database includes more than 20,000 interconnected datasets, each detailing a process-level life cycle inventory.

To calculate CO_2 emissions, emission factors developed by the company were employed. The net calorific value of fuels was determined through laboratory tests conducted at the plant. Information on electricity and fuel consumption at different production stages was derived from the 2021 Annual Energy Utilization Report submitted by the plant to the Department of Energy.

2.2.1. General Assumptions. The research utilizes Ecoinvent datasets to represent the extraction, production, and processing of raw materials (limestone, shale, silica), traditional fuels, and electricity. Emissions from calcination are equivalent to loss on ignition. Graywacke lacks Ecoinvent data and a SimaPro dataset was created for it. Silica and pozzolana adopt the datasets of dataset of silica sand and calcined clay due to similar properties. Residual materials like copper slag, fly ash, and waste oil have no emissions; impacts are attributed to the process where they are derived.

Transport includes conveyors, trucks, and barge ships. Simulation uses averaged global transport with max load capacity. The load factor or the occupied fraction of the load capacity used for truck and barge ships is 100%. And the emission for the return trip is equivalent to 20% of the emissions of the first trip. Emissions are EURO 4 standard to fit the Philippine setting. The average transport distance of alternative fuels and rice husks was estimated using a local map. Electricity and diesel use at the cement grinding and dispatch stage are estimated from equipment power and time. Wastewater treatment isn't included due to plant declaration. Emissions are categorized into three groups.

Table 1. Scope Emissions of Blended Cement Production

S1	S2	S3
Direct emissions from the combustion of fuels at the kiln (Traditional fuels, and other alternative fuels) Direct emissions from the calcination of limestone Direct emissions related to the extraction, production, and transportation of raw materials at the quarry. Direct emissions from mobile combustion sources (Diesel fuel used in the movement of materials inside the plant)	Indirect Emissions from Purchased Electricity	Emissions that are associated with the extraction, manufacturing, and conveyance of raw materials, fuels, and the energy procured or obtained by the company, but are not previously considered within the scope of S1 or S2.

2.3. Life Cycle Impact Assessment Methodology
This research used the standardized LCA methodology described in the ISO 14040 and ISO 14044 series. The impact category for this research were the climate change/global warming potential, measured in kgCO_2-eqv /mass of the product [17], and the Cumulative Energy Demand in order to evaluate the global energy requirement. SimaPro 9.3.0.3 software processes the life cycle inventory data collected in this research.

3. Results and Discussion

3.1. Baseline Carbon Footprint and Cumulative Energy Demand of Blended Cement
Table 2 presents the baseline data for the carbon footprint and cumulative energy demand of blended cement. The carbon footprint of one metric ton of clinker, and one metric ton of blended cement is 995 kg CO_2-eqv and 760 kg CO_2-eqv, respectively. On the other hand, the Cumulative Energy Demand of one metric ton of clinker, and one metric ton of blended cement is 4.74 x 10^3 MJ and 4.13 x 10^3 MJ, respectively.

Table 2. 2021 Baseline carbon footprint and cumulative energy demand

	Functional Unit	CF *kg CO_2-eqv*	CED *x10^3 MJ*
2021	*1 ton clinker*	995	4.74
Baseline	*1 ton blended cement*	760	4.13

3.1.1. Hotspot Analysis
The baseline results revealed play a central role in the cement production process, significantly impacting both the carbon footprint and cumulative energy demand of clinker production. In this case, coal and petroleum coke are among the top contributors to the cumulative energy demand of clinker, comprising 79% (3.74 x 10^3 MJ per ton of clinker) of the total energy demand of the process. On the other hand, the impact of electricity followed coal at 15.8% (751 MJ per ton of clinker). In terms of carbon footprint, the sum of the direct and embodied emissions of traditional fuels constitutes 37% (368.55 kg CO_2-eqv per ton of clinker) of the total carbon footprint, following calcination (535 kg CO_2-eqv per ton of clinker). Further, the environmental profile of blended cement in terms of carbon footprint and cumulative energy demand is a function of the amount of clinker in the product.

Tables 3 and 4 showed that S1 emissions constitute 87.8 % (873.54 kg CO_2-eqv per ton of clinker) of the carbon footprint of clinker, and 91.4% (695.03 kg CO_2-eqv per ton of clinker of the carbon footprint of blended cement. The S1 emissions are mostly direct emissions from fuel combustion and calcination. Thus, initiatives that reduce traditional fuel utilization could translate to the enhancement of the environmental profile of the product more effectively since it can impact both the carbon footprint and cumulative energy demand. Moreover, the industry has acknowledged the economic and ecological feasibility of using alternative fuels in cement production.

3.1.2 Global Data of Cement Production
The LCA results of cement manufacturing vary across the globe. Table 5 compares the CF and CED of cement production by studies from different countries worldwide. The energy consumption in the production of clinker, and blended cement in this research is significantly lower than the result of other studies. Also, the carbon footprint of clinker, and blended cement in this research is comparable with other studies, while some of the results are slightly higher. Other studies showed outstanding results in carbon footprints. The carbon footprint of clinker, and cement can be improved to 320 kg CO_2-eqv and 350 kg CO_2-eqv, respectively when 100% natural gas is used as the fuel source for the kiln.

IOP Conf. Series: Materials Science and Engineering **1318** (2024) 012025 doi:10.1088/1757-899X/1318/1/012025

Table 3. Baseline carbon footprint of clinker in terms of scope emissions

	Scope Emission	CF kg CO_2-eqv	%
Clinker {Production} | combustion, traditional fuels	S1	325.00	32.7%
Clinker {Production} | calcination, calcium carbonate	S1	535.00	53.8%
Mixed Waste {Holcim}| mixed waste, Recycled Content cut-off	S1	13.45	1.4%
Diesel {Transportation}| combustion	S1	0.09	0.0%
Electricity, high voltage {PH}| market for electricity, high voltage	S2	55.90	5.6%
Hard coal {Holcim}| hard coal mine operation and hard coal preparation	S3	40.80	4.1%
Limestone, unprocessed {RoW}| limestone quarry operation	S3	2.76	0.3%
Shale {RoW}| shale quarry operation	S3	0.12	0.0%
Mixed Waste {Holcim}| mixed waste, Recycled Content cut-off	S3	0.05	0.0%
Rice husk {Holcim)|rice husk, processing	S3	11.80	1.2%
Silica sand {Holcim}| production	S3	7.37	0.7%
Petroleum coke {Holcim}| market for	S3	2.59	0.3%
Copper slag {Holcim}| copper slag, Recycled Content cut-off	S3	0.18	0.0%
Heavy fuel oil {Holcim}| heavy fuel oil production, petroleum refinery operation	S3	0.10	0.0%
Diesel {Holcim}| diesel production, petroleum refinery operation	S3	0.06	0.0%
S1 Emissions, Total		873.54	87.8%
S2 Emissions, Total		55.90	5.6%
S3 Emissions, Total		65.83	6.6%
Total Emissions		**995.26**	**100.0%**

Table 4. Baseline carbon footprint of blended cement in terms of scope emissions

	Scope Emission	CF kg CO_2-eqv	%
Clinker	S1	695.00	91.4%
Diesel {Transportation}| combustion	S1	0.03	0.004%
Electricity, high voltage {PH}| market for electricity, high voltage	S2	32.80	4.3%
Limestone, crushed, for mill {RoW}| production	S3	0.23	0.03%
Graywacke	S3	0.00	0.0%
Pozzolana {Holcim}| pozzolana production	S3	31.10	4.1%
Gypsum, mineral {Holcim}| market for gypsum, mineral	S3	0.79	0.1%
Flyash {Holcim}| flyash, Recycled Content cut-off	S3	0.42	0.1%
Diesel {Holcim}| diesel production, petroleum refinery operation	S3	0.00	0.001%
S1 Emissions, Total		695.03	91.4%
S2 Emissions, Total		32.80	4.3%
S3 Emissions, Total		32.54	4.3%
Total Emissions		**760.37**	100.0%

IOP Conf. Series: Materials Science and Engineering **1318** (2024) 012025 doi:10.1088/1757-899X/1318/1/012025

Table 5. Comparison of carbon footprint and cumulative energy demand of cement production by studies from different countries

Reference	Country	Scenario	CF (kg CO_2-eqv)	CED (MJ)	Functional Unit	System Boundary
Martinez et al. (this study)	PHL	Baseline (Clinker)	995	4740	1 t clinker	Cradle-to-gate
Martinez et al. (this study)	PHL	Baseline (Blended Cement)	760	4130	1 t cement	Cradle-to-gate
Bulacan Plant	PHL	Baseline	697.5	NS	1 ton cementitious product	Cradle-to-gate
Song et al. (2016b)	CHN	Baseline (Blended Cement)	678	NS	1 t of cement	Cradle-to-gate
Song et al. (2016b)	CHN	Raw Material Substitution using Carbide Slag	667-674	NS	1 t of cement	Cradle-to-gate
Moretti and Caro (2016)	ITA	Baseline (Clinker)	960	NS	1 t clinker	Cradle-to-gate
Moretti and Caro (2016)	ITA	Baseline (OPC)	793	NS	1 t cement	Cradle-to-gate
Bushi and Meil (2014)	CAN	Baseline (OPC)	950	6620	1 t cement	Cradle-to-gate
Bushi and Meil (2014)	CAN	Portland-limestone cement	850	6020	1 t cement	Cradle-to-gate
Hossain et al. (2017)	HKG	Baseline (OPC)	980	5732	1 t cement	Cradle-to-gate
Hossain et al. (2017)	HKG	Portland Fly ash Cement	806	4790	1 t cement	Cradle-to-gate
Tun et al. (2020)	MMR	Baseline data from 8 cement plants (OPC)	872-1440	NS	1 t cement	Cradle-to-gate
Thwe et al. (2021)	MMR	Baseline (90% coal, 10% Natural gas)	1,113	NS	1 t cement	Cradle-to-gate
Thwe et al. (2021)	MMR	Coal-Natural Gas Mix	320 (100% Natural Gas); 1,067 (100% Coal)	NS	1 t clinker	Cradle-to-gate
Thwe et al. (2021)	MMR	Coal-Natural Gas Mix	335 (100% Natural Gas); 1153(100% Coal)	NS	1 t cement	Cradle-to-gate

4. Conclusion

This life cycle assessment research evaluated the environmental effects of blended cement production in Bulacan, Philippines in terms of carbon footprint and cumulative energy demand. Cement production has a significant effect on the environment. The most significant environmental emissions, particularly greenhouse gas (GHG) emissions, predominantly arise in the kiln, primarily due to the breakdown of calcium carbonate and the combustion of fuels. Also, the grinding operations in cement production are highly energy intensive. Thus, certain measures must be implemented to improve the sustainability of the cement industry.

The 2021 baseline carbon footprint of one metric ton of clinker, and one metric ton of blended cement is 995 kg CO_2-eqv and 760 kg CO_2-eqv, respectively, while the Cumulative Energy Demand of one metric ton of clinker, and one metric ton of blended cement is 4.74×10^3 MJ and 4.13×10^3 MJ, respectively. Hotspot analysis revealed that the quantity of traditional fossil fuels used in the kiln has substantially influenced both the carbon footprint and cumulative energy demand of clinker. As a result, the environmental profile of blended cement is also changed proportionately to the amount of clinker in the product.

The energy consumption observed in this research for the production of clinker, and blended cement registers a noteworthy reduction compared to findings from prior research endeavours. The notable decrease in energy consumption can be largely attributed to a strategic decrease in fossil fuel usage with the adoption of alternative fuels within the process of producing clinker. This initiative not only reflects commitment to sustainability but also underlined the potential for more efficient energy utilization.

Furthermore, the assessment of the carbon footprint of clinker, and blended cement in this research aligns comparably with the results from similar studies, with slight variation on certain instances. In terms of blended cement, the amount of clinker used in the mix and the use of alternative raw materials such as fly ash helped in minimizing the carbon footprint of the product.

Basically, this research underscores not only the reduction of the Cumulative Energy Demand through innovative fuel choices but also the capability of raw material selection strategies in integrating sustainability goals within the domain of clinker, and blended cement production.

References

[1] IEA 2021 Cement – Analysis. Retrieved from https://www.iea.org/reports/cement
[2] Ighalo J O and Adeniyi A G 2020 *Waste Dispos. Sustain. Energy* **2** 161 - 64
[3] Gursel A P, Masanet E, Horvath A and Stadel A 2014 Cem. Concr. Compos **51** 38–48
[4] Gao T, Shen L, Shen M, Chen F, Liu L and Gao L 2015 J. Clean. Prod. **103** 160–70
[5] Andrew R M 2019 *Earth System Science Data*, **11**, 1675-1710.
[6] Li C, Nie Z, Cui S, Gong X, Wang Z and Meng X 2014 J. Clean. Prod. **72** 204–11
[7] Al-Dadi M M, Hassan H E, Sharshar T, Arida H A and Badran H M 2014 J. Radioanal. Nucl. Chem. **302** 1103–17
[8] GLOBAL CEMENT AND CONCRETE ASSOCIATION 2021 GCCA concrete future - roadmap for net zero. Retrieved from https://gccassociation.org/concretefuture/wp-content/uploads/2021/10/GCCA-Concrete-Future-Roadmap-Document-AW.pdf
[9] Tun T Z, Bonnet S and Gheewala S H 2020 The Int. J. Life Cycle Assess. **25** 2106–21
[10] Thwe E, Khatiwada D and Gasparatos Clean. Environ. Syst. **2** 100007
[11] Rosyid A, Boedisantoso R and Iswara A P 2020 IOP Conf. Ser. Earth Environ. Sci. **506** 012024.
[12] Hossain M U, Poon C S, Lo I M C and Cheng J C P 2017 Resour. Conserv. Recycl. **120** 199–208
[13] NDC REGISTRY 2021 United Nations Framework Convention on Climate Change. Retrieved from https://unfccc.int/sites/default/files/NDC/2022-06/Philippines%20-%20NDC.pdf
[14] ISO 14040:2006 2006 Environmental Management - Life Cycle Assessment - Principles and Framework. Geneva, Switzerland: International Organization for Standardization.
[15] ISO 14044:2006 2006 Environmental Management - Life Cycle Assessment - Requirements and Guidelines. Geneva, Switzerland: International Organization for Standardization.
[16] Gao T, Shen L, Shen M, Liu L and Chen F 2016 J. Clean. Prod. **112** 553–65
[17] IPCC, 2021. Climate Change 2021: The Physical Science Basis. Contribution of Working Group I to the Sixth Assessment Report of the Intergovernmental Panel on Climate Change, eds. V. Masson-Delmotte et al. Cambridge University Press, Cambridge, United Kingdom and New York, NY, USA, In press.

Acknowledgement

The authors express their sincere gratitude to the Engineering Research and Development for Technology (ERDT) program for its generous support and funding, and to Holcim Philippines for their valuable collaboration. They also appreciate University of the Philippines-Los Baños for providing essential resources and fostering a conducive research environment, leading to the successful completion of this research.

Removal of Bunker Oil in Contaminated Seawater Using Biochar from Chicken Bones Via Slow Pyrolysis

K M Decembrano[1], M A K Hamor[1], D Sison[1], J Binay[1], E Roque[1], R V Rubi[1] and J Olay[1]

[1]Chemical Engineering Department, College of Engineering, Adamson University
Email: jerry.olay@adamson.edu.ph

Abstract. An oil spill is the unintended or deliberate release of petroleum hydrocarbons into the environment, particularly the marine ecosystem. The disseminated oil can either evaporate or form a surface slick, disperse in the water, or submerge and deposit in the sediments. One of the standard methods is the use of sorbents. Pyrolysis is one of the simplest methods for converting biomass, and it involves heating organic material in the absence of oxygen. The natural polymeric ingredients of the process are bio-oil (condensable vapours), char (solid fraction), and non-condensable gases. Chicken bones were used to produce biochar via slow pyrolysis at varying operating temperature (400°C, 450°C and 500°C) and residence time (45min, 60min and 75min) to acquire the Chicken Bones Biochar (CBB). The CBB was used to obtain Chicken Bones Activated Biochar (CBAB) after it was subjected to KOH Activation Process. For the yield obtained, the lowest temperature and shorter residence time acquired the highest yield for both CBB and CBAB. The characterization of the products was carried out by using Fourier Transform Infrared Spectroscopy (FTIR) and Scanning Electron Microscopy (SEM). The products were evaluated after it was subjected to a bunker oil contaminated seawater and the adsorption capacity was determined with varying contact time (40min, 50min and 60min). A high absorbance of the CBB and CBAB indicated the feasibility of the chicken bone as an adsorbent material.

Keywords: Oil Spill, Sorbents, Bunker Oil, Seawater, Slow Pyrolysis, Chicken Bones, KOH Activation, Fourier Transform Infrared Spectroscopy, Scanning Electron Microscopy

1. Introduction

An oil spill is the unintended or deliberate release of petroleum hydrocarbons into the environment. The wind and the current can carry an oil spill over the marine system. The disseminated oil can either evaporate or form a surface slick, disperse in the water, or submerge and deposit in the sediments [1]. Various procedures include booms, skimmer-based mechanical techniques, in-situ burning, and sorbents used to clean up oil spills along shorelines and offshore. In addition, physical measures, such as booms and skimmers, are the most utilized for cleaning up large-scale oil spills [3][4]. One of the easiest and standard methods is the use of sorbents. There is an increasing demand for sorbents that are inexpensive, simple to use, and kind to the environment to clean up spills [1]. However, the sorption effectiveness of natural-based materials is lower than synthetic materials, even though natural-based materials are inexpensive, abundant, kind to the environment, and enable the correct usage of waste. One of their primary drawbacks is their poor oleophilic and hydrophobic characteristics [5]. Thus, pyrolysis was introduced to this study.

Pyrolysis is one of the methods for converting biomass, and it involves heating material in the absence of oxygen. The natural polymeric ingredients of the process are bio-oil, char, and no condensable gases [2]. Biochar is utilized as a carbon sink, soil amendment, and adsorbent. The chemical and physical qualities can be characterized and dictated by the technologies and feedstock used to produce it. Assessing this physiochemical property is critical for the environmental use of biochar, particularly in terms of its adsorption capability. As validated substantial toluene removal from water using biochar and activated biochar using equilibrium and kinetic experiments. In addition, chicken bones are a well-known food waste that is abundant and regularly found in residential areas and trash markets [6]. However, these animal bones have calcium carbonate and carbon as their primary component making them an accessible and effective source of raw material for activated carbon [7]. This study investigates removal of bunker oil in contaminated seawater using biochar from chicken bones via slow pyrolysis.

2. Materials and Methods

2.1 Preparation of Raw Materials

The raw chicken bone was collected from local restaurants. The materials were washed and cleaned to remove the excess dirt and oil, flesh, and other contaminants. After being washed with distilled water until it achieved a neutral condition, it was dried at 80◦C. Then crushed and was placed in a corundum crucible to proceed with the heating process.

2.2 Preparation of Biochar

A sample was heated in a tube furnace at a rate of 10°C per minute to temperatures of 400°C, 450°C, and 500°C while a flow of 99.99% pure nitrogen was maintained at 100 millilitres per minute. With varying temperatures, the carbonization step was carried out for 40, 50, and 60 minutes, respectively. The carbonized sample was placed in beaker containing 2 mol/L of HCl and was stirred. The samples were dried at 80°C after being cleaned with distilled water. The obtained sample CBB.

2.3 Activation of Biochar

For the preparation of activated biochar, CBB and KOH were mixed at a 3:1 mass in grams ratio. The substance was put through the second phase of pyrolysis, which lasted an additional hour in a tube furnace at a final temperature of 500°C and contained 100 millilitres per minute of nitrogen flow. After being cooled to room temperature in a flow of 99.999% pure nitrogen, the active substance was washed and dried at 80°C using distilled water. The prepared product of activated biochar was named CBAB (chicken bone-activated biochar modified using potassium hydroxide).

2.4 Characterization

For the characterization, a scanning electron microscope (SEM) for imaging analysis was utilized by Thermo ScientificTM Phenom XL. The FTIR spectra were obtained by employing a Perkin Elmer Fourier transform infrared with a frequency range of 4000–1000 cm−1 to collect the data.

2.5 Adsorption Experiment: Standard Calibration

Mixtures of bunker oil and seawater were prepared at different concentrations. The solution then oscillated in a water shaker for 30 min with 243 r/m. Then, using the UVVIS, the adsorption of the solution with different concentration were recorded to form a calibration curve [8]. A linear standard calibration formula was formed. The coefficients were then gathered by fitting the measured adsorbance with known oil concentrations. Coefficient of determination or R^2 (0.9988) indicates that the linear relationship between adsorbance and concentration of an adsorber of electromagnetic radiation fitted the calibration curve well [8]. In the context of regression, the measurement provides information that the model or the regression line was well-suited for the actual data. The points were obtained from the highest peak adsorption peak from the wavenumber 209.15. The volume of oil was recorded at 2, 4, 6, 8, 10, and 12 using the unit mL and it was mixed with sea water. Using the FTIR, the adsorbance were recorded and plotted in a linear graph to acquire the linear formula and coefficient. The formula and volume graph are y=03154x-0.3353.

2.6 Preparation of Bunker Oil Contaminated Seawater

The experiment was performed in batch mode using flasks as the container with 90 mL of water added 10 mL bunker oil. The flasks were sealed with aluminium foil and placed in a thermostatically controlled water bath, oscillating at a 243 r/min frequency at a constant temperature for 30 min.

2.7 Experiment and Evaluation of Sorption Capacity

The CBB and CBAB were submerged in the contaminated seawater. The adsorbent dosage used in this experiment is specifically 0.2 grams, and the contact time of the adsorbent was 40min, 50min, and 60min. This experiment added a mixture of 0.2 g of chicken bone-activated biochar or chicken bone biochar to the bunker fuel-contaminated seawater. Each sample was placed in the thermostatic water bath for the second time with 243 r/min for 40, 50, and 60 min. The same procedure was carried out as a blank set-up on an additional flask containing the same bunker fuel-contaminated seawater. However, it did not contain biochar. Therefore, following the process, and the resulting absorbance was then collected and measured using UV-vis spectroscopy with the wavelength set to 209.15 nm^{-1} and same number was used for the determination of absorbance.

3. Results

3.1 Yield

Figure 1 shows the biochar yield as a function of pyrolysis temperatures ranges from 400°C to 500°C with a 50°C interval and residence times ranging from 45 to 75 minutes with a 15-minute interval. The yield at 400°C with 45 min resident time were 13.06%, 11.97% for 400°C and 60 min, and 9.37% for 400°C and 75 min. Extending the residence period under high-temperature conditions mostly altered the surface and interior structure of biochar while contributing little to biochar output [9].

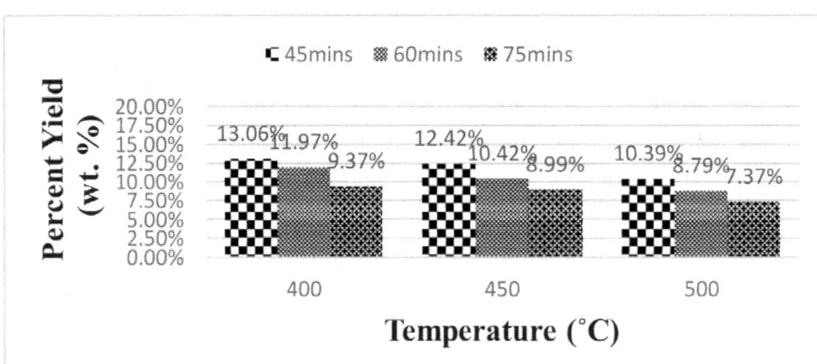

Figure 1. Percentage yields of CBB

Figure 2 shows Activated Biochar Yield as Function of Temperature and Residence Time; the yield of activated biochar varies as a function of pyrolysis temperatures that range from 400°C to 500 degrees Celsius with a 50°C interval and residence lengths that range from 45 minutes to 75 minutes with a 15-minute interval. Yield of 83.92% was achieved at 400°C and 45 minutes, 62.42% at 400°C and 60 minutes, and 55.06% at 400°C and 75 minutes. Other studies have shown that when process temperature increased, process yield decreased. The breakdown of organic components as the process temperature rose caused this dependency. It was preferable to have a greater process yield at a lower temperature

IOP Conf. Series: Materials Science and Engineering **1318** (2024) 012026 doi:10.1088/1757-899X/1318/1/012026

[10]. Using the temperature of 500°C, the percentage yield with 45 minutes resident time was 79.47%, 63.83% at 60 minutes, and 63.14% at 75 minutes. As observed, same with 400°C, the yield % is higher with lesser resident time. Meanwhile, 60 minutes-75 minutes at 450°C has a small difference in yield. At 500°C, the yield result was 59.78% at 45 minutes, 41.44% at 60 minutes, and 37.06% at 75 minutes.

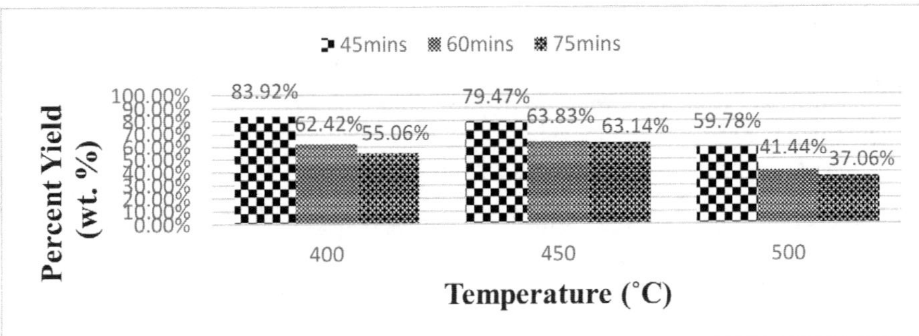

Figure 2. Percentage yields of CBAB

3.2 Characterization

Fourier Transform Infrared Spectroscopy results were obtained from the three products: CBB, CBAB and AC Bamboo that underwent two different conditions. The first condition was before the adsorption process and the other condition was after the product was submerged in the bunker oil contaminated sea water. Difference in the conditions were also recorded.

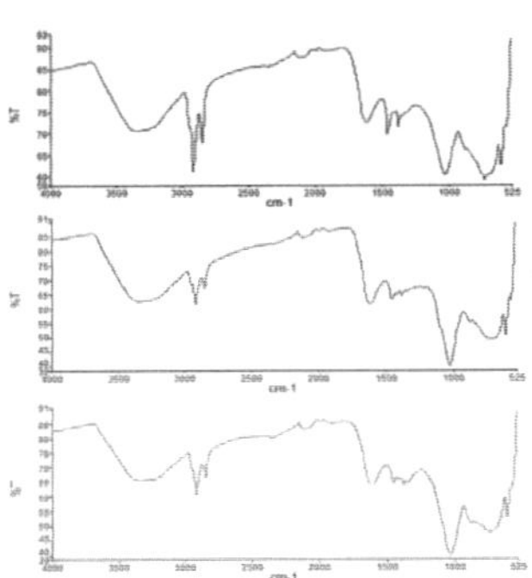

Figure 3. CBB

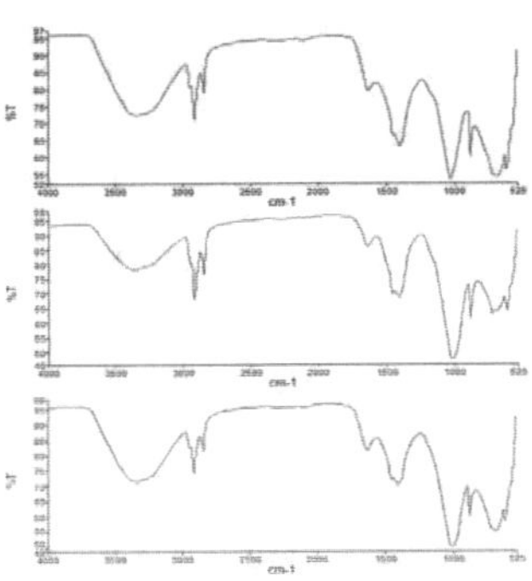

Figure 4. CBB

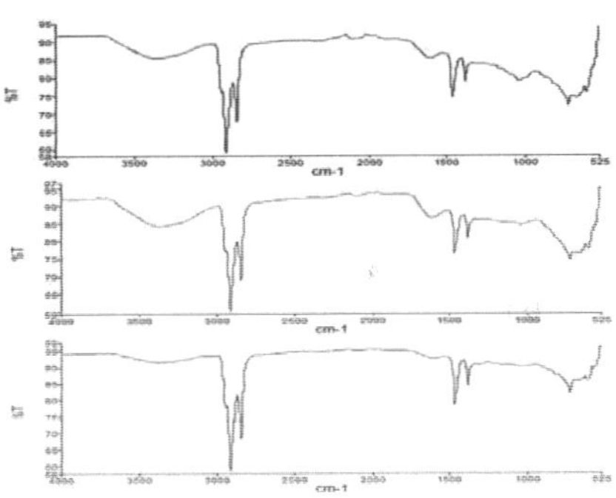

Figure 5. CBB, CBAB and Bamboo AC

Figure 3 shows the %Transmittance of CBB at 500 ˚C after adsorption at 40min, 50min, and 60min, respectively. The graph for 40min contact time had the highest transmittance of 92.90 at 525 cm-1, while the lowest transmittance was at 40.00 found on the graph for 50min and 60min contact time at 1100 cm-1. The maximum peak of transmittance and minimum transmittance after adsorption was lower than before the adsorption. Moreover, the maximum and minimum transmittance were observed at different contact times and wavelengths. The maximum and minimum transmittance of CBB before adsorption were found at 1800 cm-1 wavelength at 450˚C for 40min contact and 990 cm-1 wavelengths at 500˚C for 60min contact time, respectively. In comparison, the maximum and minimum transmittance of CBB after adsorption were found at 525 cm-1 wavelengths at 40min and 1100 cm-1 wavelength at both 50min and 60min, respectively. As shown in Figure 4, the %Transmittance of CBAB at 500 ˚C after adsorption at different temperature. The graphs for 50min and 60 contact times had the highest transmittance of 98.00 at 525cm-1, while the graph for 40min had the lowest of 46.00 at 1100 cm-1. In comparison with Figure 3, the maximum peak of 97.85 after adsorption was almost the same as that of 98.00 before the adsorption. However, the minimum after the adsorption of CBAB was higher than before its adsorption;

IOP Conf. Series: Materials Science and Engineering **1318** (2024) 012026 doi:10.1088/1757-899X/1318/1/012026

the transmittance before adsorption was 46.10 while after adsorption was 46.00. Moreover, the peaks were observed at the same wavelength. The peaks before and after the adsorption were at 1800 cm-1 wavelength. In addition, the minimum transmittance of CBAB before adsorption was found at 990 cm-1 and 1100 cm-1 after adsorption. Figure 5 displayed the %Transmittance of Commercial Carbon at 500 ˚C after adsorption and the graphs for 60 contact time have the highest transmittance of 98.00 at 525 cm^{-1}, while the graph for 40min has the lowest transmittance of 58.00 at 2900 cm^{-1}.

Figure 6. SEM CBB

Figure 7. SEM CBAB

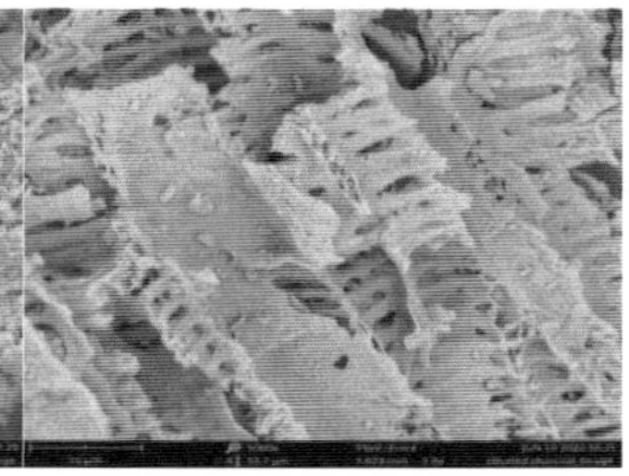

Figure 8. AC Bamboo

Figures 6 and 7 show the SEM imaging results of a dried Biochar sample from Chicken Bone both (inactivated and activated). Figure 8 shows the Commercial Activated Carbon from Bamboo which was the control sample. The parameters that were used on this sample was the CBB obtained from 500°C temperature and 45 min contact time. The microstructure of CBB powder was characterized by a fibrous structure and a surface with an irregular arrangement. The pores were homogeneous in size and loose. After KOH activation, the biochar surface exhibited an exfoliated and layered structure, as well as numerous depressions, dents, and cracks, which contributed to the increase in surface area. This could have been caused to the etching of the biochar's surface by KOH [11]. For CBAB, the parameters that were used on this sample was the CBAB obtained from 500°C temperature and 45 min contact time. The image showed a very fine microstructure of the CBAB. The alteration created a difference in the structure and resulted to finer sizes.

3.3 Adsorption Capacity and %Removal Using Different Contact Time

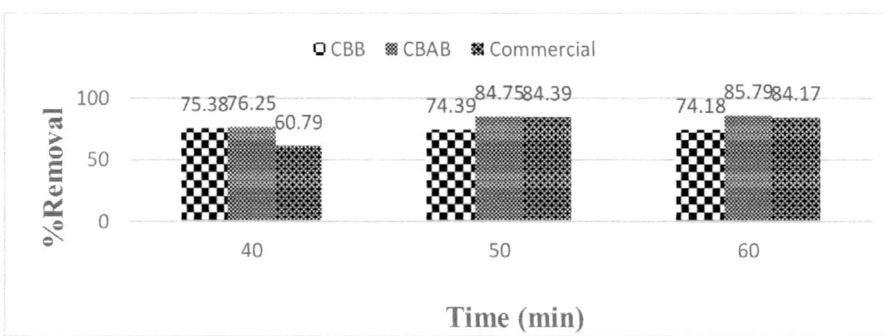

Figure 9. Summary of Removal using Different Contact Time

Figure 9 shows the summary of removal using different contact times. After the samples were submerged for three different contact times—40 minutes, 50 minutes, and 60 minutes—using the result or product with the optimal percent removal from the calculation, the products used and compared were recorded. The UV-Vis data recording was then used to compute the percentage removal of CBB, CBAB, and commercial activated carbon from bamboo.

IOP Conf. Series: Materials Science and Engineering **1318** (2024) 012026 doi:10.1088/1757-899X/1318/1/012026

4. Discussion

4.1 Yield

The CBB yield at 450°C temperature, there was a yield of 12.42% with 45 mins, 10.42% for 60 min, and 8.99% for 75 min. As observed with the results for 400°C, the percentage yield decreased as resident time increased. High boiling and less volatile substances breakdown slowly at above 400°C, whereas volatiles are completely released to gradually produce aromatic and heteroaromatic compounds, resulting in a slow yield drop [9], 500°C was the highest temperature in the study. The yield at 45 min is 10.39%, 8.79% at 60 min, and 7.37% at 75 min. It was shown that as the residence time increases, the biochar yield falls. This impact was also predicted since as residence periods increase, the reaction becomes longer, and more volatiles are produced [2]. The temperature has been used to simulate the production of energy fuels such as bio-oil, char, and biogas, with yield and energy content as responses [12]. The investigation revealed that the proportions of the various biomass had no significant effect on product yield and energy recovery. The simulation result of energy recovery shows a significant effect of temperature, related to the results of our study that the optimal biochar yield was achieved at 450°C (10.92 wt.%) at 60 mins.

The highest yield of 83.93% resulted was achieved at 400°C and 45. The lowest yield of 37.06% was obtained at 500°C for 75 minutes. It was observed that the yield drops with higher temperatures and longer time. In the activation process, the KOH melts into liquid form as it activates with the biochar with respect to the raising temperature. However, in this process, if the biochar and KOH was not properly mixed the activation will not be completely successful. Thus, creating a crystallized KOH. In this study, the microporous and mesoporous architectures of activated biochar are manipulated by manipulating the interaction between the pre-carbonization drying conditions and the carbonization temperature in the KOH activation process.

Activated biochar yields a greater mass of products than biochar, albeit with a similar declining trend as time passes. Also, a small difference is seen during the changes in temperature. In a study, most biomass residues are suitable for pyrolysis and all the operating parameters play an important role in the yield of products and their characterization [13]. Also, the production temperature significantly affects biochar chemical properties, and that activation increases sorption [14].

4.2 Characterization

These figures illustrate the FTIR spectra of CBB and CBAB. There was limited significance variance between the chicken bone biochar produced before and after the alteration, nor was there a shift in the position of the primary absorption peak. The bonds matched the O–H bending vibrations of carboxylic, hydroxyl, phenolic, and water groups [15]. The adsorption peak matched the bending and stretching vibrations of–OH and C=O, respectively. The peaks described above were discovered the samples, even though the band intensity of CBAB was lower. The C–H vibration in the methyl group was responsible for the weak bands that were observed in the CBAB spectrum. However, such a peak was not seen for the CBB sample, which indicates that the organic materials dissolved entirely following the KOH alteration. Moreover, the modification of chicken bone biochar did not significantly alter the functional groups on the surface. However, the adsorption capacity of modified biochar on bunker oil was improved to some extent, which the increase in specific surface areas. The FTIR Results were bound to give the best hit result, hit spectrum, hit description, and hit correlation of the samples.

Scanning Electron Microscopy (SEM) results showed an elaborated and stronger imaging of the microstructure. CBB was characterized by a fibrous structure, pores were homogenous in size and loose. CBAB had an exfoliated and layered structure and has a larger surface area and pore volume. The modification made it more suitable for the biochar to lock the oil through the surface and avert spill back incidents. The commercial biochar had more compact layers and larger pores than the fabricated products. In general, the physicochemical properties of generated and commercially accessible biochar are ideally suited for use as adsorbents and higher absorption in activated goods.

4.3 Adsorption Capacity and %Removal Using Different Contact Time

The results were obtained showing that there is no significant difference in the %Removal. It showed that the highest adsorption based on the UV-VIS spectrometer is the CBB in both 50 and 60 mins. However, using the calculation the %Removal is higher with lower adsorption. This is proved by the significance of these results. A trend was observed during the 50 and 60 mins %removal. It was observed that the Commercial and CBAB started to increase their adsorption and removal rate. The products still had the capacity to adsorb oil while the CBB declining graph indicates that at 50mins contact time it already achieved its optimal adsorption. It was also shown that in 50min to 60 min the increasing trend of CBAB and Commercial Activated Carbon slowed down and just made a very slight difference. According to the observations, these adsorbents were found to have low porosity and a limited surface area. Chemical activation using metal hydroxide reagents, such as potassium hydroxide (KOH), is, nevertheless, the most successful approach for producing sludge-based adsorbents with a high surface area (more than 1800 m2/g) [16]. In addition, it was stated that the performance of biochar would achieve optimum levels and that the adsorbents would regress or spill over. Thus, it is important to understand the appropriate contact duration for adsorption.

5. Conclusion

According to the results of the study, chicken bones can create biochar through a slow pyrolysis process. The following can be observed:

Using the slow pyrolysis method, chicken bone biochar (CBB) and chicken bone activated biochar (CBAB) generated from chicken bone waste were effectively prepared and have huge potential as an adsorbent for the removal of bunker oil from oily seawater. The CBB yield was recorded highest with 13.06% when the temperature was 400°C and the residence time was 45mins. The lowest yield of 7.37% was obtained from the 500°C and 60mins. It has been demonstrated that biochar yield decreases as temperature increases but increases as residence time increase. For the modified biochar, the CBAB highest and lowest yield were, 83.93% and 37.06%, respectively. Like the CBB, the modified biochar yields the highest in 400°C and 45mins and the lowest in 500°C and 60mins environment. Correspondingly, the difference in yield of CBAB and CBB was 70.87% for the highest yield and 29.69% for the lowest. This difference placed emphasis that the modification will yield more than 5 times the biochar mass because of the KOH activation process. Using various operating temperatures and resident time, statistical evidence suggests that temperature has a more significant impact than time.

Fourier Transform Infrared Spectroscopy (FTIR) before the adsorption experiment had a visible adsorbance band or the fingerprint region above the 1,500cm-1 in the infrared spectrum. The results after the product had adsorbed oil from the adsorption experiment demonstrated evident changes in behaviour and vertex of the spectrum. The adsorption peak matched the bending and stretching vibrations of the -OH and C=O, respectively. For the Scanning Electron Microscopy (SEM) results have given an elaborated and stronger imaging of the microstructure. The results obtained after the utilizations of FTIR and SEM to examine the CBB, CBAB, and commercial bamboo activated biochar. The physicochemical properties the products are ideally suited for use as adsorbents and higher absorption capacity was obtained in activated goods.

The contact time affects the adsorption performance of biochar. The activated biochar for bunker oil, and optimal experimental conditions were determined with the following parameters: temperature = 500 °C, resident time =45 min, and contact time = 60 min. It was also observed that among the three samples (CBB, CBAB, and Commercial AC Bamboo), CBAB 500 45 obtained the highest removal and adsorption at 60 mins contact time. The research results revealed that the real adsorption process of the as-prepared CBAB might be regarded as multi-layered sorption. The highest and lowest %removal was 85.79% and the lowest is 60.79%. The highest removal was from CBAB and the lowest was from commercial biochar. The pattern and %removal of the commercial increased at 50mins surpassing the CBB adsorption. The low cost and high efficiency of the raw material may have an excellent potential for the removal of oil pollutants found in water. CBB and CBAB adsorbents can now be utilized in productive, practical, and environmentally friendly ways.

Acknowledgements

Our sincerest gratitude to our family, who has been there from the start to give advice and financial and emotional support. Adamson University and all for their insightful comments and making helpful suggestions. Above all, the researchers want to thank the Almighty for giving us wisdom, strength, good health, never-ending blessings, happiness, and protection during this time of pandemic restrictions.

References

[1] Doshi B, Repo E, Heiskanen J P, Sirviö J A and Sillanpää M 2018 Sodium salt of oleoyl carboxymethyl chitosan: A sustainable adsorbent in the oil spill treatment *Journal of Cleaner Production* Volume **170** pp. 339-50
[2] Ronsse F, Bai X, Prins W and Brown R C 2012 Secondary reactions of levoglucosan and char in the fast pyrolysis of cellulose Environmental Progress & Sustainable Energy **31**(2) pp. 256-60
[3] Paulauskiene T, Jucikė I, Juščenko N and Baziuke D 2014 The Use of Natural Sorbents for Spilled Crude Oil and Diesel Cleanup from the Water Surface *Water, Air, & Soil Pollution* Volume **225** pp. 1-12
[4] Paulauskienė T and Jucikė I 2015 Aquatic oil spill cleanup using natural sorbents *Environmental Science and Pollution Research International*, Volume 22 (no.19):14874-81
[5] Zamparas M, Tzivras D, Dracopoulos V and Ioannides T 2017 Application of Sorbents for Oil Spill Cleanup Focusing on Natural-Based Modified Materials: A Review *Molecules* Volume **25**(no. 19):4522
[6] Silvani L, Vrchotová B, Kaštánek P, Demnerova K, Pettiti I and Papini M P 2017 Characterizing Biochar as Alternative sorbent for Oil Spill remediation *Scientific Reports* Volume **7**(no. 1) :43912
[7] Cruz M A E, Ruiz G CM, Faria A N, Zancanela D C, Pereira L S, Ciancaglini P and Ramos A P 2016 Calcium carbonate hybrid coating promotes the formation of biomimetic hydroxyapatite on titanium surfaces *Applied Surface Science* Volume **370** pp. 459-68
[8] Nisar A, Mamat A S, Hatim M I, Aslam M S and Ahmad M S 2017 Antioxidant and Total Phenolic Content of Catharanthus roseus Using Deep Eutectic Solvent. *Recent Advances in Biology and Medicine* Volume **3** pp. 7-10
[9] Sun Y, Gao B, Yao Y, Fang J, Zhang M, Zhou Y, Yang L 2014 Effects of feedstock type, production method, and pyrolysis temperature on biochar and hydrochar properties *Chemical Engineering Journal* Volume **240** pp. 574–78
[10] Wystalska K and Kwarciak-Kozłowska A 2021 The Effect of Biodegradable Waste Pyrolysis Temperatures on Selected Biochar Properties *Materials 2021(Basel)* Volume **14**(no. 7): 1-21
[11] Dehkhoda F, Lee C M M, Medina J and Brooks A J 2018 The Growth Hormone Receptor: Mechanism of Receptor Activation, Cell Signaling, and Physiological Aspects *Frontiers in Endocrinology Volume* **9**
[12] Hanif I, Raza S M F, Gago-De-Santos P and Abbas Q 2019 Fossil fuels, foreign direct investment, and economic growth have triggered CO2 emissions in emerging Asian economies: Some empirical evidence *Energy, Elsevier* Volume **171** pp. 493–501
[13] Vermaat J E, Wagtendonk A, Brouwer R, Sheremet O, Ansink E, Brockhoff T and Hering D 2015 Assessing the societal benefits of river restoration using the ecosystem services approach *Hydrobiologia* Volume **769**(no. 1) pp. 121–135
[14] Zhang Y, Xiao X, Wu X, Zhou S, Zhang G, Qin Y and Dong J 2017 Global gross primary production from vegetation photosynthesis model for 2000-2016, *Scientific Data* Volume **4**(no. 1):170165

IOP Conf. Series: Materials Science and Engineering **1318** (2024) 012026 doi:10.1088/1757-899X/1318/1/012026

[15] Cai L, Zhang Y, Zhou Y, Zhang X, Ji L, Song W, Zhang H and Liu J 2019 Effective Adsorption of Diesel Oil by Crab-Shell-Derived Biochar Nanomaterials *Materials (Basel)* Volume **12** (no. 2): 236

[16] Van Zwieten L, Kimber S, Morris S, Chan K Y, Downie A, Rust J and Cowie A 2009 Effects of biochar from slow pyrolysis of papermill waste on agronomic performance and soil fertility *Plant and Soil* Volume **327**(no.1–2) pp. 235–46

Environmental Sustainability Assessment of Banana Peel Briquetting

M V Torres[1], J M O Aberilla[1,*]

[1] Department of Chemical Engineering, University of the Philippines Diliman

* Corresponding author: joaberilla@up.edu.ph

Abstract. Continued population growth is expected to further strain the energy-food nexus, such as the increase of waste generation by 2050. Proper waste management is thus essential to manage the future scenarios. Banana is one of the highly produced crops globally, and especially in the Philippines, where one of its most notable wastes are the banana peels. It is estimated that 1.67 million metric tons of banana peel waste is generated from banana production in the Philippines. Despite the potential of BP across other applications, these technologies are not yet implemented on an industrial scale. While their potential as a raw material for various products has been researched extensively, the environmental sustainability of these technologies is yet to be measured. One of the technologies for utilizing waste biomass is briquetting. Briquettes are processed densified biomass which can be used as a substitute for conventional solid fuels, especially in household cooking. This study aims to report for the first time the environmental performance of industrial-scale banana peel briquetting from collected empirical data. Life cycle assessment is conducted through openLCA with ReCiPe 2016 Midpoint (H) being the impact assessment method. The results indicate that the production of briquetting performs better environmentally compared to the equivalent production of charcoal. The carbon footprint of producing briquettes from banana peels is 0.01 kg CO2-eq/MJ, over 95% lower than conventional charcoal. The briquettes from banana peels also indicate lower environmental impacts in categories such as agricultural land occupation, and fossil depletion potential. Contribution analysis of the system indicates that the use of binder contributes most to the impacts of briquetting production. Possible use of different binders or the development of different binders is expected to improve environmental performance. Alternative energy schemes for drying can also improve environmental performance.

1 Introduction

Sustainable solutions encompass good environmental performance but also take economic feasibility into account (1). Agriculture is one of the highly affected industries by climate change. Some crops are predicted to become extinct on a basis that its effect would continue to be observed in the future (2). While being affected by climate change, agricultural production-consumption cycle (including agricultural and food waste generation) contribute significant to global greenhouse gas (GHG) emissions of about one-third annually which is a significant contribution for a single sector (3).

IOP Conf. Series: Materials Science and Engineering **1318** (2024) 012027 doi:10.1088/1757-899X/1318/1/012027

Banana is one of the crops widely grown in the Philippines (4). It is also one of its known exports contributing largely to Philippine economy. The production-consumption cycle of banana generates banana peel (BP) as one of its wastes. Mostly, BP ends up in landfills where they end up underutilised despite of its potential to be used across a different number of applications which includes biomaterials, energy, and fuel(5) .

Biomass can be utilised by its conversion to, but not limited to, bioethanol (6), biomethane (7) and briquettes (8). One of the methods which can be utilised is biomass densification. It is a process in which biomass is subjected under pressure which compacts the biomass giving a more concentrated energy density when the biomass is subjected to combustion (9). In its current state, the environmental impacts of its production are currently not known.

This study contributes to the literature by reporting for the first time the life cycle impacts of banana peel briquetting. Environmental impacts of banana peel briquettes will be estimated using a life cycle assessment. The results of the current study could be used to guide future process designs to improve waste utilization

2. Methodology

Sustainability assessment of banana peel briquetting is done through an LCA. An LCA study is done under the iterative process which is shown directly in the following framework (Figure 1).

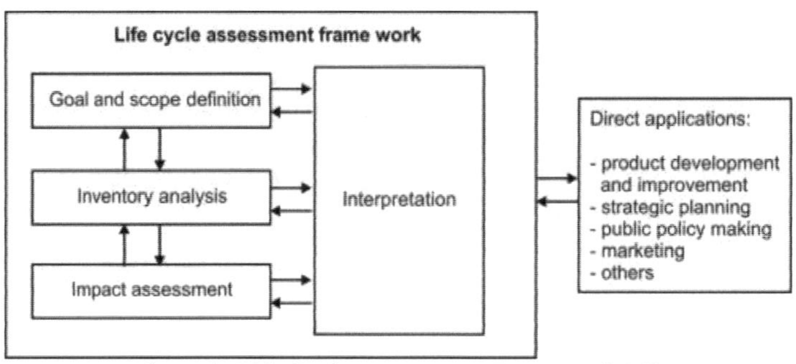

Figure 1. Life Cycle Assessment Framework(10)

The succeeding sections detail the following steps done in the conduct of LCA on this study. The fourth step of LCA on this study will be discussed as a parametric sensitivity analysis where the results can be interpreted as improvement analysis and uncertainty analysis.

2.1 Goal and Scope Definition

The goal and scope definition of this study is performed with the motivation introduced in the previous section. The main goal of this study is to estimate the environmental performance of banana peel briquetting (BPB), a form of biomass densification, on an industrial scale (11). Biomass densification is a novel technology which allows biomass to be used as a source of energy, fuel, or heat by applying converting them into a coal-like structure. Findings on any study could be used in improving future process designs for BPB

This study defines the scope to be a gate-to-gate basis. Transport of the raw material are not covered in this study. The system boundary is shown in Figure 2. Breakdown of binder recipe is detailed at Table 1.

IOP Conf. Series: Materials Science and Engineering **1318** (2024) 012027 doi:10.1088/1757-899X/1318/1/012027

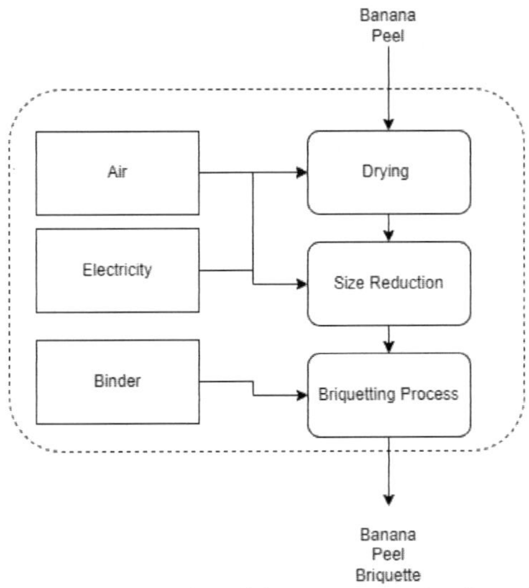

Figure 2. System Boundaries of the Banana Peel Briquetting Study

The current study does not cover the construction phases and use phases and will be focused mainly on the impacts of the process. This allows for the environmental impacts to analysed on the theme of sustainable process design. To contextualize the value of environmental impacts that will be reported, a functional unit of 1 MJ of fuel produced for power is selected. To evaluate the potential environmental benefits of BPB, the results will be compared to those of conventional solid fuels, namely coal and peat. Possible improvements on the entire product system defined in Figure 2 would be investigated as a form of Improvement Analysis (Scenario Analysis).

2.2 *Life Cycle Inventory*

Using the functional unit mentioned above, the life cycle inventory data are derived from existing studies and scale-up methodology as prescribed in literature (12). Laboratory scale data are obtained from previous studies (9,13). Background data such as water use and the use of different materials required for biomass densification are taken from ecolnvent (14), with consideration to the Philippine locality. For the energy use in the process, the current electricity mix is modelled. The data gathered are inputted to and inventoried by the software SimaPro (15). The Philippine electricity mix used in modelling for energy related process is shown in
Table 2.

Table 1. Life Cycle Inventory Data

Drying	Flow	Reference, ecoInvent
Input		
Banana peel	1 kg	
Nitrogen (in air)	8.26 kg	
Oxygen (in air)	2.20 kg	
Electricity	27 kJ	Philippine Electricity Mix as shown in Table 2. Computed from the method of Piccino(12).
Output		
Dried banana peels(8,12,16)	0.87 kg	

Water (in air)	0.13 kg	
Size Reduction		
Input		
Dried banana peels	0.87 kg	
Electricity	0.042 kJ	Philippine Electricity Mix as shown in Table 2. Computed from the method of Piccino(12).
Output		
Crushed banana peels	0.87 kg	
Briquetting Process		
Input		
Crushed banana peels	0.87 kg	
Lubricant (Coconut oil)(8,12)	0.10 g	Coconut Oil, market for (GLO)
Water(8,12)	0.76 g	Water decarbonised, market for (GLO)
Starch(8,12)	4 g	Potato Starch, market for (GLO)
Output		
Briquette	0.91 kg	

Table 2. Philippine Electricity Mix 2020

	Amount (TJ)	% Contribution
Oil	2.47	2.43%
Natural Gas	19.5	19.17%
Coal	58.18	57.19%
Geothermal	10.76	10.58%
Hydro	7.19	7.07%
Wind	1.03	1.01%
Solar	1.34	1.32%
Biomass	1.26	1.24%
Total	101.73	100%

2.3 *Life Cycle Impact Assessment (LCIA)*

Inventory which was collected based on the functional unit defined in section 2 and 2.2 are translated to environmental impacts through the life cycle impact assessment (LCIA) step. This study utilized the ReCiPe 2016 (H) impact categories. Impact potentials that will be reported are shown in the following table (Table 3).

Table 3. ReCiPe 2016 Midpoint (H) Impact Categories

Impact Category	Abbreviation	Unit
Global Warming Potential	GWP	$kg\ CO_2\ eq$
Stratospheric Ozone Depletion Potential	POFP	$kg\ CFC-11\ eq$
Ionising Radiation Potential	IRP	$kBq\ Co-60\ eq$

Ozone Formation Potential, Human Health	OFPhh	$kg\ NO_x\ eq$
Fine Particulate Matter Formation Potential	PMFP	$kg\ PM\ 2.5\ eq$
Ozone Formation Potential, Terrestrial Ecosystem	OFPte	$kg\ NO_x\ eq$
Terrestrial Acidification Potential	TAP	$kg\ SO_2\ eq$
Freshwater Eutrophication Potential	FEP	$kg\ P\ eq$
Marine Eutrophication Potential	MEP	$kg\ N\ eq$
Terrestrial Ecotoxicity Potential	TEP	$kg\ 1{,}4 - DCB\ eq$
Freshwater Ecotoxicity Potential	FETP	$kg\ 1{,}4\ DCB\ eq$
Marine Ecotoxicity Potential	METP	$kg\ 1{,}4\ DCB\ eq$
Human Carcinogenic Toxicity Potential	HETPcc	$kg\ 1{,}4\ DCB\ eq$
Human non-carcinogenic toxicity Potential	HETPnc	$kg\ 1{,}4\ DCB\ eq$
Land Used Potential	LUP	$m^2 a\ crop\ eq$
Mineral Resource Scarcity Potential	MRP	$kg\ Cu\ eq$
Fossil Resource Scarcity Potential	FRP	$kg\ oil\ eq$
Water Consumption Potential	WDP	m^3

3. Results and Discussion

This section details the life cycle environmental impacts resulting from the impact assessment in 20 impact categories. The following discussion is presented in terms of the key processes in the production of banana peel briquettes which are: (1) drying of BP, (2) size reduction of the banana peel, (3) briquetting. The results for BPB are then compared with conventional solid fuels to determine its environmental feasibility as a replacement. Lastly, potential improvements in the process of BPB production are investigated to obtain insights for future process development.

3.1 Life cycle environmental impacts of banana peel briquettes

The environmental impacts of BPB production are shown in Figure 3. The key contributors of the environmental impact previously identified can be easily identified through the contribution analysis plot shown in Figure 4. There are three processes which contribute significantly to the gate-to-gate environmental hotspots of the production of banana peel briquettes: (a) electricity or energy need which is applicable in all the processes shown, (b) binder use (in briquetting process), (c) water use in briquetting.

The use of coal in the current Philippine Electricity Mix Contributes the most to the environmental impacts where it is known in literature to be a significant contributor in GWP. In the process, use of coal contributes to about 80.5% to GWP100a. Similarly, binder use contributes to about 96% to MEP and 96.9% for FEP. Water use, which contributes significantly to the WDP is about 94%. The need for potato starch, which amounts to 4.61% of the briquette's mass, in the process recipe calls for the production of the potato crop as its key raw material. Run-off emissions during the farming stage from pesticides such as dithiocarbamates are known to adsorb into particulates and sediments. In this case, the emissions of pesticides can reduce aquatic ecosystem quality (17). Similarly, run-off emissions (i.e., N-and-P

IOP Conf. Series: Materials Science and Engineering **1318** (2024) 012027 doi:10.1088/1757-899X/1318/1/012027

emissions) are known to contribute to the available nutrient present in marine ecosystems. Excess nutrients in water encourages algal bloom which competes with the available dissolved oxygen versus the need to sustain aquatic life ultimately leading to eutrophication (18).

3.2 Comparison of BPB and Solid Fuel

The relative comparison of the environmental impacts of BPB with an equivalent amount of peat and coal is shown on Figure 3.

Based on results summarised in Figure 3, banana peel briquettes have lower environmental impacts than the two alternative product systems in all the impact categories. BPB has the lowest environmental impact in 18 of the 20 impact indicators. The MEP and FETP of BPB are between that of coal and peat. Demand for potato starch used in the production of the briquette binder results in a higher environmental impact relative to peat and coal.

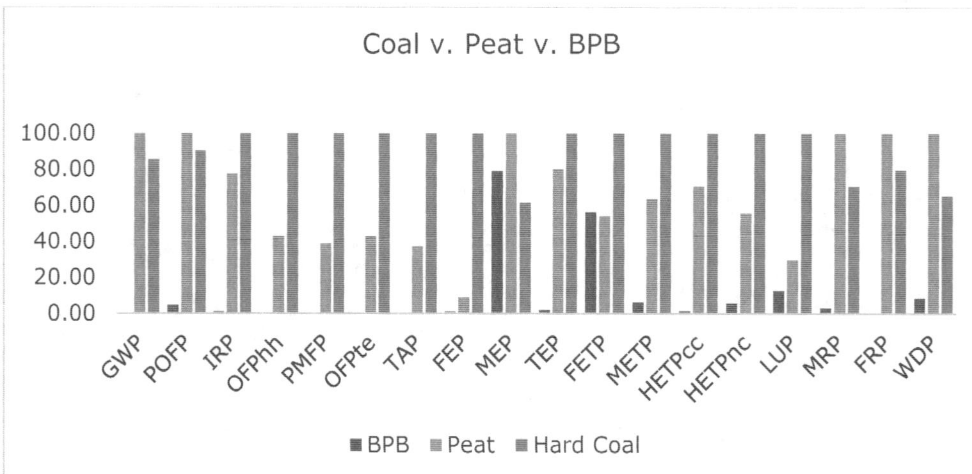

Figure 3. Relative Results of BPB versus other alternatives

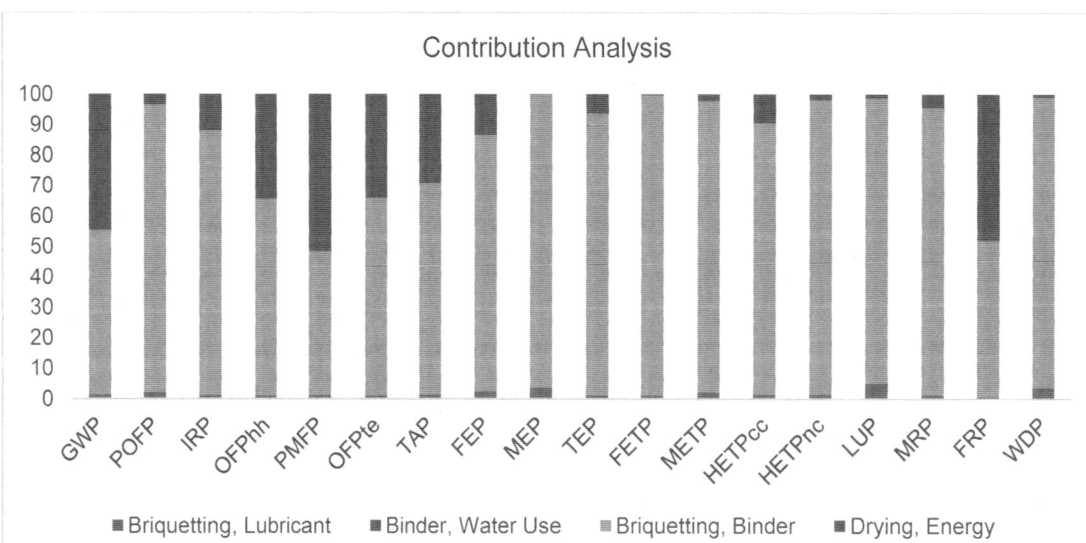

Figure 4. Contribution Analysis for Banana Peel Briquetting

3.3 Sensitivity and Improvement Analysis

In section 3.1, it was shown that there are key processes and flows involved in the manufacture of BPB that contribute mainly to its environmental impact. To inform process design, assessment is performed on potential feasible solutions which could favourably improve the environmental performance of the BPB process. Two process modifications are evaluated: increasing renewable energy use, and choice of binder.

Since one of the life cycle hotspots in BPB production is the electricity consumption, an alternative scenario for the power source is investigated. The Philippines has a clean energy target by 2040 (19) which considers increased share of renewable energy (Table 4). Using this electricity mix (shown in Figure 5), it is found that the performance of BPB has improved on: GWP by 8.6%, ODP by 6.22%, and FEP by 9.2%. In the contrary, less than 5% percent change can be seen on WDP by 0.14%, FDP by 0.6%, and METP by 2.27%.

Table 4. Philippine Energy Targets (BAU 2040)

	Amount (TJ)	% Contribution
Oil	0.28	0.08%
Coal	146.86	40.30%
Natural Gas	89.72	24.62%
Geothermal	16.18	4.44%
Hydro	51.55	14.15
Solar	53.06	14.56
Biomass	1.63	0.45%
Subtotal	364.4	100%

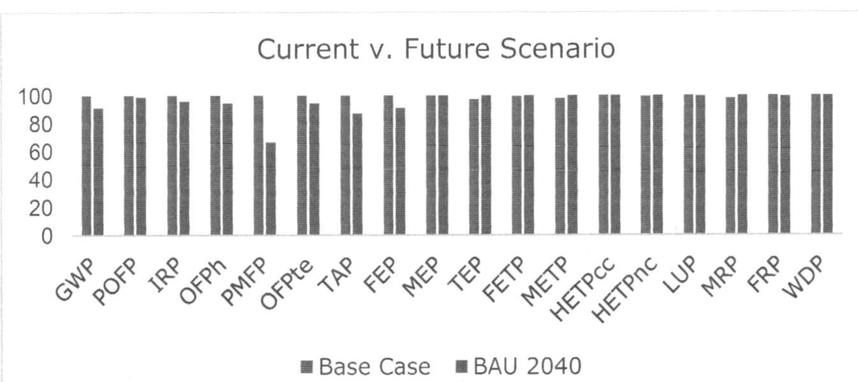

Figure 5. Improvement Analysis on Renewable Energy on BPB Production

The binder used in the densification of banana peels was also found to contribute significantly to the life cycle environmental impacts. A potential modification to the process is to consider the use of alternative binders such as maize-based options (8). Assuming a 1:1 substitution ratio for the binder, the results (Figure 6) indicate that the use of maize-based options is better than using a potato starch binder. Generally, potato production requires more intense inputs as compared to the production of maize grain (20).

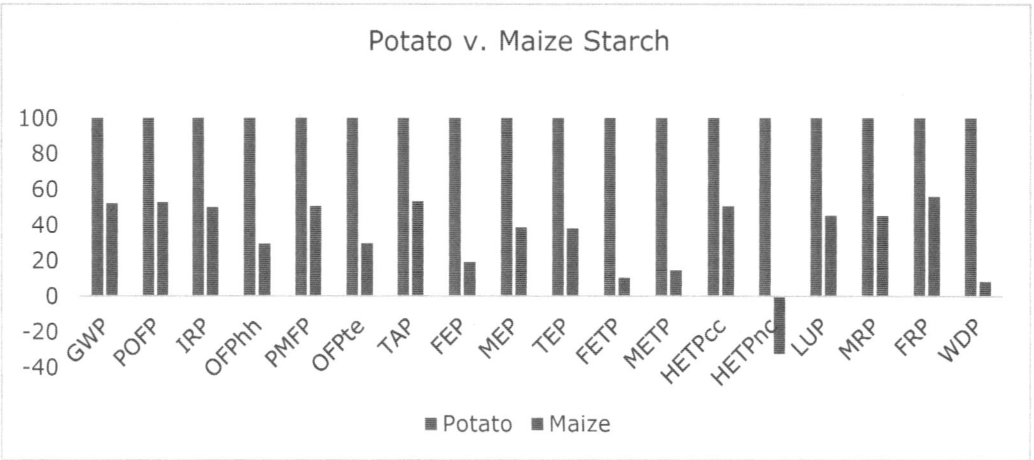

Figure 6. Improvement Analysis on use of starch

4. Conclusion and Recommendation

This study has estimated the life cycle impacts of banana peel briquette production through biomass densification. It was found that the key hotspots of the briquetting process are power consumption, binder for densification, and water consumption. In terms of the environmental performance of banana peel briquettes compared to coal and peat, it was shown that banana peel briquette has a lower environmental impact. The banana peel briquette performs better in all the environmental impacts except in the FETP and MEP versus that of peat. This study also identified potential improvements that could be easily adapted in the production process. Integration of the higher renewable energy is predicted to improve the environmental performance of banana peel briquetting by 0.6-8.6%. Use of maize starch is shown to be better than potato starch as a binder requirement from the briquetting process. The result of this work has shown that possible improvements could be made to the environmental performance of banana peel briquetting. This work could be used to inform the future process design of banana peel biomass densification. It can also act as an extension for literature on environmental impact assessment of biomass utilization. Recommendations for future study could be focused on the techno-economic assessment of BPB to further test the economic sustainability of the said process. This allows for possible improvements which could not be identified through an environmental assessment.

References

1. Bakshi BR. Toward Sustainable Chemical Engineering: The Role of Process Systems Engineering. Annu Rev Chem Biomol Eng. 2019 Jun 7;10:265–88.
2. Varma V, Bebber DP. Climate change impacts on banana yields around the world. Nat Clim Change. 2019 Sep 2;9(10):752.
3. Poore J, Nemecek T. Reducing food's environmental impacts through producers and consumers. Science. 2018 Jun 1;360(6392):987–92.
4. Major Fruit Crops Quarterly Bulletin | Philippine Statistics Authority | Republic of the Philippines [Internet]. [cited 2024 Oct 18]. Available from: https://psa.gov.ph/content/major-fruit-crops-quarterly-bulletin
5. Padam BS, Tin HS, Chye FY, Abdullah MI. Banana by-products: an under-utilized renewable food biomass with great potential. J Food Sci Technol. 2014 Dec;51(12):3527–45.
6. Gebregergs A, Gebresemati M, Sahu O. Industrial ethanol from banana peels for developing countries: Response surface methodology. Pac Sci Rev Nat Sci Eng. 2016 Jan 1;18(1):22–9.

7. Housagul S, Sirisukpoka U, Boonyawanich S, Pisutpaisal N. Biomethane Production from Co-digestion of Banana Peel and Waste Glycerol. Energy Procedia. 2014 Jan 1;61:2219–23.
8. Karimibavani B, Sengul AB, Asmatulu E. Converting briquettes of orange and banana peels into carbonaceous materials for activated sustainable carbon and fuel sources. Energy Ecol Environ. 2020 Jun 1;5(3):161–70.
9. Vaish S, Sharma NK, Kaur G. A review on various types of densification/briquetting technologies of biomass residues. IOP Conf Ser Mater Sci Eng. 2022 Mar;1228(1):012019.
10. 14:00-17:00. ISO. 2014 [cited 2023 Jun 2]. ISO 14040:2006. Available from: https://www.iso.org/standard/37456.html
11. Bjørn A, Laurent A, Owsianiak M, Olsen SI. Goal Definition. In: Hauschild MZ, Rosenbaum RK, Olsen SI, editors. Life Cycle Assessment: Theory and Practice [Internet]. Cham: Springer International Publishing; 2018 [cited 2024 Oct 18]. p. 67–74. Available from: https://doi.org/10.1007/978-3-319-56475-3_7
12. Piccinno F, Hischier R, Seeger S, Som C. From laboratory to industrial scale: a scale-up framework for chemical processes in life cycle assessment studies. J Clean Prod. 2016 Nov 1;135:1085–97.
13. Wilaipon P, Trirattansirichai K, Tangchaichit K. Moderate Die-Pressure Banana-Peel Briquettes.
14. ecoinvent v3.8 - ecoinvent [Internet]. 2021 [cited 2023 May 23]. Available from: https://ecoinvent.org/the-ecoinvent-database/data-releases/ecoinvent-3-8/
15. SimaPro [Internet]. [cited 2023 May 23]. SimaPro | LCA software for informed change-makers. Available from: https://simapro.com/
16. Ogunlade I, Akinmade AO, Ogunlade debayo O, Popoola OK, Ogunlade I, Akinmade AO, et al. Comparative study of chemical composition and evaluation of the In-Vitro antioxidant capacity of unripe and ripe banana species (Musa Sapientum) biowastes. Int J Agric Sci Food Technol. 2021 Feb 15;7(1):061–6.
17. Janz D. Dithiocarbamates. In: Wexler P, editor. Encyclopedia of Toxicology (Second Edition) [Internet]. New York: Elsevier; 2005 [cited 2023 Jun 20]. p. 86–8. Available from: https://www.sciencedirect.com/science/article/pii/B0123694000003549
18. Tabatabaie SMH, Murthy GS. Cradle to farm gate life cycle assessment of strawberry production in the United States. J Clean Prod. 2016 Jul 20;127:548–54.
19. Department of Energy Philippines [Internet]. [cited 2023 May 31]. Available from: https://www.doe.gov.ph/pep?withshield=1
20. Würdinger E. Kunststoffe aus nachwachsenden Rohstoffen: Vergleichende Ökobilanz für Loose-fill-Packmittel aus Stärke bzw. Polystyrol: Endbericht. 2002. 489 p.

IOP Conf. Series: Materials Science and Engineering **1318** (2024) 012028 doi:10.1088/1757-899X/1318/1/012028

Effect of Fatty Acid on Melting and Solidification of Erythritol

Daiju Wada, Xudong Tang, Manabu Tokushige, and Junichi Ryu*

Graduate School of Engineering, Chiba University, Chiba 263-8522 Japan

*Corresponding Author: jryu@chiba-u.jp (Junichi Ryu)

Abstract

One of the solutions to energy problems is the reuse of waste heat. One way to reuse waste heat is to use latent heat storage materials. Erythritol, of which melting point is the temperature range between 100 and 200°C, has attracted much attention as latent heat storage materials. However, as usual erythritol shows the large supercooling, its supercooling degree must be decreased to use for latent heat storage material. To achieve this, we have investigated the mixed erythritol with fatty acids as addition agents. Samples were prepared by physically mixing of erythritol and fatty acids such as (lauric acid, myristic acid, palmitic acid, and stearic acid). In this study, differential scanning calorimeter (DSC) was used (to evaluate the sample temperature program: 30→140→0→30°C, heating rate: 5°C/min, atmosphere: N_2 gas). The degree of supercooling was calculated from the DSC curve. And also, a camera was used to capture and discuss the freezing process. The erythritol physically mixed with fatty acids showed a higher freezing point and lower degree of supercooling than pure erythritol. The mixing myristic acid with erythritol significantly lowers the supercooling degree. The erythritol sample with palmitic acid or stearic acid did not show as low supercooling degree as that with myristic acid. Additionally, with lauric acid or myristic acid, the erythritol sample solidified simultaneously, while sample solidified separately with palmitic acid or stearic acid added. In summary, it was found that physical mixing of fatty acids and erythritol decreased in the supercooling degree. Furthermore, the crystallization temperatures of the sample with lauric acid or myristic acid were altered by changing the fatty acids.

1. Introduction

We are facing with severe environmental problems such as energy issues, global warming, air pollution, and so on. Especially, the depletion of energy resources is imminent because our energy consumption mainly depends on fossil fuels [1,2]. Additionally, modern industry exhausts a huge amount of heat as an unused energy which is made by burning fossil resources. Therefore, to make effective energy usage, heat storage materials are attracting much attention, because these can reduce using fossil resources [3]. Thermal storage materials are classified into three type heat storage materials such as sensible, latent, and chemical ones. Among them, the latent heat storage materials are larger gravimetric energy density than the sensible ones and have been more established technology than the chemical ones [4]. Erythritol is promising as a latent heat storage material due to the large latent heat (~340 J/g) [5, 6]. However, it has the disadvantage of large supercooling for its practical implementation [7]. This disadvantage makes the range of its usage narrow. To mitigate supercooling, the nucleating agents such as carbon nanoparticles are added in the erythritol [8, 9]. However, these nucleating materials require a lot of time for their preparation.

In our lab, we are adding zeolites to D-mannitol to mitigate supercooling and discuss polymorphs in D-mannitol. However, the problem of supercooling still exists and has not been done for erythritol [10,11]. In this study, we investigated the effects of adding four fatty acids (lauric, myristic, palmitic, and stearic acids) as the nucleating agents in erythritol on the supercooling. They are expected to promote to solidify erythritol and decrease in the supercooling degree because the fatty acids are solidified at higher temperatures than the freezing point of erythritol.

2. Experimental

2.1 Sample preparation

Erythritol (Ery; >97.0%) and fatty acid such as lauric acid (LA; >98.0%), myristic acid (MA; >98.0%), palmitic acid (PA; >95.0%) and stearic acids (SA; >95.0%) were purchased from FUJIFILM Wako Chemical Corp. Erythritol and fatty acid were physically mixed in air at room temperature.

2.2 Evaluation of thermophysical properties

Thermophysical properties of the samples were measured with about 5 mg in an Al pan (⌀5mm) by differential scanning calorimetry (DSC: DSCvesta, Rigaku Corp., Japan) with calibrating the Al_2O_3 powder (temperature program: 30→140→0→30°C, 5°C/min, N_2 gas atmosphere). The supercooling degree ΔT of sample was evaluated using the DSC curve. As shown in Fig. 1, the peak of the melting process in the DSC curve is the melting point (T_M), the starting point of the solidification process in the DSC curve is the freezing point (T_S), and the temperature difference between them is defined as the supercooling degree ΔT (Eq. 1).

IOP Conf. Series: Materials Science and Engineering **1318** (2024) 012028 doi:10.1088/1757-899X/1318/1/012028

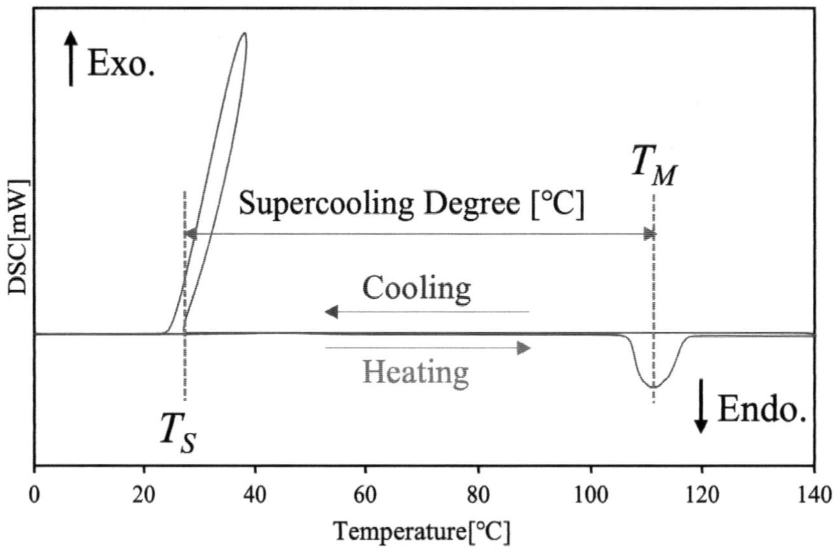

Fig.1 DSC curve of pure erythritol with Al lid.

$$\Delta T = T_M - T_S \quad (1)$$

3.Results and discussion

3.1 Supercooling Mitigation of Erythritol by Adding Fatty Acids

Figures 2 and 3 show typical DSC curves of erythritol and various fatty acids. Pure erythritol was melted at around 120°C and solidified at around 10°C, indicating a large supercooling phenomenon (supercooling degree (ΔT): 108.1°C). The observed heat of melting (ΔH_M) and heat of solidification (ΔH_S) of erythritol were $\Delta H_M = -250$ [J/g] and $\Delta H_S = 104$ [J/g], respectively. Figure 3 showed that all fatty acids were solidified at higher temperatures than that of erythritol.

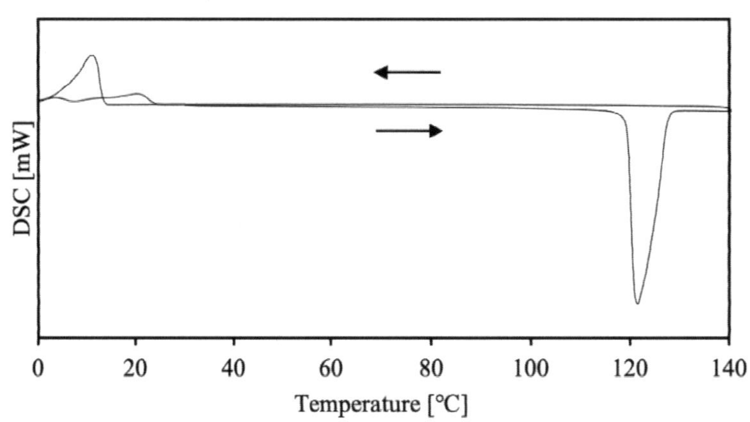

Fig.2 DSC curve of pure erythritol.

IOP Conf. Series: Materials Science and Engineering **1318** (2024) 012028 doi:10.1088/1757-899X/1318/1/012028

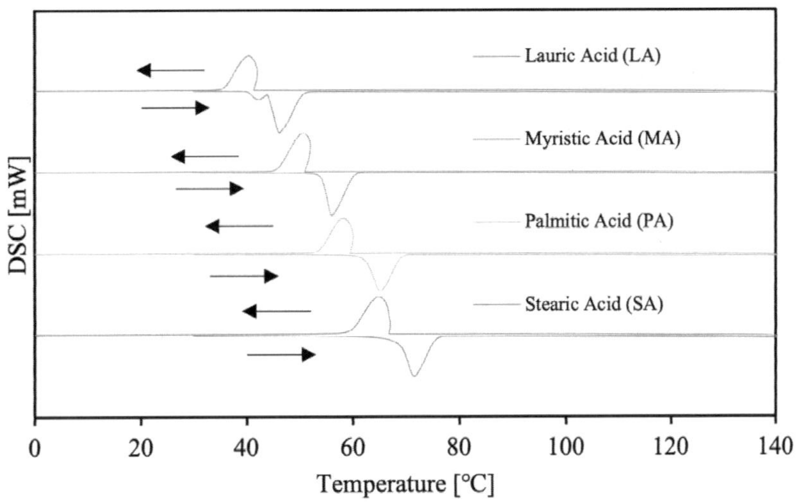

Fig.3 DSC curves of various fatty acids.

Solidification behaviors of pure erythritol (Fig. 4) and the erythritol sample with fatty acid (Fig. 5). While pure erythritol was solidified from the wall at lower temperature than 10°C, the sample with fatty acid (SA) started to be solidified from 70°C. Since the freezing point of SA is 70°C, it was suggested that the SA particles were solidified and dispersed in the erythritol melt. And then, it started to be solidified on SA particles as the growth nuclei. This consideration was supported by the results of DSC measurements.

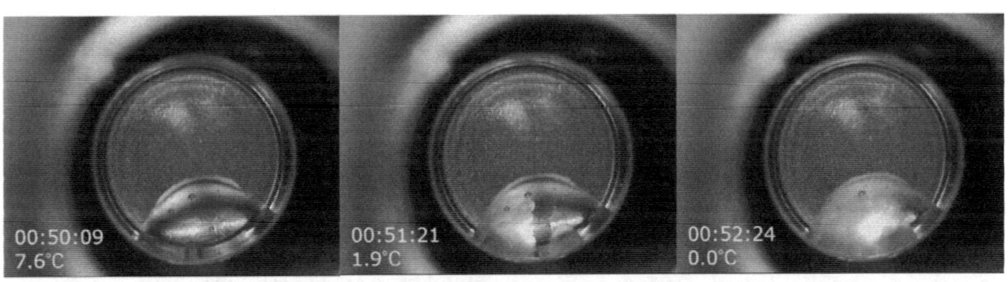

Fig.4 Solidification of pure erythritol. (Al pan : ⌀5mm)

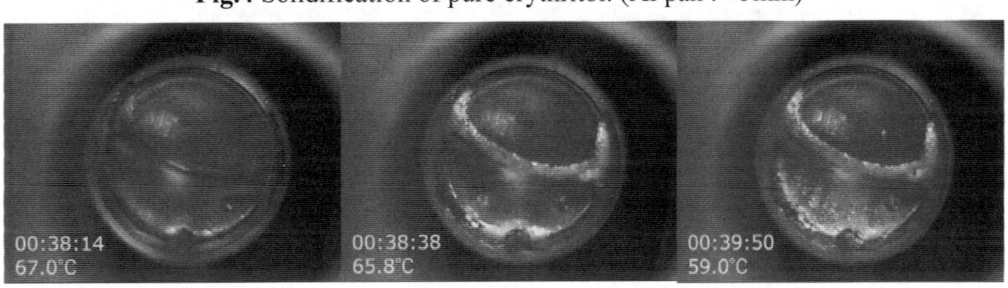

Fig.5 Solidification of erythritol sample with fatty acid (Ery-10wt% SA). (Al pan : ⌀5mm)

IOP Conf. Series: Materials Science and Engineering **1318** (2024) 012028 doi:10.1088/1757-899X/1318/1/012028

Figure 6 shows the DSC curves of the erythritol samples with 10 *wt*% of fatty acids, confirming that the supercooling in all samples were mitigated by adding fatty acids. Furthermore, the samples containing MA or PA showed the most effective supercooling mitigation (Fig. 7). The values of latent heats (ΔH_M and ΔH_S) for each sample are summarized in Table 1. Large latent heats of erythritol were observed, and these materials are expected to be practically implemented.

Figure 8 shows DSC curves for samples with different amounts of MA, confirming that the supercooling in all samples were mitigated by adding the MA. The supercooling degree was decreased with larger amount of MA until 10 *wt*% and slightly changed at larger than 10 *wt*% (Fig. 9).

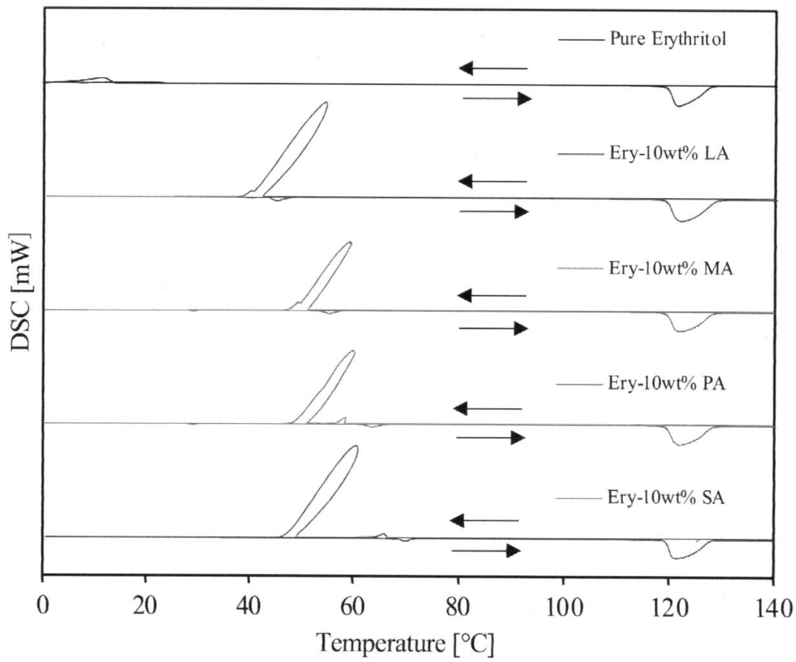

Fig.6 DSC curves of erythritol sample with 10 *wt%* of fatty acids.

Table 1 Latent heats of each sample

Sample	ΔH_M (Fatty Acid) [J/g]	ΔH_M (Erythritol) [J/g]	ΔH_S (Fatty Acid) [J/g]	ΔH_S (Erythritol) [J/g]
Ery-10wt% LA	−13	−248	–	181
Ery-10wt% MA	−15	−278	–	212
Ery-10wt% PA	−14	−260	14	190
Ery-10wt% SA	−14	−275	14	195

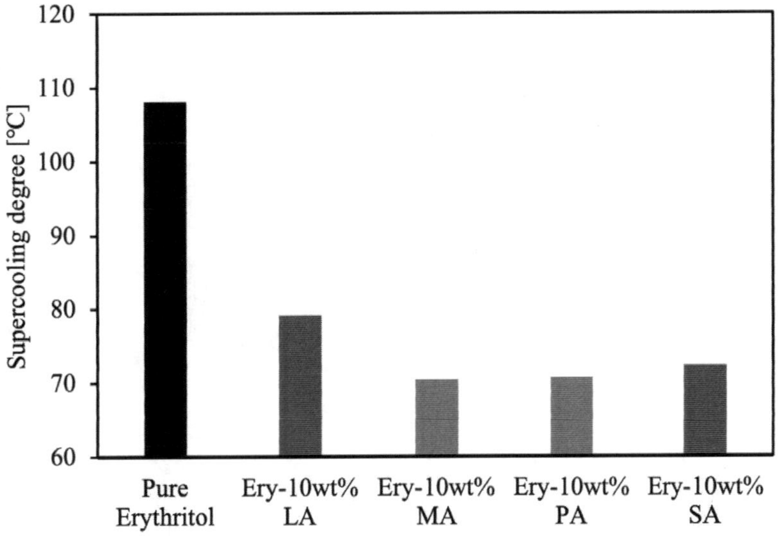

Fig.7 Supercooling degree of erythritol with 10 *wt%* of fatty acids.

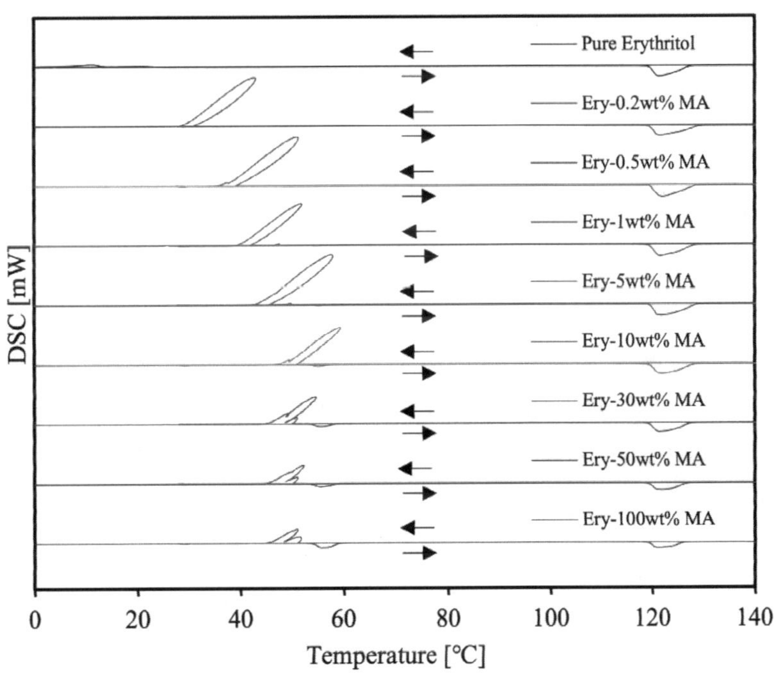

Fig.8 DSC curves of the erythritol with various amounts of MA.

IOP Conf. Series: Materials Science and Engineering **1318** (2024) 012028 doi:10.1088/1757-899X/1318/1/012028

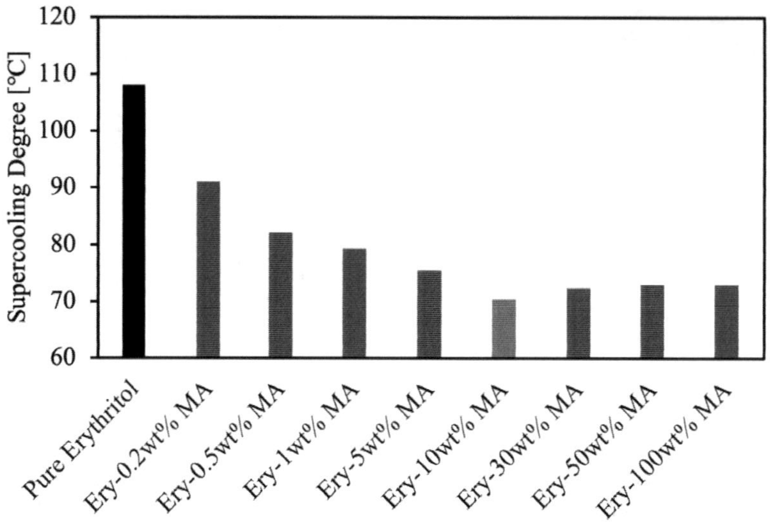

Fig.9 Supercooling degree of the erythritol with various amounts of MA.

4. Conclusion

We have investigated the mixed erythritol with fatty acids (lauric, myristic, palmitic, and stearic acids) as the nucleating agents to decrease the supercooling degree. Firstly, solidification behaviors of erythritol and various fatty acids were observed. Pure erythritol was melted at around 120°C and solidified at around 10°C, indicating a large supercooling phenomenon (supercooling degree: 108.1°C). All fatty acids were solidified at higher temperatures than that of erythritol. Secondary, while pure erythritol was solidified at lower temperatures than 10°C, the erythritols with fatty acids started to be solidified from around 50°C. Since the freezing points of the fatty acids are around 60°C, it was suggested that the particles of fatty acids were solidified and dispersed in the erythritol melt. This nucleation phenomena of fatty acids mitigated the supercooling of erythritol. The mixture of myristic or palmitic acids with erythritol significantly lowers the supercooling degree. The supercooling degree was decreased with larger amount of myristic acids until 10 *wt*% and slightly changed at larger than 10 *wt*%. This result suggested that the addition of 10 *wt*% is the most suitable for the supercooling mitigation of erythritol. When lauric and myristic acids were added, the entire samples with fatty acids also solidified near the freezing point of the fatty acids. Large latent heats of erythritol were observed, and these materials are expected to be practically implemented.

Acknowledgements

This study was supported by the Iwatani Naoji Foundation. The authors gratefully acknowledge the support provided by the institutions.

References

[1] T. Yoshio, "The Present State and Problems of the Japanese Energy Structure" IATSS Review, 38 (2013) 82–89.

[2] F. Martins, C. Felgueriras, M. Smitkova, and N. Caetano, "Analysis of Fossil Fuel Energy Consumption and Environmental Impacts in European Countries" Energies., 12 (2019) 964.

[3] I. Dincer and M.A. Rosan, "Thermal energy storage – systems and applications", Wiley, (2002).

[4] G. Antoni, M. Medrano, I. Martorell, A. Lázaro, P. Dolado, B. Zalba, and L.F. Cabeza "State of the art on high temperature thermal energy storage for power generation. Part 1 – Concepts, materials and modellization", Renew. Sust. Energ., 14 (2010) 31–55.

[5] N. Takahiro, N. Okinaka, and T. Akiyama, "Impregnation of porous material with phase change material for thermal energy storage" Mater. Chem. Phys., 15 (2009) 846–850

[6] S. Shile, S. Wu, C. Guo, J. Liang, Q. Yang, G. Xu, and J. Deng, "The effects of modified carbon nanotubes on the thermal properties of erythritol as phase change materials" Energy Convers. Manag., 157 (2018) 41–48.

[7] M. Yuan, F. Ye, and C. Xu, "Supercooling study of erythritol/EG composite phase change materials" Energy Procedia, 158 (2019) 4629–4634.

[8] J.L. Zeng, L. Zhou, Y.F. Zhang, S.L. Sun, Y.H. Chen, L. Shu, L.P. Yu, L. Zhu, L.B. Song, and Z. Cao, "Effects of some nucleating agents on the supercooling of erythritol to be applied as phase change material" J. Therm. Anal. Calorim., 129 (2017) 1291–1299.

[9] S. Shen, S. Tan, S. Wu, C. Guo, J. Liang, Q. Yang, G. Xu, and J. Deng, "The effects of modified carbon nanotubes on the thermal properties of erythritol as phase change materials" Energy Convers. Manag., 157 (2018) 41–48.

[10] X. Tang and J. Ryu, "Reduction in the Degree of Supercooling of the PCM D-mannitol Using Zeolite as a Nucleating Agent", Journal of the Japan Institute of Energy., 102 (2023) 41-50.

[11] X. Tang and J. Ryu, "Polymorphism of d-Mannitol in Zeolite-Added d-Mannitol-Based Phase Change Materials and Its Effect on Supercooling Degree", International Journal of Thermophysics., (2023) 44 -58.

IOP Conf. Series: Materials Science and Engineering **1318** (2024) 012029 doi:10.1088/1757-899X/1318/1/012029

Production characteristics in initial pyrolysis of lignin based on reactive molecular dynamics simulations

R Kameda[1], Y Sakurai[1,2], M Hiratsuka[3] and J Kobayashi[3]

[1] Graduate School of Engineering, Kogakuin University, 2665-1 Nakano-machi, Hachioji-shi, Tokyo, 192-0015, Japan
[2] Technical research institute, R & D Headquarters, Hitachi Zosen Corporation, 2-11, Funamachi 2-chome, Taisho-ku, Osaka 551-0022, Japan
[3] School of Engineering, Kogakuin University, 2665-1 Nakano-machi, Hachioji-shi, Tokyo, 192-0015, Japan

ad22501@g.kogakuin.jp

Abstract. Carbonization, direct liquefaction, and gasification, which are based on pyrolysis reactions, have been proposed as efficient technologies for energy recovery from biomass resources. However, modeling pyrolysis reactions of biomass with complex molecular structures is challenging, and numerical analysis methods that consider product compositions have not yet been established. In this study, molecular dynamics simulations were performed using LAMMPS, with the force field parameterized using ReaxFF. A molecular model based on Adler softwood lignin was created. The atomic trajectories were analyzed to identify the products of the pyrolysis process, and changes over time in inorganic gas, organic gas (C1-C4), light tar (C5-C14), heavy tar (C15-C39), and char (C40+) during the initial pyrolysis of lignin were investigated. The distribution of compositions formed and the formation behavior of typical low-molecular-weight components (such as formaldehyde and methanol) and inorganic gases from lignin were evaluated. The characteristics of the products in the initial pyrolysis of lignin, as analyzed from the molecular dynamics simulations, were based on the functional groups in the molecular model of lignin. Furthermore, similarities with the product compositions identified in experiments were also confirmed.

1. Introduction

Renewable energy, a non-fossil energy source, is gaining attention as a means to mitigate global warming. Biomass resources are becoming increasingly valuable due to their ability to provide a stable energy supply independent of climate conditions, and they are the only organic resources that can be regenerated. Direct combustion, carbonization, and gasification have been proposed as methods for energy recovery from biomass resources. However, direct combustion has low thermal efficiency, and the quality of the fuel biomass leads to fluctuations in the energy recovery rate. Pyrolysis gasification, on the other hand, aims to convert biomass resources into gases such as hydrogen and carbon monoxide for utilization [1]. Lignin, along with cellulose and hemicellulose, constitutes a major component of biomass, accounting for 95% of the total. Cellulose and hemicellulose have a linear polymerized molecular structure composed of sugars, while lignin is characterized by its aromatic nuclei. The pyrolysis of lignin occurs over a wide temperature range of 473-1073 K, with different pyrolysis reactions taking place at different temperatures [2]. During primary pyrolysis, the functional groups of the aromatic nuclei are released, and ether bonds between aromatic nuclei are cleaved. Secondary pyrolysis results in the transformation of the aromatic nuclei structure. As the products of primary pyrolysis contain radicals, these highly reactive primary pyrolysis products polymerize as secondary pyrolysis progresses, leading to complex product compositions. Experimental studies can help determine the final production volume and product composition. However, understanding the microscopic phenomena of reaction pathways, generation of intermediate products, and their characterization poses challenges.

Reaction molecular dynamics has garnered attention as a method for numerically analyzing complex pyrolysis reactions of polymers that occur sequentially and in parallel. It is a computer simulation method that examines the static and dynamic changes of atoms, molecules, and their structures on a microscopic scale. Molecular dynamics methods rely on experimental semi-empirical potential parameters to describe the breaking and forming of interatomic bonds. In conventional methods, these parameters remain unchanged throughout the calculation, making it impossible to represent chemical reactions such as bond cleavage and formation using force fields. To address this limitation, van Duin et al. developed the reactive force field (ReaxFF), which enables the modeling of chemical reactions [3].

In this study, we employ ReaxFF in reaction molecular dynamics simulations to elucidate the pyrolysis behavior of lignin. Furthermore, we evaluate the distribution of product composition and examine the formation behavior of typical small molecules.

2. Modeling methodology

2.1 Lignin model construction

The process of lignin formation is believed to involve three steps: enzymatic generation of phenoxy radicals, radical coupling, and stabilization of the resulting radical coupling entities. These reactions occur iteratively and lead to the polymerization and formation of lignin. This process results in the creation of various aromatic nuclei and their bonding patterns. Several models have been proposed to describe the chemical structure of lignin, including the softwood lignin model by Freudenberg and the hardwood lignin model by Nimz [4][5]. In this study, the softwood lignin model developed by Adler was utilized [6].

Molecular modeling of lignin was conducted using the WinmostarTM software, a versatile molecular dynamics tool [7]. A molecule consisting of 397 particles was constructed, referring to the schematic diagram of the lignin model used in Zhang et al.'s study on lignin pyrolysis [8]. Figure 1 depicts the lignin model with 397 particles. The primary functional groups in the aromatic nuclei of this model are hydroxy, carbonyl, and methoxy groups. The bonding patterns between aromatic nuclei encompass ether bonds (α-O-4, β-O-4, 4-O-5, β-O-β), carbon-carbon bonds (β-1, β-2, 5-5), and mixed bonds (α-O-4 & β-5, β-β & γ-O-α). Using this molecule as the fundamental unit, a larger molecule with 1,187 particles was generated by employing the software's polymerization function. The particle count in the system was further increased to 10,683 ($C_{4320}H_{4824}O_{1539}$) by including nine molecules with a particle count of 1,187. The simulation cell size was adjusted to yield an initial density of 0.5 g/cm^3. To validate the initial positions of atoms, the bond lengths within the created molecular models were compared to typical interatomic distances. Figure 2 illustrates the lignin molecular model, and Table 1 presents the bonding patterns and their quantities in the model.

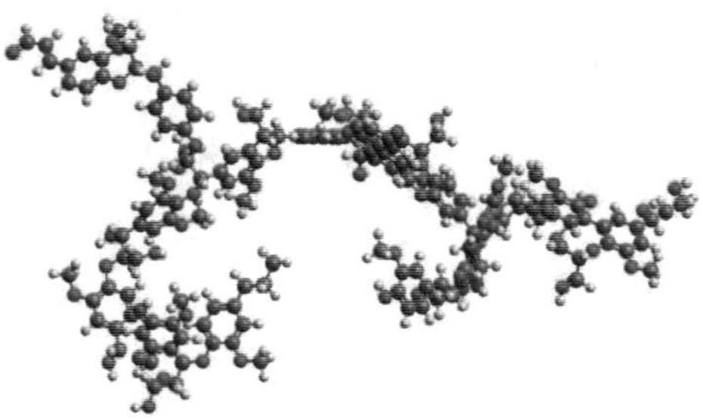

Figure 1. Snap shot of lignin molecular model

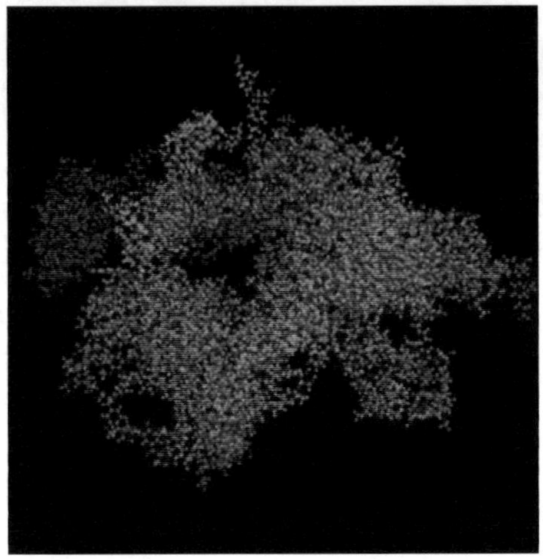

Figure 2. Snap shot of lignin molecular model with 10,683 particles

Table 1. Linkages and their amounts in the lignin model with 10,683 atoms

Linkage	Number	Linkage	Number
α-O-4	81	β-1	27
β-O-4	162	β-2	27
4-O-5	27	5-5	27
β-O-β	18	α-O-4 &β-5	27
		β-β &γ-O-α	27

2.2 Simulation strategy and details

The lignin molecule created in section 2.1 is an artificially constructed molecular model. Therefore, it is necessary to bring the structure to a thermodynamically equilibrium state. To achieve this, equilibration calculations were conducted using LAMMPS, a classical molecular dynamics program. First, energy minimization calculations were carried out to obtain the structure with the minimum energy from the initial configuration. Subsequently, equilibration calculations were performed on the lignin molecular model using the NPT ensemble, which maintains constant particle number, temperature, and pressure. The temperature was set to 298 K, the pressure to 1 atm, and the simulation time to 100 ps.

Table 2 presents the simulation conditions employed in this study. Following the equilibration calculations, the NVT ensemble was utilized for calculations at temperatures of 1273 K, 1473 K, 1673 K, and 1873 K. The simulation duration was set to 1000 ps, with a time step of 0.25 fs. ReaxFF, a reactive force field capable of representing the breaking and forming of chemical reactions between atoms, was employed in the simulations. For the potential parameter set, we utilized ReaxFF CHO, developed by Chenoweth et al. for simulating hydrocarbons based on previous investigations [9].

Table 2. Simulation conditions

Ensemble	Potential set	Time step [fs]	Number of steps	Simulation time [ps]	Temperature [K]
NVT	ReaxFF_CHO	0.25	4,000,000	1,000	1273, 1473, 1673, 1873

2.3 Trajectory analysis

The output data obtained from the reaction molecular dynamics simulation comprises atomic coordinates and physical quantities, but does not include information about the interatomic bonds. Therefore, interatomic distances were calculated based on the atom coordinates, and appropriate bond distances were determined by referring to typical covalent bond lengths. In this study, we developed a Python program specifically for this analysis. The determination of bond distances involved analyzing histograms of bond lengths. When defining the molecules, we ensured that each atom did not exceed the maximum number of covalent bonds according to theoretical considerations. Additionally, we verified that the number of atoms determined by the analysis matched the initial structure.

3. Results and discussion

3.1 Pyrolysis generation behavior related to simulation temperature

The pyrolysis behavior of lignin was investigated for the three states of matter: gas, liquid, and solid. To classify the products, a commonly used method based on carbon number in lignin pyrolysis was employed [10]. Gas products were defined for carbon numbers 1-4, light tar for carbon numbers 5-13, heavy tar for carbon numbers 14-39, and Char for carbon numbers 40 and above. Figures 3a, 3b, 3c, and 3d depict the temperature-dependent pyrolysis product behavior for each product category. The vertical axis in each figure represents the mass fraction relative to the initial lignin mass.

Figure 3a illustrates the formation behavior of gas products. At 1273 K, the yield at 1000 ps was 5.94 wt%, while it increased to 18.7 wt% at 1873 K. The gas yield showed an upward trend with increasing temperature. The formation of gas components was observed immediately after the simulation began, indicating that gas components are formed directly from the initial lignin structure during pyrolysis.

Figure 3b demonstrates the formation behavior of light tar components. The yields at 1273 K and 1473 K exhibited a monotonically increasing trend, with final yields of 10.4 wt% and 23.0 wt%, respectively. However, at 1673 K and 1873 K, the yields increased until around 400 ps and then stabilized.

Figure 3c presents the formation behavior of heavy tar components. At 1273 K, the yield monotonically increased, reaching 25.4 wt% at 1000 ps. For 1473 K, the yields increased until around 500 ps, and for 1673 K, until around 300 ps, after which they tended to stabilize. At 1873 K, the yield increased until approximately 100 ps and then started to decrease. The maximum yield observed was 51.4 wt% at 124 ps, while the yield at 1000 ps was 42.0 wt%. This decrease in yield can be attributed to the re-pyrolysis of the generated heavy tar, which decomposes into light tar and gas.

Figure 3d illustrates the formation behavior of the char component. The final char yields were 57.9 wt% at 1273 K and 19.6 wt% at 1473 K. The pyrolysis of char was more pronounced at higher temperatures. Additionally, the yield was 9.94 wt% at 1673 K and 8.65 wt% at 1873 K. The char yield stabilized after 200 ps at 1673 K and after 50 ps at 1873 K, indicating that the molecular structure of char becomes stabilized above 1673 K.

IOP Conf. Series: Materials Science and Engineering **1318** (2024) 012029 doi:10.1088/1757-899X/1318/1/012029

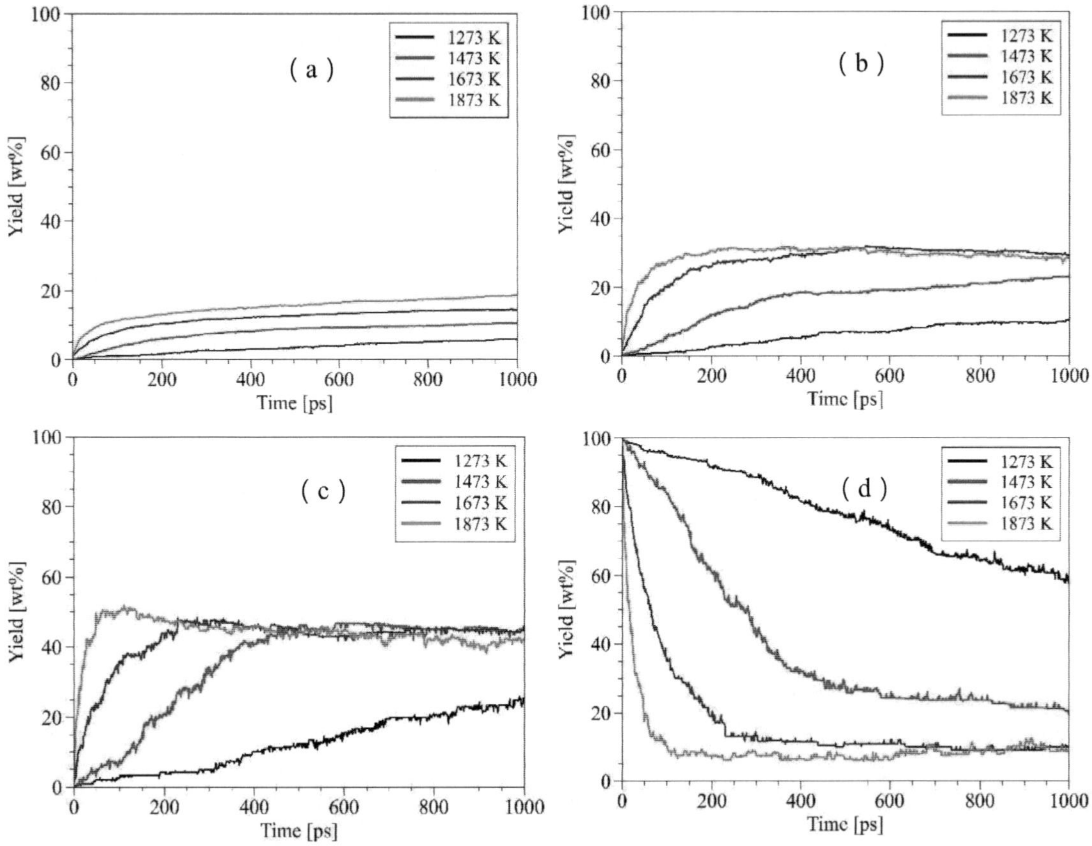

Figure 3. The overall picture of the product evolution of the four temperatures ((a) Gas, (b) Light tar, (c) Heavy tar, (d) Char)

3.2 Evaluation of char molecular structure by elemental ratio

The elemental ratios of the char obtained at each simulation temperature were calculated. For each char molecule, the H/C and O/C ratios were determined, and the average values were computed across all molecules. Figure 4 illustrates the temporal changes in the elemental ratios. It can be observed that both the H/C and O/C ratios decrease as the temperature increases. This trend suggests that the oxygen-containing functional groups in the aromatic nuclei undergo decomposition during thermal degradation. Notably, the primary functional groups in lignin comprise hydroxy and methoxy groups. The decomposition of these oxygen-containing functional groups becomes more pronounced at higher temperatures, leading to a greater reduction in H/C and O/C ratios. Furthermore, hydrogen radicals gradually combine with the decomposed oxygen-containing functional groups in the aromatic nuclei. As the molecular structure of the char becomes more stable, the char formation behavior depicted in Figure 3d tends to exhibit increased stability.

IOP Conf. Series: Materials Science and Engineering **1318** (2024) 012029 doi:10.1088/1757-899X/1318/1/012029

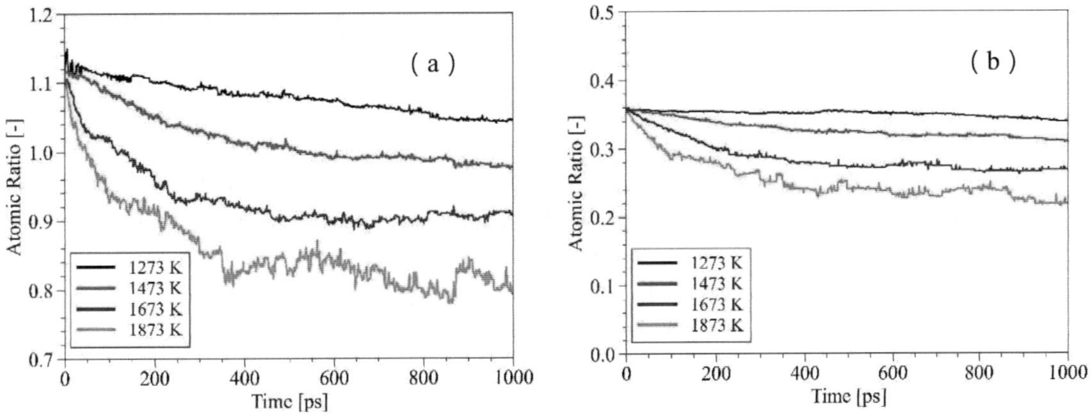

Figure 4. Effect of pyrolysis temperature on atomic ratio in char ((a) H/C, (b) O/C)

3.3 Generation behavior of major low-molecular-weight products

In the pyrolysis of lignin, our focus was on the abundant formation of low-molecular-weight products. We investigated the temporal changes in the production of these molecules and elucidated their formation pathways. The predominant low-molecular-weight products were water and formaldehyde. Figure 5 illustrates the elucidation of water and formaldehyde generation behavior. In terms of the formation pathway, water may arise from the dehydrogenation of char and the desorption of hydroxy groups present in the initial lignin structure. As discussed in section 3.1, the initial stage of lignin pyrolysis involves the desorption of functional groups from the aromatic nuclei. The subsequent step involves the cleavage of bonds between aromatic nuclei. Formaldehyde is primarily formed through the desorption of functional groups, particularly methoxy groups, making it a significant product in the early stage of pyrolysis.

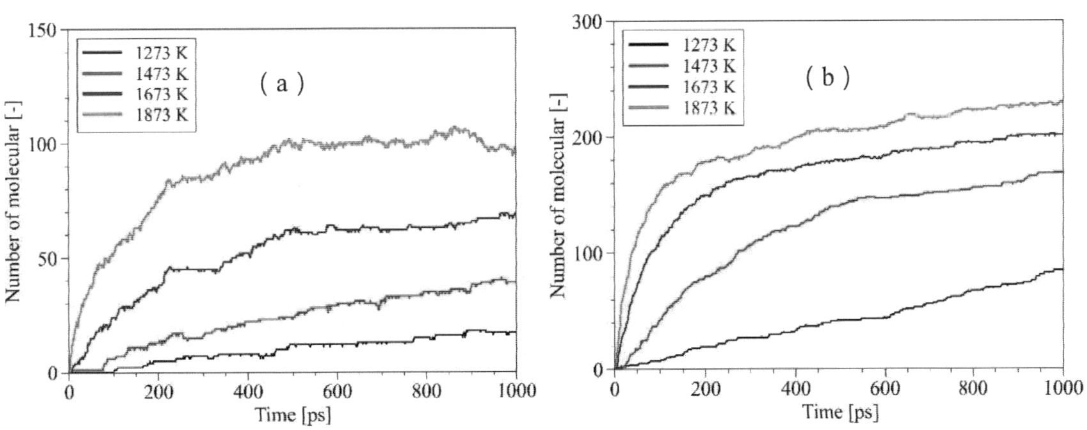

Figure 5. Time evolution of species from lignin model pyrolysis ((a) water, (b)formaldehyde)

4. Conclusion

Reaction molecular dynamics simulations using ReaxFF were conducted to investigate the pyrolysis behavior of lignin, a complex process primarily attributed to the presence of aromatic nuclei in its molecular structure. The results revealed that higher temperatures led to decreased char yields, while gas and tar yields increased. Low molecular weight products were generated either directly from the initial lignin structure or through further pyrolysis of existing pyrolysis products. The primary factor contributing to the decrease in H/C ratio was the cleavage of functional groups within the aromatic core. Water and formaldehyde were identified as the predominant low molecular weight products, originating from the cleavage of functional groups within the aromatic nucleus and the breaking of bonds between aromatic nuclei.

Reference

[1] Okada T, Okumura Y and Okazaki K 2009 *Journal of the Japan Institute of Energy* **88** 301-9
[2] Asmadi M and Kawamoto H 2011 *Journal of analytical and Applied Pyrolysis* **92** 417-25
[3] Adri van Duin C T, Siddharth D, Francois L and William G A 2001 *The Journals of Physical Chemistry* **105** 9396-409
[4] Freudenberg K and Neish A C 1968 *Constitution and Biosynthesis of Lignin* **2**
[5] Nimz H 1974 *Angewandte Chemie International Edition in English* **13** 313-21
[6] Adler E1977 *Wood Science and Technology* 33 169-218
[7] Winmostar V11 X-Ability Co. Ltd., Tokyo, Japan, 2022
[8] Zhang T, Li X, and Guo L 2017 *Langumuir* **33** 11646-57
[9] Chenoweth K, Adri van Duin C T and Willism G A 2008 *Journal of analytical and Applied Chemistry* **112** 1040-53
[10] Chen C, Zhao L, Wang J and Lin S 2017 *Industrial & Engineering Chemistry Research* 56 12276-88

Acknowledgments
A part of this research was supported by Hitachi Zosen Corporation.

IOP Conf. Series: Materials Science and Engineering **1318** (2024) 012030 doi:10.1088/1757-899X/1318/1/012030

Ranking Energy Storage Technologies Using Analytical Hierarchy Process (AHP) and VIKOR

Ron David M. Cruzat, Jerico R. Fojas, Ernest Andre S. Villena, Marcus O. Uy, Joseph R. Ortenero and Raymond R. Tan

Department of Chemical Engineering, De La Salle University, 2401 Taft Avenue, Manila 0922, Philippines

ron_cruzat@dlsu.edu.ph, jerico_fojas@dlsu.edu.ph, joseph.ortenero@dlsu.edu.ph, raymond.tan@dlsu.edu.ph

Abstract. Due to increasing global average temperatures, it is imperative to increase investment in renewable energy and energy storage technologies to reduce the negative impact to the environment. As such, improvements associated with renewable energy intermittency and variability is a priority. One solution is through the use of energy storage technologies (ESTs) to improve the reliability and stability of power supplies. ESTs come in various selections, each with its own application and attributes. Since no EST is universally applicable, utilization of MADM methods helps in narrowing selected technologies in a specific application. In this study, a hybrid AHP and VIKOR method is utilized for stationary power applications. AHP is used for determining weights associated with each EST attribute from surveyed experts in academe and industry. The attributes efficiency and cycle life obtained the highest weights of 0.2384 and 0.2322 respectively. VIKOR is used in ranking ESTs based on available data gathered. Supercapacitor is ranked as best alternative due to its cycle efficiency, storage lifetime, power capacity, and power density. Subsequent rankings are flywheel, pumped-hydro, and lithium-ion battery. Technology readiness level and specific application details could be incorporated in further studies to improve ranking. This study demonstrates the potential of hybrid AHP-VIKOR in guiding researchers and decision makers select the appropriate EST given competing performance in different criteria.

1. Introduction

Global warming is a phenomenon that has been getting much traction in the past decades, being an emerging threat to current and future generations. Human activities are closely linked to global warming, specifically, the use of fossil fuels which is used to supply energy demands contribute to 89% of global CO_2 emissions which raised the average global temperature by 0.3°C to 1°C [1]. This makes it the single biggest contributor to global temperature increase. Fossil fuels are used primarily for three reasons: (1) efficacy when used as fuels; (2) low cost compared to alternative sources of energy; (3) logistically well-established [2]. An eventual switch to renewable sources should be prioritized to mitigate global warming and its effects. The emergence of renewable energy through the efforts of international organizations such as the UN and the actions undertaken by numerous countries has been sanctioned to promote a net zero-carbon alternative post-industrial revolution. However, the feasibility of such a proposal dictates the need for investment and innovation in the energy sector. It is also necessary to listen to financial and engineering concerns to understand the challenges associated with renewable energy [3].

Energy storage systems involve technologies that harness the electrical energy provided by various sources of energy, which is then converted to a form that is suitable for storage [4]. For renewable energy sources like wind and solar energy, energy must be collected and stored until a time of use, then must have the ability to release the energy from containment. Through energy storage systems, peak shaving, and load leveling is achieved, in which a balance is met between supply and demand, with the added benefit of evening out the discrepancies in renewable power from distribution generations [5]. Energy storage technologies (ESTs) are extensively utilized through different applications in the industrial and transportation sectors. It provides advantages within energy systems such as increased energy generation and economic performance from renewable energy sources [6]. Energy storage technologies can be categorized into a multitude of classifications based on system field applications, energy densities, power density, storage capacity, cost, efficiency, and lifetime [7]. Examples of energy storage classifications are chemical energy storages (hydrogen storage), electrochemical energy storage (batteries), electrical energy storages (capacitors, supercapacitor), mechanical storages (flywheel, pumped hydro, compressed air), and thermal energy storages (sensible heat, latent heat). Evaluating ESTs can be a challenge due to the complex criteria to consider in determining the performance of a specific technology. Since multiple technologies are being evaluated according to their technical and economic performance, the methodology must be able to take these attributes into account in its decision-making process.

Multi-attribute decision-making (MADM) models achieve all these conditions and is a forefront methodology in numerous studies ranking ESTs [8]. MADM excels in solving problems involving a finite number of alternatives and a set of complex attributes by creating a hierarchical structure for the evaluation of the problem and using it to find the best solution. This study combines the two MADM methods: the analytical hierarchy process (AHP) and the multi-criteria optimization and compromise solution method, VlseKriterijumska Optimizacija I Kompromisno Resenje (VIKOR) in the ranking of ESTs. These were chosen because the combination of these methodologies allows for the creation of a procedure that achieves a high level of accuracy, certainty, and reliability in its results. An AHP-VIKOR model also accounts for the complex nature of the raw data of ESTs, being a popular model in ranking problems such as the ranking of maritime transportation [9], and fertilizer production systems [10]. Both share the same trait of their evaluation criteria being subjected to complex economic, technological, and environmental criteria. While the ranking of ESTs using MADM has already been done previously [11, 12], none have ranked them using an AHP-VIKOR model.

Different ESTs are used in various applications, with each technology defined by its unique specifications, with no single technology dominating all technical and economic aspects. As such, a systematic process such as the AHP-VIKOR model would be useful to decision-makers in selecting the most appropriate EST for stationary power applications. This model would determine the weights of the attributes based on relative importance using AHP, then utilize these attributes to rank each specific EST using VIKOR. In terms of ranking, the development of a computational methodology for ranking ESTs under the criteria of each technical (energy capacity, power capacity, energy density, power density) and economic (power capital cost, energy capital cost) attributes using hybrid MADM methods is prioritized.

While the ranking of ESTs has been the topic of numerous studies; limited research is available involving a hybrid AHP-VIKOR methodology focused on stationary electrical applications. This paper addresses that research gap by using these models to obtain relevant weights of the technologies' attributes and subsequently rank ESTs to determine the best alternative for stationary power applications. Data for the attributes was obtained from the review published by Koohi-Fayegh and Rosen (2020) [4], with the paper's ancestral references being used as the data repository. This was done to get a more direct source of data, with this specific review paper only being used as a library.

The ESTs covered in this study are the following: hydrogen, Li-ion, Lead acid, Ni-Cd, capacitor, supercapacitor, flywheel, pumped-hydro, compressed air, sensible heat, and latent heat. These ESTs are broadly classified under chemical (hydrogen), electrochemical (Li-ion, Lead acid, Ni-Cd), electrical (capacitor, supercapacitor), mechanical (flywheel, pumped-hydro, compressed air), and thermal (sensible heat, latent heat) sectors. The various technical specifications/attributes used to evaluate each EST were based on the available data presented in the review by Koohi-Fayegh and Rosen (2020): power capacity, power density, energy density, cycle efficiency, storage lifetime, power capital cost, energy capital cost. These ESTs were chosen for their detailed review discussions and technological maturity in industrial applications. The data were treated as a non-Fuzzy set. Finally, there is large uncertainty in the pairwise comparison part of AHP since determining which technical specification attribute is more important relative to other attributes is highly subjective, which leads to consulting experts at academe and industry levels.

2. Methodology

The data relies heavily on the values collected for EST attributes presented in the review done by Koohi-Fayegh and Rosen (2020). Important technical specifications for ESTs such as power density and cycle efficiency were gathered from the references and tables within the review. Questions such as "Is power density more important than cycle efficiency?" need to be answered in the criteria due to the complexity of the numerous specifications of ESTs. It soon became evident that traditional single-attribute decision-making cannot be used in this study. Thus, this study settled on using a MADM methodology utilizing AHP and VIKOR for the judging criteria.

AHP was chosen because it can be used in a wide variety of criteria since it is up to the decision maker to customize the relative importance of each attribute. This makes it suitable for the study data as EST attributes are complex in nature and cannot be easily compared with each other. AHP allows for the establishment of the evaluation criteria by determining the criteria weights of each attribute. Once the criteria are established, another MADM methodology is needed to evaluate the storage technologies, which can be done using VIKOR [13]. The method evaluates each alternative by creating a multicriteria ranking index based on how close each one is to the optimal solution, and subsequently ranks them based on their index value [14]. The way the MADM methods were used to achieve the study goals is illustrated in **Figure 1**.

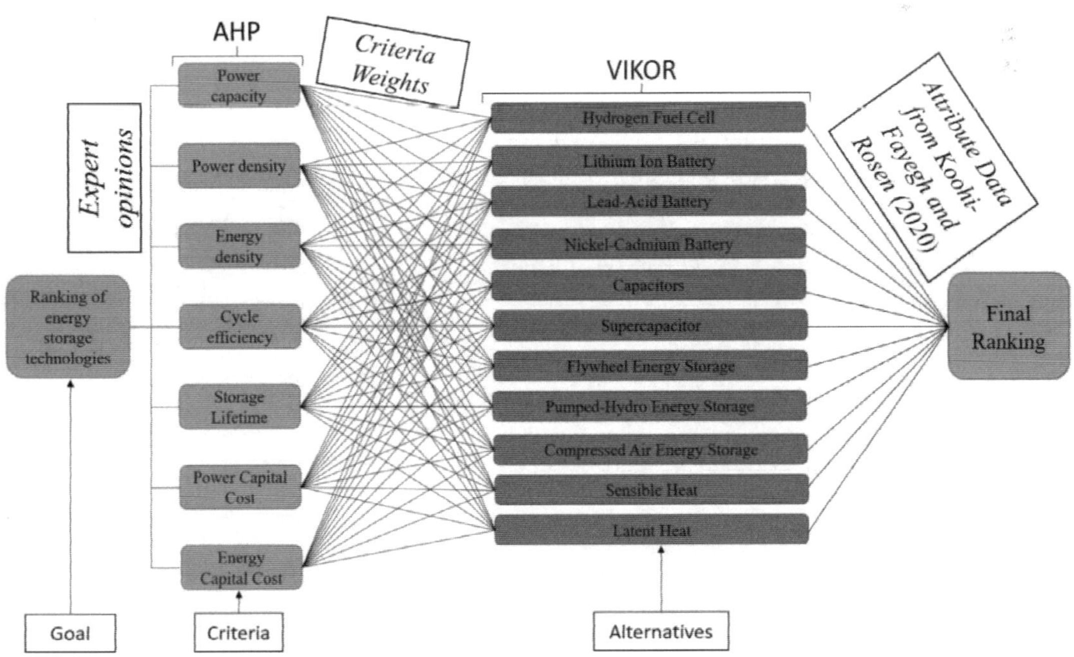

Figure 1. MADM Hierarchy structure for the ranking of ESTs

The finished methodology is divided into two distinct steps, the determination of AHP criteria weights via the distribution of questionnaires to experts, the checking of the consistency of the gathered weights data, and the ranking of ESTs using VIKOR. This process is summarized in **Figure 2**. Data from experts is needed since a consistent opinion is needed on the relative importance of each technical attribute. In addition, all calculations were done using Excel®.

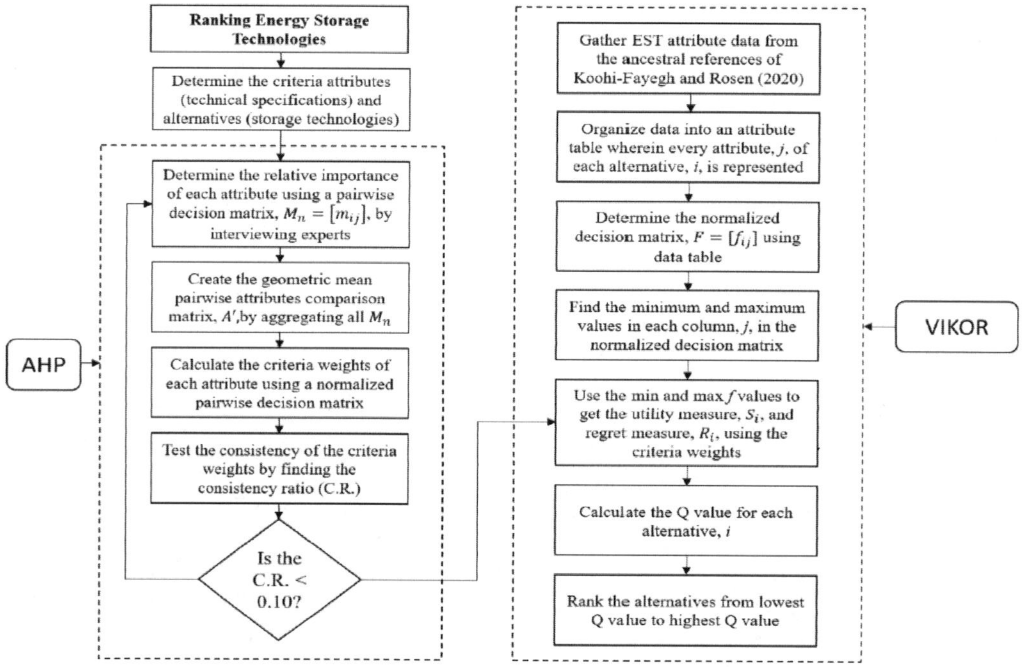

Figure 2. Methodology flowchart for the ranking of ESTs using AHP and VIKOR

2.1. Determination of criteria weights using Analytical Hierarchy Process (AHP)

The first step is to establish the evaluation criteria by determining the criteria weights, $\boldsymbol{w}$, associated to each criterion, $\boldsymbol{C}$, using AHP. This was done by creating a pairwise decision matrix, $\boldsymbol{M}$, where each technical specification was compared with each other, as seen in Equation (1). In each pairwise comparison, $\boldsymbol{m_{ij}}$, a value of scale of relative importance was assigned to each pairwise comparison following the fundamental AHP ratio scale, as seen in **Table 1**.

$$M = \begin{bmatrix} m_{11} & \cdots & m_{1j} & \cdots & m_{1n} \\ \vdots & & \vdots & & \vdots \\ m_{i1} & \cdots & m_{ij} & \cdots & m_{in} \\ \vdots & & \vdots & & \vdots \\ m_{n1} & \cdots & m_{nj} & \cdots & m_{nn} \end{bmatrix} \quad (1)$$

Table 1. Saaty AHP fundamental ratio rating scale [15][16]

Ratio Scale Value	**Description**
1	Equal Importance
3	Moderate importance
5	Essential or strong importance
7	Very strong importance
9	Extreme importance
2,4,6,8	Intermediate values between two adjacent judgements
Reciprocals (12, 13..., 19)	If activity ***i*** has one of the above numbers assigned to it when compared to activity j, then j has the reciprocal value when compared with i.

For the determination of the values for the sample pairwise comparison table, an AHP questionnaire was created wherein the respondents provided educated guesses on the values of each pairwise comparison, m_{nm}. The goal is to send the questionnaire to multiple experts and aggregate their data to determine a geomean value for each pairwise comparison. This subsequent average pairwise comparison table, ***A'***, will then be the main data that will undergo the AHP process. This methodology defines an EST expert as someone who has or is currently working with energy storage, preferably with one to two years of experience working in the energy storage sector. As such, the group reached out to researchers who published review studies covering various ESTs as well as field engineers, project managers, and researchers in R&D working in ACEN Philippines and San Miguel Corporation. They were contacted through the social platforms of Scopus, LinkedIn and ResearchGate, as well as a few personal contacts working in the energy storage sector. After every pairwise comparison in the matrix was assigned a value, the next step is to calculate the criteria weights, w_i, using Equation (2) by finding the average of the normalized values for each row of a normalized matrix, ***K***, as seen in Equation (3).

$$w_i = \frac{\sum_{j=1}^{j} m_{ij}}{j} \tag{2}$$

$$K = \begin{bmatrix} m_{11}/T_1 & \dots & m_{1j}/T_j & \dots & m_{1n}/T_n \\ \vdots & & \vdots & & \vdots \\ m_{i1}/T_1 & \dots & m_{ij}/T_j & \dots & m_{in}/T_n \\ \vdots & & \vdots & & \vdots \\ m_{n1}/T_1 & \dots & m_{nj}/T_j & \dots & m_{nn}/T_n \end{bmatrix} \tag{3}$$

2.2. *Testing the consistency of calculated weights*

After the criteria weights of each attribute were calculated, the next step was to test the consistency of the calculated weights, which gives the accuracy of the weight values. It was calculated by finding the consistency index (C.I) by finding λ_{max}, using Equation (4), from a ***M*** x ***w*** matrix, as seen in Equation (5). The consistency ratio (C.R.) is then subsequently calculated using Equation (6), utilizing C.I. and R.I, with the latter's value corresponding to the number of attributes present, which is seen in **Table 2**. The goal was to find a value of C.R. < 0.10 in order to maintain consistency and accuracy [16].

$$\lambda_{max} = \frac{\sum_{j=1}^{j} \lambda_j}{j} \tag{4}$$

$$M \times w = \begin{bmatrix} m_{11} & \cdots & m_{1j} & \cdots & m_{1n} \\ \vdots & & \vdots & & \vdots \\ m_{i1} & \cdots & m_{ij} & \cdots & m_{in} \\ \vdots & & \vdots & & \vdots \\ m_{n1} & \cdots & m_{nj} & \cdots & m_{nn} \end{bmatrix} \begin{bmatrix} w_1 \\ w_2 \\ \vdots \\ w_j \end{bmatrix} = \begin{bmatrix} WS_1 \\ WS_2 \\ \vdots \\ WS_j \end{bmatrix} \tag{5}$$

$$C.R. = \frac{C.I.}{R.I.} = \frac{\lambda_{max} - n}{(R.I.) * (n-1)} \tag{6}$$

2.3. Evaluation and ranking of ESTs via VIKOR

Once the consistency of the criteria weights was verified, the values were entered into the VIKOR method in the evaluation criteria of the process. Using technical data for each ESTs, a normalized decision matrix, $\boldsymbol{F = [f_{ij}]}$, was created where each value, $\boldsymbol{I_{ij}}$ corresponds to the performance of an EST alternative, $\boldsymbol{i}$, with respect to its technical/economical specification attribute, $\boldsymbol{j}$, as seen in Equation (7).

$$f_{ij} = \frac{I_{ij}}{\sqrt{\sum_{i=1}^{m} (I_{ij})^2}} \tag{7}$$

The next step after the normalized decision matrix was to calculate the utility value, $\boldsymbol{S_i}$, and the regret value, $\boldsymbol{R_i}$. The forms of these equations changes depending on whether the attribute in question is a beneficial attribute or a non-beneficial attribute, as seen from Equation (8) to Equation (11). Afterwards, the VIKOR index value (Q) is calculated using Equation (12), with the ranking corresponding to a decreasing index value.

For beneficial,

$$S_i = \sum_{j=1}^{n} \left[w_j * \frac{(f_j)_{max} - f_{ij}}{(f_j)_{max} - (f_j)_{min}} \right] \tag{8}$$

$$R_i = max \left[w_j * \frac{(f_j)_{max} - f_{ij}}{(f_j)_{max} - (f_j)_{min}} \right] \tag{9}$$

For non-beneficial,

$$S_i = \sum_{j=1}^{n} \left[w_j * \frac{f_{ij} - (f_j)_{min}}{(f_j)_{max} - (f_j)_{min}} \right] \tag{10}$$

$$R_i = max \left[w_j * \frac{f_{ij} - (f_j)_{min}}{(f_j)_{max} - (f_j)_{min}} \right] \tag{11}$$

VIKOR Index,

$$Q_i = v \left[\frac{S_i - (S_i)_{min}}{(S_i)_{max} - (S_i)_{min}} \right] + (1 + v) \left[\frac{R_i - (R_i)_{min}}{(R_i)_{max} - (R_i)_{min}} \right] \tag{12}$$

IOP Conf. Series: Materials Science and Engineering **1318** (2024) 012030 doi:10.1088/1757-899X/1318/1/012030

3. Results and Discussion

After gathering sufficient data from 9 experts in the academe and industry, the data was tabulated and aggregated. Using the AHP method, the data from the experts was then used to calculate the attribute criteria weights, which is shown in **Table 2**. It can be observed that the most important attributes are cycle efficiency and storage lifetime.

Table 2. Calculated Criteria Weights

	Criteria Weights (w_n)
Power Capacity	0.1001
Power Density	0.0695
Energy Density	0.1139
Cycle Efficiency	0.2384
Storage Lifetime	0.2322
Power Capital Cost	0.1001
Energy Capital Cost	0.1059

Using the following AHP criteria weights the consistency index and consistency ratio of the calculated criteria was obtained, having a value of 0.03 and 0.0233 respectively, which is summarized in **Table 3**. Since the consistency ratio is less than the maximum consistency of 0.1, this affirms that the responses gathered from the nine experts were consistent.

Table 3. AHP Consistency Ratio

λ_{max}	Consistency Index (C.I.)	Random Index (R.I.)	Consistency Ratio (C.R.)
0.0300	0.0300	1.2500	0.0233

Following the verification of the criteria weights, data for all ESTs were gathered and is an essential component for ranking the ESTs using VIKOR. The data for the EST attributes were obtained from references from Koohi-Fayegh and Rosen (2020), with its data being tabulated in **Table 4**.

Table 4. EST Attribute Data [4]

	PC (MW)	VPD (kW/m^3)	VED (kWh/m^3)	CE (%)	SL (cycles)	PCC ($/kW)	ECC ($/kWh)
HFC	50	500	3,000	50	1,000	10,000	20
LIB	0.1	360	750	95	15,000	4,000	2,500
PAB	20	400	90	90	2,000	900	1,100
NCB	40	141.05	150	90	10,000	1,500	3,500
CPC	0.05	100,000	10	70	50,000	400	1,000
SCPC	0.3	45,000	35	99	1,000,000	800	94,000
FWES	0.25	2,000	80	95	200,000	1,000	6,000
CAES	5,000	0.12	1.33	87	60,000	5,288	291.2
SH	300	10	20	89	30,000	1,500	50
LH	60	0	120	90	0	7,900	50

This data is then processed into the VIKOR method, with two categories being assigned to the chosen criteria in terms of beneficial and non-beneficial. All the ESTs except the latent and sensible heat are complete in the data. Both latent and sensible heat ESTs are part of thermal ESTs which are generally difficult to obtain data from as most thermal ESTs are used for heating rather than energy

IOP Conf. Series: Materials Science and Engineering **1318** (2024) 012030 doi:10.1088/1757-899X/1318/1/012030

storage. Most notably among the ESTs, the supercapacitor had the largest value for the most important weight criteria which are storage lifetime and cycle efficiency.

Among the criteria weights the power capital cost and energy capital cost at a high value will not be beneficial in selecting EST and as such the only two non-beneficial criteria. Using the VIKOR method to select an EST based on equal maximum group utility and individual regret using a *v*-value of 0.5 the ranking was determined in **Table 5**. In the final ranking, the top four ESTs are supercapacitors, flywheel, pumped-hydro, and Li-ion respectively, whilst the bottom three consists of sensible heat, lead acid, and hydrogen.

Table 5. Final VIKOR Ranking

EST	Qi	Rank
Hydrogen	1.0000	11
Li-ion	0.6672	4
Lead Acid	0.7056	9
Ni-Cd	0.7044	8
Capacitors	0.6701	5
Supercapacitors	0.0000	1
Flywheel	0.3826	2
Pumped-Hydro	0.5679	3
Compressed Air	0.6735	6
Sensible Heat	0.8137	10
Latent Heat	0.6810	7

3.1. Sensitivity Analysis

Sensitivity analysis is done on the effect of differing *v*-values used in the VIKOR process. In the ranking, the VIKOR process was done using a *v*-value of 0.5 which gives the ranking that equally compromises between utility and regret. This can be seen in **Figure 3**, in which a change in *v*-value can change the ranking of those ranked 4th to 9th. A higher *v*-value gives more precedence to utility while a lower *v*-value gives more precedence to regret.

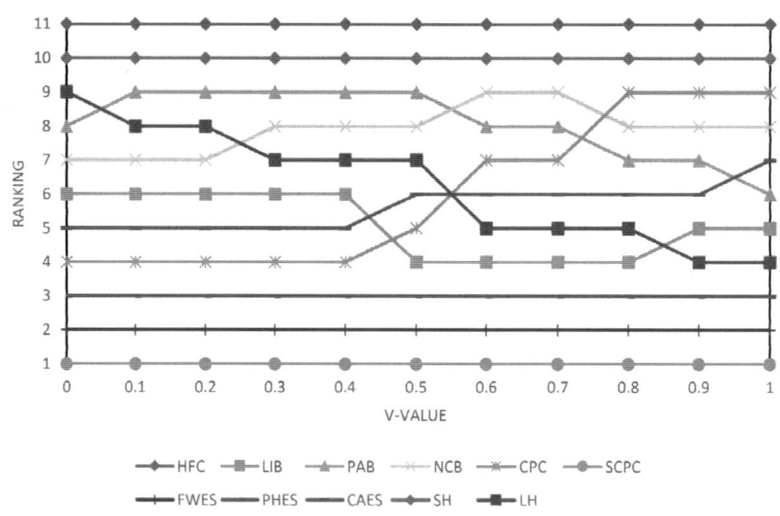

Figure 3. EST ranking under different *v*-values

IOP Conf. Series: Materials Science and Engineering **1318** (2024) 012030 doi:10.1088/1757-899X/1318/1/012030

Further sensitivity analysis is done by determining the effects of the power capacity attribute and power density attribute on the final ranking. The effect of power capacity is shown in **Figure 4**, where it can be inferred that the power capacity attribute has a large effect on the ranking, indicating that it is an attribute of importance when considering ESTs. However, when it comes to supercapacitors and flywheels, it is not a driving factor since their rankings hardly changed.

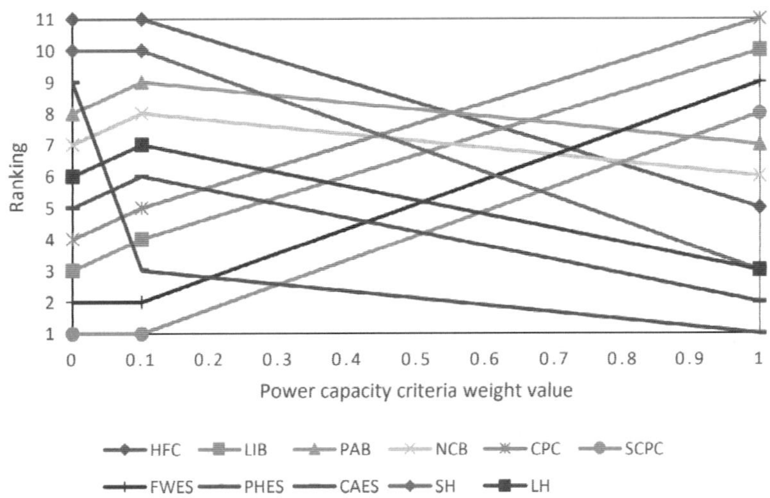

Figure 4. Effect of power capacity criteria weight on EST Ranking

On the other hand, when analyzing the effect of power density, drastic changes in ranking are observed notably in the capacitor EST, which is summarized in **Figure 5**. Since the ranking of capacitors increases rapidly the more important power density is, it can be concluded that power density is the biggest strength of capacitors. Like power capacity, power density has little effect on the top two ESTs, indicating that power density is not a driving factor as to why supercapacitors and flywheel are ranked so highly.

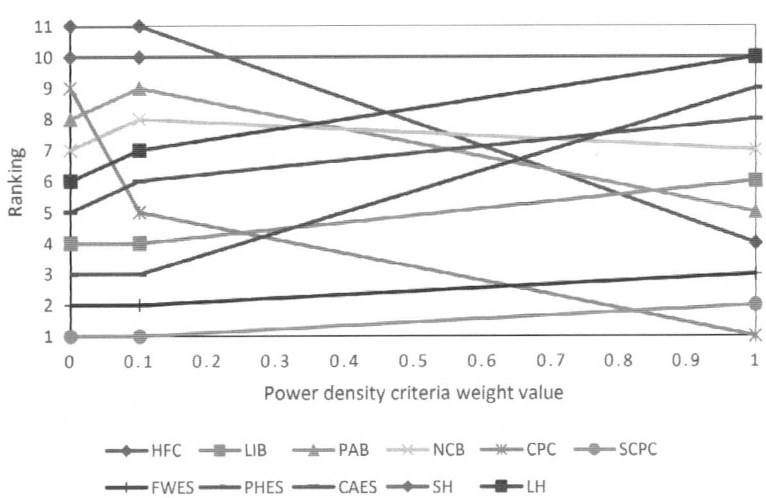

Figure 5. Effect of power density criteria weight on EST Ranking

3.2. Ranking Analysis

When looking at the final ranking and looking deeper into each EST, their ranking starts to make sense. Supercapacitors (SCPC) are ranked as the best overall EST under the methodology's criteria. This can be attributed to supercapacitors having the highest cycle efficiency as well as the longest storage lifetime. However, its disadvantages include its high self-discharge rate, low cycle duration, and limited use in large-scale applications. The next best is flywheel storage (FWS), which are characterized by their high cycle efficiency at 95% as well as their high storage lifetime at 200,000 cycles. These specifications are due to flywheel being technologically mature, with further developments being slow, albeit still showcases potential. Furthermore, pumped hydro storage (PHS) and lithium-ion batteries (LIB) are ranked third and fourth respectively. Pumped hydro is ranked highly due to its large volume, long storage lifetime, and its maturity, with the disadvantages only being its potential for improvement is low as it is already a mature EST. Lithium-ion is one of the most developed ESTs with commercial success at low-scale energy storage. Despite its use, lithium-ion technology has yet to fully mature with improvements. These four ESTs are the overall best ESTs, with their places being easily swapped around depending on the application.

The middle tier of ESTs consists of capacitors (CPC), compressed air energy storage (CAES), latent heat (LH), nickel-cadmium batteries (NCB), and lead-acid batteries (PAB) as the fifth to ninth places respectively. Capacitors are suitable for applications that do not require a high energy density frequency. Unlike supercapacitors, capacitors are affected by their size which affects their costs and use for stationary energy storage applications. Compressed air is similar to pumped-hydro storage as it requires geological formations to use. In terms of technological maturity, it is considered to be developed, however its rarity makes its potential limited. Its cycle efficiency and storage lifetime are considered middle. Latent heat is part of thermal-based ESTs which are used primarily in heating applications. Its rarity impedes its ranking as power density and storage lifetime are not noted. Nickel-cadmium main characteristic is its high energy density which is reliable in small-scale applications. However, it is impeded by its low power and high energy capital costs. Lead acid batteries are the oldest battery energy storage system making it one of the most technologically mature ESTs in the market. However, its attributes are considered average at best.

Lastly, the bottom tier ESTs consist of sensible heat (SH), and hydrogen fuel cells (HFC) as the worst-ranked ESTs. Sensible heat is a thermal ESTs that require high temperatures for energy storage applications. As with Latent Heat, its data rarity for power density and storage lifetime impedes its ranking. It also has the second-highest power capital cost. Hydrogen fuel cells are ranked as the worst EST, with it having the highest capital cost and low cycle efficiency of the rest of the ESTs in the study. However, hydrogen fuel cells show potential as their low technological maturity can lead to future development of this EST to become a more viable option in the future. In the end, these two ESTs might not be good at stationary electrical applications, however, their advantages can make them much viable options for other non-electrical energy applications or even in specific short-term electrical energy applications.

4. Conclusion

This study used a hybrid AHP and VIKOR methodology for ranking eleven ESTs for stationary power applications. Each chosen EST has been researched for its attributes, which were used for the VIKOR data. The weights for AHP have been measured based on the chosen aspects of each surveyed expert where they were asked to determine which attribute is more important for an EST to have. Nine experts agreed to participate in the study, and based on their data the criteria weights were calculated, with cycle efficiency and storage lifetime being regarded as the two most important attributes in an EST. Then, using attribute data from Koohi-Fayegh and Rosen (2020), ESTs were ranked via the VIKOR process with the parameters set to a v-value of 0.5, and the two cost attributes (PCC, ECC) set as the non-beneficial attribute, whilst the rest (PC, VPD, VED, CE, SL) is set as the beneficial attribute. The final VIKOR ranking showed that supercapacitor is ranked as the best option based on the educated opinions surveyed from contacted experts, with flywheel technologies and pumped hydro storage ranking in second and third respectively. The worst-performing energy storage technologies are sensible heat storage and hydrogen fuel cells given the parameters and the preferences of the experts. Additionally, the rankings of the top and bottom three alternatives are stable and insensitive to changes made in VIKOR parameters. The methodology can be adjusted such that the weights for the attribute can be changed if the opinion of the expert changes or additional experts give weightings to the attributes. This study shows that AHP in conjunction with VIKOR can guide researchers and decision makers in selecting the appropriate energy storage technology given multiple alternatives and criteria. The findings of the thesis suggest that this methodology can be used in determining the best alternative EST for stationary power applications. Other criteria such as geographical limitation are hard to quantify which makes implementation difficult for a certain scenario.

References

[1] Clientearth. (2020, November 11). Fossil fuels and climate change: the facts. Clientearth Communications. Retrieved from https://www.clientearth.org/latest/latest-updates/stories/fossil-fuels-and-climate-change-the-facts/Another reference

[2] Cheek, L. (2020, April 10). 3 Reasons Why We Are Still Using Fossil Fuels. Retrieved from https://wordpressua.uark.edu/sustain/3-reasons-we-are-still-using-fossil-fuels/

[3] ÓhAiseadha, C., Quinn, G., Connolly, R., Connolly, M., & Soon, W. (2020). Energy and Climate Policy—An Evaluation of Global Climate Change Expenditure 2011–2018. Energies, 13(18), 4839. MDPI AG. Retrieved from http://dx.doi.org/10.3390/en13184839

[4] Koohi-Fayegh, S., & Rosen, M. A. (2020). A review of energy storage types, applications and recent developments. Journal of Energy Storage, 27. Elsevier Ltd. https://doi.org/10.1016/j.est.2019.101047

[5] Fathima, A. H., & Palanisamy, K. (2018). Renewable systems and energy storages for hybrid systems. Hybrid-renewable energy systems in microgrids: Integration, developments and control, 147–164. https://doi.org/10.1016/B978-0-08-102493-5.00008-X

[6] Al Shaqsi, A. Z., Sopian, K., & Al-Hinai, A. (2020). Review of energy storage services, applications, limitations, and benefits. Energy Reports, 6, 288–306. https://doi.org/10.1016/J.EGYR.2020.07.028

[7] Guney, M. S., & Tepe, Y. (2017). Classification and assessment of energy storage systems. Renewable and Sustainable Energy Reviews, 75, 1187–1197. https://doi.org/10.1016/J.RSER.2016.11.102

[8] Mohaghar, A., Moradi-Moghadam, M., Khanmohammadi, E., & Molavi, M. (2013). Applying Interval VIKOR and Fuzzy AHP Methods for Supplier Selection. New York Science Journal, 6(6). http://www.sciencepub.net/newyork

[9] Soner, O., Celik, E., & Akyuz, E. (2017). Application of AHP and VIKOR methods under interval type 2 fuzzy environment in maritime transportation. Ocean Engineering, 129, 107–116. https://doi.org/10.1016/J.OCEANENG.2016.11.010

[10] Shahnazari, A., Pourdej, H., & Kharage, M. D. (2021). Ranking of organic fertilizer production from solid municipal waste systems using analytic hierarchy process (AHP) and VIKOR models. *Biocatalysis and Agricultural Biotechnology*, 32, 101946. https://doi.org/10.1016/J.BCAB.2021.101946

[11] Ortenero, J. R., & Tan, R. R. (2021). Ranking Energy Storage Technologies with VIKOR. *Chemical Engineering Transactions*, *88*, 151-156. https://doi.org/10.3303/CET2188025

[12] Ortenero, J. R., Choi, A. E. S., & Promentilla, M. A. B. (2022). A Multi-Agent Decision-Making Model for the Ranking of Energy Storage Technologies. *Chemical Engineering Transactions*, *94*, 1141-1146. https://doi.org/10.3303/CET2294190

[13] Chatterjee, P., & Chakraborty, S. (2016). A comparative analysis of VIKOR method and its variants. *Decision Science Letters*, 5(4), 469–486. https://doi.org/10.5267/j.dsl.2016.5.004

[14] Tzeng, G.-H., and Huang, J.-J. (2011). Multiple Attribute Decision Making: Methods and Applications. *CRC Press/Taylor & Francis*.

[15] Coyle, G. (2004). THE ANALYTIC HIERARCHY PROCESS (AHP). *Practical Strategy*.

https://training.fws.gov/courses/references/tutorials/geospatial/CSP7306/Readings/AHP-Technique.pdf

[16] Saaty, R. W. (1987). The analytic hierarchy process—what it is and how it is used. *Mathematical Modelling*, 9(3–5), 161–176. https://doi.org/10.1016/0270-0255(87)904

IOP Conf. Series: Materials Science and Engineering **1318** (2024) 012031 doi:10.1088/1757-899X/1318/1/012031

Sulfur oxidation of ferrate synthesized from sludge in the aspect of desulfurization: A parametric study

M M Haboc[1], N P Dugos[1], A E S Choi[1] and M W Wan[2]

[1]Department of Chemical Engineering, De La Salle University, 2401 Taft Ave., Manila 0922, Philippines

[2]Department of Environmental Engineering, Chia Nan University of Pharmacy and Science, Tainan 71710, Taiwan

angelo.choi@dlsu.edu.ph, peterwan@mail.cnu.edu.tw

Abstract. The emission of sulfur in the atmosphere poses a significant threat to human health and the environment. To address this issue, stringent regulations have been implemented to limit the sulfur content in diesel, and novel desulfurization technologies are being developed. One notable technology is oxidative desulfurization (ODS), which employs oxidants to transform sulfur compounds into their corresponding sulfones, which are relatively easier to recover. The application of high-shear mixing in ODS has been studied to increase sulfur-to-sulfone conversion by creating smaller droplets and reducing mass transfer resistance. This research investigates the application of potassium ferrate derived from drinking water treatment sludge (DWTS), in the mixing-assisted oxidative desulfurization (MAOD) of a dibenzothiophene (DBT) model fuel. Potassium ferrate was synthesized using the wet oxidation method. The study evaluated the effects of ferrate concentration (400 to 600 ppm), agitation speed (4,400 to 10,800 rpm), and temperature (40 to 60 °C) on the efficiency of DBT conversion. The results revealed that 493.2 ppm DBT conversion was achieved at 550 ppm Fe(VI) concentration, 7,600 rpm agitation speed, and 50 °C temperature. Notably, increasing Fe(VI) concentration, agitation speed, and temperature had significant effects on sulfur reduction. This study demonstrates the potential of using potassium ferrate derived from DWTS in MAOD for effective desulfurization and discusses insights into the effects of operating conditions to enhance desulfurization efficiency. Ultimately, the study contributes to the development of environmentally friendly and cost-effective desulfurization technologies.

1. Introduction

Organosulfur compounds in fuels are undesirable due to their negative impacts. Sulfur causes corrosion problems in piping and significantly affects the efficacy of catalytic converters. When emitted, sulfur oxide gases react with water, ultimately, causing acid rain which can damage infrastructure and vegetation [1]. Stringent regulations have been set in the USA to limit sulfur emissions to 10 ppm and 15 ppm in gasoline and diesel, respectively [2]. With this, eliminating organosulfur compounds is a crucial stage in the production of environmentally friendly fuels.

In the petroleum industry, the most common desulfurization method is hydrodesulfurization (HDS). HDS is effective in removing thiols, sulfides, and disulfides, however, it is ineffective in completely removing heterocyclic thiophenic compounds, such as dibenzothiophene (DBT) and benzothiophene (BT) [3]. Oxidative desulfurization (ODS) is another technology that has gained significant attention in recent years. The advantage of ODS over HDS is its requirement for mild operating conditions, such as low reaction temperature and pressure [4]. A novel assisted ODS technology called mixing-assisted oxidative desulfurization (MAOD) has been studied to increase the efficiency of oxidation reactions [5]. This technique employs high-shear mixing to increase the formation of emulsions and increase oxidant dispersion at the liquid-liquid interface [6]. Following the oxidation of the sulfur compounds to their corresponding sulfones, the relatively increased polarity is utilized to remove the sulfur compounds by adsorption or extraction [7].

Common oxidants used in MAOD include H_2O_2, O_2, and Fe(VI). Among these, Fe(VI) has the highest reduction potential of 2.2 V in acidic media [8]. Fe(VI) is also known to be environmentally friendly due to its reaction product, Fe(III) [9]. The use of commercial Fe(VI) in MAOD has resulted in 93.68% DBT reduction for model fuel and 93.15% for diesel oil [10]. However, commercial Fe(VI) is costly to acquire. Drinking water treatment sludge (DWTS) is a potential source of Fe(III) and consequently, Fe(VI). A previous study was able to synthesize Fe(VI) from DWTS using the wet-oxidation method and apply it in an ultrasound-assisted oxidative desulfurization system [11].

With this, the novelty of the paper lies in the application of Fe(VI) derived from DWTS in the MAOD of a DBT model fuel. Furthermore, this paper highlights the parametric analysis of MAOD factors through factor variation and trend observations. Specifically, the MAOD factors include Fe(VI) concentration, agitation speed, and temperature.

2. Experimental

2.1. Materials

Dibenzothiophene ($C_{12}H_8S$; 98% purity) was obtained from Tokyo Chemical Industry Co. Ltd.). Chemicals such as toluene (C_7H_8), tetraoctylammonium bromide ($C_{32}H_{68}BrN$, 98% purity), glacial acetic acid (CH_3COOH), sodium hydroxide pellets (NaOH), and potassium hydroxide pellets (KOH) were purchased from Sigma-Aldrich. Nitric acid (HNO_3, 69%) was obtained from Merck. Sodium hypochlorite (NaOCl, 6-12%) was acquired from Nihon Shiyaku. Drinking water treatment sludge (pellet form) was obtained from Changhua No. 3 Water Purification Plant (Changhua, Taiwan).

2.2. Instrumental analysis

The MAOD experiments were performed using a high-shear mixer (IKA T-25 Digital Ultra-Turrax mixer). An electric control unit from Panchum Scientific Corporation was used to maintain temperature control. A UV-Vis spectrophotometer (Hach DR 5000) was utilized to measure Fe(VI) concentration. The measurement of DBT before and after oxidation was analyzed using a gas chromatograph (GC, Agilent 7890A Gas Chromatograph) with fused-silica capillary column HP-5ms with 0.25 μm film thickness. The oven temperature for DBT analysis was set at an initial of 200 °C for 1 min, heated at 20 °C/min to 280 °C and kept constant for 1 min.

2.3. Fe(VI) synthesis

To synthesize Fe(VI) from drinking water treatment sludge, 10 g sludge was dissolved in 10 mL of 2 M HNO_3 and mixed for 1.5 hr at 100 rpm. In a separate beaker, 100 g NaOH and 200 mL NaOCl were stirred in a cold-water bath. The mixtures were mixed at 500 ppm for 1 hr. Afterwards, 250 mL of saturated KOH was added to the solution and mixed for 30 min at the same speed. After the mixing stage, the solution underwent centrifugation. The dark purple liquid was collected and used fresh in the oxidation experiment. The concentration of the Fe(VI) oxidant was measured using a UV-Vis spectrophotometer at 510 nm.

IOP Conf. Series: Materials Science and Engineering **1318** (2024) 012031 doi:10.1088/1757-899X/1318/1/012031

2.4. MAOD methodology

DBT was dissolved in toluene to create a 500-ppm sulfur model fuel. Pre-determined amounts of the model fuel, phase transfer agent, and Fe(VI) were added to a 400-mL beaker. The temperature was controlled with a water bath. Acetic acid was added to the mixture until pH 5 was achieved. The mixture was agitated for 30 min, forming an emulsion. The emulsion is separated by means of centrifugation, where the oil phase was recovered for analysis.

2.5. Trend analysis

The study focused on three factors, namely, Fe(VI) concentration, agitation speed, and temperature. To determine each parameter effect in MAOD, experimental runs were performed by varying one parameter while maintaining the rest of the parameters constant. The Fe(VI) concentration varied from 400 to 600 ppm, with 50-ppm intervals. The agitation speed was tested from 4,400 to 10,800 rpm using 1,600-rpm intervals. Lastly, the temperature of the reaction was observed from 40 to 60 °C under 5 °C intervals.

3. Results and Discussion

3.1. Effect of Fe(VI) concentration

Figure 1 shows an increase in sulfur conversion from 466.7 ppm to 493.2 ppm as Fe(VI) concentration is increased from 400 to 550 ppm. The presence of more Fe(VI) ions results in a higher oxidation activity to convert DBT to DBT sulfone (DBTO), especially in acidic environments [12]. However, as Fe(VI) concentration increases from 550 to 600 ppm, the sulfur conversion decreased slightly to 489.5 ppm. This may be attributed to the reaction of Fe(VI) with water to produce $Fe(OH)_3$, consequently increasing the pH of the system [13]. In this case, the increase in Fe(VI) concentration may not necessarily lead to an increase in sulfur conversion.

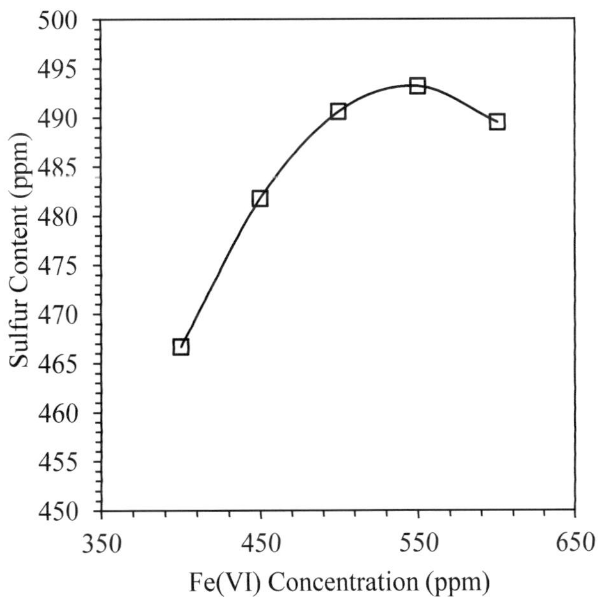

Figure 1. Effect of Fe(VI) concentration in sulfur conversion.

3.2. Effect of agitation speed

Figure 2 presents the trend of increasing agitation speed with sulfur conversion. An increased oxidation state from 485.0 to 491.3 ppm was observed from 4,400 to 9,200 rpm. The sulfur conversion was lowered to 490.6 ppm when the agitation speed increased to 10,800 rpm. Elevated agitation speed can enhance the oxidation rate by promoting the formation of uniformly shaped droplets. The increased number of droplets results in a larger interfacial area and reduces mass transfer resistance [10]. In the case of elevated agitation speed, the formation of whirlpools may occur, hindering further oxidation activity [14].

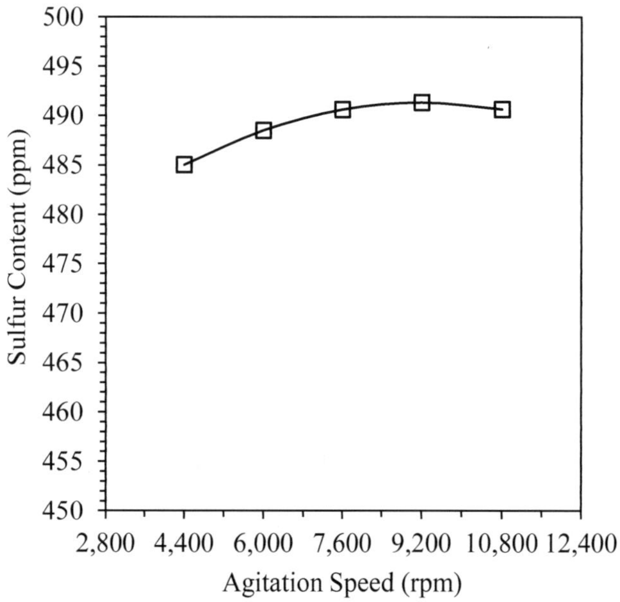

Figure 2. Effect of agitation speed in sulfur conversion.

3.3. *Effect of temperature*

Figure 1c illustrates the increase of sulfur conversion from 458.5 to 492.7 ppm when the temperature is increased from 40 to 55 °C. This is due to the increase of molecular kinetic energy that causes an increase in the frequency of molecular collisions. Increasing the temperature also decreases interfacial tension between the oxidant and oil phases [15]. However, further increasing the temperature to 60 °C causes the sulfur conversion to lower to 485.6 ppm due to the increased decomposition reaction of Fe(VI) with water [16].

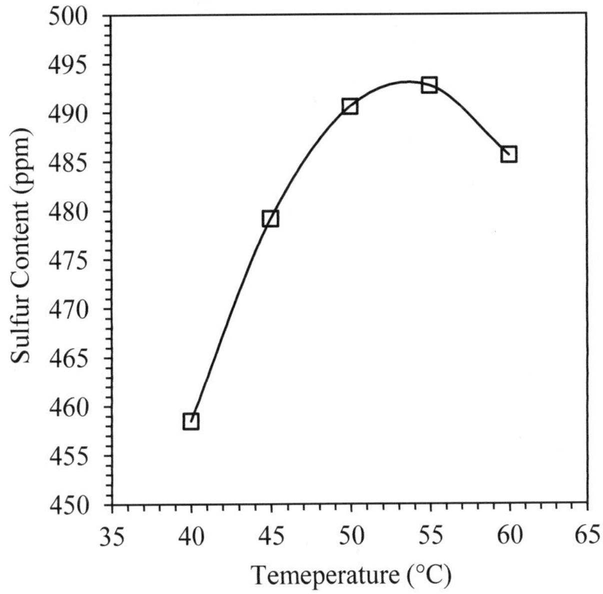

Figure 3. Effect of temperature in sulfur conversion.

4. Conclusion

In this study, results indicated that Fe(VI) derived from drinking water treatment sludge is effective in oxidizing DBT to DBTO in a MAOD system. Trend plots illustrated that increasing the Fe(VI) concentration, agitation speed, and temperature results to an increase to sulfur conversion. However, increasing Fe(VI) concentration and temperature to elevated levels may reduce the sulfur conversion due to the increased decomposition of Fe(VI). Agitation was observed to improve sulfur conversion as more uniformly shaped droplets are formed. From the experimental runs, the highest sulfur conversion of 493.2 ppm DBT to DBTO was achieved at 550 ppm Fe(VI) concentration, 7,600 agitation speed, and 50 °C.

Acknowledgments

The authors give acknowledgment to the Ministry of Science and Technology (MOST 111-2221-E-041-002-MY3) and the Department of Science and Technology (DOST) of the Philippines — Engineering Research and Development for Technology (ERDT) program for providing financial support.

References

[1] Srivastava V C 2012 An evaluation of desulfurization technologies for sulfur removal from liquid fuels *RSC Adv.* **2** 759–83
[2] U.S. EPA 2014 Control of Air Pollution From Motor Vehicles: Tier 3 Motor Vehicle Emission and Fuel Standards *United States Environ. Prot. Agency*
[3] Wan M W, de Luna M D G, Golosinda L R, Futalan C M, Phatai P and Lu M C 2019 Oxidative desulfurization of dibenzothiophene via high-shear mixing with phosphotungstic acid: the influence of calcination temperature on kinetics and catalytic activity *Clean Technol. Environ. Policy* **21** 1459–69
[4] Zhang X, Shi Y and Liu G 2016 Direct preparation of [(CH3)3NC16H33]4Mo8O26 and its catalytic performance in oxidative desulfurization *Catal. Sci. Technol.* **6** 1016–24
[5] Lu M C, Agripa M L, Wan M W and Dalida M L P 2014 Removal of oxidized sulfur

compounds using different types of activated carbon, aluminum oxide, and chitosan-coated bentonite *New pub Balaban* **52** 873–9
[6] Choi A E S, Roces S, Dugos N and Wan M W 2016 Oxidation by H2O2 of bezothiophene and dibenzothiophene over different polyoxometalate catalysts in the frame of ultrasound and mixing assisted oxidative desulfurization *Fuel* **180** 127–36
[7] Choi A E S, Roces S, Dugos N and Wan M W 2016 Mixing-assisted oxidative desulfurization of model sulfur compounds using polyoxometalate/H2O2 catalytic system *Sustain. Environ. Res.* **26** 184–90
[8] Jiang J-Q 2007 Research progress in the use of ferrate(VI) for the environmental remediation *J. Hazard. Mater.* **146** 617–23
[9] Sharma V K 2002 *Potassium ferrate VI : an environmentally friendly oxidant* vol 6
[10] Choi A E S, Roces S, Dugos N, Futalan C M and Wan M-W 2016 Optimization analysis of mixing-assisted oxidative desulfurization of model sulfur compounds using commercial ferrate(VI) *Desalin. Water Treat.* **57** 17616–23
[11] Arcega A A, Futalan C M, Dalida M L P and Wan M-W 2020 Ultrasound-assisted oxidative desulfurization of organosulfur compounds using ferrate(VI) derived from drinking water treatment sludge *15th International Symposium on East Asian Resources Recycling Technology, EARTH 2019*
[12] Choi A E S, Roces S, Dugos N, Futalan C M, Lin S S and Wan M W 2014 Optimization of ultrasound-assisted oxidative desulfurization of model sulfur compounds using commercial ferrate (VI) *J. Taiwan Inst. Chem. Eng.* **45** 2935–42
[13] Liu S, Wang B, Cui B and Sun L 2008 Deep desulfurization of diesel oil oxidized by Fe (VI) systems *Fuel* **87** 422–8
[14] Huang D, Wang Y J, Yang L M and Luo G S 2006 Chemical Oxidation of Dibenzothiophene with a Directly Combined Amphiphilic Catalyst for Deep Desulfurization *Ind. Eng. Chem. Res.* **45** 1880–5
[15] Lu M-C, Biel L C C, Wan M-W, de Leon R and Arco S 2014 The Oxidative Desulfurization of Fuels with a Transition Metal Catalyst: A Comparative Assessment of Different Mixing Techniques *Int. J. Green Energy* **11** 833–48
[16] Sharma V K 2011 Oxidation of inorganic contaminants by ferrates (VI, V, and IV)-kinetics and mechanisms: A review *J. Environ. Manage.* **92** 1051–73

Effect of Polar Aprotic Solvents on the Physical, Chemical, and Mechanical Properties of Cellulose Acetate Bioplastic Films

M F U Rizon, A K M Cabigting, K S Sanchez, and R J I Tamargo*

Advanced Materials and Organic Synthesis (AMOS) Laboratory, Department of Chemical Engineering, College of Engineering, University of the Philippines Diliman, Quezon City, Philippines 1101
*ritamargo@up.edu.ph

Abstract. There has been a growing interest in bioplastics as a solution to increasing greenhouse gases and the pile up of waste caused by fossil-based non-biodegradable plastics. However, the use of harmful solvents in their synthesis may continue to be detrimental to the environment. A novel bio-based solvent, Cyrene^{TM}, is seen as an alternative to toxic polar aprotic solvents. With this, the study aimed to compare the properties of cellulose acetate (CA) bioplastic films using acetone, dimethylformamide (DMF), and Cyrene^{TM} as solvents and glycerol as the plasticizer. The following tests were used to characterize the synthesized bioplastics: Fourier-Transform Infrared Spectroscopy (FTIR), Scanning Electron Microscopy (SEM), density, moisture content, biodegradability, tensile strength, Young's Modulus, and elongation at break. Cyrene^{TM} films were found to have significantly high moisture content and elongation at break. A weight loss of 29% in biodegradation was recorded for Cyrene^{TM} films without glycerol over the course of 23 days. The properties of the Cyrene^{TM} films show that it may be applicable for mulching films and seedling containers in agricultural applications due to their high affinity for water. It can be concluded that the study of Cyrene^{TM} solvent for the synthesis of CA films and its properties contributes to greener synthesis methods for bioplastics.

1. Introduction

In 2018, 359 million metric tons of plastic were produced globally [1]. The cradle-to-grave assessment of plastic shows its detrimental effects to human health and the environment. In its production, the extraction of fossil fuels for plastic resins releases large amounts of greenhouse gases. Additionally, leaching of unreacted monomers, additives, and plastic debris from consumption and disposal can affect consumer health [2].

To reduce the dependence on fossil-derived polymers, bioplastics made from biomass-derived polymers have been widely studied. However, harmful polar aprotic solvents such as acetone, N,N-dimethylformamide (DMF), N-methyl-2-pyrrolidone (NMP), and dimethylsulfoxide (DMSO) have been reported in bioplastic production [3][4][5][6]. The European Chemicals Agency has included DMF, NMP and DMSO in the list of restricted substances under Registration, Evaluation, Authorisation and Restriction of Chemicals (REACH) [7].

This study aims to explore the use of Cyrene^{TM} as an alternative green solvent to replace conventional polar aprotic solvents in bioplastic production. Cyrene^{TM} is a dipolar aprotic solvent often cited as a substitute to DMF and NMP due to its low flammability, toxicity, and non-corrosive properties [8]. The synthesis of bioplastic films was based on the study of Goncalves et al. [4] using cellulose acetate (CA) as the polymer, acetone (AC) as the solvent, and glycerol as the plasticizer.

2. Methodology

2.1. Gathering of Materials

Cellulose acetate (average Mn ~50,000 by GPC) was purchased from Sigma-Aldrich. Acetone, dimethylformamide, and glycerol were purchased from DKL Lab Supplies. Cyrene^{TM} solvent was provided by Circa Group Australia.

2.2. Preparation of Bioplastic Films

CA was magnetically stirred in AC, DMF and Cyrene^{TM} solvents following a 1:10 w/v CA-to-solvent ratio until homogeneous to form the setups AC0, DMF0, and CY0. Afterwards, three more setups were made with glycerol (10% w/v relative to mixture) which was mixed in after dissolution of the polymer [4] to form setups AC10, DMF10, and CY10. A total of six setups were casted onto glass trays under ambient conditions until dry. The films were then washed with deionized water to remove excess solvents, after which drying was done at 110°C for 1 hour.

2.3. Characterization of Bioplastic Films

2.3.1. Fourier Transform Infrared Spectroscopy (FTIR). The determination of the functional groups of the bioplastic films were done using FTIR spectroscopy at the Chemical Engineering Analytical (CEAL) Laboratory, Department of Chemical Engineering, University of the Philippines Diliman. Scans were taken at a wavelength range of 400-4000 cm^{-1} with 32 scans.

2.3.2. Scanning Electron Microscopy (SEM). SEM analysis was performed in the CEAL Laboratory. Analysis of the bioplastics' surface morphology was done at 500x magnification with an accelerating voltage of 5.00 kV.

2.3.3. Density. The mass of 4x4 cm films were obtained through an analytical balance after which the films' thicknesses at five random points were measured through a digital calliper [6]. Film density (g/cm^3) was then calculated using the equation below.

$$\text{Density} = \frac{\text{mass}}{(\text{area})(\text{average thickness})} \tag{1}$$

2.3.4. Moisture Content. The initial weights (w_0) of 3x3 cm films were obtained, after which drying of the films at 105°C for 3 hours was performed [7]. The films were then weighed again to determine the final weight (w_f). Moisture content was then calculated as weight loss using the equation below.

$$\text{Weight Loss (\%)} = \frac{w_0 - w_f}{w_0} \times 100 \tag{2}$$

2.3.5. Biodegradability. Three replicates sized 3x3 cm were prepared per bioplastic setup. The films were first dried in an oven at 70°C for 24 hours after which they were weighed to determine the initial weight (w_0). The films were then buried 10 cm deep in soil for 23 days under ambient conditions [6]. Their masses were recorded in four random days within this time period. The samples were oven dried at 70°C for 3 hours prior to measuring the final weight (w_f). Determination of the films' capability to degrade was done measured through weight loss similar to Equation 2.

IOP Conf. Series: Materials Science and Engineering **1318** (2024) 012032 doi:10.1088/1757-899X/1318/1/012032

2.3.6. Mechanical Properties. Tensile strength, Young's Modulus, and elongation at break were determined with a Universal Testing Machine (UTM) in the CEAL Laboratory. Bioplastic films sized 10x2.5 cm were used. The machine was operated at a speed of 1 mm/s up until the films split into two [4].

3. Results and Discussion

3.1. FTIR

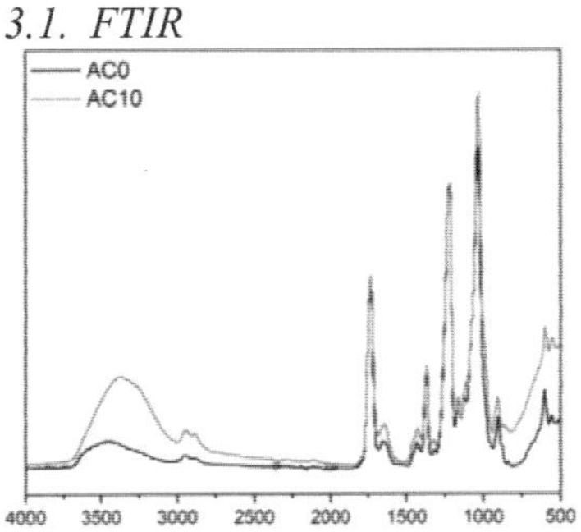

Figure 1. Absorbance of films made from AC.

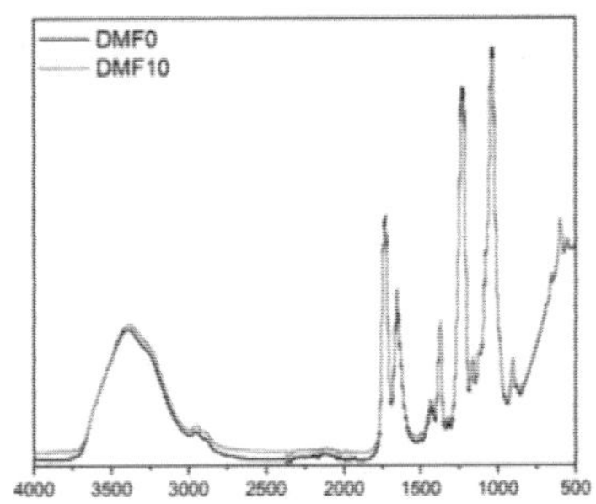

Figure 2. Absorbance of films made from DMF.

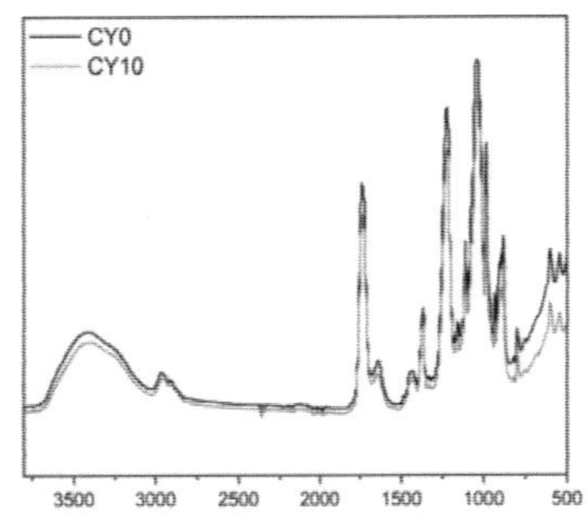

Figure 3. Absorbance of films made from Cyrene^{TM}.

The figures above show comparisons between the films with and without glycerol for their respective solvents. Increased peaks near 3400 cm^{-1} for cellulosic -OH elongation, 2931 cm^{-1} for C-H bond elongation, and 1230 cm^{-1} and 1027 cm^{-1} for steric carbonyl elongation have been observed for films with glycerol [4]. This difference is most visible for AC films. DMF films, on the other hand, have nearly identical absorbances at some wavelengths (1230 cm^{-1} and 1027 cm^{-1}). However, a decrease in the peaks for CY10 compared to CY0 is observed, contrary to the expected trend. The interactions between the functional groups of Cyrene^{TM} and glycerol may have caused the consistent decrease in peaks, but further studies are needed to support this outcome.

3.2. SEM

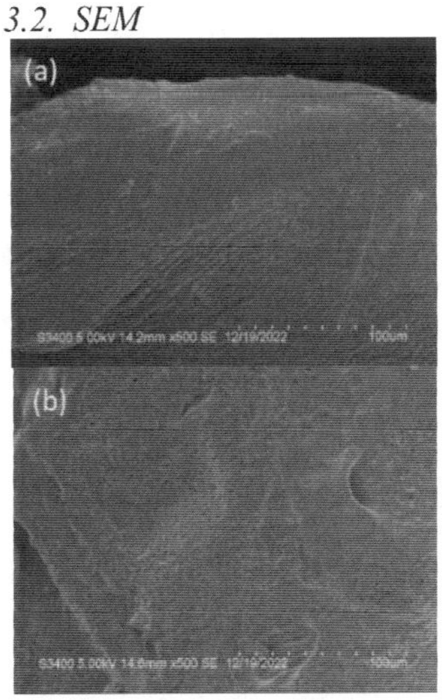

Figure 4. SEM images for AC0 (a) and AC10 (b).

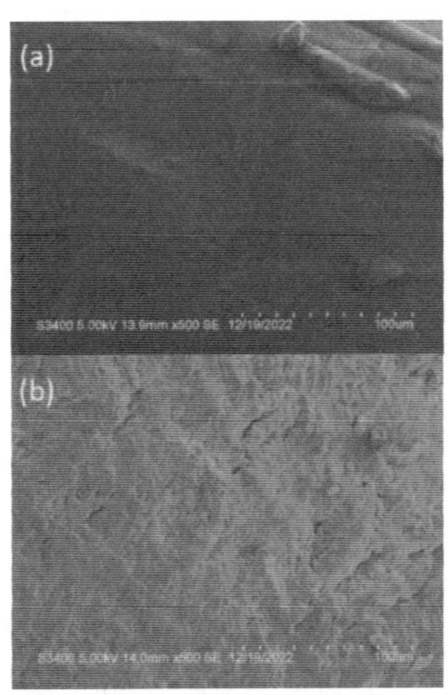

Figure 5. SEM images for DMF0 (a) and DMF10 (b).

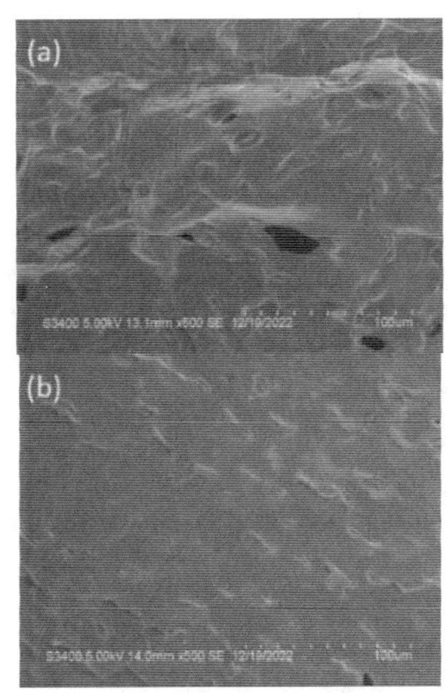

Figure 6. SEM images for CY0 (a) and CY10 (b).

Figures 4 to 6 show the SEM images of the films at 500x magnification. Both AC films show roughness, but the film with glycerol also shows porous texture, similar to the study of Goncalves et al. [4]. The difference in texture suggests modification to the polymer network of CA by glycerol. The micropores is also expected to cause less tensile strength for glycerol films [4].

Meanwhile, the DMF film shows significant roughness when glycerol is added. This is backed. By Caicedo et al. wherein roughness glycerol concentration is directly proportional [9]. The more homogeneous surface seen in DMF0 implies better mechanical properties as well due to the absence of tears [9].

Lastly, rough surfaces are also observed for films made in Cyrene™ [10]. The voids present in the CY0 films may be a result of the polymer agglomerating more in some regions [9]. On the other hand, irregular agglomeration has also been avoided in the CY10 film due to addition of glycerol [9].

3.3. Density

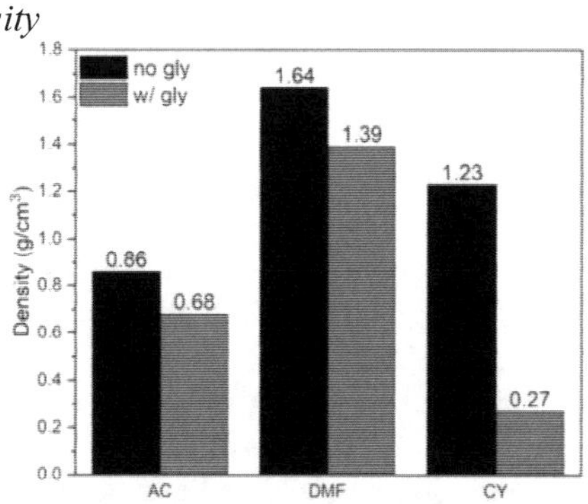

Figure 7. Densities of films.

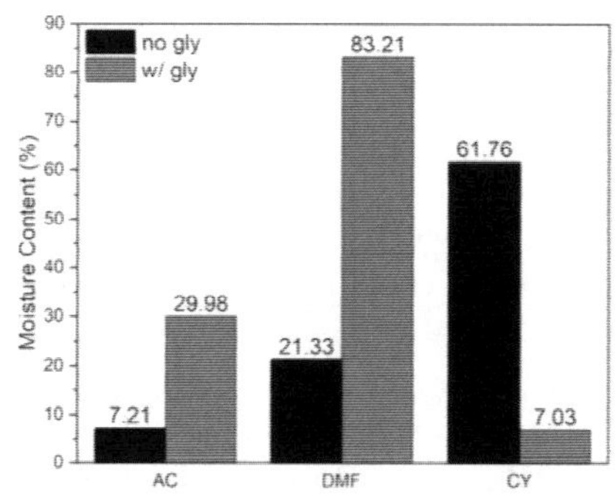

Figure 8. Moisture content of films.

The average density of each film is shown in Figure 7. Glycerol films are less dense than their pure film counterparts due to glycerol thickening the films, which in effect, increases their volume as well. Plasticizers can restructure the polymer to create more free volume [12]. In addition, glycerol is hydrophilic, so films with glycerol are expected to hold more moisture resulting in greater thickness [13][14]. Glycerol also helps in the swelling of the polymeric network through its hydroxyl groups forming H-bonds with the polymer, replacing polymer-polymer bonds [15][16].

Among the solvents, AC and DMF gave the lowest and highest densities, respectively. The low densities of AC films can be explained by the volatility of its solvent as it quickly leaves the CA film. Meanwhile, the high moisture content of DMF films also observed in this study (Figure 8) may also be a cause for high density.

The low density of the CY10 is expected due to the porous structure of Cyrene™ films with plasticizer [17]. Plasticizer additive is described to act as a non-solvent to give pores that trap air pockets in the membrane. The presence of air contributes to the low density of the films.

3.4. Moisture Content

Table 1. Hansen solubility parameters of different solvents and difference from water.

	δ_d [MPa$^{1/2}$]	δ_p [MPa$^{1/2}$]	δ_h [MPa$^{1/2}$]	Difference from water [MPa$^{1/2}$]
Water [23]	15.5	16.0	42.4	-
Acetone [23]	15.5	10.4	7.0	35.8
DMF [23]	17.4	13.7	11.3	31.4
Cyrene™ [24]	18.9	12.4	7.1	36.1

$$R_{ij} = \sqrt{4(15.5 - \delta_d)^2 + \left(16.0 - \delta_p\right)^2 + (42.4 - \delta_h)^2} \quad (3)$$

As suggested by the hydrophilic property of glycerol, films with plasticizer hold more moisture [18]. It is also shown in Figure 8 that films made with Cyrene™ and DMF hold the highest moisture content.

Table 1 shows the Hansen solubility parameters of each solvent and their difference with water (computed based on Equation 3). A lower difference with water results in higher affinity with water, thus a higher moisture content [19]. Since DMF has the closest parameters to water, films made with DMF are expected to hold the highest moisture as seen with DMF10. Cyrene™ films also show high moisture content, as expected due to their hydrophilic properties [11]. Meanwhile, the low moisture content of the acetone films can be explained by the volatility of acetone. Most of the moisture content shown by AC0 and AC10 films may come from the low moisture absorption of the polymer alone [20].

3.5. Biodegradability

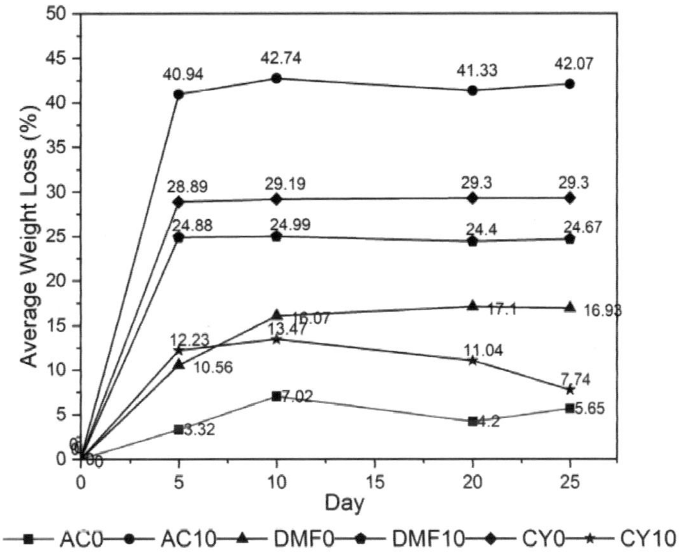

Figure 9. Average weight loss of each bioplastic setup.

Shown on Figure 9, there was a significant decrease in weight up until the 8th day across all set-ups. For the films without glycerol, CY0 showed the highest biodegradability, followed by DMF0, then AC0. Cyrene™ is a bio-based solvent thus it is expected to be fully biodegradable [21][22]. As for DMF and acetone, since the films were dried before each measuring, it is possible that most of the solvent has already left the films, leading to higher polymer-polymer interactions [23], resulting in lower biodegradability.

The addition of a plasticizer for the acetone and DMF set-ups also accelerated biodegradation which is in accordance with literature [6]. However, the opposite trend was observed for the Cyrene™ set-ups which may be due to insufficient drying. Given that the CY10 samples showed high water absorbency, sufficient drying until constant weight would have resulted in higher degradation results.

3.6. Mechanical Properties

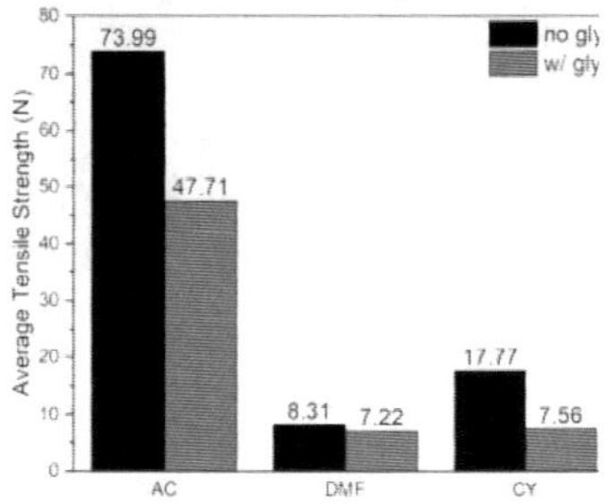

Figure 10. Tensile strength of films.

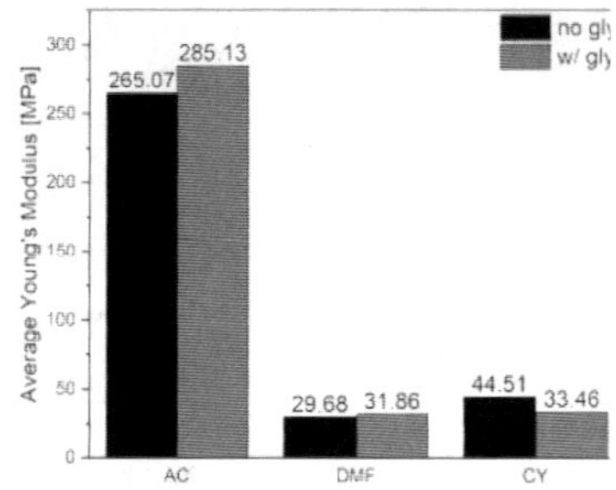

Figure 11. Young's Modulus of films.

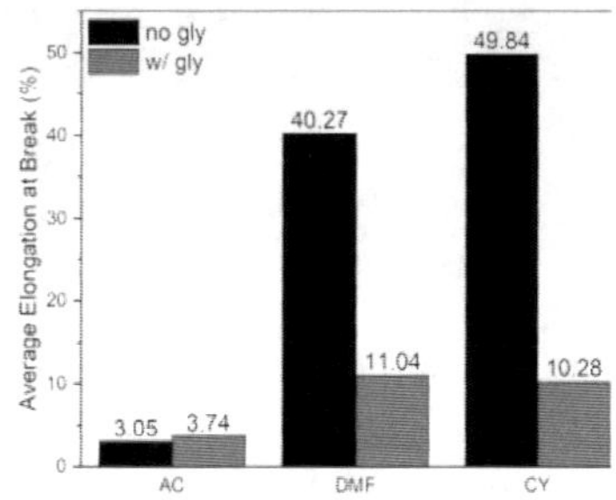

Figure 12. Elongation at break of films.

3.6.1. Tensile Strength. Figure 10 shows a decrease in tensile strength was observed when glycerol was added per solvent set-up since glycerol can weaken the hydrogen bonds in the polymer matrix as well as contribute to pore formation [4]. However, Goncalves et al. [4] reports that tensile strength increases up until 20% glycerol concentration in their study. The decrease observed at 10% glycerol concentration may be attributed to the lack of interfacial interaction between CA and glycerol because of the presence of pores.

Across solvent set-ups, acetone set-ups had the highest tensile strength followed by Cyrene^{TM} and then DMF. Significant differences were observed for AC0 when compared to the tensile strengths of DMF0, DMF10, and CY10. There was also a significant difference between AC10, DMF10, and CY10. The decrease in tensile strength may be due to the solvent properties wherein acetone has a high volatility thus having less moisture as opposed to Cyrene^{TM} and DMF which have hydrophilic properties [11][23]. Moisture can then disrupt the hydrogen bonding between CA and glycerol, leading to swelling and being more vulnerable to external forces.

3.6.2. Young's Modulus. Setups with glycerol are expected to have a lower Young's Modulus due to weakened polymer interactions caused by glycerol [4]. However, Figure 11 shows that this was only observed for CY0 and CY10. Despite the increasing trend in Young's Modulus for the acetone and DMF setups, the deviation is statistically insignificant between AC0 and AC10, DMF0 and DMF10, as well as CY0 and CY10 thus having negligible effect on the stiffness when glycerol is added.

The Young's Modulus values between solvent set-ups also observed a similar trend to that of tensile strength. There were significant differences between AC0, DMF0, DMF10, and CY10 as well as between AC10, DMF0, DMF10, and CY10. The high moisture content for Cyrene^{TM}- and DMF-based bioplastics resulted to extra hydrogen bonding and swelling because of water thereby affecting the other existing hydrogen bonds in the polymer matrix.

3.6.3. Elongation at Break. From Figure 12, there was an increase in elongation between AC0 and AC10 since glycerol adds flexibility [4]. However, a decrease was observed for Cyrene^{TM} and DMF set-ups which may be due to the high moisture content for DMF10 and CY10. The presence of water increases swelling and weakens the hydrogen bonds hence the results for DMF10 and CY10.

For the effect of the type of solvent, the set-ups dissolved in acetone had a lower value compared to the setups dissolved in Cyrene^{TM} and DMF. This is due to the stiffness of the bioplastic wherein a higher Young's Modulus was observed for the acetone setups thus is not as flexible compared to the Cyrene^{TM} and DMF setups.

4. Conclusion

Bioplastic films made from cellulose acetate were successfully synthesized using Cyrene^{TM} as the solvent which were also characterized and compared with bioplastics using acetone and DMF as solvents. The FTIR spectra verified the functional groups as well as the interaction of the bioplastic components. The SEM images showed increasing porosity when glycerol was added as a plasticizer. For density, a decrease was observed upon the addition of glycerol since it loosens the polymer matrix. Moisture content revealed bioplastics' water affinity due to hydroxyl and ester groups. All bioplastics showed weight losses in the biodegradability test, confirming their biodegradability. Lastly, the obtained mechanical properties of the bioplastics showed higher tensile strength and Young's Modulus for acetone-based bioplastics while higher elongation for the DMF and Cyrene^{TM}-based bioplastics.

The synthesized bioplastics can be efficient for agricultural applications particularly as mulching films and seedling containers because of their high affinity to water and their capability to degrade.

It is recommended to conduct additional analysis of the solvents' properties and their impact on CA bioplastics to better explain the trends. Additionally, varying the glycerol concentration, utilizing other plasticizers, or adding additives can be explored to improve the bioplastic's versatility.

Acknowledgements
This work was primarily funded by the UP System Balik PhD Program (OVPAA-BPhD-2020-01). Furthermore, the authors acknowledge the Office of the Chancellor of the University of the Philippines Diliman, through the Office of the Vice Chancellor for Research and Development, for funding support through the PhD Incentive Award. The researchers also express their gratitude to the College of Engineering Office of the Associate Dean for Student Affairs (UPD COE ADSA) and the Department of Science and Technology Science Education Institute (DOST-SEI) through the Engineering Research and Development for Technology (ERDT) and the Merit Scholarship Program for the additional funding support.

References

[1] Okoffo E D, Donner E, McGrath S P, Tscharke B J, O'Brien J W, O'Brien S, Ribeiro F, Burrows S D, Toapanta T, Rauert C, Samanipour S, Mueller J F and Thomas K V 2021 Plastics in Biosolids from 1950 to 2016: A function of global plastic production and consumption *Water Research* **201** 117367

[2] Azoulay D, Villa P, Arellano Y, Gordon M, Moon D, Miller K and Thompson K 2019 ed A Kistler (Center for International Environmental Law)

[3] Cindradewi A W, Bandi R, Park C-W, Park J-S, Lee E-A, Kim J-K, Kwon G-J, Han S-Y and Lee S-H 2021 Preparation and characterization of cellulose acetate film reinforced with cellulose nanofibril *Polymers* **13** 2990

[4] Gonçalves S M, dos Santos D C, Motta J F, Santos R R, Chávez D W and Melo N R 2019 Structure and functional properties of cellulose acetate films incorporated with glycerol *Carbohydrate Polymers* **209** 190–7

[5] Kim Y and Jung C 2022 Reaction mechanisms applied to starch modification for biodegradable plastics: Etherification and esterification *International Journal of Polymer Science* **2022** 1–10

[6] Nigam S, Das A K and Patidar M K 2021 Synthesis, characterization and biodegradation of bioplastic films produced from *Parthenium Hysterophorus* by incorporating a plasticizer (PEG600) *Environmental Challenges* **5** 100280

[7] European Chemicals Agency 2022 Substances restricted under reach *ECHA*

[8] Kong D and Dolzhenko A V 2022 Cyrene^{TM}: A bio-based sustainable solvent for organic synthesis *Sustainable Chemistry and Pharmacy* **25** 100591

[9] Caicedo C, Díaz-Cruz C A, Jiménez-Regalado E J and Aguirre-Loredo R Y 2022 Effect of plasticizer content on mechanical and water vapor permeability of maize starch/PVOH/chitosan composite films *Materials* **15** 1274

[10] Martins da Costa J C, Lima Miki K S, da Silva Ramos A and Teixeira-Costa B E 2020 Development of biodegradable films based on purple yam starch/Chitosan for Food Application *Heliyon* **6**

[11] Wibisono Y, Noviani V, Ramadhani A T, Devianto L A and Sulianto A A 2022 Eco-friendly forward osmosis membrane manufacturing using dihydrolevoglucosenone *Results in Engineering* **16** 100712

[12] Sanyang M L, Sapuan S M, Jawaid M, Ishak M R and Sahari J 2015 Effect of plasticizer type and concentration on physical properties of biodegradable films based on sugar palm (*Arenga pinnata*) starch for food packaging *J Food Sci Technol* **53** 326–36

[13] Ahmadi R, Kalbasi-Ashtari A, Oromiehie A, Yarmand M-S and Jahandideh F 2012 Development and characterization of a novel biodegradable edible film obtained from psyllium seed (*Plantago ovata Forsk*) *Journal of Food Engineering* **109** 745–51

[14] Teixeira S C, Silva R R, de Oliveira T V, Stringheta P C, Pinto M R and Soares N de 2021 Glycerol and triethyl citrate plasticizer effects on molecular, thermal, mechanical, and barrier properties of cellulose acetate films *Food Bioscience* **42** 101202

[15] Dick M, Costa T M, Gomaa A, Subirade M, Rios A de and Flôres S H 2015 Edible film production from Chia seed mucilage: Effect of glycerol concentration on its physicochemical and mechanical properties *Carbohydrate Polymers* **130** 198–205

[16] Wypych G 2017 Effect of plasticizers on properties of plasticized materials *Handbook of Plasticizers* 209–332

[17] Tomietto P, Russo F, Galiano F, Loulergue P, Salerno S, Paugam L, Audic J-L, De Bartolo L and Figoli A 2022 Sustainable fabrication and pervaporation application of bio-based membranes: Combining a polyhydroxyalkanoate (PHA) as biopolymer and Cyrene^{TM} as Green Solvent *Journal of Membrane Science* **643** 120061

[18] Ghasemlou M, Khodaiyan F and Oromiehie A 2011 Physical, mechanical, barrier, and thermal properties of polyol-plasticized biodegradable edible film made from Kefiran *Carbohydrate Polymers* **84** 477–83

[19] Fink J K 2007 Emulsions, dispersions and foams *Special Publications* 53–4

[20] Khoshtinat S, Carvelli V and Marano C 2021 Moisture absorption measurement and modelling of a cellulose acetate *Cellulose* **28** 9039–50

[21] Dolzhenko A V 2021 Cyrene^{TM} *Green Sustainable Process for Chemical and Environmental Engineering and Science* 205–18

[22] Marathianos A, Liarou E, Hancox E, Grace J L, Lester D W and Haddleton D M 2020 Dihydrolevoglucosenone (cyreneTM) as a bio-renewable solvent for cu(0)wire-mediated reversible deactivation radical polymerization (RDRP) without external deoxygenation *Green Chemistry* **22** 5833–7

[23] Ferrarezi M M, Rodrigues G V, Felisberti M I and Gonçalves M do 2013 Investigation of cellulose acetate viscoelastic properties in different solvents and microstructure *European Polymer Journal* **49** 2730–7

[24] Milescu R A, Zhenova A, Vastano M, Gammons R, Lin S, Lau C H, Clark J H, McElroy C R and Pellis A 2021 Polymer chemistry applications of Cyrene and its derivative Cygnet 0.0 as safer replacements for polar aprotic solvents *ChemSusChem* **14** 3367–81

IOP Conf. Series: Materials Science and Engineering **1318** (2024) 012033 doi:10.1088/1757-899X/1318/1/012033

Rare earth elements enrichment and extraction potential of bauxite deposits in Samar, Philippines

J A Gabo-Ratio[1], L A Arenque[1], B Payot[1], M D Balela[2], M C Vegafria[3], J Guzman[4], P C Manalo[5], D K Tungpalan[2]

[1]Rushurgent Working Group, National Institute of Geological Sciences, University of the Philippines, Diliman, Quezon City, Philippines

[2]Department of Mining, Metallurgical and Materials Engineering, University of the Philippines, Diliman, Quezon City, Philippines

[3]College of Engineering and Technology, Mindanao State University – Iligan Institute of Technology, Iligan City, Philippines

[4]Marcventures Mining and Development Corporation, Makati City, Philippines

[5]Graduate School of International Resource Sciences, Akita University, Akita, Japan

jgratio@nigs.upd.edu.ph

Abstract. Rare earth elements (REE) play an integral part of modern life, from smartphones to satellites. Dubbed 'critical metals', the demand for REE has significantly increased, driven by the growth in global production for green technologies such as e-vehicles, solar panels, wind turbines, and other energy-economic infrastructure. Therefore, there is a need to explore alternative REE sources to ensure continued development of emerging green technologies. A possible alternative REE source is bauxite, which is soil that forms from intense weathering of aluminum oxide-rich rocks exposed to tropical climate. This paper investigates the REE enrichment and extraction potential from bauxite deposits in Paranas, Samar, Philippines. The sampling survey included collection of rock and soil on surface exposures, test pits, and drill cores from accessible portions of the bauxite mineral reservation site. The soil and parent rock samples collected were then subjected to various geochemical and mineralogical analyses. The main ore minerals of aluminum are gibbsite and boehmite; with minor goethite, hematite, and magnetite. Investigation of the geochemical composition of the bauxite reveals a total REE content of up to ~300 ppm, which is one of the highest REE and critical metal content among geological deposits in the country. These results will provide inputs in the design of a green and economical process to recover REE from bauxites. If the bauxite deposits prove to be a valuable REE resource, this study will help maximize the economic potential of the mineral resource in the Philippines and contribute to an economically efficient and environment-friendly way to produce e-tech elements for the country.

1. Introduction

Rare earth elements (REE) are the 15 lanthanides (lanthanum to lutetium), scandium and yttrium. This group of elements has a wide variety of applications and have become pervasive in many energy-economic technologies and infrastructure such as e-vehicles, solar panels, and wind turbines [1]. Exploration activity for REE has increased greatly in recent years and some of these metals were identified as critical metals – essential metals that are prone to increased supply risk [2]. The REE supply chain is restricted to several Chinese mining districts, although REE enrichment has been reported in a wide range of geological settings [3]. Therefore, there is a need to explore alternative REE sources to ensure continued development of emerging technologies. To minimize the supply risks of these critical metals, shallower ore deposits associated with weathering are considered as possible sources.

One of the deposits with potential as REE source is bauxite, which is the main source of aluminum. Bauxite deposits are paleo-weathering clay-like products of various aluminum-bearing igneous and carbonate-sedimentary rocks [4]. The major element composition of bauxite is dominated by elevated SiO_2, TiO_2, Al_2O_3 and Fe_2O_3 [5]. It is characterized by an aggregate composition where the oxide and/or hydroxide minerals of aluminum, iron, and titanium make up more than half of the total content. Aside from aluminum, bauxite has recently been tapped as an alternative source of REE [6]. Bauxite mines in selected countries have determined that producing REE as by-product of large-scale aluminum production is cost-effective and results to a smaller environmental footprint [7].

This study investigates the bauxite deposit in Paranas municipality in Samar Island being developed by Alumina Mining Philippines Inc. to determine its field characteristics, mineralogy, and composition and evaluate their potential for REE enrichment. The results hope to contribute to value addition of Philippine bauxite, which holds great potential for future aluminum and REE production.

2. Bauxite deposits in the Philippines

In the Philippines, bauxite deposits are reported in Bucas Grande Island, Surigao del Norte [8] and in various portions of Samar Island [9], both in the central-eastern portion of the archipelago (Figure 1a). The ferruginous bauxite in Bucas Grande Island was believed to have formed from in situ weathering of mafic and ultramafic rocks. According to [8], the Bucas Grande bauxite is characterized by gibbsite and kaolinite with hematite and goethite. This type of bauxite is termed "lateritic" bauxites, which constitute 85% of the world's bauxite deposits [10]. However, majority of the bauxite deposits identified in the Philippines are identified in Samar Island, which prompted the establishment of a bauxite mining reservation under a 1977 Presidential Proclamation (Figure 1b). The bauxites in Samar Island are closely associated with the limestone sinkholes in the area. The sinkholes provided good accumulation sites for bauxite deposition, from the probable weathering of nearby aluminum-rich volcanic and clastic units [9]. This type of bauxite formed on carbonate rock are termed as "karst bauxites" [4].

Among the bauxite deposits in Samar, the deposit in Paranas being developed by Alumina Mining Philippines Inc. (AMPI) is the most well-explored with numerous test pit and drill core data. It is currently in the permitting stage of development and is gearing for production in the coming years. Two types of bauxite were identified in the area: lateritic bauxite (Figure 2a) from the in-situ weathering of mafic sedimentary rocks; and karst bauxites (Figure 2b) weathered from basalts then deposited on limestone depressions in between karst hills. Of the two, the karst bauxite in the southwest portion contains higher aluminum grades (38-45 wt. %) compared to the lateritic bauxite (35-38 wt.%) in the northwest portion of the area. The southwest area is dominated by limestone peaks and discontinuous bauxite deposits in the lower elevation sinkhole basins (<200 masl). The higher elevations (up to 450 masl) in the northeast portion is dominated by sedimentary strata that transitions to the lateritic bauxite weathering profiles.

The bauxites appear as homogenous orange-brown clayey soil. In some portions, black manganese particles are present. In the upper portion of the soil profile, the manganese becomes more abundant and occurs as coarse-grained nodules (Figure 2c). The minerals identified from X-ray diffraction analyses of representative samples reveal that the brown-orange soil is dominantly gibbsite ($Al(OH)_3$), boehmite ($AlO(OH)$), goethite ($FeO(OH)$), lepidocrocite (γ-$FeO(OH)$), cristobalite (SiO_2), and anatase (TiO_2) (Figure 3).

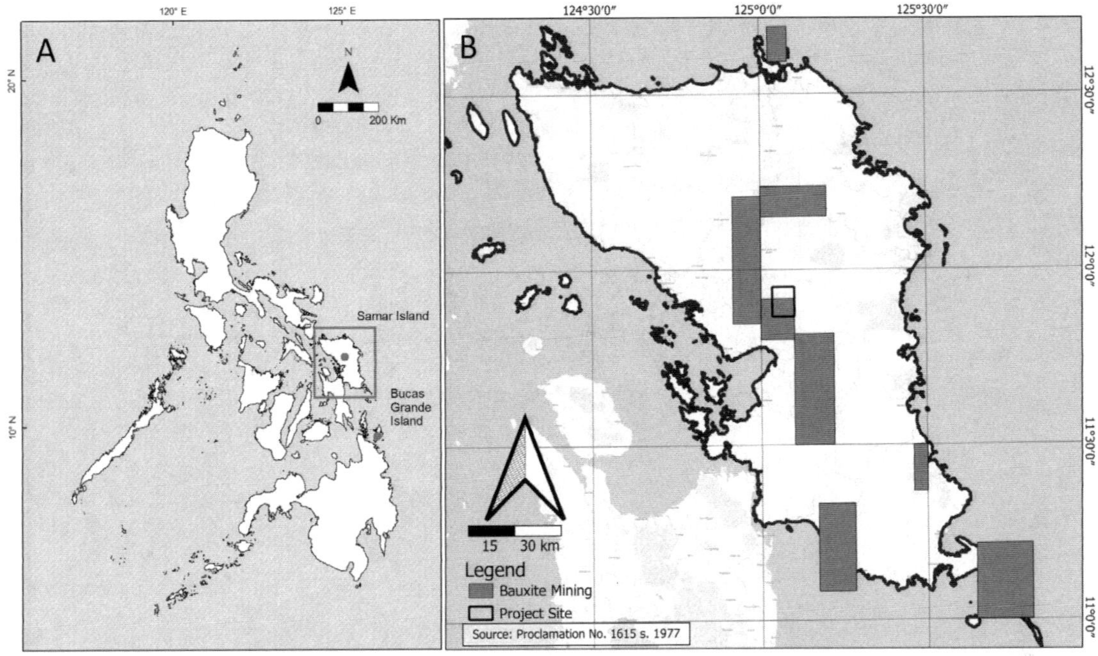

Figure 1. A. Bauxite in the Philippines has only been identified in Bucas Grande and Samar. B. Areas in pink are declared part of the Samar Bauxite Mineral Reservation. The black box outlines the location of the study area in Paranas.

Figure 2. Representative bauxite profiles in Paranas: lateritic bauxite exposure (left); test pit of karst bauxite test located in limestone depressions (center); black manganese nodules in the extracted from the first meter of the karst bauxite test pit.

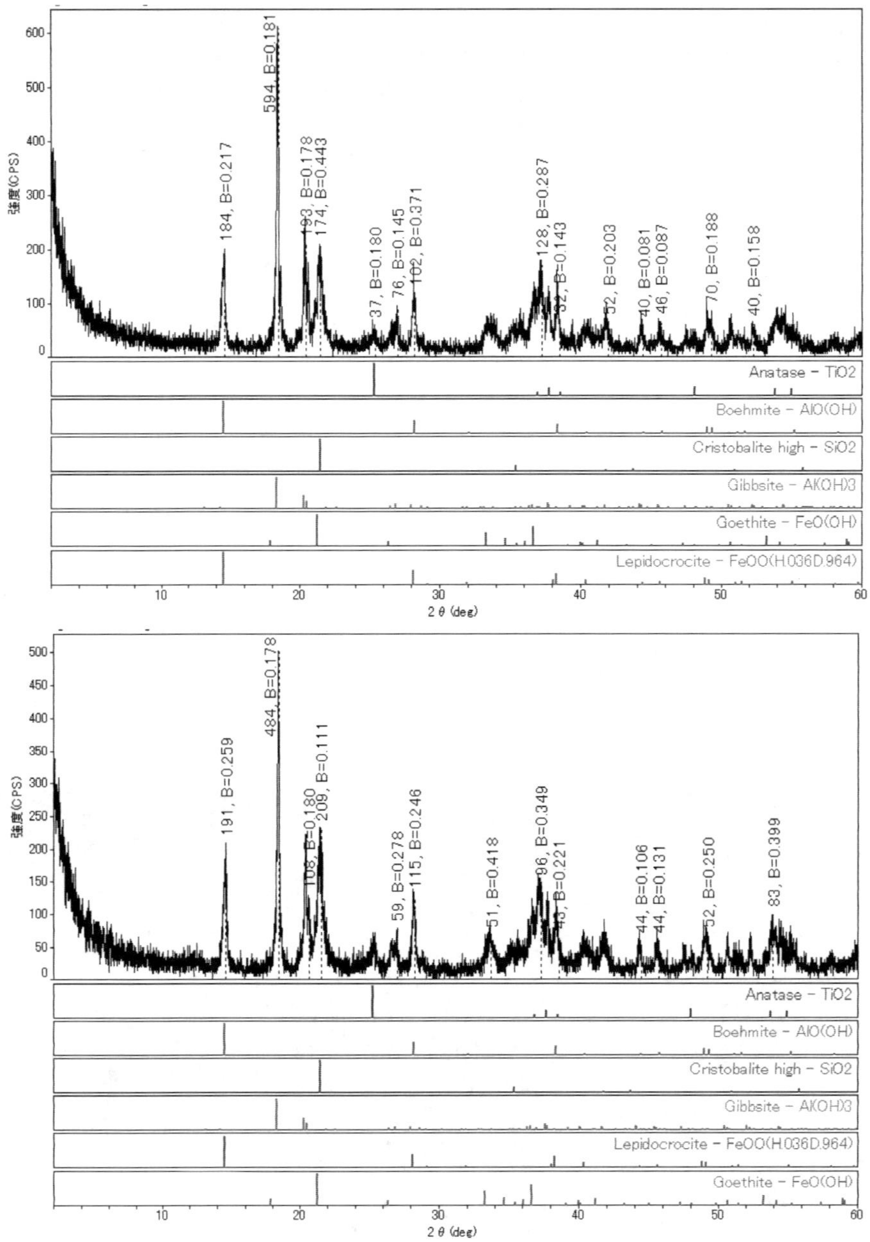

Figure 3. Representative X-ray diffraction peaks of samples from the karst bauxite test pit. Minerals identified include gibbsite, boehmite, goethite, lepidocrocite, cristobalite, and anatase.

3. REE enrichment and extraction potential in bauxites

Geochemical analysis of a representative test pit data provides insights into the elemental distribution within the shallow depth (up to four meters from the surface) of karst bauxite (Table 1). The major oxide composition displays minimal variability within the first four meters, except for manganese (Mn) which exhibits enrichment within the initial meter (3605 ppm), followed by a relatively constant concentration in the subsequent second to fourth meters (2255-2300 ppm). This phenomenon can be attributed to the abundant presence of Mn nodules near the surface, as evidenced by field observations (Figure 2). Similarly, concentrations of rare earth elements (REE) maintain consistency throughout the first four meters. The cumulative content of light REE, spanning from lanthanum to samarium (LREE), totals 221 ppm, while heavy REE (HREE), europium to lutetium, reach levels of up to 59 ppm. These findings closely parallel outcomes reported in the bauxites from the study conducted by [11].

Table 1. Geochemical composition of a representative test pit from the karst bauxite area in Paranas.

Major Oxides (wt.%)	TP 3-1	TP 3-2	TP 3-3	TP 3-4
SiO_2	5.33	6.02	6.41	6.65
Al_2O_3	44.89	44.15	44.3	44.33
Fe_2O_3	24.36	25.35	24.99	24.76
MgO	0.2	0.19	0.23	0.2
CaO	0.1	0.09	0.09	0.09
Na_2O	0.01	0.01	0.01	0.03
K_2O	0.02	0.01	0.01	<0.01
MnO	0.47	0.3	0.3	0.3
P_2O_5	0.625	0.64	0.633	0.602
TiO_2	2	2.07	2.03	2.02

Trace Elements (ppm)	TP 3-1	TP 3-2	TP 3-3	TP 3-4
Co	70.8	69.1	68.5	67.7
Sc	25	22	22	39
Y	21.1	16.8	17.5	18.7
La	56.9	52.8	52.9	53.9
Ce	73.4	73.5	71.9	71
Pr	13.84	13.97	12.84	12.85
Nd	62.2	56.7	56.1	56.3
Sm	14.7	13.1	12.4	11.9
Eu	4	3.6	3.4	3.3
Gd	15.5	14.3	13.1	12.4
Tb	2.27	2.04	1.98	1.91
Dy	15.1	13.8	13	12.6
Ho	3.3	3	2.8	2.7
Er	9.2	8.5	7.9	7.3
Tm	1.2	1.1	1	0.9
Yb	7.7	7.2	7.1	6.2
Lu	1.19	1.12	0.94	0.97

Diverse mechanisms of REE fractionation and redistribution have been documented in other bauxite regions, encompassing processes such as surficial weathering, supergene transformations, and diagenesis (e.g., [12, 13, 14]). Furthermore, factors like the pH and Eh conditions of the depositional environment, the presence of REE-bearing minerals, and the occurrence of organic complexes within the original source material can profoundly influence the mobilization and fixation of REE within the bauxite (e.g., [5, 13, 15]). During bauxite formation, both LREE and HREE demonstrate a degree of mobility (e.g., [5, 16]). Notably, HREEs tend to be leached during the bauxitization process [17] and exhibit an association with titanium oxides [16], often accumulating near the bedrock [18]. Conversely, LREEs exhibit enrichment in bauxite horizons rich in iron (Fe) due to the affinity of goethite scavenging [19, 20]. Within the bauxite, REE typically manifest in authigenic and diagenetic carbonate and phosphate minerals [14] or adhere to the surfaces of clay or diaspore minerals [21].

Further studies on the modal mineralogy of bauxite in Paranas, Samar will hopefully shed light on the specific mineral phases responsible for REE mineralization. To determine REE concentration and enrichment behavior in bauxite deposits, various geochemical and mineralogical techniques still need to be conducted [e.g., 13, 17]. Microscopic bauxite mineralogical features are usually identified using a scanning electron microscope (SEM) and electron probe microanalyzer (EPMA) [e.g., 12, 13] while REE mineral chemistry is detected with laser-ablation techniques [14].

In general, the REE concentration in the ore typically ranges around a few hundred parts per million (ppm). As a result, a preliminary concentration stage is necessary to enhance the enrichment of REE minerals, rendering them suitable for subsequent processing [22]. This preliminary step involves a combination of techniques including gravity concentration, magnetic separation, electrostatic separation, and froth flotation. Flotation is particularly favored for treating fine-grained or complex ores.

But even with heightened REE concentrations in the ore, a significant portion of these elements cannot be fully dissolved into aqueous solution at ambient temperatures due to their robust mineral associations [23]. To liberate the REE, the ore or concentrate must undergo intense leaching conditions to enhance REE extraction. Various leaching methods can be categorized into four main groups: alkaline, acid, chlorination, and ion exchange [24]. All these methods utilize aqueous solutions for leaching. Typically, these groups involve a roasting pre-treatment followed by subsequent leaching.

A recent advancement in extractive metallurgy, known as solvometallurgy, presents a solution that minimizes water consumption while enhancing metal selectivity. It employs non-aqueous solutions to extract metals from ores. Its notable advantages over conventional hydrometallurgy include reduced water and energy consumption, diminished acid requirements, elevated metal selectivity during leaching, and the use of biodegradable and recyclable solvents [25]. Future work will attempt to utilize the mineralogical and geochemical data of the Paranas bauxites to guide the design of possible leaching techniques to extract the REE.

4. Conclusions

Rare earth elements (REE) have become an integral part of green technologies and have become critical metals, such that alternative supply sources are being explored. A potential source of REE is bauxite, which is abundant in Samar Island. In Paranas, the high-grade bauxites are the karst bauxites, which contains 38-45 wt.% Al_2O_3. Manganese nodules are present in the orange-brown clayey bauxite soil that consists of gibbsite, boehmite, goethite, cristobalite, and anatase minerals. Analysis of a representative test pit indicates consistent geochemical composition for the first 4 meters of the shallow subsurface. The LREE content is more abundant (221 ppm) compared to the HREE (59 ppm), which is attributed to goethite scavenging of the REE in the topsoil during meteoric water leaching. The minerals possibly hosting the REE phases are authigenic and diagenetic carbonate and phosphate minerals clay surfaces. Future work will identify these specific mineral phases to aid in the attempts to design solvometallurgical extraction processes. These results will hopefully contribute to REE exploration and value addition of bauxite in the country.

5. References

[1] Wall F 2014 Rare earth elements *Critical Metals Handbook* ed Gunn G (Nottigham: John Wiley & Sons, Ltd) pp 312-339

[2] Habib K, Hansdóttir S T and Habib H 2020 Critical metals for electromobility: Global demand scenarios for passenger vehicles, 2015–2050 *Resour Conserv and Recycl* **154** 104603

[3] Chakhmouradian A R and Wall F 2012 Rare earth elements: minerals, mines, magnets (and more) *Elements* **8** pp 333-340

[4] Gow N N and Lozej G P 1993 Bauxite *Geosci Can* **20** pp 9-16

[5] Mongelli G, Boni M, Buccione R and Sinisi R 2014 Geochemistry of the Apulian karst bauxites (southern Italy): Chemical fractionation and parental affinities *Ore Geol Rev* **63** pp 9-21

[6] Deady E A, Mouchos E, Goodenough K, Williamson B J and Wall F 2016 A review of the potential for rare-earth element resources from European red muds: Examples from Seydisehir, Turkey and Parnassus- Giona, Greece *Mineral Mag* **80** pp 43-61

[7] Borra C R, Pontikes Y, Binnemans K and Van Gerven T 2015 Leaching of rare earths from bauxite residue (red mud) *Miner Eng* **76** pp 20-27

[8] Esguerra F B 1967 Genesis of ferruginous bauxite deposits in Bucas Grande Island, Surigao Mineral Reservation, Mindanao *The Phil Geol* **13** pp 71-98

[9] Jagolino R B 1976 Bauxite deposits of Samar *Journ Geol Soc Phil* **30** pp1-17

[10] Bárdossy G 1982 *Karst Bauxites* Elsevier Scientific Vol 14

[11] Gibaga C R L, Samaniego J O, Tanciongco A M, Quierrez R N M, Montano M O, Gervasio J H C, Reyes R C G and Peralta M J V 2022 The rare earth element (REE) potential of the Philippines. *Journ Geochem Explo* **242** 107082

[12] Mordberg L E, Stanley C J and Germann K 2001 Mineralogy and geochemistry of trace elements in bauxites: the Devonian Schugorsk deposit, Russia *Mineral Mag* **65** pp 8-101

IOP Conf. Series: Materials Science and Engineering **1318** (2024) 012033 doi:10.1088/1757-899X/1318/1/012033

[13] Liu X, Wang Q, Zhang Q, Feng Y and Cai S 2012 Mineralogical characteristics of the superlarge Quaternary bauxite deposits in Jingxi and Debao counties, western Guangxi, China *J Asian Earth Sci* **52** pp 53-62
[14] Gamaletsos P N, Godelitsas A, Filippidis A, Pontikes Y 2019 The rare earth elements potential of Greek bauxite active mines in the light of a sustainable REE demand *J Sustain Metall* **5** pp 20-47
[15] Li Z, Din J, Xu J, Liao C, Yin F, Lu T, Cheng L and Li J 2013 Discovery of the REE minerals in the Wulong–Nanchuan bauxite deposits, Chongqing, China: Insights on conditions of formation and processes *Journ Geochem Explor* **133** pp 88-102
[16] Yuste, A., Bauluz, B., Mayayo, M. J. (2017). Origin and geochemical evolution from ferrallitized clays to karst bauxite: An example from the Lower Cretaceous of NE Spain. *Ore Geol Rev* **84** pp 67-79
[17] Abedini A, Azizi M R and Calagari A A 2019 REE mobility and tetrad effects in bauxites: an example from the Kanisheeteh deposit, NW Iran. *Acta Geodyn et Geomater* **193** pp 11- 26
[18] Maksimovic Z and Pantó G Y 1991 Contribution to the geochemistry of the rare earth elements in the karst-bauxite deposits of Yugoslavia and Greece *Geoderma* **51** pp 93-109
[19] Radusinović S, Jelenković R, Pačevski A, Simić V, Božović D, Holclajtner-Antunović I, Životić D 2017 Content and mode of occurrences of rare earth elements in the Zagrad karstic bauxite deposit (Nikšić area, Montenegro) *Ore Geol Rev* **80** pp406-428
[20] Mameli P, Mongelli G, Oggiano G, Dinelli E 2007 Geological, geochemical and mineralogical features of some bauxite deposits from Nurra (Western Sardinia, Italy): insights on conditions of formation and parental affinity. *Int J Earth Sci* **96** pp 887-902
[21] Wang Q, Deng J, Liu X, Zhang Q, Sun S, Jiang C and Zhou F 2010 Discovery of the REE minerals and its geological significance in the Quyang bauxite deposit, West Guangxi, China. *J Asian Earth Sci* **39** pp 701-712
[22] Zhang J, Zhao B and Schreiner B 2016 Rare earth beneficiation and hydrometallurgical processing In: *Separation Hydrometallurgy of Rare Earth Elements* Springer, Cham.
[23] Habashi F 2013 Extractive metallurgy of rare earths *Can Metall Q* **52** pp 224-233
[24] Gupta C K and Krishnamurthy N 1992 Extractive metallurgy of rarc carths *Int Mater Rev* **37** pp 197-248

Acknowledgments
The authors acknowledge the financial support provided by the Department of Science and Technology – Philippine Council for Industry, Energy and Emerging Technology Research and Development for the project 'Rare earth elements in bauxite: Characterization and extraction of a potential resource'. The authors wish to thank the staff of Alumina Mining Philippines Inc. and the graduate students at the University of the Philippines National Institute of Geological Sciences for the logistical support.

IOP Conf. Series: Materials Science and Engineering **1318** (2024) 012034 doi:10.1088/1757-899X/1318/1/012034

Production of Defragmentable Bioplastic using Starch Extracted from *Annona Muricata* (Soursoup) Seed Reinforced with Bentonite Nanoclay

A V Melendres[1], S Q Angles[1], E P G Gaspar[1], J F Saliendra[1], M M S Buenaventura[1], G Sobrepeña[1], J G Olay[1], R V C Rubi[1]

[1]Department of Chemical Engineering, College of Engineering, Adamson University, 900 San Marcelino St., Ermita, Manila 1000, Philippines

E-mail:ariel.melendres@adamson.edu.ph

Abstract. Seeds of Annona Muricata, often called soursop or guyabano were used as a source of starch for the production of defragmentable bioplastic. The researchers used mixture of water and starch content of 10 and 20 grams respectively from Annona Muricata (soursop) seeds with various concentration (20%, 25%, 30%) of glycerol (plasticizer) and reinforced with three different concentrations (0%, 3%, 6%) of bentonite nanoclay (filler). The characterization of the bioplastic was carried out by determining the effect of variation of the ratio between the quantity of starch from soursop seeds, bentonite nanoclay, and glycerol. The starch-based defragmentable bioplastic was evaluated with its water retention, degradation, and mechanical properties. The result showed that the tensile strength improved significantly with the addition of bentonite nanoclay and the elongation at break with the addition of glycerol. Same way in water retention test, where the presence of the bentonite nanoclay increased the strength of the bioplastic with its water solubility. In soil burial test, a higher level of glycerol concentration increased the rate of the bioplastic degradation. By adding the bentonite nanoclay as a filler, the rate of bioplastic degradation decreased compared to the bioplastic without filler. In shelf-life test, adding glycerol as plasticizer improved the shelf life of bioplastics. A high concentration of glycerol served as an anti-fungal for bioplastics..

1. Introduction

Bioplastic is a biodegradable polymer derived from renewable resources that can be used to alleviate the world's plastic waste crisis, which is strangling the planet and contaminating the air (Folino et al., 2020). Plastic degradability is one of today's most serious environmental issues, as large amounts of plastic garbage take years to degrade (Alshehrei, 2017). Even everyday plastic bags can take up to 1,000 years to degrade, while plastic bottles can take up to 450 years to decompose (LeBlanc, 2021). In order to prevent plastic pollution, the industry is still looking for new ways to reduce plastic usage. By far the most effective alternative to plastic degradability is recycling (Baker, 2018; Villando et al., 2021). Approximately 400 million tons of fossil fuel based polymer waste is generated globally every year, where only about 9% is recycled and reused (Haider et al., 2019). As a result, bioplastic has become a viable option (Gibbens, 2018). The use of biodegradable polymers derived from natural polysaccharides, primarily starch, aids in the reduction of non-degradable substances (Kuruvila et al., 2018). The goal of this study is to explore the applicability of *Annona Muricata* seed as a source of starch for the production of defragmentable bioplastics. *Annona Muricata,* also known as soursop is native to South and North America's hottest regions, but it is now extensively spread across the world's tropical and subtropical regions such as Central and South America, Western Africa, and Southeast Asia, including the Philippines. The seed comprises at around 4% of the total weight of the fruit, seeds of Annona Muricata are flat and firm, and they contain something which may be utilized to make paint or pesticide (Fasakin et al., 2008). Soursop seed has a carbohydrate content of 67.85 percent (Onyechi et al., 2014). Starch is made up of two types of a-D-glucose polymers: amylose, which is a straight particle, and amylopectin, which is a thoroughly distributed atom. The ratio of amylose to amylopectin changes depending on the starch source (Melendres, et al., 2021). With more than 65% carbohydrate, *Annona Muricata* (soursop) seed can be utilized as a source of starch to produce defragmentable bioplastics.

Researchers on bioplastics are mostly focused on the utilization of starch from rice, wheat, corn, and potatoes (Marichelvam et al., 2019). The development of starch-based bioplastic was influenced by the factor which is to produce better product quality, abundant, cheap, and renewable resource (Othman et al., 2023). It contains additives such as plasticizers and compatibilizers to improve process ability, water-resistance and tear strength (Cherkaev & Leelavanichkul, 2012; Alhanish, & Abu, 2021). Bioplastics require reinforcement, such as nanoclay, to improve their mechanical characteristics. The use of nanoclay for bioplastics was investigated by Wahyuningtiyas & Suryanto, 2018). Nanoclay when incorporated in bioplastics can provide a greater improvement of the property (Zamrud et al.,2021). It can also speed up the decomposition of starch-based bioplastics, as well as change and improve the structure of bioplastics, allowing them to higher tensile strength and less water retention.

Despite substantial research on the use of starch for packaging applications, there are no study of starch based on *Annona Muricata* seed reinforced with bentonite nanoclay. Hence, in this research, starch extracted from soursop seeds combined with glycerol (plasticizer) were reinforced by bentonite nanoclay in order to produce a defragmentable bioplastic film. This study aimed to test the effectiveness of the bentonite nanoclay to the defragmentable bioplastic produced from the starch of *Annona Muricata* for its mechanical, physical, and biodegradability properties.

2. Materials and Methods

2.1. Reagents and Raw Materials

Pure glycerol (C3H8O3) was purchased from Alyson's Chemical Enterprises Inc., Quezon City. For the starch test, Iodine –Potassium Iodide (KI) Reagent was mixed along with distilled water which was purchased at local convenience store. Bentonite nanoclay with an average particle size of one nanometer was obtained from the Guangzhou Peak Chemical Technology Co. Ltd, China. The guyabano (*Annona Muricata*) was purchased at Daraitan Tanay, Rizal. The analytical balance (Sartorius BSA 2245-CW), lab mixer (EURO-ST 40 D 5000), water bath with agitation (VS – 310SWO), and air-drying oven (1600 HAFO Series) used in the production of the bioplastic were all available in the Adamson University Chemical Laboratory.

2.2. Methodologies

2.2.1. Preparation of Bioplastics

The soursop (*Annona Muricata*) seeds of 1.6 kilograms were washed with distilled water and decorticated using blender to extract the starch. The extracted starch was dried at 60°C and filtered through a 100 mesh strainer. Bentonite nanoclay solution was made by dissolving bentonite in 20%, 25% and 30% glycerol solution to produce each 0, 3 and 6% bentonite. The bentonite nanoclay solution was placed in a lab mixer for 20 minutes. After mixing, the solution was withdrawn and mixed with the extracted starch. The solution was then heated and repeatedly agitated until it became gelatinized. (80°C). The gelatinized solution was poured into the required mold, dried in a 50°C oven for 6 hours, and then allowed to cool at room temperature. The bioplastic film produced was removed from the mold and ready for testing.

2.2.2. Water Retention

The bioplastic film samples were immersed in distilled water for about five minutes. The samples were then placed in test tubes and centrifuge for about ten minutes. After centrifugation, the bioplastic film samples were transferred to 250 mL beakers and weighed to obtain the wet mass (W). The bioplastic film samples were then dried in a vacuum oven for an hour at 100-110oC. After the drying process, the bioplastic samples were placed in the desiccator for cooling. Finally, the bioplastic film samples including the container were weighed and it was recorded as dry mass (D). The determination of percent water retention (R) of bioplastic was calculated using equation below.

$$R = \left(\frac{W - D}{W}\right) x\ 100$$

2.2.3. Tensile Strength and % Elongation
The tensile test was performed using a tensile tester (Model-61008431) from Milageto. The initial surface area of the bioplastic film was calculated by measuring its length and height. The tensile tester was attached to the center of the bioplastic film and weight load was added until the bioplastic film ruptured and separated. The maximum force is equivalent to the highest load added until rupture. The tensile strength is calculated using the following equation.

$$\textbf{Tensile Strength} = \frac{\textbf{Maximum Force}}{\textbf{Initial Surface Area}}$$

The length of the bioplastic film at break and the length of the bioplastic film before being pulled by the tensile tester can be used to calculate percent length. The elongation at break can be calculated using equation below.

$$\%\ \textbf{Elongation} = \left(\frac{\textbf{Final Length} - \textbf{Initial Length}}{\textbf{Initial Length}}\right) \textbf{x100}$$

2.2.4. Biodegradability Test
The soil burial degradation test was used to determine bioplastic's biodegradability. To check the biodegradability, the bioplastic films samples were cut into pieces with a size of 10 mm x 10 mm and weighed before being buried at a depth of 10 cm. The weight of the samples was measured after 12 days. The samples were taken out and examined for signs of defragmentation. The equation below the biodegradability of the polymer sample.

$$\textbf{Degree of Defragmentation}\ (\%) = \frac{\textbf{Final Mass} - \textbf{Initial Mass}}{\textbf{Initial Mass}}$$

The shelf life test was determined by placing the sample in an environment with low oxygen or humidity of 45-60% RH. The trial lasted for three months.The bioplastic samples were examined after 12 weeks. The degree of degradation was calculated based on the initial mass and final mass.

3. Results and Discussions

Pictures of the soursop (Annona Muricata) seeds after washing with distilled water is shown in figure 1-a. Starch is not very soluble in water, thus, a slurry is produced after decorticating the seed using a blender to extract the starch. Figure 1-b shows the extracted starch from the soursop seed. Using different concentrations of glycerol at various bentonite nanoclay loading and following the procedure outlined in the methodologies, film of bioplastic is produced as shown in figure 1-c.

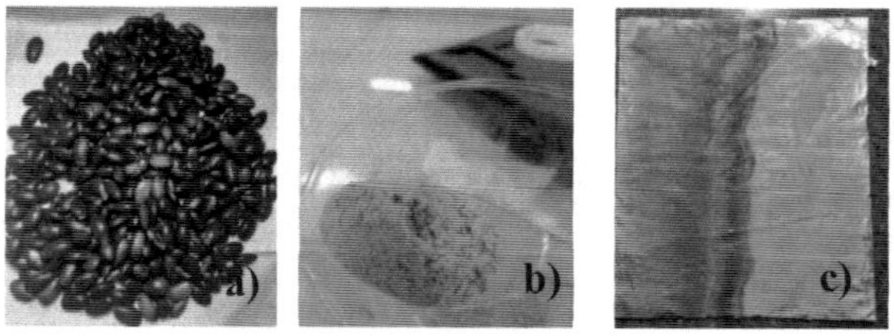

Figure 1. a) Seeds of *Annona Muricata,* b) extracted starch, c) Bioplastic film.

Table 1 shows the results of tensile strength at various concentrations of starch, glycerol and bentonite nanoclay added to the defragmentable bioplastic film. According to the test results, adding bentonite nanoclay to the defragmentable bioplastic film improved its physical, mechanical, and degrading capabilities. The defragmentable bioplastic film sample with the greatest nanoclay content (6%) exhibited the best tensile strength but a coarser film surface. Because bentonite nanoclay has

IOP Conf. Series: Materials Science and Engineering **1318** (2024) 012034 doi:10.1088/1757-899X/1318/1/012034

hydrophobic qualities, the film's water resistance was enhanced, resulting in rough and hard bioplastic films. The defragmentable bioplastic samples with no nanoclay (0 percent) concentration had the lowest tensile strength and flexibility when compared to the defragmentable bioplastic with bentonite nanoclay. It showed that the highest tensile strength produced was in sample 8, which has the highest concentration of bentonite nanoclay (6%). This finding is similar to the study of Zia et al., (2011). Samples 2 & 4 and samples 5 & 8 had the same composition of glycerol and starch but different concentrations of bentonite nanoclay, and the result showed that higher bentonite nanoclay concentration tend to have higher tensile strength. In general, the tensile strength of the defragmentable bioplastic films can also be affected by differences in glycerol composition. The hydrophilic nature of glycerol tends to decrease the tensile strength of the polymer.

Table 1. Tensile strength of bioplastic

Sample	Concentration of Glycerol (wt%)	Concentration of Bentonite Nanoclay (wt%)	Amount of Starch (grams)	Tensile (MPa)
1	20	0	10	0.01750
2	20	3	10	0.01415
3	20	3	20	0.01325
4	20	6	10	0.02250
5	25	0	20	0.00044
6	25	3	10	0.01290
7	25	3	10	0.01490
8	25	6	20	0.03050
9	30	0	10	0.01300
10	30	3	20	0.01500
11	30	6	10	0.01550

Table 2 shows the effects of varying concentration of bioplastic components to the elongation at break of defragmentable bioplastic film samples. It can be seen that sample 11, which had the highest amount of glycerol, had also the highest percentage of elongation in the bioplastic film produced (30% elongation). This was because increasing the plasticizer concentration raised the viscoelastic response of the plastic polymer chain, which rai sed the elastic property and elongation value of the product (Sofiah et al., 2019). Sample 2 and 7, as well as Sample 3 and 10 had the same concentration of bentonite nanoclay and starch but different values of glycerol. In both cases, samples, with higher amount of glycerol have higher % elongation. Glycerol reduces the stiffness and improve the flexibility of the plastics by increasing the mobility of the polymer bonding (De Roza et al. 2012). The percentage for elongation values showed a decrease with increasing concentration of bentonite nanoclay.

Table 2. Elongation at break of bioplasti

Sample	Concentration of Glycerol (wt%)	Concentration of Bentonite Nanoclay (wt%)	Amount of Starch (grams)	Elongation (%)
1	20	0	10	6.3300
2	20	3	10	12.500
3	20	3	20	9.1650
4	20	6	10	8.7100
5	25	0	20	25.000
6	25	3	10	18.375
7	25	3	10	20.000
8	25	6	20	9.4000
9	30	0	10	22.000
10	30	3	20	25.335
11	30	6	10	30.000

Table 3 presents how various concentration of bentonite nanoclay added to defragmentable bioplastic films affects its water solubility. It can be seen that sample 5, which had the highest composition of starch (20g) and no amount of bentonite nanoclay (0%) showed the highest water solubility (57.7%).

Table 3. Percentage water retention

Sample	Concentration of Glycerol (wt%)	Concentration of Bentonite Nanoclay (wt%)	Amount of Starch (grams)	Water Retention (%)
1	20	0	10	38.391
2	20	3	10	31.797
3	20	3	20	38.901
4	20	6	10	21.013
5	25	0	20	57.713
6	25	3	10	29.798
7	25	3	10	31.581
8	25	6	20	31.999
9	30	0	10	38.974
10	30	3	20	41.009
11	30	6	10	21.246

The solubility of the bioplastic films increased steadily as the starch content in the film-forming solution increased, the glycerol concentration also had an effect on the films' solubility. This could be due to the interactions between amylose, amylopectin molecules, and intramolecular bonds of starch characteristics. Table 4 presents the data of various compositions added to the defragmentable sample film and how glycerol concentration affects the physical property by the used of soil burial test. The biodegradability (disintegration) of the defragmentable bioplastic film was expressed by mass retention of the bioplastic film after 12 days of soil burial test as shown in the table. The defragmentation of bioplastics was assessed utilizing soil burial testing as a parameter of an environmentally friendly material to determine the defragmentation level of plastic film in the environment. The samples were buried at a depth of 10 cm in loam soil, and the percentage of defragmentation by weight loss was determined. Result showed that the highest degradation were achieved on sample 10, which had the highest concentration of glycerol (30%) and had a result of 70.5% weight loss. In a study similar to that of Acquavia et al., (2021), a higher level of glycerol concentration accelerated the rate of defragmentable bioplastic decomposition. This was due to the hydrophilic nature of glycerol in relation to water. Furthermore, because water was the medium in which most bacteria live in the soil, glycerol has the potential to absorb water quickly. As a result, the higher the glycerol content, the more water may pass

Table 4. Biodegradability of bioplastic

Sample	Concentration of Glycerol (wt%)	Concentration of Bentonite Nanoclay (wt%)	Amount of Starch (grams)	Final Weight Loss aftr 12 days(%)
1	20	0	10	45.288
2	20	3	10	47.048
3	20	3	20	46.252
4	20	6	10	41.396
5	25	0	20	60.908
6	25	3	10	45.647
7	25	3	10	52.288
8	25	6	20	55.576
9	30	0	10	69.407
10	30	3	20	70.433
11	30	6	10	61.283

through the defragmentable bioplastic's structure causing faster degradation. The inclusion of bentonite nanoclay, on the other hand, slows down the pace of defragmentation. This was due to the hydrophobic qualities of bentonite nanoclay which would prevent water absorption. This was shown in sample 4, which had the highest concentration of bentonite nanoclay (6%), but only 20% glycerol which showed the least deterioration with a degradation rate of only 41.396%.

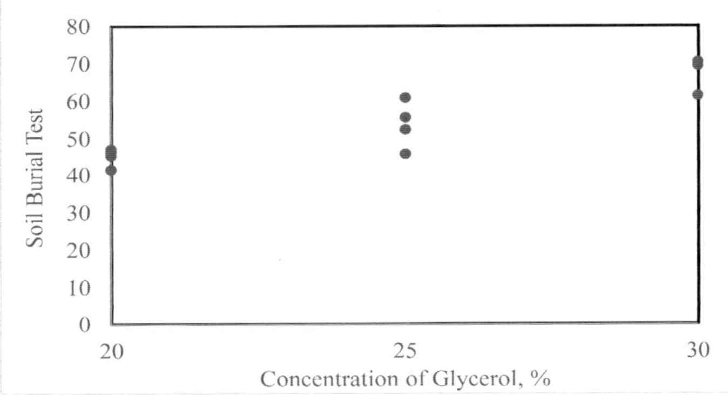

Figure 2. Bioplastic's biodegradability using soil burial test at various glycerol concentrations.

The results on burial test based on the data Table 4 data is shown in figure 2 as scattered graph. In this graph, increasing soil burial test resulted with increasing concentration of glycerol in general.

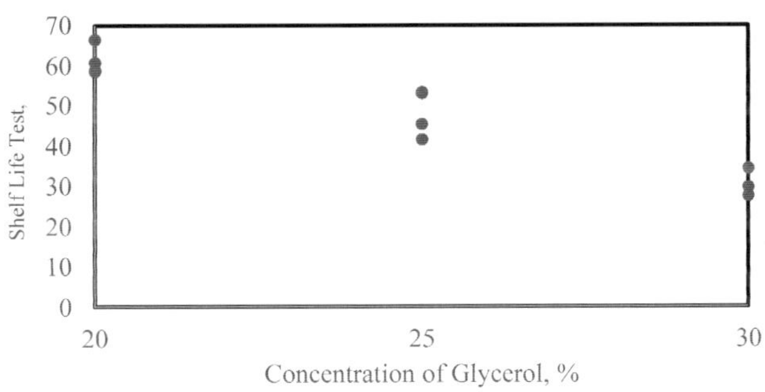

Figure 3. Bioplastic's shelf life test as % degradation.

As explained above, the effect of increasing the concentrations of bentonite showed a slow down on the degradation of bioplastic. Figure 3 shows the effect of glycerol concentration to the shelf life of defragmentable bioplastic films. Shelf life is expressed in terms of % degradation after 12 weeks at high moisture content environment. It showed a downward trend on the graph showing lesser degradation at high glycerol concentration which acts as anti-fungal. The dotted points in the graph are the degradation of defragmentable bioplastic after 12 weeks. The downward trend shows improving shelf life as the concentration of glycerol increases.

4. Conclusion

Varying starch, bentonite nanoclay, and glycerol concentrations resulted in different physical properties of defragmentable bioplastics in terms of texture, appearance, and characteristic. To detect the presence of starch, extracted starch from soursop were analyzed and the data were for producing the bioplastic. The addition of bentonite nanoclay to the defragmentable bioplastic film increased its tensile strength significantly. Bentonite nanoclay enhanced the tensile strength of defragmentable bioplastic films, including those containing glycerol. Glycerol enhances the elasticity of the defragmentable bioplastic film, reducing the film stiffness. As the glycerol concentration in the defragmentable bioplastic film increases, so does the elongation, demonstrating a direct relationship between the two variables. The defragmentable bioplastic films with the highest starch and glycerol content made the films more soluble in water. Adding more glycerol to the defragmentable bioplastics increased defragmentation; however, adding bentonite nanoclay as a filler slowed down the defragmentation process.

5. References

[1] Acquavia MA, Pascale R, Martelli G, Bondoni M, Bianco G. Natural polymeric materials: A solution to plastic pollution from the agro-food sector. *Polymers*. 2021 Jan 4;13(1):158.

[2] Agustin MB, Ahmmad B, Alonzo SM, Patriana FM. Bioplastic based on starch and cellulose nanocrystals from rice straw. *Journal of Reinforced Plastics and Composites*. 2014 Dec;33(24):2205-13.

[3] Alhanish A, Abu Ghalia M. Developments of biobased plasticizers for compostable polymers in the green packaging applications: *A review. Biotechnology Progress*. 2021 Nov;37(6):e3210.

[4] Alshehrei F. Biodegradation of synthetic and natural plastic by microorganisms. *Journal of Applied & Environmental Microbiology*. 2017;5(1):8-19.

[5] Cherkaev A, Leelavanichkul S. An impact protective structure with bistable links. *International Journal of Damage Mechanics*. 2012 Jul;21(5):697-711.

[6] da Rosa Zavareze E, Pinto VZ, Klein B, El Halal SL, Elias MC, Prentice-Hernández C, Dias AR. Development of oxidised and heat–moisture treated potato starch film. *Food chemistry*. 2012 May 1;132(1):344-50.

[7] Fasakin AO, Fehintola EO, Obijole OA, Oseni OA. Compositional analyses of the seed of sour sop, Annona muricata L., as a potential animal feed supplement. *Scientific Research and Essay*. 2008 Oct 1;3(10):521-3.

[8] Folino A, Karageorgiou A, Calabrò PS, Komilis D. Biodegradation of wasted bioplastics in natural and industrial environments: *A review. Sustainability*. 2020 Jul 27;12(15):6030.

[9] Gibbens S. What you need to know about plant-based plastics. *National Geographic*. 2018 Nov.

[10] Haider TP, Völker C, Kramm J, Landfester K, Wurm FR. Plastics of the future? The impact of biodegradable polymers on the environment and on society. *Angewandte Chemie International Edition*. 2019 Jan 2;58(1):50-62.

[11] Iamareerat B, Singh M, Sadiq MB, Anal AK. Reinforced cassava starch based edible film incorporated with essential oil and sodium bentonite nanoclay as food packaging material. *Journal of food science and technology*. 2018 May;55:1953-9.

[12] Kuruvilla S, Sadana R, Montesinos EV, Beard J, Vasdeki JF, de Carvalho IA, Thomas RB, Drisse MN, Daelmans B, Goodman T, Koller T. A life-course approach to health: synergy with sustainable development goals. *Bulletin of the World Health Organization*. 2018 Jan 1;96(1):42.

[13] Leblanc R. The decomposition of waste in landfills: A story of time and materials. *The Balance Small Bussiness*. 2019 Oct.

[14] Marichelvam MK, Jawaid M, Asim M. Corn and rice starch-based bio-plastics as alternative packaging materials. *Fibers*. 2019 Apr 9;7(4):32.

[15] Melendres AV, Cruz RP. Absorption of water vapor using superabsorbent polymer composite material. InKey Engineering Materials 2020 (Vol. 858, pp. 129-139). *Trans Tech Publications Ltd.*

[16] Melendres A, Cruz RV, Espenilla MB, Monsada A. Absorbency and Corrosion Inhibition Property of Polyamide Epoxy and Superabsorbent Polymer Composite Material. *Chemical Engineering Transactions*. 2021 Jun 15;86:1297-302.

[17] Onyechi UA, Ibeanu VN, Eme P, Ugwuanyi CU. Nutrient and photochemical composition of Annona muricata seed flour (soursop). *J. dietitian Assoc. Nig*. 2014;5(4):46-55.

[18] Othman SA, Nasir NN, Azman NF. Natural Fibre for Composite Structural Application. *Structural Integrity and Monitoring for Composite Materials*. 2023 Jan 28:165-78.

[19] Aznury M. Mechanical properties of bioplastics product from Musa paradisica formatypica concentrate with plasticizer variables. InJournal of Physics: Conference Series 2019 Feb 1 (Vol. 1167, No. 1, p. 012048). *IOP Publishing*.

[20] Vilando AC, Rubi RV, Lacsa FJ. Utilization of low-cost waste materials in wastewater treatments. InIntegrated and hybrid process technology for water and wastewater treatment 2021 Jan 1 (pp. 99-119). *Elsevier.*

[21] Wahyuningtiyas NE, Suryanto H. Properties of cassava starch based bioplastic reinforced by nanoclay. *Journal of Mechanical Engineering Science and Technology* (JMEST). 2018 Aug 30;2(1):20-6.

[22] Zamrud Z, Ng WM, Salleh HM. Effect of bentonite nanoclay filler on the properties of bioplastic based on sago starch. InIOP Conference Series: Earth and Environmental Science 2021 May 1 (Vol. 765, No. 1, p. 012009). *IOP Publishing.*

[23] Zuber M, Zia KM, Mahboob S, Hassan M, Bhatti IA. Synthesis of chitin–bentonite clay based polyurethane bio-nanocomposites. *International Journal of Biological Macromolecules*. 2010 Aug 1;47(2):196-200.

Zinc-imidazole-based metal-organic framework nanosheet membrane for H_2/O_2 separation

Shunsuke Tanaka[1,2], Yuka Kimura[1], Kojiro Fuku[1,2], Naoki Ikenaga[1,2] and Keizo Nakagawa[3,4]

[1]Department of Chemical, Energy and Environmental Engineering, Kansai University, Osaka, Japan
[2]Organization for Research and Development of Innovative Science and Technology, Kansai University, Osaka, Japan
[3]Graduate School of Science Technology and Innovation, Kobe University, Kobe, Japan
[4]Research Center for Membrane and Film Technology, Kobe University, Kobe, Japan

E-mail: shun_tnk@kansai-u.ac.jp

Abstract. Metal-organic framework (MOF) nanosheets are promising candidates for molecular sieve because of their structural diversity and minimized mass transfer barrier. However, design of appropriate MOF nanosheets and preparation of high-performance MOF nanosheet-based membranes, especially for gas separation, remains great challenges. Structural degradation may simultaneously occur with conventional exfoliation method, which has hindered its widespread application in high-performance molecular sieve membrane preparation. Even if nanosheets could be stacked, grain boundaries would form between the nanosheets, which could be applied to liquid separation but not to gas separation. To address these challenges, we applied bottom-up growth of Zn_2(benzimidazole)$_4$ nanosheet to fabricate defect-free membranes. Zn_2(benzimidazole)$_4$ is composed of benzimidazole-zinc tetrahedral units and layers are connected by van der Waals interaction. At first, the Zn_2(benzimidazole)$_4$ nanosheets were deposited onto a porous support to prepare gutter layer. Next, Zn-based amorphous layer was coated on the gutter layer and crystallized by benzimidazole vapor treatment. Highly oriented Zn_2(benzimidazole)$_4$ nanosheet based membranes were fabricated by anisotropically controlling crystallization. The prepared Zn_2(benzimidazole)$_4$ nanosheet-based membranes show separation performance in hydrogen purification with H_2/O_2 ideal separation factor of 11.6 and H_2 permeance of 5650 GPU.

1. Introduction

The COVID-19 pandemic and growing international demand have caused wholesale energy prices to skyrocket worldwide. With an energy self-sufficiency rate of only about 12%, Japan will increasingly need to utilize renewable energy sources. Hydrogen does not emit CO_2 when burned, and is attracting attention as an energy source for a new era toward carbon neutrality. In order to use hydrogen as a truly clean energy source, hydrogen generated from water splitting using renewable energy sources such as solar and wind power must be efficiently separated and purified from oxygen. Membrane separation is expected to be an energy-saving method for separating mixed gases. In this study, we focused on metal-organic frameworks (MOFs) as a membrane material [1–3]. MOFs are crystalline microporous and inorganic-organic hybrid materials, which can be self-assembled from metal ions/clusters and organic ligands. Among the many MOFs, zeolitic imidazolate frameworks (ZIFs) are undoubtedly the most extensively studied because of their facile synthesis coupled with their structural stability [4]. ZIF-7 has a sodalite topology, which comprises of interconnected six-membered rings with an inner cage size of 5.6 Å and pore aperture of 2.9 Å [5,6]. Zn_2(benzimidazole)$_4$, with the same composition as ZIF-7 and layered structure, is a promising nanosheet precursor for hydrogen sieving [7,8]. Therefore, if membranes are fabricated from nanosheets of Zn_2(benzimidazole)$_4$, they are expected to be applied as hydrogen separation membranes with high permeability and selectivity. Several methods have been reported to exfoliate layered MOF crystals by ultrasonic or soft physical methods to form nanosheets, but structural degradation during the exfoliation process is a serious problem [9,10].

In this study, we developed a bottom-up method of nanosheet membrane formation in which zinc-based amorphous layer is applied on top of an intermediate gutter layer of Zn_2(benzimidazole)$_4$ nanosheets fabricated by conventional interface synthesis, followed by crystallization of zinc-based layer by supplying benzimidazole vapor. The concept of our method is to anisotropically control the crystallization of Zn_2(benzimidazole)$_4$ during crystallization of the zinc-based amorphous layer by using surfactant. In addition, the separation performance of the resulting membranes was investigated.

2. Experimental Section

2.1. Chemicals
All the commercially available chemicals were used without any further purification. Zinc nitrate hexahydrate, benzyltrimethylammonium chloride, ethylene glycol monomethyl ether, 2-ethanolamine, methanol and dichloromethane were purchased from Wako Pure Chemical Industries. Benzimidazole was purchased from Sigma-Aldrich, Japan.

2.2. Synthesis of Zn_2(benzimidazole)$_4$
0.67 mmol of zinc nitrate hexahydrate and 9.3 mmol of benzimidazole were dissolved in 60 mL of a mixed solvent of ethylene glycol monomethyl ether and methanol (V : V = 1 : 2). The solution was stirred for 24 h at room temperature. After stirring, the solution was centrifuged and separated into the liquid and solid phases. The product was thrice-washed with methanol and vacuum dried overnight at 40 °C.

2.3. Synthesis of Zn_2(benzimidazole)$_4$ nanosheet gutter layer and zinc-based amorphous layer
0.2 mol/L dichloromethane solution of benzimidazole was dropwise spread on the liquid surface of 0.4 mol/L aqueous solution of zinc nitrate. The nanosheet formed on the liquid surface was transferred to the surface of anodic aluminum oxide (AAO) substrate. The product was dried at 100 °C.

2.4. Conversion of zinc-based amorphous layer to Zn_2(benzimidazole)$_4$ nanosheet
1.1 mmol of zinc nitrate hexahydrate and 0.27 mmol of benzyltrimethylammonium chloride were dissolved in 5 mL of 2-methoxyethanol. The resulting solution was cast onto the AAO substrate with Zn_2(benzimidazole)$_4$ nanosheet gutter layer. The AAO substrate with zinc-based amorphous layer and nanosheet gutter layer was placed separately with solid-phase benzimidazole in a Teflon-lined stainless-steel autoclave. The autoclave was sealed and heated at 180 °C for benzimidazole vapor deposition. The conversion of zinc-based amorphous layer to Zn_2(benzimidazole)$_4$ nanosheet was conducted with deposition time for 2 h. Finally, the resulting Zn_2(benzimidazole)$_4$ nanosheet membrane was washed with methanol for several times and vacuum dried at 40 °C.

2.5. Characterization
Powder X-ray diffraction (PXRD) patterns were recorded on a RIGAKU MiniFlex600. CuKα (λ = 1.5418°) was used as an X-ray tube at 30 kV and 15 mA. Thermogravimetric analysis (TGA) was conducted using Shimadzu DTG-60H. Field emission scanning electron microscopy (FESEM) images

of membrane were obtained using a Hitachi High-Tech S-4800. The measurement was performed at an acceleration voltage of 1.0–5.0 kV and an emission current of 10 μA.

The single gas permeation of membranes was measured at 35 °C using a constant-volume apparatus. The membranes were sealed on a stainless-steel permeation cell with an effective test area of 0.79 cm^2. The feed pressure of each test gas is 100 kPa. The molar flow rate, N_i, of the permeate was calculated from the linear pressure rise. Membrane permeance, P_i (mol/($m^2 \cdot s \cdot Pa$)), was calculated as follows;

$$P_i = \frac{N_i}{A\Delta P}$$

where ΔP is the pressure difference of upstream and downstream of the membrane, and A is the test area. The ideal separation factor was calculated with following equation.

$$\alpha = \frac{P_i}{P_j}$$

3. Results and Discussion

3.1. Structure and thermal stability of bulk Zn_2(benzimidazole)$_4$

As shown in Figure 1 left, PXRD pattern confirms that our bulk product is isostructural to Zn_2(benzimidazole)$_4$ (CCDC ref.#675375). TGA performed on Zn_2(benzimidazole)$_4$ revealed the structural stability up to 400 °C (Figure 1 right). In addition, TGA shows weight loss of about 75% in the temperature range 400–600 °C, indicating that structure completely collapsed and ZnO residue is left.

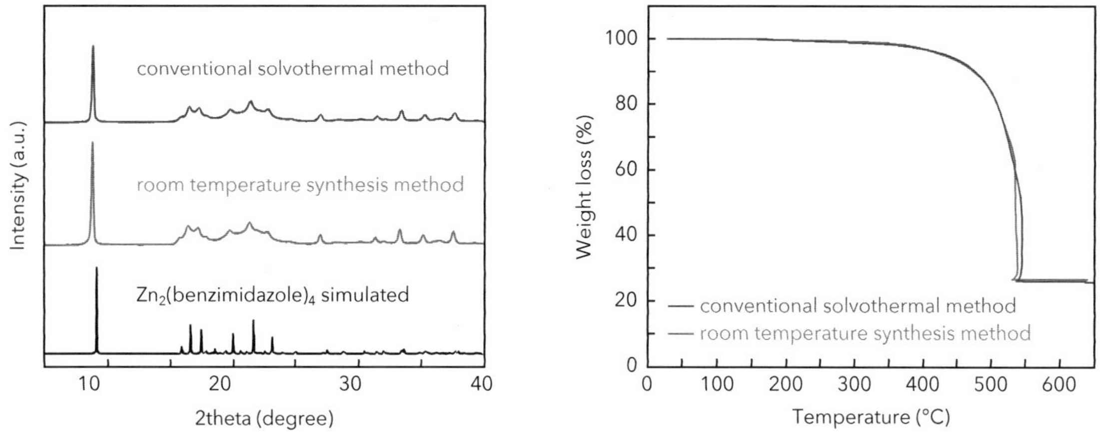

Figure 1. PXRD patterns (left) and TGA curves (right) of the bulk Zn_2(benzimidazole)$_4$ products prepared by the conventional solvothermal method and our room temperature synthesis method.

3.2. Zn_2(benzimidazole)$_4$ layer on AAO substrate with and without gutter layer

PXRD patterns of the membranes prepared on AAO substrate are shown in Figure 2 left. PXRD pattern of Zn-based layer shows broad diffraction. On the other hand, after Zn-based layer was exposed to benzimidazole vapor at 180 °C, sharp diffraction peaks appeared in the PXRD pattern. The PXRD shows 1st, 2nd and 3rd order diffraction peaks, indicating that the product possesses a layered structure. Ordered diffraction corresponding to the (002), (004) and (006) planes of the bulk Zn_2(benzimidazole)$_4$ was observed, indicating that the structure of the layer is strongly oriented along the *c* axis (Figure 2 right). PXRD pattern confirms that the zinc-based layer prepared on AAO substrate both with and without gutter layer after benzimidazole vapor deposition are layered structure, respectively. On the other hand, when the Zn_2(benzimidazole)$_4$ gutter layer was not coated, the broad diffraction around 25° still remained after benzimidazole vapor deposition, indicating that the membrane contains crystalline structure and amorphous region.

Figure 3 shows FESEM images of top-view and cross-section of Zn_2(benzimidazole)$_4$ layer prepared on AAO substrate. When the Zn_2(benzimidazole)$_4$ gutter layer was not coated, zinc-based amorphous layer penetrated into the pores of AAO substrate, resulting in failure to fabricate thin nanosheet membranes on the AAO surface. On the other hand, when the Zn_2(benzimidazole)$_4$ nanosheets synthesized by the conventional interfacial synthesis method were transferred onto the AAO substrate for the fabrication of gutter layer, the pores of AAO were covered by the gutter layer, preventing the zinc-based amorphous layer from seeping into the AAO.

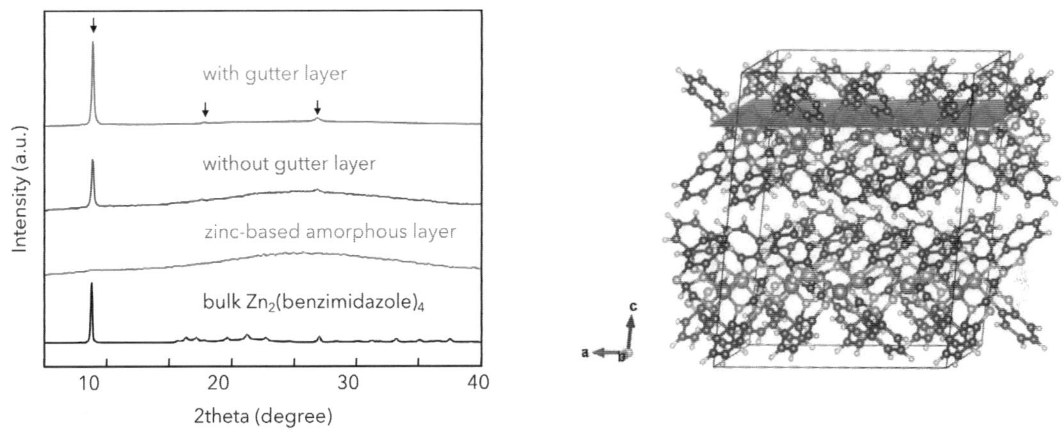

Figure 2. PXRD patterns of the membranes prepared on AAO substrate (left) and schematic illustration of *c* axis oriented layered Zn_2(benzimidazole)$_4$ visualized using VESTA [11] (right).

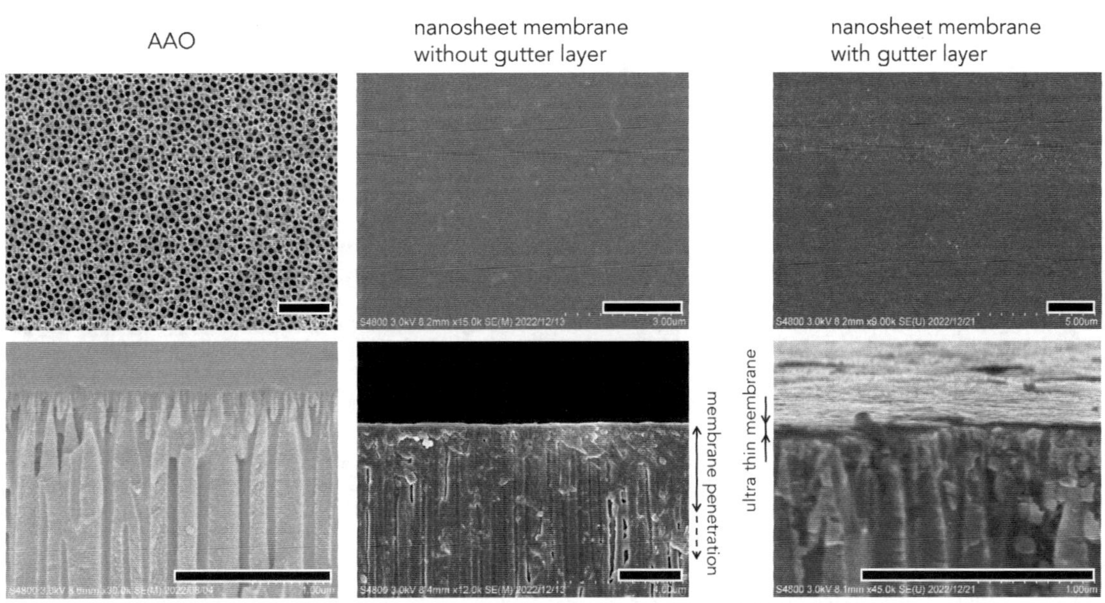

Figure 3. FESEM images of the nanosheet membranes prepared on AAO substrate.

3.3. Gas permeation properties

Figure 4 shows the permeance of hydrogen and ideal separation factor of H_2/O_2 through the Zn_2(benzimidazole)$_4$ nanosheet membranes. The AAO substrate itself shows the highest gas permeance compared to the others, but the ideal separation factor ($\alpha = 2.7$) is smaller than the Knudsen diffusion selectivity ($\alpha_{Kn} = 4$). In contrast, the ideal separation factor of Zn_2(benzimidazole)$_4$ nanosheet membranes without gutter layer exceeds the Knudsen diffusion selectivity. However, the permeance was significantly decreased. This is because the nanosheet membrane did not form thinly on the AAO substrate surface, but rather penetrated inside the substrate, resulting in thicker membrane.

On the other hand, the ideal separation factor of the Zn_2(benzimidazole)$_4$ nanosheet membrane fabricated on the gutter layer was 11.6 and its hydrogen permeance reached 5650 GPU (1 GPU = 3.35×10^{-10} mol/(m^2·s·Pa)). This is due to the formation of extremely thin membrane on the AAO substrate surface, as well as the epitaxial growth of Zn_2(benzimidazole)$_4$ from the gutter layer and the anisotropic crystal growth.

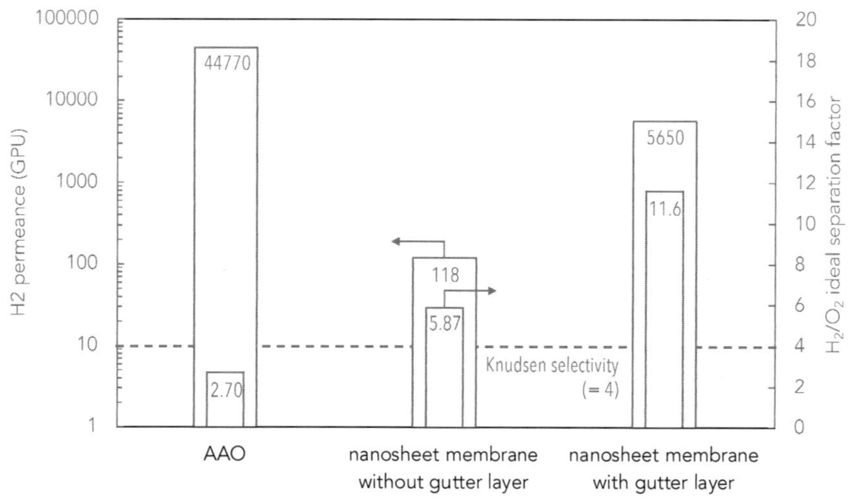

Figure 4. Comparison of hydrogen permeance and ideal separation factor of H_2/O_2 for Zn_2(benzimidazole)$_4$ nanosheet membranes.

4. Conclusions

In this study, high-quality Zn_2(benzimidazole)$_4$ nanosheet membrane was successfully grown on the AAO substrate with gutter layer. The use of a smooth gutter layer of nanosheets was effective in producing ultrathin nanosheet membranes. Nanosheet membranes on gutter layer showed excellent gas permeation properties. Bottom-up synthesis of nanosheet membranes, as proposed in this study, can provide improved reliability of membrane preparation and improved separation performance.

Acknowledgments

This work was supported by the Kansai University Fund for Supporting Outlay Research Centers, 2021, JKA and its promotion funds from KEIRIN RACE (Grant No. 2023M-412).

References

[1] G. Férey, Hybrid porous solids: past, present, future, *Chem. Soc. Rev.* **37** (2008) 191–214.
[2] H. Furukawa, K. E. Cordova, M. O'Keeffe, O. M. Yaghi, The chemistry and applications of metal-organic frameworks, *Science* **341** (2013) 1230444.
[3] S. L. Qiu, M. Xue, G. S. Zhu, Metal-organic framework membranes: from synthesis to separation application, *Chem. Soc. Rev.* **43** (2014) 6116–6140.
[4] K. S. Park, Z. Ni, A. P. Cote, J. Y. Choi, R. D. Huang, F. J. Uribe-Romo, H. K. Chae, M. O'Keeffe, O. M. Yaghi, Exceptional chemical and thermal stability of zeolitic imidazolate frameworks, *Proc. Natl. Acad. Sci. USA* **103** (2006) 10186–10191.
[5] Y. S. Li, F. Y. Liang, H. G. Bux, W. S. Yang, J. Caro, Zeolitic imidazolate framework ZIF-7

based molecular sieve membrane for hydrogen separation, *J. Membr. Sci.* **354** (2010) 48–54.

[6] A. Arami-Niya, G. Birkett, Z. H. Zhu, T. E. Rufford, Gate opening effect of zeolitic imidazolate framework ZIF-7 for adsorption of CH_4 and CO_2 from N_2, *J. Mater. Chem. A* **5** (2017) 21389–21399.

[7] P. Zhao, G. I. Lampronti, G. O. Lloyd, M. T. Wharmby, S. Facq, A. K. Cheetham, S. A. T. Redfern, Phase transitions in zeolitic imidazolate framework 7: The importance of framework flexibility and guest-induced instability, *Chem. Mater.* **26** (2014)1767–1769.

[8] R. M. Xu, M. P. Jian, Q. H. Ji, C. Z. Hu, C. C. Tang, R. P. Liu, X. W. Zhang, J. H. Qu, 2D water-stable zinc-benzimidazole framework nanosheets for ultrafast and selective removal of heavy metals, *Chem. Eng. J.* **382** (2020) 122658.

[9] Y. Peng, Y. S. Li, Y. J. Ban, W. S. Yang, Two-dimensional metal-organic framework nanosheets for membrane-based gas separation, *Angew. Chem., Int. Ed.* **56** (2017) 9757–9761.

[10] C. L. Tan, X. H. Cao, X. J. Wu, Q. Y. He, J. Yang, X. Zhang, J. Z. Chen, W. Zhao, S. K. Han, G. H. Nam, M. Sindoro, H. Zhang, Recent advances in ultrathin two-dimensional nanomaterials, *Chem. Rev.* **117** (2017) 6225–6331.

[11] K. Momma, F. Izumi, VESTA 3 for three-dimensional visualization of crystal, volumetric and morphology data, *J. Appl. Crystallogr.* **44** (2011) 1272–1276.

IOP Conf. Series: Materials Science and Engineering **1318** (2024) 012036 doi:10.1088/1757-899X/1318/1/012036

Density functional theory (DFT) analysis of the effect of oxygen vacancies in TiO_2 for the binding of oxides

C J N Lee[1], C F L Mangulabnan[1] and J A M Mendoza[1]

[1]School of Chemical, Biological, and Materials Engineering and Sciences, Mapua University, Intramuros, Manila, Philippines 1002

Email: jammendoza@mapua.edu.ph

Abstract. Challenges in the environment, particularly the rise in atmospheric pollution, entail the development of new methods to assist in sustainable development. Earlier studies have proven that oxygen vacancies (OVs) occur as the most prevalent defect in transition metal oxides, which play an essential role in gaseous pollutant degradation. Titanium dioxide (TiO_2), a naturally occurring semiconductor, is widely employed in pollutant removal. It is significant to examine the effects of oxygen vacancies on the adsorption characteristics of surfaces since O deficiencies are common on metal oxide surfaces. To date, earlier studies have centred on the reaction mechanism of TiO_2, leaving the observation of OVs largely unexplored. The purpose of this study was to measure the effects of OV in TiO_2 on the binding of oxides. In this case, the CP2K software performed DFT calculations to determine the effect of OVs on the overall binding of oxides and identify the TiO_2 energy correlated with computational analysis. As a result, having OVs in the anatase TiO_2 structure decreased the formation energy, thereby increasing its adsorption affinity for oxide binding. Therefore, it is significant to study the OVs in TiO_2 and to further modify TiO_2 to improve oxide binding for the removal of pollutants.

1. Introduction

Emerging pollutants emitted by active pharmaceutical ingredients and personal care products, as well as their potential impact on public health, have recently become a source of concern [1]. With the rapid development of global industry, many harmful substances have been released into the natural environment, resulting in a range of serious pollution [2]. At a time when environmental pollution is becoming highly critical, titanium dioxide (TiO_2) plays an increasingly important role in wastewater purification, inorganic membranes, catalyst support, and other applications [3], given its low cost and vast source, and safety for the environment and humans [4].

Structural defects pose a challenge in various energy conversion technologies [5], which substantially convert solar energy into chemical energy like the reduction of pollutants or removal of toxic compounds [6]. Oxygen vacancy (OV) is the most common type of defect in transition metal oxides (TMOs), and it can alter the intrinsic properties of TMOs [7]. Such defects are formed when an oxygen atom is eliminated from the lattice and significantly affect the photocatalytic performance [5]. Defect engineering is a wise, practical, and controllable method to adjust the performances of function materials [7].

Titanium dioxide (TiO_2) recently has attracted a great deal of attention in different fields, such as renewable energy and pollutant removal, due to its low price and amphoteric nature. Many studies have pointed out that phase composition and oxygen vacancies play a certain role in promoting the performance of TiO_2 [3].

Strategies for synthesizing defective TiO_2 with oxygen vacancies have been investigated, whereas the importance of OVs has been recognized. As a result, many efforts have been conducted to improve the performance of TiO_2 in applications. To date, earlier studies centered on the reaction mechanism of the photocatalytic reactions, while the observation of oxygen vacancies remains unexplored mainly [8].

IOP Conf. Series: Materials Science and Engineering **1318** (2024) 012036 doi:10.1088/1757-899X/1318/1/012036

The effect of oxygen vacancy on the stability of TiO_2 has been ignored, which limits its application under certain conditions [3].

This study seeks to measure the effect of oxygen vacancies in TiO_2 for the binding of oxides using the density functional theory (DFT). We are to calculate the effect of OV in TiO_2 for the binding of common pollutants (NO, NO_2, CO, H_2O, and CO_2) correlated with computational analysis. Here, we are to differentiate the lowest energy formation difference of TiO_2 with and without oxygen vacancy. The difference in energy formation would symbolize the energy adsorption of TiO_2. Since OV in oxides is extensively used to manipulate vital material properties, this study will focus on analyzing the computational description of oxygen vacancies in TiO_2 under the parameters available to DFT. The CP2K software will simulate our DFT calculations.

Since oxygen defects are widely present on the surfaces of metal oxides, it is very important to study the effects of OV on the adsorption characteristics of actual surfaces in a normal working environment [9]. Although methods for predicting defect formation energies have progressed, understanding of the intrinsic material properties that govern defect energy of the system remains unclear. Studies that directly investigate the impact of these OVs have been relatively constrained and are often contradictory, thereby limiting our understanding of the TiO_2 performance [5]. The analysis of the oxygen vacancies of TiO_2 could be an efficient approach to an increased adsorption capability of TiO_2. The findings of this study could potentially serve as a guide for future environmental studies, particularly in improving oxide binding for pollutant removal.

The present work will focus on the effect of OVs in TiO_2 for the binding of oxides. It will be limited to the optimization of the crystalline structure of TiO_2 with oxygen vacancy. The unit cell of Anatase will be used for this study. A computational method will be used to investigate the effect of the oxygen vacancies in TiO_2. The results of this study can be validated through an experimental validation system. However, in our case, the study will be interpreted based on literature studies.

2. Methodology

2.1. Characterization

Among the phases of TiO_2, the Anatase has been the widely used TiO_2 polymorph. To further understand the structural properties of the anatase TiO_2, DFT calculations will be carried out. In reference to the work of Hinuma et al, anatase phase has a tetragonal structure with the space group of 141− I41/amd as shown in Table 1. The standard setting of the space group number will indicate the surface orientation of the phase [8]. Furthermore, the optimized lattice constants of the bulk structures for the anatase phase are a = b = 3.858 Å and c = 9.557 Å [8]. However, the lattice constants utilized for Anatase will be obtained from Table 1. The supercell to be used is 2 × 1 × 2 for Anatase in perfect and defective states [10].

The Gaussian and plane waves (GPW) method and its implementation in Quickstep, which is part of the freely available program package CP2K, will measure the total energy, crystal configuration, band structure, and partial density of orbitals (PDOS) [10].

Table 1. Crystallographic properties of TiO_2 [1].

Properties	Rutile	Anatase	Brookite
Crystal Structures	Tetragonal	Tetragonal	Orthorhombic
Lattice constants (A°)	a = 4.5936	a = 3.784	a = 9.184, b = 5.447

IOP Conf. Series: Materials Science and Engineering **1318** (2024) 012036 doi:10.1088/1757-899X/1318/1/012036

	c = 2.9587	c = 9.515	c = 51.54
Molecule (cell)	2	2	4
Space groups	$P4_2/mnm$	$I4_1/amd$	Pbca
Volume/molecule	31.2160	34.061	32.172
Density (g/cm^3)	4.13	3.79	3.99
O-Ti-O bond angle	81.2°	77.7°	77.0°-105°
	90.2°	92.6°	

2.2. Computational Method

In this study, software from CP2K served as our simulator for our DFT calculations to study the effects of OVs in TiO_2. CP2K is an open–source quantum chemistry and solid–state physics software tool that provides a broad framework for various modelling methods, including DFT. It can efficiently and accurately execute atomistic simulations of solid state, liquid, molecular, periodic, material, crystal, and biological systems [11].

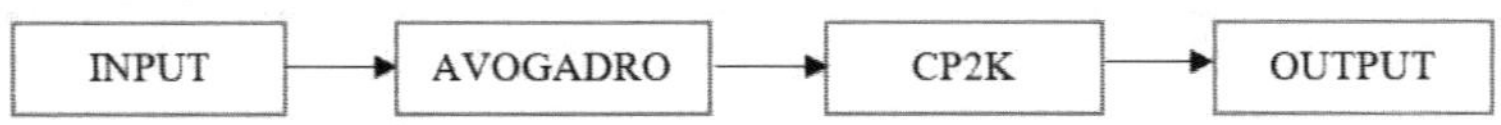

Figure 1. Flow diagram of the methodology.

Aside from that, we also used a third-party software, Avogadro, to accompany CP2K for our graphical user interface since CP2K does not have any graphical input. As shown in figure 1, the procedure began with importing the crystalline structure of Anatase TiO_2 unit cell. The file of the structure (figure 2) is found inside the database of Avogadro, which served as our input interface for this simulation. Next, we build the supercell of the Anatase structure to simulate the crystal slab of TiO_2 shown in figure 3. Afterward, we copied the atomic coordinates of the structure for the input file for the running of geometry optimization of the structure of TiO_2.

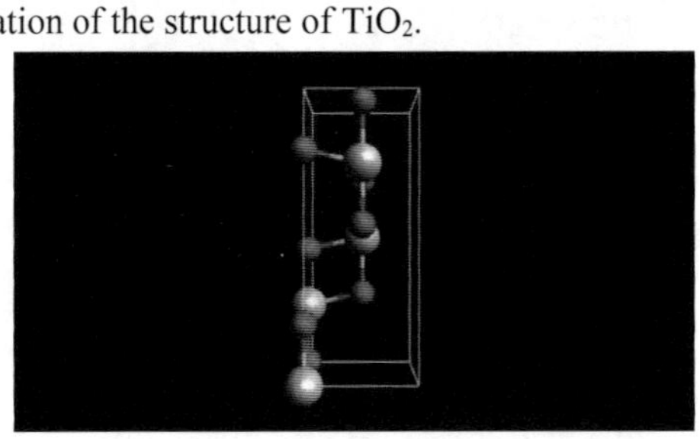

Figure 2. Crystalline structure of Anatase TiO_2. The white spheres represent the titanium dioxide molecule, while the red represents the oxygen atom.

After the simulation of the geometry optimization, CP2K would provide an output file of the coordinates of the atoms, that would serve as the basis atomic coordinates for the rest of the simulation.

The removal of oxygen atoms served as the OVs in the study. This section was done with 2 or 3 oxygen vacancies depending on each compound.

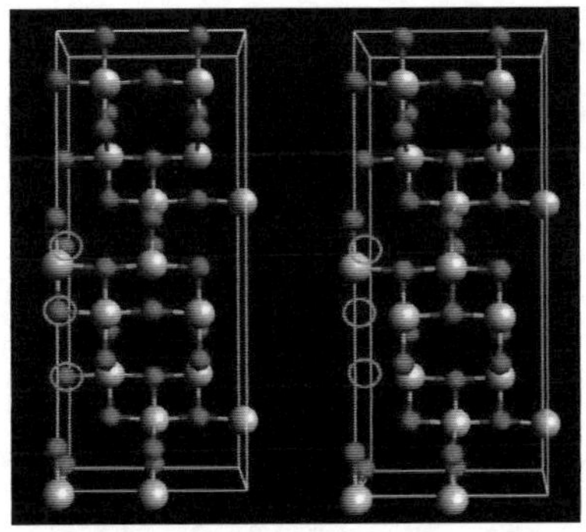

Figure 3. Structure of TiO_2 slab (left) and TiO_2 with oxygen vacancies (right). The white spheres represent the titanium dioxide molecule, while the red represents the oxygen atom.

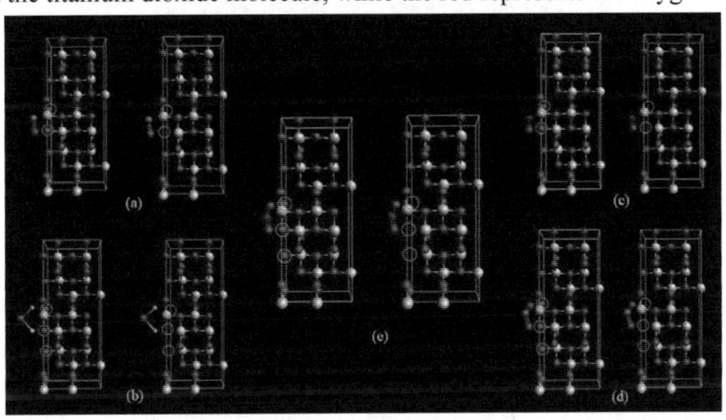

Figure 4. Structural models in the DFT calculations. (a) TiO_2- CO, (b) TiO_2- H_2O, (c) TiO_2-NO, (d) TiO_2-CO_2, (e) TiO_2-NO_2. The white spheres represent the titanium dioxide molecule, while the red represents the oxygen atom. The blue circles indicate the oxygen atoms that were removed to create structure with OVs.

Then, NO, NO_2, CO, CO_2, and H2O were added on the side of the surface of TiO_2, respectively, as shown in figure 4. Afterwards, oxygen vacancies were added near the surface where the added molecules are also close by. It should be noted that these molecules are introduced to the system separately to collect the adsorption affinities of TiO_2 with varying molecules supplied to the system. The file is then exported to the CP2K simulator, where the energy of formation in the OVs is calculated. We used notepad to read the output file and evaluate the simulation results. The adsorption energy was calculated using the equation (1),

$$\Delta E_{ads} = \sum Eproducts - \sum Ereactants, \quad (1)$$

where $E_{product}$ is the energy of formation of the TiO_2 with oxides, and $E_{reactant}$ is the summation of the crystal structure of TiO_2 plus the energy of formation of the oxides [12,18].

2.3. Electronic Structure methods in CP2K

Our primary interest in this study is to analyze the OVs of TiO_2 in adsorbing oxides. For this study, the total energies of each molecule are calculated using density functional theory (DFT) through the CP2K. DFT is commonly used in solid-state physics for its high reliability. An earlier study has shown that comparing DFT-calculated formation energy to those obtained experimentally has the same accuracy.

Table 2. Potentials used for each molecule.

Element	Code
Ti	GTH-BLYP-q12
O	GTH-BLYP-q6
N	GTH-BLYP-q5
C	GTH-BLYP-q4
H	GTH-BLYP-q1

Pseudopotentials are known for the reduction of basis set the size to speed up the calculation. The Goedecker, Teter, and Hutter (GTH) pseudopotentials were used to run the calculation under the GWP approach [13]. Notably, the potentials provided in Table 2 show the code for each molecule, which may be found online [13]. The value beside q represents the number of valence electrons used by each pseudo potential [13]. The pseudopotentials of GTH are particularly well suited for application with Quickstep [13]. Also, GTH was used because Quickstep only supports GTH pseudopotentials [13]. Furthermore, the Becke, Lee, Yang, and Parr (BLYP) were employed in our study among the several optimized parameter configurations for the exchange-correlation potentials [13]. **Table 3.** Basis Set used for each element in the system.

Element	Code
Ti	DZV-GTH-PADE
O	DZVP-ALL-PADE-NEW
N	DZVP-ALL-PADE
C	DZVP-ALL-PADE-NEW
H	DZVP-ALL-PADE

Basis sets, on the other hand, are noted for their systematic increase in reliability. It has been assumed that smaller basis yields higher error to accept accurate calculation [14]. Gaussian basis sets are offered in various forms, including DZVP, TZVP, TZV2P, QZV2P, and QZV3P; nonetheless, suitable basis sets must be optimized for use alongside GTH pseudo potentials [15]. An appropriate set of Gaussian functions must be defined as a basis set for each Quickstep calculation, and it has been found that the quality of the DZVP basis set is adequate for most applications as it yields more accurate results compared with other smaller basis sets [14]. It was also constructed for DFT calculation as its purpose, and it exhibits significantly smaller (BSSE) than other basis sets of similar size. [14]. With that said, the basis sets shown in Table 3 were the ones used in our study [13]. It can be noticed that the basis set used for Ti was DZV instead of DZVP because with CP2K, there was no basis set given for Ti under DZVP. However, the DZV basis set is also under DZVP basis set in CP2K.

The potentials and basis sets considered in our analysis were both common types prior to earlier studies. Also, considering the elements in our simulation, particularly those with transition metals like Titanium, it is the simplest we could use. Other basis sets and pseudopotentials are more complex and may take longer to generate a result.

3. Results and Discussion

3.1. Influence of Oxygen Vacancies

The surface defects have been considered as charge carrier traps and adsorption sites where the charge transfer to adsorbed species can prevent the e–h recombination. Here, bulk defects only act as charge carrier traps where e–h recombine [16]. When the crystal structure of TiO_2 was modified to simulate the oxygen vacancies with added molecules in the supercell, the results in Figure 4 were observed.

The TiO_2 anatase crystal structure has obtained an -1439.5044 kJ/mol formation energy. Adding oxygen vacancies to the structure increased the formation energy of the TiO_2 crystal structure from 1407.0384 kJ/mol for two OVs in the structure and, -1391.1436 kJ/mol for three OVs in the structure. These values would serve as the basis for the calculation of the energy absorption using the equation given.

Table 4. Energy Adsorption for Different Systems with and without OVs

System	ΔE_{ads} w/o OVs (kJ/mol)	ΔE_{ads} w/ OVs (kJ/mol)
TiO_2-NO	0.6412	0.0318
TiO_2-NO_2	-3.5272	-4.1310
TiO_2-CO	1.8493	1.2576
TiO_2-CO_2	-3.3601	-4.0100
TiO_2-H_2O	0.7164	0.2834

From the molecules added, it can be noted that the TiO_2-NO_2 compound has the lowest energy formation difference of -4.1310 kJ/mol, shown in Table 4. NO_2 is strongly bound by oxygen vacancies on a TiO_2 surface that is defective. Compared to the same system with any defects which resulted in 3.5272 kJ/mol energy adsorption. The adsorption capabilities in some molecules are modified by defects associated with oxygen vacancies, favoring dissociative over molecular adsorption. The dissociative adsorption of the compound would be enhanced by oxygen vacancies on the TiO_2 surface [16]. One of the reasons NO_2 has the lowest E_{ads} is its higher reactivity with oxidation; with NO_2, both O and N, or both molecules, can bond with metals [17]. With the DeNOx operation, NO_2 could undergo partial ($NO_2 \rightarrow NO + O$) or complete dissociation ($NO_2 \rightarrow N + O_2$) [17].

Furthermore, the addition of WO_3 in the TiO_2 system with oxygen vacancy showed a similar trend in data. This is because WO_3 is a semiconductor that has a lower band gap of 2.6 kJ/mol than TiO_2 with 3.20 kJ/mol, which can adsorb more light energy which translated to positive results [18]. Kato et al found that oxygen vacancies in WO_3 primarily serve as recombination sites. Similarly, with Mendoza, J. A. (2017), TiO_2 bounded with WO_3 without defects or oxygen vacancy resulted in energy adsorption of 3.18 kJ/mol, compared to the same system with one OV in the TiO_2 structure results in an energy adsorption of -7.45 kJ/mol [19].

This contrasts with other photocatalysts such as TiO_2 and $SrTiO_3$, where the carrier lifetime increases with increasing oxygen vacancy concentration. These differences can be attributed to the difference in the distance between oxygen vacancies [20]. In relation to the study of Wang et al, they found that replacing W in WO_3 with another metal of the same valency, lowers the band gap by pushing the conduction band edge to the lower side. However, the addition of lower valency metal such as Ti, causes a rise in the band gap owing to the conduction band edge moving higher [21].

The results were consistent across all oxide molecules used in the study, wherein adding oxygen vacancies in TiO_2, resulted in decreased energy adsorption for the TiO_2. Table 4 also shows the results gathered for NO, 0.6412 kJ/mol for a system without defects and 0.0318 kJ/mol for a system with oxygen vacancies. It may be noticed that the NO system does not have a negative result compared to NO_2, which is mainly due to the reactivity of the molecules in which NO_2 has higher electron affinity [17]. However, it is ideal to have a negative result for energy adsorption, but having a positive result is also acceptable as the values decreased when having oxygen vacancies in the structure. It should be noted, however, that dissociation adsorption of the adsorbates is frequently followed by re-oxidation of the TiO_2 surface. As a result, before we can analyze the dissociation adsorption characteristics of reduced TiO_2, we must first stabilize the oxygen vacancies [16].

Furthermore, CO, CO_2, and H_2O systems follow the trend observed by the researchers for this study, in which having defects in the surface of TiO_2 decrease the system's energy adsorption while increasing TiO_2 adsorption affinity. Similarly, with NOx, COx molecules have a significant difference in their energy adsorption, in which the CO_2 system is more negative -4.0100 kJ/mol, compared to 1.2576 kJ/mol for CO energy adsorption. The researchers concluded that having more oxygen atoms in the free molecules being adsorbed affects the adsorption of oxides in TiO_2, as the results show that NO_2 and CO_2 are both on the negative side of the number scale, whereas NO, CO, and H_2O, which each have one oxygen atom, are on the positive side. Nevertheless, our result is consistent with the study of Jiang et al, wherein it has been found that subsurface OVs further enhance the binding of CO_2 molecules to the surface. The OV-enhanced CO_2 adsorption not only accelerates the photocatalytic reaction but also allows the reaction to be carried out at low CO2 concentrations [22]. Furthermore, Pan et al stated that the presence of OVs promotes CO_2 adsorption and activation on ß-Ga_2O_3 (100) whereas OV provided active sites for CO_2 adsorption and activation [23].

The researchers were able to synthesize the structure of Anatase TiO_2 with oxygen vacancy in binding oxides such as NOx, COx, and H_2O favorably, which makes TiO_2 a good adsorbent.

4. Conclusion

As industrialization advances, pollutants are becoming an urgent environmental concern. This study reviews recent developments in the applications of TiO_2 containing oxygen vacancies for the binding of oxides, particularly NOx. Even though understanding of the role of oxygen vacancies in TiO_2 is still in its early stage, it has been discussed that defective TiO_2 is advantageous for a wide range of applications. The study found that having some oxygen vacancies in the anatase TiO_2 structure improves energy adsorption, as well as the adsorption affinity. Also, TiO_2 is most effective in adsorbing NO_2 molecules, followed by an efficient adsorption of CO_2, NO, H_2O, and lastly, CO, which could signify that these compounds can be adsorbed by TiO_2 and lead to a more efficient dissociation reaction.

References

[1] A.A. Esmailpour Valmazouyi, UNDERSTANDING THE ROLE OF OXYGEN VACANCIES IN CATALYTIC OZONATION PROCESS TO DEGRADE ORGANIC POLLUTANTS, (2021) https://doi.org/10.26190/UNSWORKS/24484

[2] S. Zhao, Y. Yang, F. Bi, Y. Chen, M. Wu, X. Zhang, G. Wang, Oxygen vacancies in the catalyst: Efficient degradation of gaseous pollutants, Chem. Eng. J. 454 (2023) 140376 https://doi.org/10.1016/J.CEJ.2022.140376

[3] Z. Tian, S. Du, X. Cheng, J. Zhang, F. Li, Z. Chen, Y. Lv, Y. Zhu, G. Liu, The Role of Oxygen Vacancy in Anatase to Rutile Transformation of TiO2, Cryst. Growth Des. 22 (2022) 6852–6856 https://doi.org/10.1021/ACS.CGD.2C01062/ASSET/IMAGES/MEDIUM/CG2C01062_0005.GIF

[4] A. Trapalis, N. Todorova, T. Giannakopoulou, N. Boukos, T. Speliotis, D. Dimotikali, J. Yu, TiO2/graphene composite photocatalysts for NOx removal: A comparison of surfactant-stabilized graphene and reduced graphene oxide, Appl. Catal. B Environ. 180 (2016) 637–647 https://doi.org/10.1016/j.apcatb.2015.07.009

[5] S. Selim, E. Pastor, M. García-Tecedor, M.R. Morris, L. Francàs, M. Sachs, B. Moss, S. Corby, C.A. Mesa, S. Gimenez, A. Kafizas, A.A. Bakulin, J.R. Durrant, Impact of Oxygen Vacancy Occupancy on Charge Carrier Dynamics in BiVO4 Photoanodes, J. Am. Chem. Soc. 141 (2019) 18791–18798. https://doi.org/10.1021/jacs.9b09056

[6] P. Gupta, V. Rathore, Materials Today : Proceedings Study of TiO 2 material : A photocatalyst for contrary pollutants, Mater. Today Proc. (2021) https://doi.org/10.1016/j.matpr.2020.12.1198

[7] Z. Liu, Y. Huang, Y. Cai, X. Wang, Y. Zhang, Y. Guo, J. Ding, W. Cheng, Oxygen Vacancy Enhanced Two-Dimensional Lithium Titanate for Ultrafast and Long-Life Bifunctional Lithium Storage, ACS Appl. Mater. Interfaces. 13 (2021) 18876–18886. https://doi.org/10.1021/ACSAMI.1C02962/SUPPL_FILE/AM1C02962_SI_001.PDF

[8] Y. Hinuma, T. Toyao, T. Kamachi, Z. Maeno, S. Takakusagi, S. Furukawa, I. Takigawa, K.I. Shimizu, Density Functional Theory Calculations of Oxygen Vacancy Formation and Subsequent Molecular Adsorption on Oxide Surfaces, J. Phys. Chem. C. 122 (2018) 29435–29444 https://doi.org/10.1021/acs.jpcc.8b11279.

[9] Q. Zhang, F.W. Tang, Z. Zhao, Z.R. Nie, X.Y. Song, Surface modification of tungsten oxide by oxygen vacancies for hydrogen adsorption, J. Mater. Sci. Technol. 117 (2022) 23–35 https://doi.org/10.1016/J.JMST.2021.10.048

[10] M. Elahifard, M.R. Sadrian, A. Mirzanejad, R. Behjatmanesh-Ardakani, S. Ahmadvand, Dispersion of defects in TiO2 semiconductor: Oxygen vacancies in the bulk and surface of rutile and anatase, Catalysts. 10 (2020) https://doi.org/10.3390/catal10040397

[11] about [CP2K Open Source Molecular Dynamics], (n.d.). https://www.cp2k.org/

[12] X. Wang, X. Yang, General hydrogen gas sensing mechanism on SnO2 (110) surface based on DFT simulations, Surf. Sci. 721 (2022) https://doi.org/10.1016/j.susc.2022.122064

[13] T.D. Kühne, M. Iannuzzi, M. Del Ben, V. V. Rybkin, P. Seewald, F. Stein, T. Laino, R.Z. Khaliullin, O. Schütt, F. Schiffmann, et al. CP2K: An electronic structure and molecular dynamics software package - Quickstep: Efficient and accurate electronic structure calculations, J. Chem. Phys. 152 (2020) https://doi.org/10.1063/5.0007045

[14] J. Hostaš, J. Řezáč, Accurate DFT-D3 Calculations in a Small Basis Set, J. Chem. Theory Comput. 13 (2017) 3575–3585 https://doi.org/10.1021/ACS.JCTC.7B00365/ASSET/IMAGES/LARGE/CT-2017003653_0003.JPEG

[15] J. Vandevondele, M. Krack, F. Mohamed, M. Parrinello, T. Chassaing, J. Hutter, Quickstep: Fast and accurate density functional calculations using a mixed Gaussian and plane waves approach, Comput. Phys. Commun. 167 (2005) 103–128 https://doi.org/10.1016/J.CPC.2004.12.014

[16] X. Pan, M. Yang, N. Zhang, Defective TiO2 with oxygen vacancies: Synthesis, properties and photocatalytic applications, (2019) https://doi.org/10.1039/c3nr00476g

[17] A. Rodriguez, T. Jirsak, G. Liu, J. Hrbek, J. Dvorak, N. York, S. Road, S. Diego, R. V May, Chemistry of NO 2 on Oxide Surfaces : Formation of NO 3 on TiO 2 (110) and NO 2 T O Vacancy Interactions, 2 (2001) 9597–9605

[18] J. Li, C. Guo, L. Li, Y. Gu, K. BoK-Hee, J. Huang, Experimental and DFT studies of WO3–CuO p-n heterojunctions for enhanced photoelectrochemical performance, J. Phys. Chem. Solids. 168 (2022) 110801 https://doi.org/10.1016/J.JPCS.2022.110801

[19] J.A. Mendoza, J.A. Mendoza, 지도교수 강 주 현 동국대학교 대학원 건설환경공학과, (2017).

[20] K. Kato, Y. Uemura, K. Asakura, A. Yamakata, Role of Oxygen Vacancy in the Photocarrier Dynamics of WO3Photocatalysts: The Case of Recombination Centers, J. Phys. Chem. C. 126 (2022) 9257–9263 https://doi.org/10.1021/ACS.JPCC.2C01662/SUPPL_FILE/JP2C01662_SI_001.PDF

[21] R. Ameta, M.S. Solanki, S. Benjamin, S.C. Ameta, Photocatalysis, Adv. Oxid. Process. Wastewater Treat. Emerg. Green Chem. Technol. (2018) 135–175 https://doi.org/10.1016/B978-0-12-810499-6.00006-1

[22] W. Jiang, H. Loh, B.Q.L. Low, H. Zhu, J. Low, J.Z.X. Heng, K.Y. Tang, Z. Li, X.J. Loh, E. Ye, Y. Xiong, Role of oxygen vacancy in metal oxides for photocatalytic CO2 reduction, Appl. Catal. B Environ. 321 (2023) 122079 https://doi.org/10.1016/J.APCATB.2022.122079

[23] Y.X. Pan, C.J. Liu, D. Mei, Q. Ge, Effects of hydration and oxygen vacancy on CO2 Adsorption and activation on β-Ga2O3(100), Langmuir. 26 (2010) 5551–5558 https://doi.org/10.1021/LA903836V/SUPPL_FILE/LA903836V_SI_003.MPG

Acknowledgements

First and foremost, we would like to honor and thank God, the Almighty, for providing us with countless blessings, knowledge, and opportunities, allowing us to finally complete the research.

This endeavour would not have been possible without the unwavering support of our parents and siblings, who have been there for us throughout this journey and are always willing to help.

We would also like to express our deepest appreciation to our thesis adviser, Dr. Joseph Albert M. Mendoza, who guided and supported us throughout this research. Your insights and guidance helped us refine our study and provided quality work.

Apart from our efforts, the success of this research depends largely on the encouragement and guidelines of many others. We would like to use this opportunity to extend our heartfelt appreciation to everyone who contributed to the successful completion of this research.

IOP Conf. Series: Materials Science and Engineering **1318** (2024) 012037 doi:10.1088/1757-899X/1318/1/012037

Production of few-layer graphene and boron nitride nano onions via lamp ablation

J. He[1,2,+], H. Zhang[1,2,+], W. Zhang[3,4], J. M. Gordon[5], H. T. Chua[1,*]

[1] Department of Chemical Engineering, The University of Western Australia, Perth, WA 6009

[2] Ablano Pty. Ltd., 5 Lindsay Street, Perth, WA 6000, Australia

[3] School of Materials Science and Engineering, Taiyuan University of Technology, Taiyuan 030024, China

[4] Energy Research Institute at NTU (ERI@N), Nanyang Technological University, 1 Clean Tech Loop #06-04, Singapore, 637141

[5] Department of Solar Energy and Environmental Physics, Blaustein Institutes for Desert Research, Ben-Gurion University of the Negev, Sede Boqer Campus 8499000, Israel

[+] Equal first author

[*] Corresponding author: huitong.chua@uwa.edu.au

Abstract.
Few-layer graphene, and separately, boron nitride nano onions have been synthesised by a novel lamp ablation method which is a one-step, high-yield process that is catalyst-free, devoid of toxic reagents, with short reaction times, and potentially scalable. The precursor for the few-layer graphene is graphite flake, while that of boron nitride nano onions is hexagonal boron nitride powder. Product nanostructures are characterised by high-resolution transmission electron microscopy (HRTEM), energy-dispersive X-ray spectroscopy (EDS) and electron energy loss spectroscopy (EELS). We also report on our commercialisation trajectory to translate promising research findings in the laboratory to the market. For the few-layer graphene product, it is presently on a pre-commercial pilot plant phase.

Keywords
few-layer graphene, boron nitride, nano onions, lamp ablation.

1.Introduction

Hexagonal boron nitride (h-BN) has attracted a great deal of attention due to its outstanding physicochemical properties, such as high thermal stability, impressive oxidation and corrosion resistance, large specific surface area, low dielectric constant, and a wide bandgap of 5–6 eV [1-5]. Furthermore, the biocompatibility and non-toxicity of h-BN materials, ensuing from their structural stability and chemical inertness, present vast possibilities for bioapplications [6]. Similarly, graphene presents its own set of remarkable properties, such as exceptional mechanical strength, high thermal and electrical conductivity, and unique optical characteristics, making it a revolutionary material in various fields[7-11]. Few-layer graphene (FLG), consisting of only a handful of layers, is of particular interest due to its promising optoelectronic properties and compatibility with various substrates.

However, the synthesis of these fascinating materials, particularly h-BN nano-onions (BNNOs) and FLG, presents significant challenges. While various methods such as pulsed laser ablation [12-14], chemical vapour deposition [15, 16], electron beam irradiation [17], and pyrolysis [18] have been employed, they often necessitate complex procedures, catalysts, templates, or the use of toxic reagents. Even so, the synthesis of h-BN nano-onions has remained particularly elusive, with reported yields and product purity often being sub-optimal [19].

In this paper, we present a breakthrough in the scalable and repeatable synthesis of BNNOs and FLG using the innovative method of lamp ablation (please refer to our previous works [20, 21] for details of the apparatus and operations). This strategy uses readily available h-BN powder and graphite as the only precursors, making the method non-toxic, chemically stable, simple, and rapid. Additionally, the use of lamp ablation has the potential to not only increase the quality and yield of the products but also reduce the complexity and safety risks associated with their synthesis. This opens up new horizons in the large-scale production of these promising nanomaterials and their incorporation into next-generation applications.

2. Results

2.1. Material characterisation and properties of BNNOs

Figure 1 presents characteristic results derived from the product synthesised over a 2-hour ablation. The high-yield production of BNNOs is highlighted in Figs. 1(A) and 1(B), while Fig. 1(C) visually demonstrates the hollow structure of a ~90 nm BNNO composed of approximately 30 layers. The composition of the product is unveiled through elemental mapping (Figs. 1(B), 1(C), and 1(D)) and energy-dispersive X-ray spectroscopy (EDS) (Fig. 1(D)). The hollow BNNOs solely contain boron and nitrogen, with the presence of elemental carbon and copper originating from the carbon film supporting the material and the TEM copper grid, as further reinforced by Fig. 1(D).

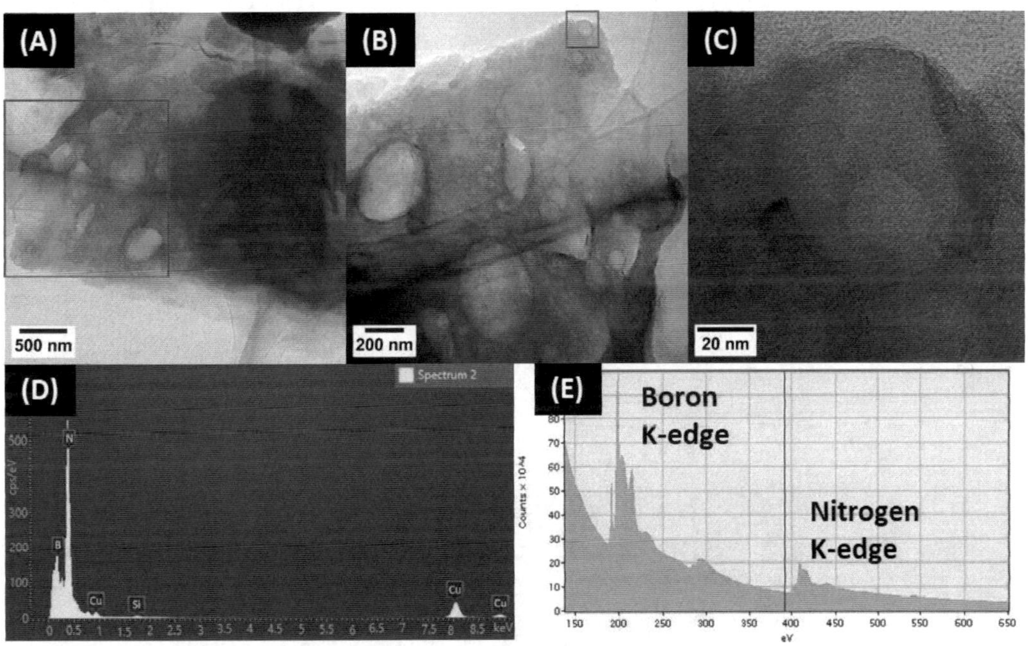

Figure 1. TEM analysis of the ablated product. (A) A representative cluster of BNNOs following a 2-hour ablation. (B) Progressive high-resolution depictions of the area defined by the large red square in (A). (C) HRTEM image of the BNNO encased by the small red square in (B). (D) EDS spectrum characterising the BNNOs. (E) EELS spectrum obtained from a collection of BNNOs.

To evaluate the B/N ratio in BNNOs, Electron Energy Loss Spectroscopy (EELS) was employed. EELS provides a more precise determination of the B/N ratio than EDS, particularly when lightweight elements like boron are implicated. This is because EDS calculations are more prone to interference from background signals and sample thickness [12]. The EELS also confirmed the bandgap of the synthesised BNNOs. As demonstrated in Fig. 1(E), the B and N K edges are discernible at 188 and 401 eV, respectively [22]. The distinct π^* and σ^* peaks at the B and N K edges signify the typical sp^2 bonding configuration, further verifying the h-BN structure within the BNNOs[23]. The observed C edge, located at 284 eV, can be ascribed to the carbon film on the TEM copper grid. A separate examination of the EELS spectrum revealed that the atomic ratio of B/N was 1:1 [24].

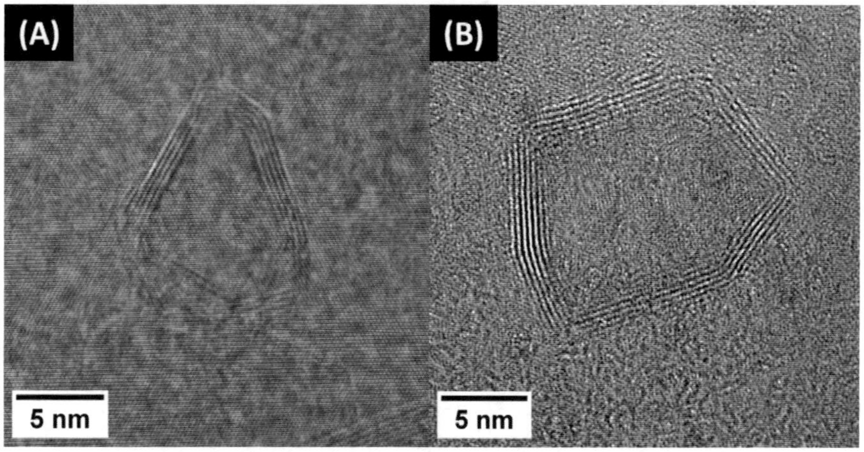

Figure 2. HRTEM images for representative small BNNOs (with polyhedral cross-sections).

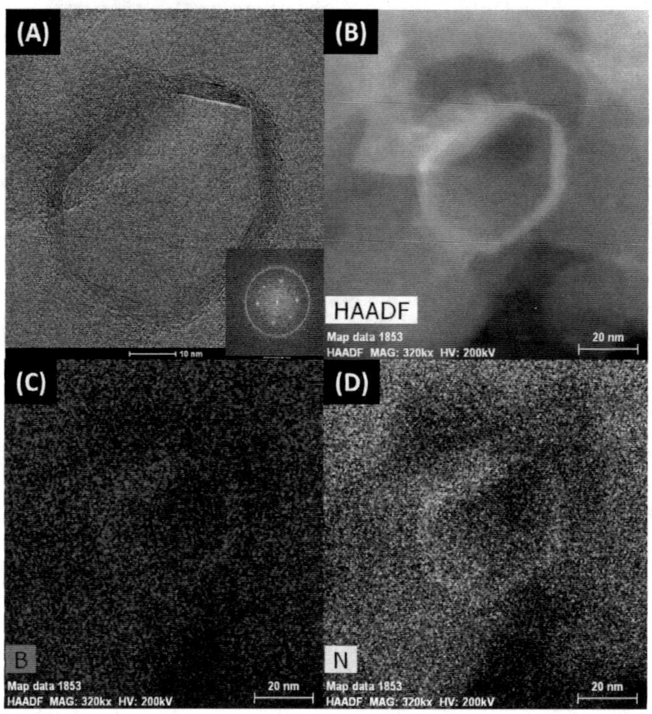

Figure 3. (A) A representative HRTEM image of a BNNO, with an inset showing the FFT. (B) HAADF image. (C) and (D) Elemental mapping derived from the HAADF image, showcasing the distribution of boron and nitrogen, respectively, as labelled.

The HRTEM images displayed in Figs. 2(A) and 2(B) feature the high crystallinity of the hollow BNNOs. These smaller BNNOs typically consist of around 4-7 crystalline layers and display a diameter of ~10 nm. The BNNOs depicted by TEM (Fig. 1) reveal a wide range of diameters, ranging from 6–

600 nm. However, it should be mentioned that the use of the term "nano" may be disputed for particles reaching as large as 600 nm in size.

The HRTEM image presenting a single BNNO along with the attendant fast Fourier-transform (FFT) is illustrated in the high-resolution TEM image of Fig. 3(A), with the HAADF image depicted in Fig. 3(B). The relative thickness can be inferred from the intensity of the image. The nano-onion shown in the HAADF image is made up exclusively of boron and nitrogen atoms, a conclusion corroborated by the elemental maps depicted in magenta and yellow, respectively, in Figs. 3(C) and 3(D). The FFT analysis was used to measure an interlayer spacing of 0.335 nm (± 2%), a figure that aligns with the typical (002) d-spacing of h-BN when accounting for the margin of error, with c/2 = 0.3336 nm for h-BN as determined from XRD (COD ID: 1010602).

Moving forward, the synthesis methods for BNNOs will be refined and improved in order to increase the yield and achieve industrial-scale production. This advancement is essential for realising the full potential of BNNOs in various applications, and will involve the optimisation of the manufacturing processes. By enhancing the yield and scalability, it is possible to make the production of BNNOs more economically viable and thus pave the way for their broader adoption in industries.

2.2. Characterisation and properties of few-layer graphene

The SEM images displayed in Figs. 4(A) suggest a multitude of identifiable few-layer graphene formations, in the presence of silica nano-fibres, while 4(B) depicts a pristine grade of partial FLG product (with the unconverted material being graphite precursor) obtained through an improved synthesis method. The generation of such SiO_2 nanostructures is not surprising, on account of the softening of the quartz reactor in contact with the black precursor up to 2700K [25-28]. With an upgrade to the previous lamp ablation method, the synthesised FLG displays an unmatched level of purity, with effectively no impurities present.

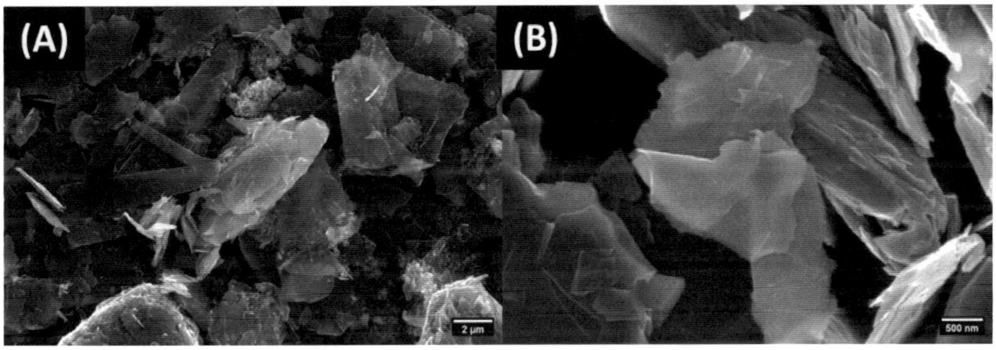

Figure 4. SEM images. (A) FLG after lamp ablation tainted with silica nano-fibres, (B) FLG obtained through an improved procedure, devoid of SiO_2 contaminants.

Figure 5 presents TEM images highlighting the few-layer graphene. The SAED pattern inset in Fig. 5(A) allows the calculation of two distinct distances, i.e. 2.15 Å and 3.48 Å, which align with the (110) and (002) graphite planes, respectively [29]. The crisp spots underscore the high crystallinity of the few-layer graphene. Fig. 5(B) presents a HRTEM image of the few-layer graphene, indicating that it is 7-8 layer thick.

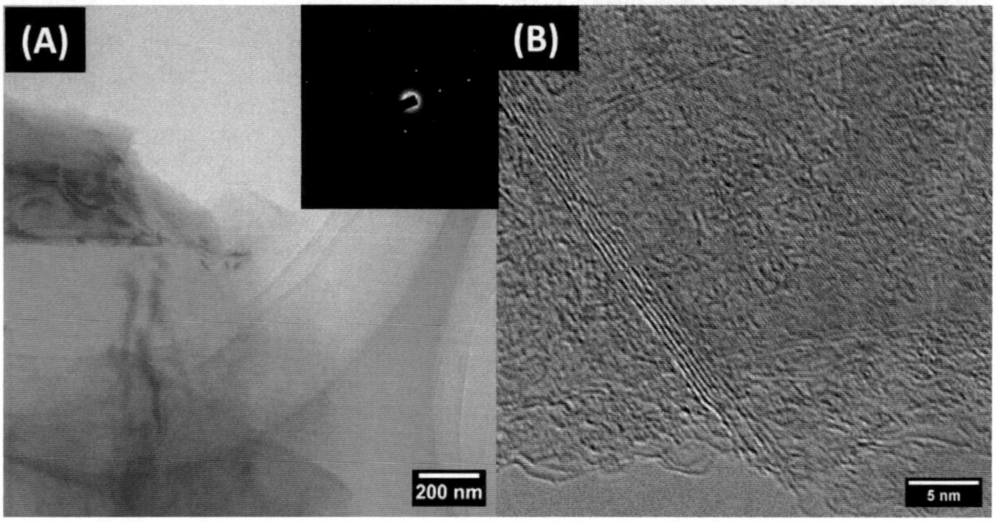

Figure 5. (A) TEM image displaying FLG with an inset SAED pattern (B) HRTEM image of FLG.

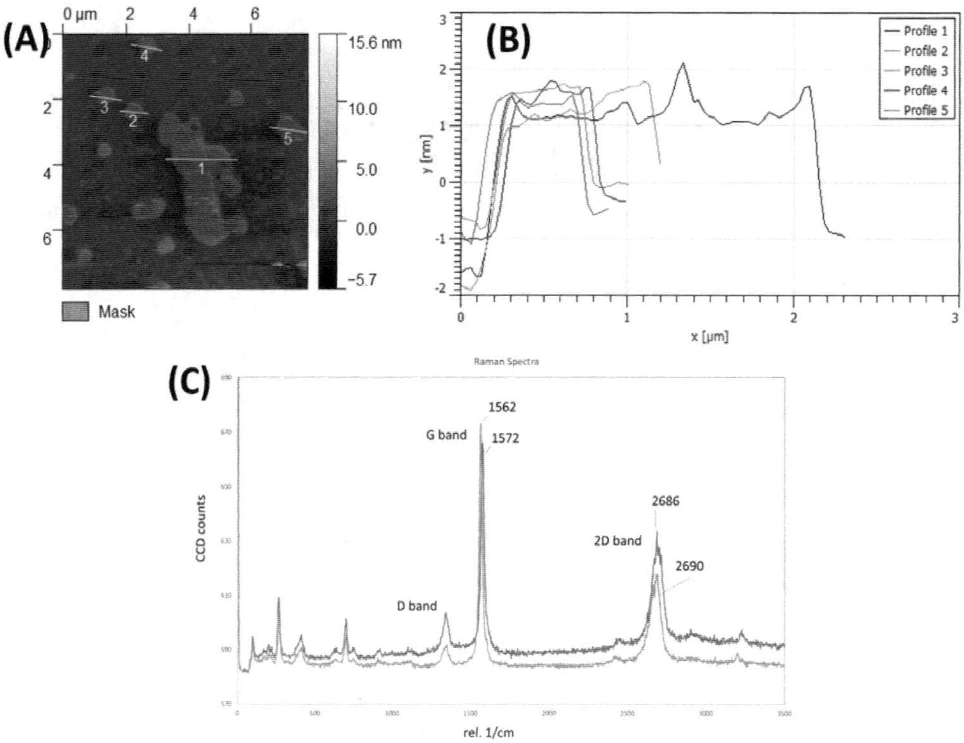

Figure 6. AFM images and corresponding line profiles of typical FLG flakes. (A) Photograph from a typical AFM scan. (B) Profiles along the lines overlaid in (A), and (C) Raman spectra of the FLG.

The tapping mode AFM images, as displayed in Fig. 6(A), along with the line profile in Fig. 6(B), demonstrate that the synthesised few-layer graphene products are, on average, ~2.0–3.0 nm thick, consistent with 7–8 layer few-layer graphene [30]. With reference to the presented Raman spectra in Fig. 6(C), the ratios of I2D/IG are recorded as 0.56 and 0.43, respectively, which closely approximate 0.5, again suggesting a layer number of 7 [31]. The number of graphene layers can also be ascertained from the positioning and contour of the 2D band. The positions of the D and 2D bands are dependent on excitation energy and typically occur at around 1350 cm^{-1} and 2700 cm^{-1}, respectively [32]. The shape of the 2D mode in the Raman spectrum serves as a primary tool for determining the number of layers in graphene samples [33-36]; the contour of the 2D peak undergoes noticeable alterations corresponding to the variations in the number of graphene layers [34-36]. The Raman spectra appear to suggest the existence of FLG consisting of less than 10 layers, as deduced from the ratio of the intensities of the 2D and G bands, as well as the frequency at which the 2D band peaks. These observations are consistent with the TEM analyses.

We acknowledge that there are companies which have commercialised solution based FLG [37], but this is the first time, to our knowledge, that one could access FLG in its dry and pristine form.

3. Conclusions

We have successfully showcased a refined synthesis method for BNNOs and FLG. The one-pot procedure is also (1) devoid of any toxic or hazardous chemicals, (2) uncomplicated, (3) holds potential for scaling, and (4) is time-efficient. The EELS and EDS confirm that the nano-onions, visible in SEM and TEM images, are composed of pure, uncontaminated BN in a 1:1 ratio. Separately, the TEM analyses, AFM and Raman spectra seem to indicate the presence of FLG with fewer than 10 layers.

Although BNNOs and dry FLG have been previously synthesised, none of the methods employed have proven capable of achieving scalability and effectively addressing the practical constraints mentioned earlier. This paper introduces a novel approach that overcomes these limitations and offers a promising solution.

The refinement in the synthesis methods for BNNOs and FLG is instrumental in increasing the yield and purity of the products. These advancements are indispensable for translating these materials into industrial-scale production, consequently creating avenues for practical applications and technological breakthroughs. Such progress is paramount in unleashing the full potential of BNNO and FLG across various domains, and in facilitating their seamless integration into contemporary technologies.

Acknowledgements

The authors gratefully acknowledge funding from Ablano Pty. Ltd. The authors also acknowledge the facilities, and the scientific and technical assistance of Microscopy Australia at the Centre for Microscopy, Characterisation & Analysis, The University of Western Australia, a facility funded by the University, State and Commonwealth Governments.

Disclosure statement

Hongyu Zhang and Jijiang He are full-time employees of, while Hui Tong Chua is a consultant to Ablano Pty. Ltd. Hongyu Zhang, Jijiang He, Weike Zhang, Jeffrey Gordon and Hui Tong Chua each holds financial interests in Ablano Pty. Ltd. Ablano Pty. Ltd. is a spin-off company of The University of Western Australia focussing on the commercial production of boron nitride nano onions and few-layer graphene.

Reference

[1] A. Zunger, A. Katzir, A. Halperin, Optical properties of hexagonal boron nitride, Physical Review B 13(12) (1976) 5560-5573.

[2] J.-G. Kho, K.-T. Moon, J.-H. Kim, D.-P. Kim, Properties of Boron Nitride (BxNy) Films Produced by the Spin-Coating Process of Polyborazine, Journal of the American Ceramic Society 83(11) (2000) 2681-2683.
[3] R.T. Paine, C.K. Narula, Synthetic routes to boron nitride, Chemical Reviews 90(1) (1990) 73-91.
[4] T.E. O'Connor, Synthesis of Boron Nitride, Journal of the American Chemical Society 84(9) (1962) 1753-1754.
[5] M. Corso, W. Auwärter, M. Muntwiler, A. Tamai, T. Greber, J. Osterwalder, Boron Nitride Nanomesh, Science 303(5655) (2004) 217-220.
[6] X. Chen, P. Wu, M. Rousseas, D. Okawa, Z. Gartner, A. Zettl, C.R. Bertozzi, Boron Nitride Nanotubes Are Noncytotoxic and Can Be Functionalized for Interaction with Proteins and Cells, Journal of the American Chemical Society 131(3) (2009) 890-891.
[7] Z. Liu, L.H.G. Tizei, Y. Sato, Y.-C. Lin, C.-H. Yeh, P.-W. Chiu, M. Terauchi, S. Iijima, K. Suenaga, Postsynthesis of h-BN/Graphene Heterostructures Inside a STEM, Small 12(2) (2016) 252-259.
[8] S.K. Tiwari, V. Kumar, A. Huczko, R. Oraon, A.D. Adhikari, G.C. Nayak, Magical Allotropes of Carbon: Prospects and Applications, Critical Reviews in Solid State and Materials Sciences 41(4) (2016) 257-317.
[9] S.K. Tiwari, R.K. Mishra, S.K. Ha, A. Huczko, Evolution of Graphene Oxide and Graphene: From Imagination to Industrialization, ChemNanoMat 4(7) (2018) 598-620.
[10] L. Liao, H. Peng, Z. Liu, Chemistry Makes Graphene beyond Graphene, Journal of the American Chemical Society 136(35) (2014) 12194-12200.
[11] Y. Yang, R. Liu, J. Wu, X. Jiang, P. Cao, X. Hu, T. Pan, C. Qiu, J. Yang, Y. Song, D. Wu, Y. Su, Bottom-up Fabrication of Graphene on Silicon/Silica Substrate via a Facile Soft-hard Template Approach, Scientific Reports 5(1) (2015) 13480.
[12] D. Golberg, A. Rode, Y. Bando, M. Mitome, E. Gamaly, B. Luther-Davies, Boron nitride nanostructures formed by ultra-high-repetition rate laser ablation, Diamond and related materials 12(8) (2003) 1269-1274.
[13] R.S. Lee, J. Gavillet, M.L. de la Chapelle, A. Loiseau, J.L. Cochon, D. Pigache, J. Thibault, F. Willaime, Catalyst-free synthesis of boron nitride single-wall nanotubes with a preferred zig-zag configuration, Physical review. B, Condensed matter and materials physics 64(12) (2001).
[14] D. Golberg, Y. Bando, M. Eremets, K. Takemura, K. Kurashima, H. Yusa, Nanotubes in boron nitride laser heated at high pressure, Applied Physics Letters 69(14) (1996) 2045-2047.
[15] C. Tang, Y. Bando, D. Golberg, Large-scale synthesis and structure of boron nitride sub-micron spherical particles, Chemical communications (Cambridge, England) 2(23) (2002) 2826-2827.
[16] C. Tang, Y. Bando, Y. Huang, C. Zhi, D. Golberg, Synthetic Routes and Formation Mechanisms of Spherical Boron Nitride Nanoparticles, Advanced Functional Materials 18(22) (2008) 3653-3661.
[17] D. Golberg, Y. Bando, O. Stéphan, K. Kurashima, Octahedral boron nitride fullerenes formed by electron beam irradiation, Applied Physics Letters 73(17) (1998) 2441-2443.
[18] C. Sun, C. Guo, X. Ma, L. Xu, Y. Qian, A facile route to prepare boron nitride hollow particles at 450°C, Journal of Crystal Growth 311(14) (2009) 3682-3686.
[19] B. Zhong, X.H. Tang, X.X. Huang, L. Xia, X.D. Zhang, C.J. Wang, G.W. Wen, Boron nitride hollow nanospheres: Synthesis, formation mechanism and dielectric property, Materials Research Bulletin 64 (2015) 61-67.
[20] S. Eshon, W. Zhang, M. Saunders, Y. Zhang, H.T. Chua, J.M. Gordon, Panorama of boron nitride nanostructures via lamp ablation, Nano Research 12(3) (2019) 557-562.
[21] H. Zhang, J. He, W. Zhang, I.E. Castelli, M. Saunders, J.M. Gordon, H.T. Chua, Rapid, one-pot, non-toxic and scalable synthesis of boron nitride nano-onions via lamp ablation, Materials Today (2023).
[22] D. Golberg, Y. Bando, M. Mitome, K. Kurashima, T. Sato, N. Grobert, M. Reyes-Reyes, H. Terrones, M. Terrones, Preparation of aligned multi-walled BN and B/C/N nanotubular arrays and their characterization using HRTEM, EELS and energy-filtered TEM, Physica. B, Condensed matter 323(1) (2002) 60-66.

IOP Conf. Series: Materials Science and Engineering **1318** (2024) 012037 doi:10.1088/1757-899X/1318/1/012037

[23] B. Zhong, Y. Wu, X. Huang, G. Wen, H. Yu, T. Zhang, Hollow BN microspheres constructed by nanoplates: Synthesis, growth mechanism and cathodoluminescence property, CrystEngComm 13(3) (2011) 819-826.

[24] A. Rubio, J.L. Corkill, M.L. Cohen, Theory of graphitic boron nitride nanotubes, Physical review. B, Condensed matter 49(7) (1994) 5081-5084.

[25] M. Levy, A. Albu-Yaron, R. Tenne, D. Feuermann, E.A. Katz, D. Babai, J.M. Gordon, Synthesis of Inorganic Fullerene-like Nanostructures by Concentrated Solar and Artificial Light, Israel journal of chemistry 50(4) (2010) 417-425.

[26] J.M. Gordon, E.A. Katz, D. Feuermann, A. Albu-Yaron, M. Levy, R. Tenne, Singular MoS2, SiO2 and Si nanostructures—synthesis by solar ablation, Journal of Materials Chemistry 18(4) (2008) 458-462.

[27] H.-b. Lu, B.C.Y. Chan, X. Wang, H.T. Chua, C.L. Raston, A. Albu-Yaron, M. Levy, R. Popowitz-Biro, R. Tenne, D. Feuermann, J.M. Gordon, High-yield synthesis of silicon carbide nanowires by solar and lamp ablation, Nanotechnology 24(33) (2013) 335603-335603.

[28] H. Lu, W.S. Woi, X. Tan, C.T. Gibson, X. Chen, C.L. Raston, J.M. Gordon, H.T. Chua, Synthesis of few-layer graphene by lamp ablation, Carbon 94 (2015) 349-351.

[29] Y. Shen, A.C. Lua, A facile method for the large-scale continuous synthesis of graphene sheets using a novel catalyst, Scientific Reports 3(1) (2013) 3037.

[30] P. Nemes-Incze, Z. Osváth, K. Kamarás, L.P. Biró, Anomalies in thickness measurements of graphene and few layer graphite crystals by tapping mode atomic force microscopy, Carbon 46(11) (2008) 1435-1442.

[31] Y. Hwangbo, C.-K. Lee, A.E. Mag-Isa, J.-W. Jang, H.-J. Lee, S.-B. Lee, S.-S. Kim, J.-H. Kim, Interlayer non-coupled optical properties for determining the number of layers in arbitrarily stacked multilayer graphenes, Carbon 77 (2014) 454-461.

[32] Y. Hao, Y. Wang, L. Wang, Z. Ni, Z. Wang, R. Wang, C.K. Koo, Z. Shen, J.T.L. Thong, Probing Layer Number and Stacking Order of Few-Layer Graphene by Raman Spectroscopy, Small 6(2) (2010) 195-200.

[33] D. Kim, Y. Hwangbo, L. Zhu, A.E. Mag-Isa, K.-S. Kim, J.-H. Kim, Dimensional dependence of phonon transport in freestanding atomic layer systems, Nanoscale 5(23) (2013) 11870-11875.

[34] A.C. Ferrari, J.C. Meyer, V. Scardaci, C. Casiraghi, M. Lazzeri, F. Mauri, S. Piscanec, D. Jiang, K.S. Novoselov, S. Roth, A.K. Geim, Raman Spectrum of Graphene and Graphene Layers, Physical Review Letters 97(18) (2006) 187401.

[35] D. Graf, F. Molitor, K. Ensslin, C. Stampfer, A. Jungen, C. Hierold, L. Wirtz, Spatially Resolved Raman Spectroscopy of Single- and Few-Layer Graphene, Nano Letters 7(2) (2007) 238-242.

[36] L.M. Malard, M.A. Pimenta, G. Dresselhaus, M.S. Dresselhaus, Raman spectroscopy in graphene, Physics Reports 473(5) (2009) 51-87.

[37] GrapheneUP, GUP® Based Intermediary Products, 2022. https://www.grapheneup.com/gup-products/. (Accessed 11 May 2023.

Exfoliation of Ti_3C_2 MXene using $(NH_4)_2TiF_6$ Etchant and Simultaneous Synthesis of Ti_3C_2 MXene/NH_4TiOF_3 Mesocrystals Hybrids

T Wang[1], L Zhu[1], W Y Zhu[1] and H kanda*,[1]
[1] Department of Materials Process Engineering, Nagoya University, Nagoya 464-8603, Japan.

*e-mail: kanda.hideki@material.nagoya-u.ac.jp

Abstract. A new single-step method is suggested to fabricate block structures consisting of NH_4TiOF_3 wrapped around Ti_3C_2 MXene (Ti_3C_2) by etching Ti_3AlC_2 MAX phase with $(NH_4)_2TiF_6$. The optimal reaction conditions for Ti_3AlC_2 MAX phase and $(NH_4)_2TiF_6$ are systematically investigated. $(NH_4)_2TiF_6$ selectively etches the Al in the Ti_3AlC_2 MAX phase and promotes promoting the mild hydrolysis reaction of $(NH_4)_2TiF_6$ to form NH_4TiOF_3 crystals. The exterior shapes of Ti_3C_2/NH_4TiOF_3 hybrids can be tuned by adjusting the reagent concentration, reaction temperature. The phase composition, morphology and photophysical properties of the Ti_3C_2/NH_4TiOF_3 hybrids are investigated using various characterization techniques. Ti_3AlC_2 MAX phase reacts with $(NH_4)_2TiF_6$ at 60 °C for 24 h to form hybrids which are surrounded by NH_4TiOF_3 crystals simultaneously with the formation of Ti_3C_2 nanosheet. The resulting Ti_3C_2/NH_4TiOF_3 are porous and finely structured, showing potential for various applications in biotechnology, sensing, and environmental fields. This method provides important insights into future applications of Ti_3C_2/NH_4TiOF_3 hybrids.

1. Introduction

Within low-dimensional nanospaces, physicochemical properties change dramatically, including changes in the phase behavior of materials [1,2]. Since its discovery in 2004, graphene has rapidly advanced the study of two-dimensional solids due to its unique properties resulting from its low dimensionality [3]. In recent years, many researchers have synthesized MXene, a non-graphene two-dimensional material containing carbon and metals such as titanium, and have attempted to apply it to sensors, photonic devices, energy storage, and catalysts [4].

First identified in 2011 by a research team at Drexel University, the MXene family of two-dimensional transition metal carbides and nitrides has attracted considerable attention in various fields due to their outstanding properties. These properties include exceptional electrical and thermal conductivity, toughness, chemical resistance, and thermal stability [5]. Ti_3C_2 MXene (Ti_3C_2) has been the most studied due to its graphene-like properties, excellent electrical and thermal conductivity, heat resistance, and chemical resistance [6]. However, the conventional method of synthesizing MXene using HF acid etchant has limited applicability due to its high toxicity and generation of putrid effluent as expressed in Eq. (1) [7]. Therefore, alternative non-hazardous synthesis methods such as $NaBF_4$–HCl [8], LiF–HCl [9], NH_4HF_2 organic solvents [10], NH_4F–HCl [11], ZnCl [12], $CuCl_2$ [13], ultrasonic etching [14], and hydrothermal [15] have been investigated.

Liquid phase deposition (LPD) is a deposition method based on the hydrolytic equilibrium reaction of metal complex ions, which is promoted by the addition of fluoride ion scavengers such as boric acid, as shown in Eq. (2-4) [16]. In our previous work, we developed an alternative synthetic method to avoid the use of conventional methods involving toxic HF acids. Instead, we used the LPD technique, in which $(NH_4)_2TiF_6$ is deposited on the surface of the fiber optic sensor, to fabricate a composite consisting of Ti_3C_2 multilayers and TiO_2 [17,18].

In this study, the feasibility of liquid phase etching and exfoliation of Ti_3AlC_2 in the presence of $(NH_4)_2TiF_6$ was demonstrated. The results show that NH_4TiOF_3 precipitates during the etching process and forms volumetrically in the Ti_3C_2 as described by Eq. (1). This is the first study on the surface microstructure of Ti_3C_2 and NH_4TiOF_3 hybrids synthesized by the LPD method.

$$Ti_3AlC_2 + 3HF \leftrightarrow Ti_3C_2 + AlF_3 + (3/2)H_2. \quad (1)$$

$$MFx^{(x-2n)-} + nH_2O \leftrightarrow MO_n + xF^- + 2nH^+, \quad (2)$$

$$H_3BO_3 + 4H^+ + 4F^- \leftrightarrow HBF_4 + 3H_2O, \quad (3)$$

$$Al + 6H^+ + 6F^- \leftrightarrow H_3AlF_6 + (3/2)H_2. \quad (4)$$

$$(NH_4)_2TiF_6 + H_2O \leftrightarrow NH_4TiOF_3 + 2HF + NH_4F. \quad (5)$$

2. Materials and methods

2.1. Materials

The Ti_3AlC_2 MAX phase (titanium aluminum carbide 312, purity ≥ 90%) was supplied by Sigma–Aldrich, Japan. $(NH_4)_2TiF_6$ and ethanol were purchased from Fujifilm Wako Pure Chemical, Osaka, Japan.

2.2. Preparation Of the Ti_3C_2/NH_4TiOF_3 Hybrids

The process for synthesizing the Ti_3C_2/NH_4TiOF_3 hybrid derived from Ti_3AlC_2 and $(NH_4)_2TiF_6$ is shown in Figure 1. $(NH_4)_2TiF_6$ (80 mL) was dissolved in deionized water and then transferred to a hot stirrer. The solution was stirred for 10 min to obtain a $(NH_4)_2TiF_6$ solution (0.5 M). Ti_3AlC_2 powder was added to the $(NH_4)_2TiF_6$ solution to form a 10 mg mL^{-1} Ti_3AlC_2 suspension. The suspension was stirred at 500 rpm for 24 h at both 25 and 60 °C, abbreviated as Ti_3C_2/NH_4TiOF_3 (25) and Ti_3C_2/NH_4TiOF_3 (60), respectively. The mixed solution was centrifuged at 2,147 × g for 10 min. The liquid at the top was removed, and the resulting Ti_3C_2/NH_4TiOF_3 powder was collected. The collected Ti_3C_2/NH_4TiOF_3 powder was washed with ethanol and water, and then passed through a membrane filter.

2.3. Characterization of the Ti_3C_2/NH_4TiOF

An X-ray diffractometer (Rigaku SmartLab, Tokyo, Japan) was used in the angle range of 5–80° to analyse the phase components of the synthesized Ti_3C_2/NH_4TiOF_3. A high–resolution field emission scanning electron microscope (FE–SEM) (JSM–7500F, JEOL, Tokyo, Japan) was used to characterize the nanostructure of the Ti_3C_2/NH_4TiOF_3. A JEM 2100Plus field emission transmission electron microscope (TEM, JEOL, Tokyo, Japan) was used for analysis. Thermogravimetric (TG)/differential thermal analysis (DTA) measurements were performed under a nitrogen atmosphere using a RIGAKU TG 8120 812000 Thermo plus (Tokyo, Japan). Ti_3C_2/NH_4TiOF_3 were heated from 25 to 1000 °C at a rate of 20 °C min^{-1}. The gas flow rate was maintained at 100 mL min^{-1}. The chemical changes in the synthesized Ti_3C_2/NH_4TiOF_3 were investigated by Fourier transform infrared spectroscopy (FT–IR).

IOP Conf. Series: Materials Science and Engineering **1318** (2024) 012038 doi:10.1088/1757-899X/1318/1/012038

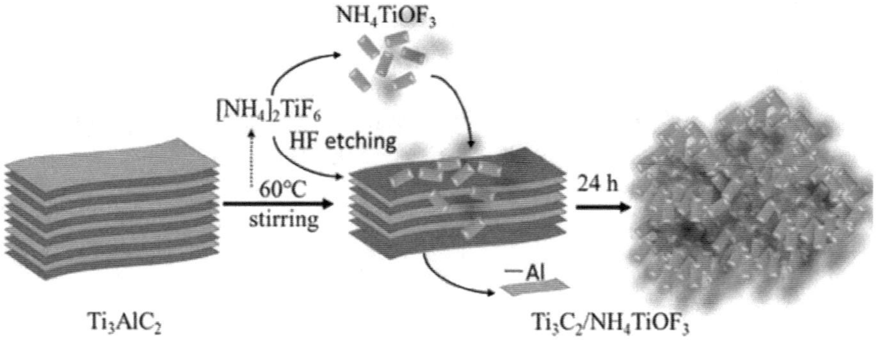

Figure 1. Schematic of the Ti_3AlC_2 exfoliation process and the subsequent formation of Ti_3C_2/NH_4TiOF.

3. Results and discussion

Figure 2(a) shows the layered structure of Ti_3AlC_2. As shown in Figure 2(b) and 2(c), a small amount of NH_4TiOF_3 crystallites and low coverage of Ti_3C_2 were observed after reaction with 0.5 M $(NH_4)_2TiF_6$ at 25 °C for 24 h. In Figure 2(d–f), the coverage of Ti_3C_2/NH_4TiOF_3 (60) was uniform when the etching temperature was increased to 60 °C. Figure 3 shows TEM images of Ti_3AlC_2 and Ti_3C_2/NH_4TiOF_3 (60). NH_4TiOF_3 crystal structures are observed on the Ti_3C_2 nanosheets.

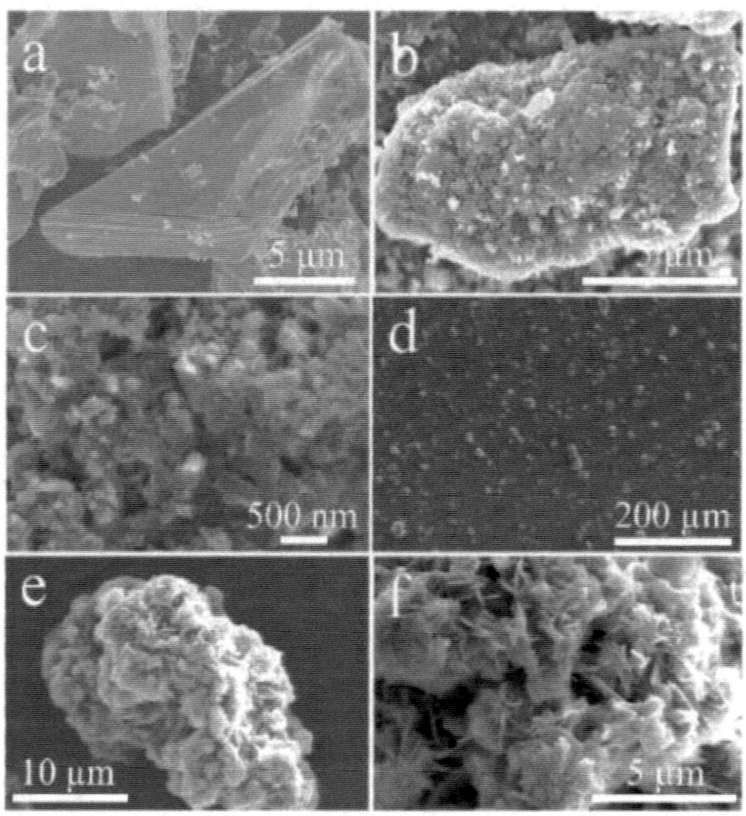

Figure 2. SEM images of (a) Ti_3AlC_2, (b, c) Ti_3C_2/NH_4TiOF_3 (25) and (d–f) Ti_3C_2/NH_4TiOF_3 (60).

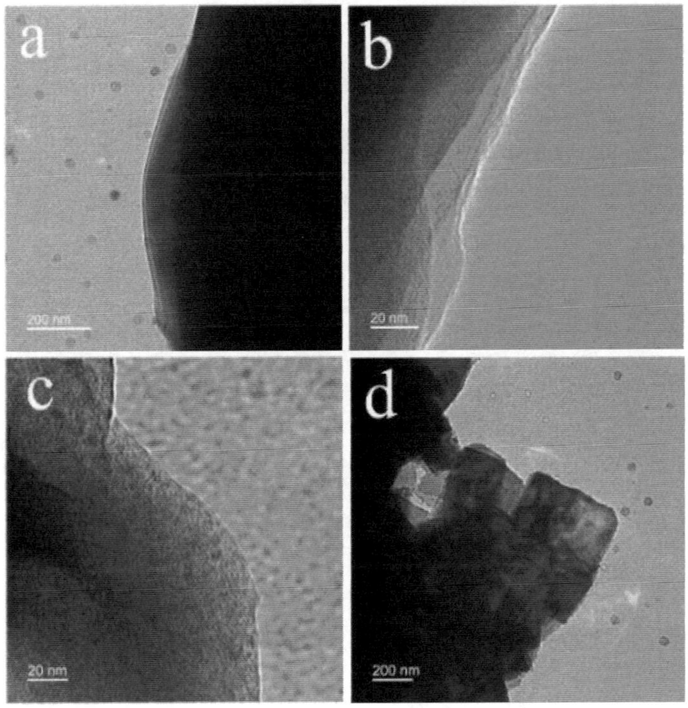

Figure 3. TEM images of (a, b) Ti_3AlC_2 and (c, d) Ti_3C_2/NH_4TiOF_3 (60).

In Figure 4(a), Ti_3AlC_2 shows the presence of its constituent elements Ti, C, O, and Al. Figure 4(b) shows uniform bonding between NH_4TiOF_3 crystals and Ti_3C_2. The formation of Ti_3C_2/NH_4TiOF_3 by exfoliation and delamination is evidenced by the presence of Ti, C and O, while Al is completely absent [19].

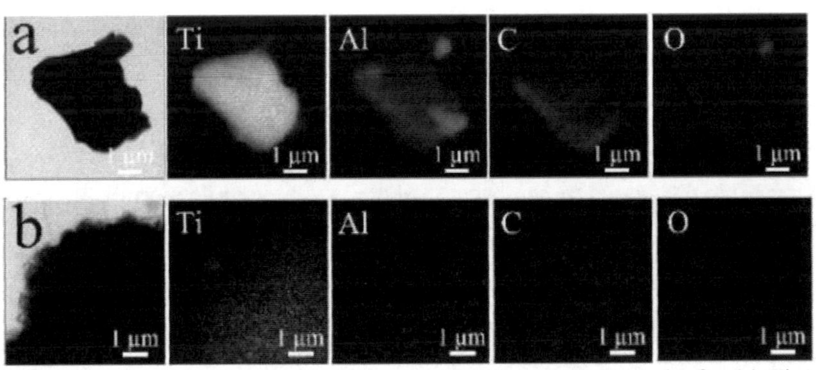

Figure 4. Energy dispersive X-ray spectroscopy (EDX) mapping results for (a) Ti_3AlC_2, and (b) Ti_3C_2 MXene/NH_4TiOF_3 (60).

The X-ray diffraction pattern (XRD) of the Ti_3C_2/NH_4TiOF_3 are shown in Figure 5(a). After etching Ti_3AlC_2 with $(NH_4)_2TiF_6$, the reflection peak of Ti_3AlC_2 at 9.6° is significantly attenuated and shifted to 9.3°. In addition, the Ti_3AlC_2 peaks at 34° and 39° are shifted to lower angles. These results confirm that the Ti_3C_2 MXene is transformed into a single layer during the etching process while maintaining the overlapping multilayer structure [20]. For Ti_3C_2/NH_4TiOF_3 (60), typical peaks at 23.4°, 33.8°, 48.1° and 56.3° were observed. These peaks corresponded to the (101) crystal plane of the synthesized NH_4TiOF_3 [21]. This indicates that the reaction of Ti_3C_2 with $(NH_4)_2TiF_6$ formed NH_4TiOF_3 crystals with specific crystal planes.

Figure 5(b) shows FT–IR spectra of Ti_3C_2 MXene/NH_4TiOF_3 and Ti_3AlC_2. In the Ti_3C_2/NH_4TiOF_3, peaks at 3189, 2901, 1700, 1411, 1150 and 442 cm^{-1} are observed, which are attributed to –OH, C–H, Ti–O, –OH, C–F and Ti–C, respectively [20]. The prominent peaks observed between 500 and 1000 cm^{-1} are attributed to Ti–O and Ti–O–Ti bonds, suggesting the incorporation of NH_4TiOF_3 crystals within the hybrid structure. The emergence of peaks associated with C–F, Ti–C, and Ti–O groups in FT-IR spectra of Ti_3C_2/NH_4TiOF_3 indicates the effective etching and exfoliation of the Ti_3AlC_2 MAX phase. These results provide evidence for the chemical changes that occur during the synthesis of the Ti_3C_2/NH_4TiOF_3, where Ti_3AlC_2 is effectively etched and exfoliated to form the Ti_3C_2/NH_4TiOF_3.

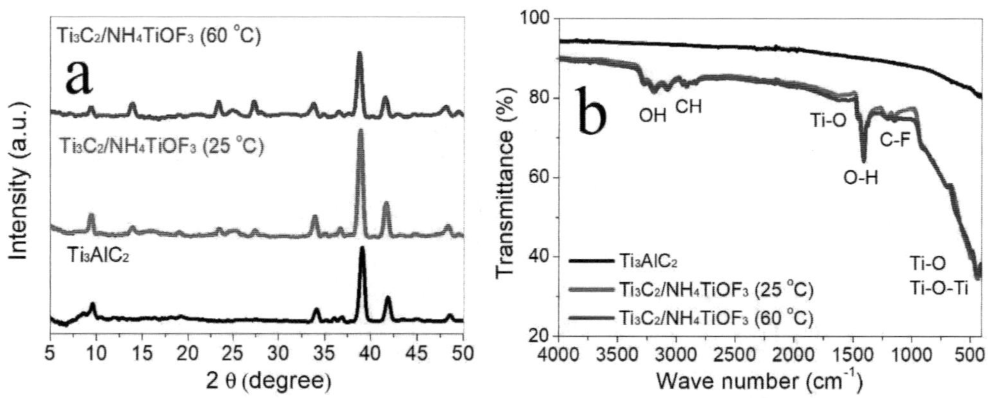

Figure 5. shows (a) the X-ray diffraction pattern and (b) the Fourier transform infrared spectrum measurement of Ti_3AlC_2 (black line) next to Ti_3C_2/NH_4TiOF_3 (25, red line) and (60, blue line).

Figure 6(a) shows the TG/DTA profile of the synthesized Ti_3C_2/NH_4TiOF_3 (25), and Figure 6(b) shows the profile of Ti_3C_2/NH_4TiOF_3 (60) from 25 °C to 1000 °C under nitrogen atmosphere. During the initial weight loss regime (25–200 °C), both samples experienced a consistent mass loss of approximately 1.5%.

This was ascribed to the elimination of both physically and chemically trapped water [22]. According to the endothermic peaks in the TG profile at temperatures in the 250–370 °C range for Ti_3C_2/NH_4TiOF_3 (25) and at temperatures in the 250–420 °C range for Ti_3C_2/NH_4TiOF_3 (60), in the second regime the weight losses of 7.9% and 11.9% for Ti_3C_2/NH_4TiOF_3 (25) and Ti_3C_2/NH_4TiOF_3 (60), respectively. This phenomenon has been attributed to the degradation of the crystallized water and the organic matter [22]. The weight for Ti_3C_2/NH_4TiOF_3 (25) decreased slightly by 0.3% for temperatures in the 370–595 °C range, which was associated with the titanium transformation from the amorphous to the anatase [23]. The weight loss observed for Ti_3C_2/NH_4TiOF_3 (60) was 2.8% in the temperature range of 420–645 °C, corresponding to the titanium transition from the anatase to the rutile [23]. At elevated temperatures (exceeding 600 °C), there was a gradual and slight increase in the weight of both Ti_3C_2/NH_4TiOF_3 samples. This increase was attributed to the selective oxidation of Ti_3C_2/NH_4TiOF_3 upon heating with nitrogen, which featured a significantly low oxygen content [24]. At these elevated temperatures, Ti_3C_2/NH_4TiOF_3 underwent reactions with previously released oxygen, resulting in the formation of oxides. An endothermic peak at about 828 °C was observed in the DTA curve of Ti_3C_2/NH_4TiOF_3 (60). This was accompanied by removal of surface bound F groups, disruption of the layer structure, disproportionation of Ti_3C_2, and formation of TiO_2 and TiC [25].

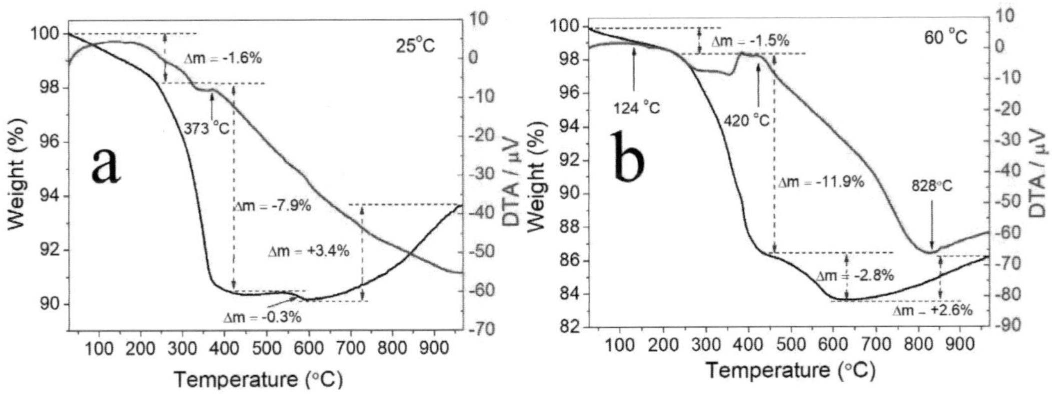

Figure 6. The TG-DTA curves of (a) Ti_3C_2/NH_4TiOF_3 (25) and (b) Ti_3C_2/NH_4TiOF_3 (60).

4. Conclusions

In summary, our research has effectively demonstrated the fabrication of Ti_3C_2/NH_4TiOF_3 through a streamlined process involving Ti_3AlC_2 exfoliation and $(NH_4)_2TiF_6$ etching. Using the LPD technique, we were able to exfoliate Ti_3C_2 while simultaneously incorporating NH_4TiOF_3 mesocrystals on its layered surface. This innovative and safe method paves the way for future applications utilizing Ti_3C_2 and the uniform integration of Ti_3C_2 with NH_4TiOF_3 mesocrystals. Our findings offer valuable understanding regarding the production of Ti_3C_2, indicating promising progress in developing MXene-based materials with improved attributes.

Acknowledgement

This research received partial funding from the Tatematsu Public Interest Foundation located in Kariya, Japan.

References

[1] Kanda H, Miyahara M, Yoshioka T and Okazaki M 2000 *Langmuir*. 35 6622
[2] Miyahara M, Kanda H, Shibao M and Higashitani K 2000 *J. Chem. Phys*. 112 9909
[3] Zhang Y, Tan Y W, Stormer H L and Kim P 2005 *Nature*. 438 201
[4] Murali G, Reddy Modigunta J K, Park Y H, Lee J H, Rawal J, Lee S Y, In I and Park S J 2022 *ACS Nano*. 16 13370
[5] Naguib M, Kurtoglu M, Presser V, Lu J, Niu J, Heon M, Hultman L, Gogotsi Y and Barsoum M W 2011 *Adv. Maters*. 23 4248
[6] Xu S, Dall'Agnese Y, Li J, Gogotsi Y and Han W 2018 *Chem. Eur. J*. 24 1855
[7] Wang S, Liu Y, Liu Y and Hu W 2023 *Chem. Eng. J*. 452 13951
[8] Li Y, Ding L, Yin S, Liang Z, Xue Y, Wang X, Cui H and Tian J 2019 *Nanomicro Lett*. 1 6
[9] Ghazaly A E, Ahmed H, Rezk A R, Halim J, Persson P O Å, Yeo L Y and Rosen J 2021 *ACS Nano*. 15 4287
[10] Natu V, Pai R, Sokol M, Carey M, Kalra V and Barsoum M W 2020 *Chem*. 6 616
[11] Wang L, Zhang H, Wang B, Shen C, Zhang C, Hu Q, Zhou A and Liu B 2016 *Electron. Mater. Lett*. 12 702
[12] Li M, Lu J, Luo K, Li Y, Chang K, Chen K, Zhou J, Rosen J, Hultman L, Eklund P, Persson P O Å, Du S, Chai Z, Huang Z and Huang Q 2019 *J. Am. Chem. Soc*. 141 4730
[13] Li M, Li X, Qin G, Luo K, Lu J, Li Y, Liang G, Huang Z, Zhou J, Hultman L, Eklund P, Persson P O Å, Du S, Chai Z, Zhi C and Huang Q 2021 *ACS Nano*. 15 1077
[14] Feng W, Luo H, Wang Y, Zeng S, Tan Y, Zhang H and Peng S 2018 *Ceram. Int*. 44 7084
[15] Peng C, Wei P, Chen X, Zhang Y, Zhu F, Cao Y, Wang H, Yu H and Peng F 2018 *Ceram. Int*. 44 18886

[16] Wang T, Korposh S, James S and Lee S W 2020 *Microchim Acta*. 187 663
[17] Wang T, Zhu L and Kanda H 2023 *Sens. Actuators B Chem*. 393 134136
[18] Wang T, Zhu L, Zhu WY and Kanda H 2023 *Heliyon*. 9 e18718
[19] Li S, Que X, Chen X, Lin T, Sheng L, Peng J, Li J and Zhai M 2020 *ACS Appl. Energy Mater*. 3 10882
[20] Liu Y, Zhang S, Su X, Xu J and Li Y 2020 *J. Mater. Sci*. 55 10339
[21] Lei B, Valluvar Oli A, Ivaturi A and Robertson N 2022 *Sustain. Energy Fuels*. 6 502
[22] Han M, Yin X, Wu H, Hou Z, Song C, Li X, Zhang L and Cheng L 2016 *ACS Appl. Mater. Interfaces*. 8 21011
[23] Zhu X, Han S, Feng W, Kong Q, Dong Z, Wang C, Lei J and Yi Q 2018 *RSC Adv*. 8 14249
[24] Arod P and Shivashankar S A 2016 *RSC Adv*. 6 65366
[25] Feng A, Yu Y, Jiang F, Wang Y, Mi L, Yu Y and Song L 2017 *Ceram. Int*. 43 6322

IOP Conf. Series: Materials Science and Engineering **1318** (2024) 012039 doi:10.1088/1757-899X/1318/1/012039

The effects of cellulose nanocrystals (CNC) and halloysite nanotubes (HNT) on selected properties of crosslinked modified tapioca starch films

R A Rañada[1,a], J L Castro[1,b], R D Manalo[2,c], H L Ong[3,d], M U Herrera[4,e] and M D L Balela[1,f]

[1]Department of Mining, Metallurgical and Materials Engineering, University of the Philippines Diliman, Quezon City, Philippines 1101
[2]Department of Forest Products and Paper Science, University of the Philippines Los Baños, Laguna, Philippines 4031
[3]Faculty of Chemical Engineering Technology, Universiti Malaysia Perlis, 02600 Arau, Perlis, Malaysia
[4]Institute of Mathematical Sciences and Physics, University of the Philippines Los Baños, Laguna, Philippines 4031
[a]raranada@up.edu.ph, [b]jlcastro3@up.edu.ph, [c]rdmanalo@up.edu.ph, [d]hlong@unimap.edu.my, [e]muherrera@up.edu.ph, [f]mlbalela1@up.edu.ph

Abstract. Crosslinked modified tapioca starch films with CNC and HNT have the potential to be good renewable and biodegradable food packaging. Starch, CNC, and HNT were used because of their availability and cheap cost. This study aimed to investigate the effect of varying concentrations of CNC and HNT on the morphology, mechanical and optical properties of crosslinked modified tapioca starch films as an initial step in developing an active food package. Both CNC and HNT have high crystallinity and a large amount of OH groups. The OH groups as well as the ion-dipole interaction in HNT increase the tensile strength of the films. The highest tensile strength and the lowest elongation were recorded in films containing 5% CNC and 5% HNT. Beyond 5%, the addition of these nanomaterials diminished the tensile strength due to agglomeration which is the weak point of the films. Below the 5%, the nanomaterials serve as good reinforcing agents which improve the tensile strength. In addition, all films have good transmission capacity for visible light, while films with HNT outperform films with CNC in UV blocking ability.

1. Introduction

The extensive usage of polyethylene and polypropylene-based plastics caused problems regarding their disposal and possible toxicity to living organisms. Hence, a need to develop and improve existing biodegradable packaging materials which are safe for both living organisms and the environment should be a priority. Natural polymers such as starch can form films that have a high potential to be a good packaging material. Several literatures have been reported regarding this topic [1-3].

The citric acid (CA) crosslinks starch molecules to enhance the film's properties [4-6]. Crosslinking is done to give rise to a more stable starch structure [7]. However, further improvement on the starch films can be done by the addition of reinforcing agents or additives, for instance, cellulose nanocrystals (CNC) and halloysite nanotubes (HNT). These nanomaterials have huge surface areas which enable them to interact more with the crosslinked starch molecules of the films [8,9]. Starch, CNC, and HNT are widely

available and low cost compared to other nanomaterials such as titanium oxide (TiO_2) nanoparticles which made them desirable. Yet, we opted to use modified tapioca starch instead of pristine tapioca starch since only a small number of studies have tackled this material. Specifically, the modified tapioca starch was a distarch phosphate wherein two molecules of starch were attached to a common phosphate group.
The objective of this study was to investigate the effect of varying amounts of CNC and HNT on the morphology, mechanical, and optical properties of crosslinked modified starch films.

2. Materials and Methodology

2.1 Materials
The CNC (Emtex) and HNT were purchased from Nanotronics and Sigma-Aldrich, respectively. The modified tapioca starch was bought from Dalkem and the glycerol (99%) was supplied by HmbG Chemicals. The anhydrous citric acid (99%) was acquired from Thermo Fisher Scientific.

2.2 Film formation with varying concentrations of CNC and HNT
The preparation of films was based on the works of Ramos et al. [6] and Owi et al. [8] with some modifications. Six grams (6 g) of modified tapioca starch, 0.6 g of anhydrous citric acid (10% based on the mass of starch), and 1.2 g of glycerol (20% based on the mass of starch) were mixed in 100 mL of distilled water. Varying amounts of CNC (1, 3, 5, and 7% based on the mass of starch) or HNT (2.5, 5, 10, 15% based on the mass of starch) were added with 50 mL of distilled water and sonicated at 50% amplitude with 2:5 (sec/sec) on-off pulse for 30 mins. The homogenized mixture of nanomaterial was blended with the previous mixture thereby diluting the final mixture to 150 mL. Then, the mixture was heated at 80 °C for 30 mins. with continuous stirring and cast onto a non-stick carbon steel mold (18 cm internal diameter). After casting, the mixture was cured at 105 °C for 15 mins. and dried at 50 °C for 18 hrs. Subsequently, the mold was placed in a desiccator for about 30 mins. Afterward, the film was removed carefully from the mold and stored. Triplicates were produced for each treatment.

2.3 Film characterization

2.3.1 Scanning Electron Microscope (SEM). The surface characteristics of the films were analyzed using JEOL 5310 scanning electron microscope under 20 000x magnification and 5 kV standard mode.

2.3.2 Film thickness. The thickness of films from different treatments was measured in triplicates using a Mitutoyo digital caliper with an accuracy of ±0.01 mm.

2.3.3 Mechanical properties. The films were cut into strips with dimensions of 15 mm height and 1 mm width. Then, the tensile properties of the films were determined using Universal Testing Machine following the standard procedure from ASTM D882-12 with a load of 1 kN and a testing speed of 10 mm/min. The testing was done in triplicates.

2.3.4 Optical properties. The films were cut into dimensions of 200 mm x 300 mm and loaded inside UV-Vis 2600i. The scanning range was from 200 to 800 nm. Then the values at 290 and 550 nm were analyzed.

3. Results and Discussion
The fabrication of crosslinked modified tapioca starch films containing varying amounts of CNC and HNT was successful. Figures 1 and 2 show the digital images of the films. In both figures, the fabricated films were colorless to slightly yellow and translucent. The surface of the films appeared to be smooth to the naked eye. Moreover, an increased amount of nanomaterials (CNC and HNT) caused turbidity of the films.

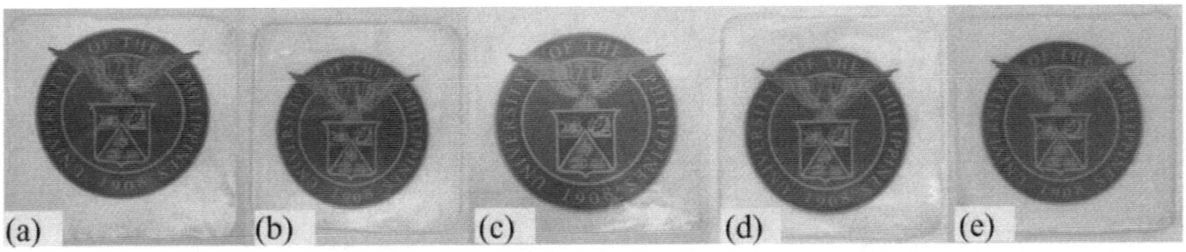

Figure 1. Digital images of crosslinked modified tapioca starch film with (a) 0%, (b) 1%, (c) 3%, (d) 5%, (e) 7% CNC.

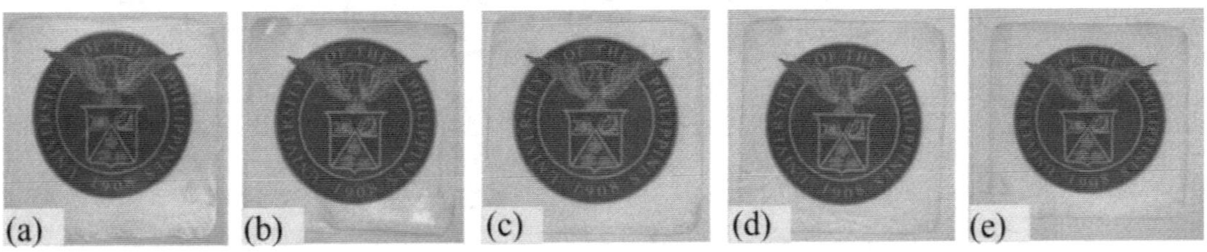

Figure 2. Digital images of crosslinked modified tapioca starch film with (a) 0%, (b) 2.5%, (c) 5%, (d) 10%, (e) 15% HNT.

Figures 3 and 4 present the SEM micrographs of modified tapioca starch film with varying amounts of CNC or HNT. In Figures 5a, 5b, 5c, and 5d, most of the films have cracking and pores/holes on the surface. These features provide weak points. On the other hand, in Figures 6a, 6b, 6c, and 6d, the surface of the films with different amounts of HNT has no cracking or pore/hole. Additionally, rough surfaces in Figure 6 are noticeable, wherein the higher amount of HNT resulted in a rougher surface.

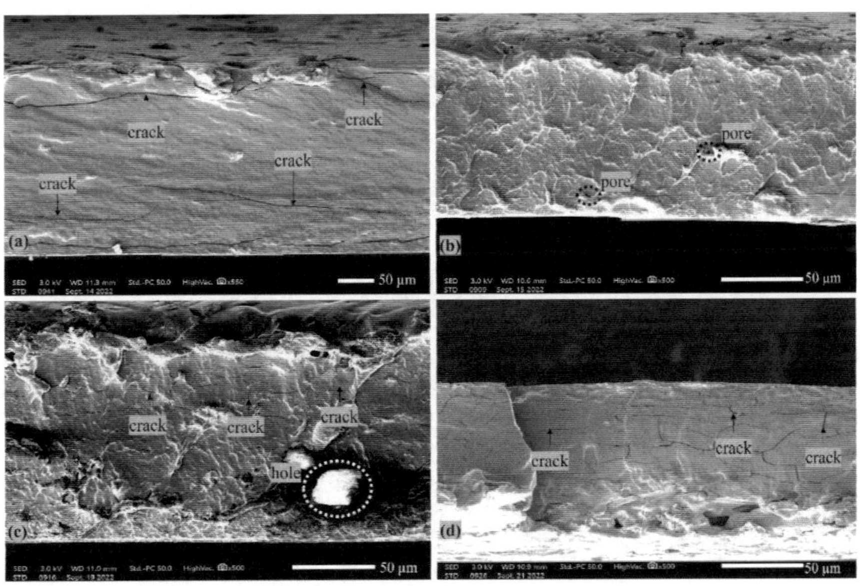

Figure 3. SEM micrographs of crosslinked modified tapioca starch film with (a) 1%, (b) 3%, (c) 5%, and (d) 7% CNC at a 20 000x magnification.

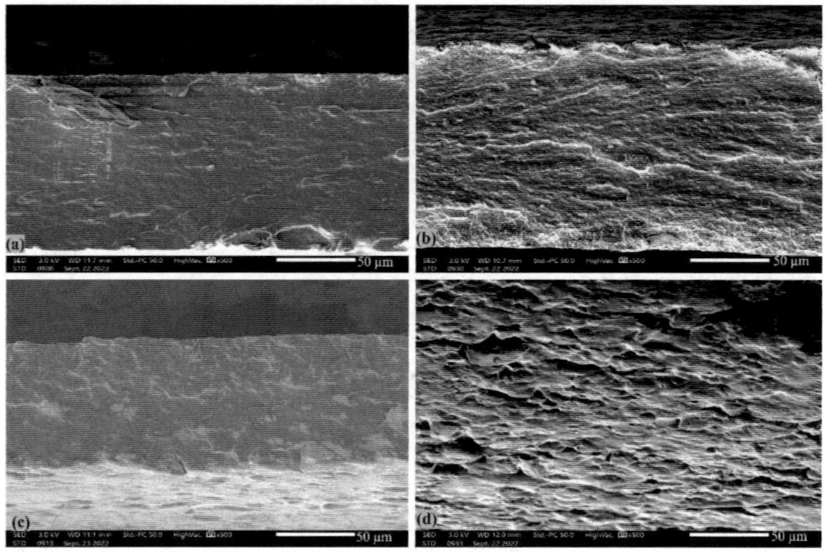

Figure 4. SEM micrograph of crosslinked modified tapioca starch films with (a) 2.5%, (b) 5%, (c) 10%, and (c) 15% HNT at a magnification of 20 000x.images

Figure 5 illustrates the effect of CNC and HNT on the thickness of modified tapioca starch films. Statistical analysis showed that the thickness of films was not affected by varying amounts of nanomaterials. On average, films with no nanomaterial additive have a thickness of 0.142 ± 0.030 mm, while films with 1, 3, 5, and 7% CNC have 0.147 ± 0.022, 0.177 ± 0.001, 0.179 ± 0.018 mm, and 0.136 ± 0.025 mm, respectively. Furthermore, films with 2.5, 5, 10, and 15% HNT have thicknesses of 0.154 ± 0.025, 0.163 ± 0.023, 0.157 ± 0.029, and 0.161 ± 0.016 mm, respectively. Possibly, CNC and HNT are nanoscale and take up only the small interstitial spaces between crosslinked modified tapioca starch molecules in the matrix film. This results in an unchanged thickness of the films even if the amounts of either CNC or HNT are increased or decreased

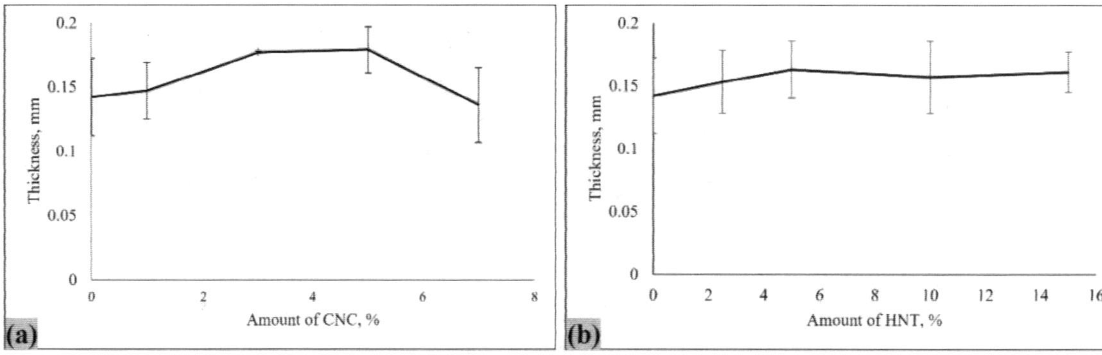

Figure 5. The thickness of crosslinked modified tapioca starch films with (a) CNC and (b) HNT.

Table 1 presents the effect of varying concentrations of nanomaterials on the mechanical properties of the films. The highest tensile strength was recorded in films containing 5% CNC (7.710 ± 0.155 MPa) and 5% HNT (9.845 ± 1.131 MPa). Compared to films without nanomaterials, the tensile strength of films with 5% CNC and 5% HNT has increased by 9.368 folds and 11.96 folds, respectively.
The addition of CNC and HNT provides extra intermolecular bonding between crosslinked modified starch molecules and the nanomaterials. In the case of CNC, each cyclic monomer has three available OH groups which can form hydrogen bonding with another OH group from the starch molecules. For HNT, substantial hydrogen bonding exists between the nanomaterial and the film [10]. Additionally, the negative charges on the outer surface of HNT can interact with the OH groups of crosslinked starch

IOP Conf. Series: Materials Science and Engineering **1318** (2024) 012039 doi:10.1088/1757-899X/1318/1/012039

molecules; thus, providing a strong ion-dipole interaction. Both hydrogen bonding and ion-dipole interaction increase the tensile strength.

The tensile strength of films increased as the amount of CNC and HNT increased until 5% of the nanomaterials was reached. Beyond 5%, the tensile strength started to decrease. This happens due to the agglomeration of the nanomaterials in the matrix of film. Below 5%, the nanomaterials serve as a good reinforcing agent. Since incorporating 5% of CNC and 5% of HNT yielded in most improved tensile strength, these amounts were considered to be the optimum.

The films containing HNT have higher tensile strength than films with CNC. This could be attributed to stronger bonds between HNT and starch molecules (ion-dipole interactions and hydrogen bonds) than between CNC and starch molecules (hydrogen bonds). Moreover, the lowest break elongation was also observed in films with 5% CNC (2.196 ± 1.178%) and 5% HNT (8.311 ± 8.067%).

Table 1. Average tensile strength and mean elongation of 10% CA-crosslinked modified tapioca starch film with varying amounts of CNC and HNT.

Nanomaterial	Amount, %	Tensile Strength, MPa	Break Elongation, %
CNC	0	0.823 ± 0.150	71.110 ± 40.889
	1	5.220 ± 0.533	10.835 ± 12.784
	3	6.070 ± 0.761	23.537 ± 19.733
	5	7.710 ± 0.155	2.196 ± 1.178
	7	4.170 ± 0.957	19.921 ± 19.624
HNT	0	0.823 ± 0.150	71.110 ± 40.889
	2.5	9.693 ± 1.824	14.773 ± 15. 182
	5	9.845 ± 1.131	8.311 ± 8.067
	10	7.028 ± 1.282	10.723 ± 9.962
	15	6.696 ± 0.343	9.355 ± 10.504

Figures 6 and 7 present the effect of varying amounts of CNC and HNT on the light transmittance of the films. At 550 nm, every single film transmits nearly all of the light. It means that the films allow visible light to pass through. This is an advantageous property of films because it enables the packed food to be easily seen. At 295 nm, generally, a higher concentration of CNC in films leads to lower light transmittance (Figure 6). Low transmittance of light at 295 nm indicates that UV light is being blocked by CNC, and an increased amount of CNC provides better blocking ability. On the other hand, films with HNT have the same light transmittance at 295 nm regardless of their amount (Figure 7). Comparing the CNC with HNT, the latter has better UV blocking ability.

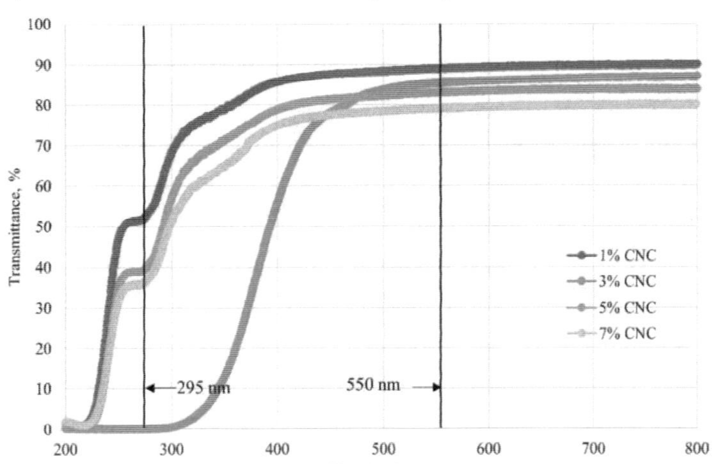

Figure 6. Ultraviolet-visible spectra of crosslinked modified tapioca starch films with varying amounts of CNC.

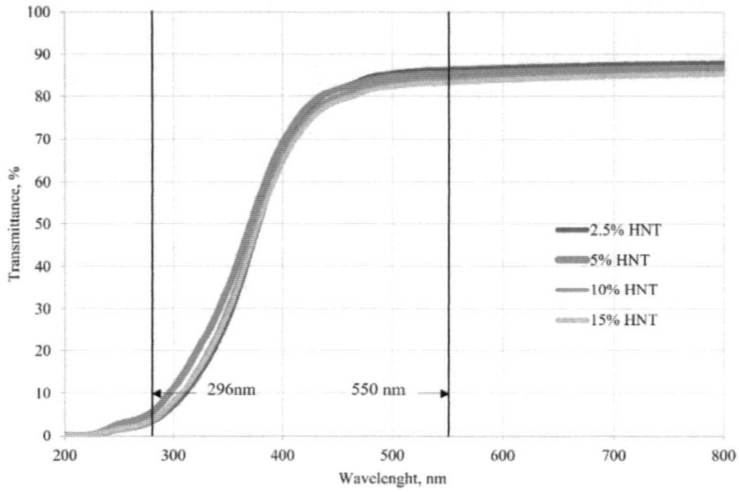

Figure 7. Ultraviolet-visible spectra of crosslinked modified tapioca starch films with varying amounts of HNT.

4. Summary and Conclusion

The CNC and HNT were incorporated into citric acid-crosslinked modified tapioca starch film. The tensile strength of films improves upon the addition of CNC and HNT up to a certain amount (5% CNC and 5% HNT). Then, the tensile strength diminishes upon further addition. Moreover, both films with CNC and HNT performed similarly in transmitting visible light. But films with HNT surpassed the performance of films with CNC in blocking the UV light.

Acknowledgment

This study is supported by a grant from the University of the Philippines System's Emerging Interdisciplinary Research (EIDR) program (OVPAA-EIDR-C09-10). The authors wish to thank Ana Rose Ramos, RCh. for assisting in the laboratory work.

References

[1] Cai J, Zhang D, Zhou R, Zhu R, Fei P, Zhu Z, Cheng S and Ding W 2021 *J. Agric. Food. Chem.* **69** 5067-5075

[2] Saiful, Helwati H and Iqbalsyah T M 2018 *Material Science and Engineering* **523** 012015

[3] Othman S, Tarmiti N, Shapi'I R, Zahiruddin S Tawakkal I and Basha R 2020 *BioRes.* **15** 3984-3998

[4] Narendra R and Yiqi Y 2009 *Food Chemistry* **118** 702-711

[5] Gerezgiger A G and Szabó T 2022 *Journal of Physics* **2315** 012036

[6] Ramos A R, Manalo R D, Ong H L, Herrera M U and Balela M D 2023 *Materials Science Forum* **1087** 29-34

[7] Golachowski A, Drożdż W, Golachowska M, Kapelko-Żeberska M and Raszewski B 2020 *Foods* **9** 1311

[8] Owi W T, Ong H L, Sam T S, Villagracia A R, Tsai C and Akil H M 2019 *Industrial Crops & Products* **139** 111548

[9] Saif M J, Asif H M and Naveed M 2018 *J. Chil. Chem. Soc.* **63**

[10] Huang D, Zhang Z, Zheng Y, Quan Q, Wang W and Wang A 2020 *Food Hydrocolloids* **101** 105471

IOP Conf. Series: Materials Science and Engineering **1318** (2024) 012040 doi:10.1088/1757-899X/1318/1/012040

Free-standing polyaniline-reduced graphene oxide (PANI-rGO) thermoelectric films via electrochemical synthesis

A M San Juan and E P Enriquez
Department of Chemistry, School of Science and Engineering, Ateneo de Manila University, Quezon City, Philippines

E-mail: ana.sanjuan@obf.ateneo.edu, epenriquez@ateneo.edu

Abstract. Heat-based energy harvesting is gaining attention as an alternative energy source. Thermoelectric materials (TEs) generate voltages in response to a thermal gradient and conductive polymer-carbon composites have properties that make them favorable for these applications. Polyaniline-reduced graphene oxide (PANi-GO) thermoelectric thin films were fabricated by the electropolymerization of PANi with simultaneous co-deposition of graphene onto fluorine-doped tin oxide (FTO). Poly(butyl acrylate) was spin-coated onto these films as backing support to make a stable free-standing film of PANi-rGO. The free-standing film was formed upon GO reduction which resulted in simultaneous delamination from the FTO electrode. Compositional and morphological data were obtained via Raman microspectroscopy and scanning electron microscopy and the TE properties were measured in terms of the Seebeck coefficient and conductivities from Hall effect measurements. The Seebeck coefficients of the films measured between 6.6 μV K^{-1} to 13.4 μV K^{-1}, whereas Hall effect conductivities ranged from 0.483 to 2.68 S cm^{-1}. The highest film power factor measured was 0.025 μW $m^{-1}K^{-2}$. Varying the GO content used in the electropolymerized film did not have a significant effect on the overall TE property though varying the reduction cycle number and voltage resulted in significant changes in the film Seebeck coefficients and conductivities.

1. Introduction

Thermoelectrics (TEs) are a class of materials that convert heat to electricity, thus making them attractive for recovering waste heat loss [1-3]. While TEs are available in industry for a variety of applications, new innovations in organic and flexible TEs have demonstrated potential use in wearable electronics, flexible devices, and self-powered electronics [4-5]. The fundamental principle underlying TEs is the Seebeck-Peltier effect. This describes what occurs when a temperature gradient (ΔT) is established along one axis of a material. The majority charge carriers are excited via the applied thermal energy on the heated end and migrate towards the cold end [6]. Due to charge carrier accumulation, a potential difference (V) is generated which is quantified by the following relationship:

$$V = \alpha \Delta T \tag{1}$$

where the quantities are related by the Seebeck coefficient α, the measure for thermopower. The thermoelectric property figure-of-merit is given by zT (2) which relates the material's electrical conductivity (σ), thermopower (α), and thermal conductivity (λ) as follows [1, 7]:

$$zT = \frac{\alpha^2 \sigma}{\lambda} T \tag{2}$$

Equation 2 shows that it is insufficient to just increase the Seebeck coefficient and conductivity. The thermal conductivity must also be optimized given its inverse relationship with zT [1, 7]. Organic TEs are gaining attention for inherently low thermal conductivity, low cost of raw materials, and greater flexibility than inorganic materials [1-2, 8]. One such material is the conductive polymer polyaniline (PANi) whose limited conductivity can be further improved by compositing it with conductive carbon-based fillers such as graphene or reduced graphene oxide (rGO). This approach has produced high-performing organic-carbon TE devices [1, 9-10]. Many PANi-based TEs are made using solution-dispersal methods, but synthesis modes for PANi-rGO TE devices require strong oxidizing and reducing agents to polymerize aniline and reduce graphene oxide [10]. This study developed a new method of fabricating polyaniline-reduced graphene oxide TE thin films via electrochemical synthesis that minimizes strong reagents. Electropolymerization was then followed by simultaneous delamination-reduction of GO to form free-standing PANi-rGO composite films [11-12]. The effect of the various fabrication steps on the TE properties was studied: the electropolymerization cycle number, GO concentration, polymer-backing, cycling potential, and number of delamination-reduction cycles.

2. Materials and Methods

2.1. Electropolymerization of polyaniline and co-deposition of graphene oxide thin film

Aniline monomer (Merck) was purified using vacuum distillation until a clear, light yellow and slightly oily liquid was obtained. Reagent grade concentrated H_2SO_4 (~18 M) was used without further purification. Graphene oxide was synthesized using a modified Hummer's method [13]. The FTO working electrodes were cleaned and prepared by successive sonication of 15 min each in the following solutions: detergent, deionized water, ethanol, deionized water, and acetone with a final rinsing with deionized water. These were then cleaned using UV-ozone treatment for 10 min.

A 3.3 mg mL^{-1} GO dispersion in water was prepared and subjected to ultrasonication for 1 h to further exfoliate GO layers. A three-electrode electrochemical cell was set up in a 50 mL beaker with a saturated calomel electrode (SCE) reference electrode, Pt wire counter electrode, and the cleaned FTO as a working electrode (25x25 mm^2). These were connected to a GAMRY Reference 3000 electrochemical workstation. To 50 mL of 1 M H_2SO_4 electrolyte, enough aniline was added to form a 0.12 M aniline solution. Varying volumes of GO dispersion was also added to this according to Table 1 below, then the electrolyte was sonicated for 1 h. Electropolymerization and co-deposition was initiated similar to what was reported by Harima et al. [14] by applying a triangular wave voltage between ± 1.1 V with a scan rate of 100 mVs^{-1} for a number of cycles for the experiment designation.

2.2. Reduction and simultaneous delamination of graphene oxide

Prior to GO reduction with simultaneous delamination, a two-fold diluted poly(butyl acrylate) [PBA] (a commercial colorless nail lacquer) was spin-coated (Laurell WS-650-23B Spin Coater) onto the dried PANi film on the FTO. Here, 350 μL of the diluted PBA was deposited on the film surface under a static spin coating program for 30s at 600 RPM. The FTO-PANi-rGO electrode was used as the working electrode while a Pt wire was the counter-reference electrode. These were immersed in a 0.5 M Na_2SO_4 electrolyte solution. Delamination-reduction was conducted at a scan rate of 100 mV s^{-1} from ±1.4 or ± 2 V depending on the experiment performed (see Table 1) [15-16]. The number of cycles and triangular wave voltage were varied according to the experimental design shown in Table 1. A stable, free-standing PBA-backed PANI-rGO was then produced and the film thickness measured using a profilometer (KLA Tencor Profilometer). The films were also characterized using Raman microspectrometry (Horiba Xplora Plus) and scanning electron microscopy (SEM, Hitachi SU1510).

2.3. Electrical measurements

To measure the Seebeck coefficient, a four-probe Seebeck coefficient measurement device was set up based on the design of Cho et al. [8, 17]. A sample mounting stage was 3D printed using polylactic acid (PLA). Two copper tubes were attached to the stage and connected to water pumps immersed in two separate temperature control thermostat baths (PolyScience AP07R-40 Refrigeration/Heating Circulating Bath and Brookfield TC-650 Water Bath). Copper thermal contact were attached to the top of the copper tubes using Arctic MX-4 thermal paste. Temperature was measured using a thermocouple

IOP Conf. Series: Materials Science and Engineering **1318** (2024) 012040 doi:10.1088/1757-899X/1318/1/012040

and the voltage obtained using a precision multimeter (YEW Model 2501A Precision Digital Multimeter). An acrylic cover was placed on top of the sample to ensure good thermal contact with the copper plate surface. The measurement set up was calibrated using Cu and Ni metal plates with mass % purity of 96.50% and 94.05%, respectively, from X-ray fluorescence measurement (Horiba XGT-7200). The sample were prepared for measurement by soldering copper wires at the edges of the test metal plate. Film conductivities and carrier mobilities in the PANi and PANi-rGO films were measured using a Hall Effect apparatus (Ecopia Hall Effect Measurement System HMS-3000). Silver epoxy was used to connect copper wires to the two ends of the material for the multimeter leads to clamp onto.

Table 1. Design of experiments (DoE) for PANi-rGO fabrication protocol.

Experiment No.	Cycle Number	GO Concentration (mL of 3.3 mg mL^{-1} GO in 100 mL electrolyte)	Reduction Potential	Reduction Cycle Number
1	100	0.2	-2	50
2	30	2	-1.4	200
3	100	2	-2	50
4	100	2	-2	200
5	30	0.2	-2	50
6	100	0.2	-1.4	200
7	30	2	-2	50
8	100	2	-1.4	50

3. Results and Discussion

3.1. Characterization of PANi-rGO films

During electropolymerization, the characteristic redox transition peaks of polyaniline for the different oxidation forms were observed. The peak positions agree with those reported in the literature for the different peaks formed in the electropolymerization of polyaniline. The *c* peak indicates the formation of the monomer cation radical for polymerization initiation. The *a-a'* redox pair is associated with the transition between the fully reduced (leucoemeraldine) and partially oxidized and protonated emeraldine salt states while the *b-b'* transition is between the emeraldine and fully oxidized pernigraniline state [11, 18]. As the leucoemeraldine and pernigraniline states are less conductive forms of PANi, to ensure that the conductive emeraldine form was obtained, the reaction was stopped at 0.25 V. A distinct peak within the range of -0.5 to -1.0 V is typically reported in the literature that corresponds to the reduction of GO in the system (Figure 1) [19]. However, this was not distinctly observed in the CV graphs, which could be due to the low loading of graphene oxide in the films.

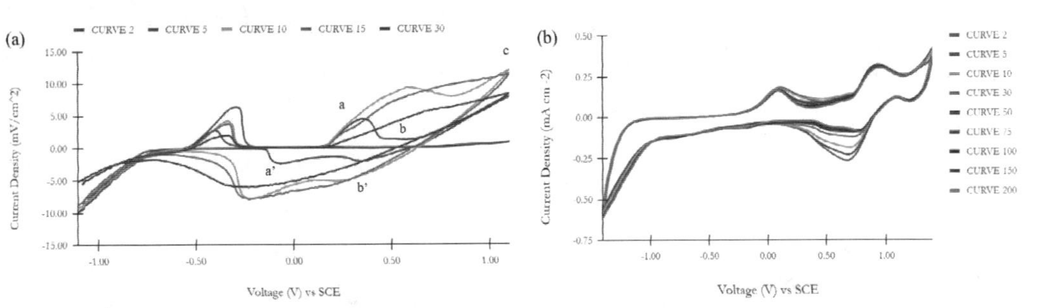

Figure 1. Cyclic voltammogram of PANi-rGO electropolymerization-codeposition (a) and delamination-reduction (b) cycles.

IOP Conf. Series: Materials Science and Engineering **1318** (2024) 012040 doi:10.1088/1757-899X/1318/1/012040

The films were demonstrated to be electrochemically stable over the delamination-reduction potential range. Stability is indicated by minimal observed change (if any) in the CVs going up to 200 cycles during the delamination-reduction step. The delaminated film had final thicknesses consistently < 40 μm including the PBA polymer backing. The films remained dimensionally stable when the preliminary PANi-rGO was grown as a pristine layer on the FTO electrode.

The Raman spectra shown in Figure 2 for pure PANi and PANi-rGO demonstrated similar profiles to those reported in the literature within the 1000 to 1600 cm^{-1} region. Benzenoid C-H bending corresponds to the peaks observed from 1160 to 1170 cm^{-1}. C-N stretching peaks occur between 1320 to 1340 cm^{-1} and quinoid C=C stretching peaks were found between 1580 to 1615 cm^{-1} [20-21]. Because graphene oxide peaks overlap with those associated with PANi, its presence is instead confirmed using the specific peak at 1481 cm^{-1} which corresponds to π-π interactions that are amplified by the GO basal plane and PANi aromatic group interactions. As no significant difference in peak height could be observed within this region, this confirms relatively low compositing of GO in the film which also explains the absence of a CV peak for GO reduction. The morphological characteristics of the films were evaluated via SEM (Figure 4). Generally, the film morphology of pure PANi (without rGO, Figure 3a) and the PANI-rGO (Figure 4b) were very similar. There was no observable difference between these two films at these image resolutions, which suggests no aggregation of GO into large phases, also given the low loading of the GO in the PANi-rGO films.

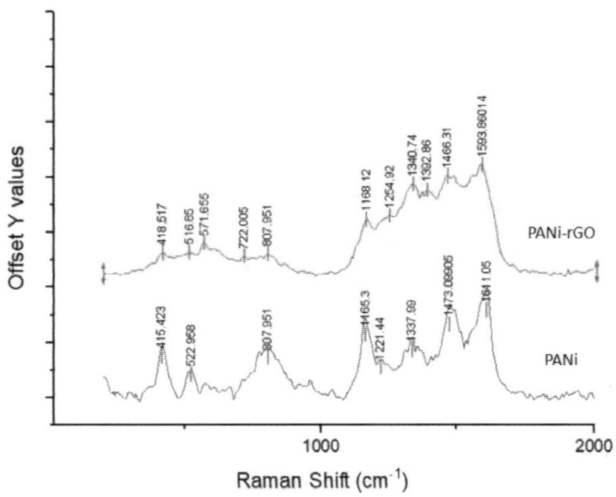

Figure 2. Raman spectra of Pure PANi and PANi-rGO from DOE experiment 5 (Table 1).

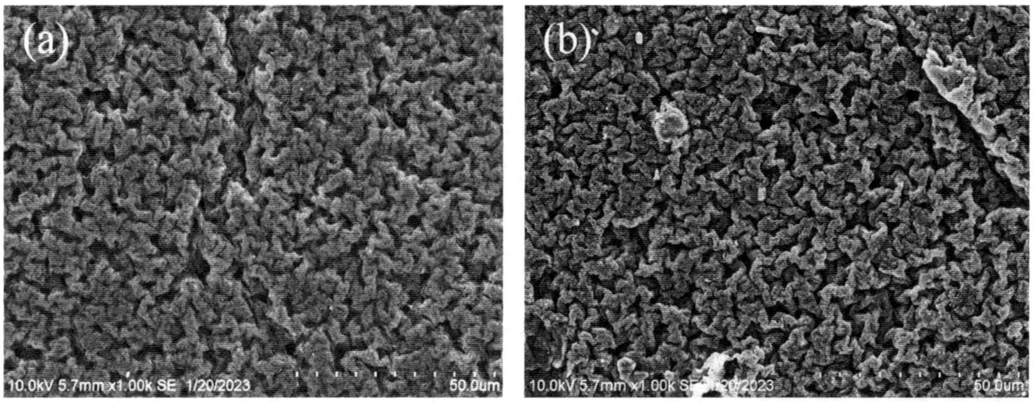

Figure 3. SEM images for a) Pure PANi thin film and b) PANI-rGO thin film (DOE Exp 2).

3.2. Calibration of Seebeck coefficient measurement device
The thermoelectric properties of the thin films were assessed using a Seebeck coefficient measurement set up. The steady-state differential measurement method was employed where a thermal gradient $\Delta T = T_2 - T_1$ was applied across the sample while keeping a constant mean temperature: $T_o = (T_1 + T_2)/2$. The slope of the best-fit line obtained from plotting the generated potential difference V vs ΔT should give the sample Seebeck coefficient [22-23]. Cu and Ni metal samples were used to calibrate the apparatus for any bias, with the Ni undergoing additional hysteresis testing. The best-fit line demonstrated a high goodness-of-fit (Figure 4a) from the Ni samples ($R^2 > 0.999$) with a measured slope of -22.7 μV K^{-1}. During hysteresis testing wherein the thermal gradients of the two sides of the sample were reversed in direction, linear behavior was retained by Ni, which indicated no measurable hysteresis in both increasing and decreasing ΔT directions. Should hysteresis occur from reversing temperature differential application direction, this would be indicative of poor thermal contact measurements. As there was neither hysteresis nor time dependence, the thermal contacts were deemed sufficient [24].

Due to the low expected Seebeck coefficient of Cu, four thin films were connected in series to increase the signal. Good linear correlation ($R^2 = 0.978$) was observed, which was lower than that for Ni. An average slope of 1.7 μV K^{-1} was obtained. Consistent bias (% deviation) were observed throughout the trials, as seen in Table 2.

Table 2. Thermoelectric properties of calibration materials (standard deviation s. d. in parentheses).

Material	Slope / μV K^{-1}	Literature Seebeck Coefficient / μV K^{-1}	% Deviation
Ni metal	−22.8 (0.1)	-19.5 [25]	17
Cu metal	1.7 (0.1)	1.83 [26]	9.6
Pure Polyaniline	10.5 (0.1)	12 [10]	13

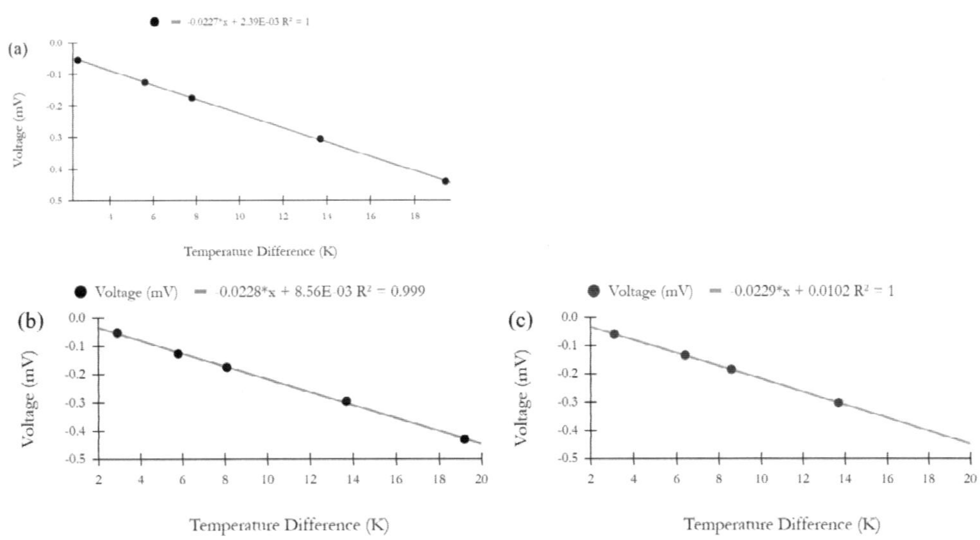

Figure 4. Voltage vs Temperature Difference graphs for a) nickel, b) hysteresis testing with increasing ΔT, and c) hysteresis testing with decreasing ΔT.

The measured slopes for the pure Ni and Cu samples were corrected to get the calibrated Seebeck coefficient, and the correction was also applied to the case of the pure PANi. For pure PANi, a high R^2

IOP Conf. Series: Materials Science and Engineering **1318** (2024) 012040 doi:10.1088/1757-899X/1318/1/012040

= 0.996 was also observed. The calibrated Seebeck coefficient values were then computed as follows: 0.93(*Slope*) + 1.41, where *Slope* is from the V versus ΔT plot.

3.3. Thermoelectric properties of the PANi-rGO films

The electrical properties were measured using a Hall-effect Apparatus where the free-standing films (Figure 5) were cut into 4x4 mm^2 squares and mounted onto glass slides with silver epoxy contacts at the corners.

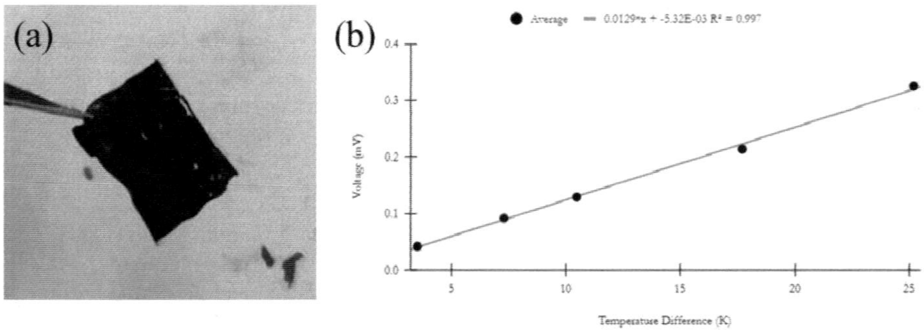

Figure 5. PANi-rGO a) Free-standing film and b) Experiment 5 SC measurement.

During Seebeck coefficient measurement, the average voltage at each temperature gradient was obtained. When the potential was plotted against the temperature gradient, a high R^2 (=0.997 to 1) was obtained from all the films. It was observed that at the highest ΔT (between 10 and 45 °C), the measured voltage had large fluctuations indicating possible reduced film stability for $\Delta T > 25$ K. All the obtained Seebeck coefficient values were positive, which is indicative of PANi being a p-type TE material.

The TE power factor (*PF*) was calculated from the Seebeck coefficient and electrical conductivity values by the Equation 3, which is frequently used to compare the TE performance of organic-carbon thin films, noting that these films have poor thermal conductivity (and hence the denominator is not included in the PF value).

$$PF = \alpha^2\sigma \tag{3}$$

Multiple experiments were performed to determine the effect of various factors on the final film TE properties. The results are given in Table 3. It was found that with the current protocols, changing the concentration of added GO in the electrolyte solution did not result in significant changes to the Seebeck coefficient and electrical conductivity. This may be attributed to low loading of GO in the composite films. This low loading is possibly due to the instability of GO at high ionic strength of the deposition electrolyte, i.e., 1 M H_2SO_4 [27-28]. The GO could have aggregated during electropolymerization, resulting in relatively low loading as the aggregated particles could not sufficiently migrate to the electrode surface to co-deposit with the PANi fibers. The challenges that this may pose can be resolved with further experimentation involving surfactants to form stable dispersions of the GO in the deposition electrolyte solution.

TE performance was instead primarily affected by changes in the final reduction potential and cycle number during the reduction-delamination procedure. It was found that a higher magnitude reduction voltage of ± 2 rather than ± 1.4 V with a lower number of cycles (50 rather than 200) generally improves the final Seebeck coefficient of the films.

As shown in Table 3, a comparable performance of 0.025 μWm^{-1}K^{-2} to that of pure PANi was observed for the case of Experiment 8 in Table 1, whereas the other measured values were lower. The lower performance of the other films is likely due to the delamination-reduction procedure, from which it was observed that applying a lower reduction potential or a greater number of reduction cycles reduced the film TE performance. For Experiment 8, the higher PF may be attributed to the relatively higher electrical conductivity of the film even with a relatively lower Seebeck coefficient.

Electrochemical methods for the fabrication of free-standing thermoelectric devices are not a frequently utilized technique. In a procedure from Harima et al. [14], a PF value of 3.6 $\mu Wm^{-1}K^{-2}$ was obtained using a two-electrode synthesis method to form PANi-rGO films. This is much higher than those obtained in the present study suggesting that further optimization of the process used in this study could still improve film's TE performance. With further compositing of the material, the Seebeck coefficient tends to increase until it levels off at a certain concentration of the dopant. The primary manipulable factor is therefore electrical conductivity, which can be increased by enhancing charge carrier mobility [5, 12, 29]. Improving the concentration of rGO incorporated into the composite films, to increase the conductivity, should ideally further improve the TE performance of these films.

Table 3. Thermoelectric properties of synthesized thin films (standard deviation s. d. in parentheses).

Experiment No.	Thickness / µm (s.d.)	Elec. Conductivity / S cm^{-1} (s. d.)	Measured Slope / µV K^{-1}	Computed Seebeck Coefficient / µV K^{-1}	Computed Power Factor / µW $m^{-1}K^{-2}$
1	16 (5)	1.392 (0.006)	9.0	9.7	0.013
2	14 (6)	0.483 (0.009)	7.1	8.0	0.003
3	12 (1)	2.077 (0.004)	8.0	8.8	0.016
3 T1	16 (3)	1.180 (0.007)	9.1	9.9	0.011
5	12 (5)	1.25 (0.01)	12.9	13.4	0.022
6	9 (3)	1.62 (0.02)	5.6	6.6	0.007
7	6.6 (0.9)	1.17 (0.03)	12.6	13.1	0.020
8	11.8 (0.8)	2.68 (0.09)	8.8	9.6	0.025
3 T2	8.5 (0.4)	2.17 (0.02)	9.18	9.95	0.021
4 T2	11.9 (0.6)	1.6 (0.2)	9.85	10.57	0.018
5 T2	11 (2)	2.14 (0.03)	7.8	8.62	0.016
Pure PANi	16 (2)	1.730 (0.004)	10.5	12	0.025

4. Conclusion

A fabrication protocol was demonstrated to produce free-standing PANi-rGO thin films that exhibited thermoelectric property. The results indicated low loading of GO in the actual composite, which was likely due to low GO concentration in the electrolyte solution (possibly due to aggregation). This in turn resulted in low co-deposition in the PANi film during electropolymerization. The measured Seebeck coefficients were primarily affected by changes in the parameters of the delamination-reduction step, with higher values obtained when a fewer number of reduction cycles (50) and higher magnitude reduction voltage (±2 V) were used. Due to low rGO loading however, the effect of the rGO concentration on film properties was not observed. Higher power factors were obtained from experiments 5 and 8 at 0.022 and 0.025 $\mu Wm^{-1}K^{-2}$, respectively which were comparable with those of pure PANi using the optimized procedure (0.025 $\mu Wm^{-1}K^{-2}$). For experiment 8, which had improved conductivity, the PF was higher than most of the films. It is recommended that improving the stabilization of GO in the electrolyte solution, e.g., by using surfactants, could improve co-deposition and further increase the conductivities of the free-standing composite films following the fabrication protocol presented in this study.

Acknowledgements

The authors would like to acknowledge L C Macaraig, J L Pauco, R Bareo, and J D Vilela for assistance in the SEM and Raman analyses, and IIID58-CHED-PCARI and DOST-PCIEERD for funding support in the Printed Electronics and Nanotechnology Laboratory. EPE would also like to acknowledge the Ateneo Research Institute for Science and Engineering (ARISE) for the research load and Senator Gil J. Puyat Endowed Professorial Chair.

References

[1] Nandihalli N, Liu CJ and Mori T 2020 *Nano Energy* **78** 105186
[2] Soleimani Z, Zoras S, Ceranic B, Shahzad S and Cui Y 2020 *Sustain. Energy Technol. Assess.* **37** 100604
[3] Wang L, Zhang Z, Liu Y, Wang B, Fang L, Qiu J, Zhang K and Wang S 2018 *Nat. Commun.* **9** 1
[4] Beltrán-Pitarch B, Pardo-Gonjal J, Powerll A, Martínez-Julián F and García-Cañadas J 2019 *J. Phy. Chem.* **123** 12608-13
[5] R, Glaudell A, Urban J, Chabinyc M and Segalman R *Nat. Rev. Mater.* **1** 1-14
[6] Rowe D 2006 *Thermoelectrics Handbook: Macro to Nano* (USA: Taylor & Francis Group) pp (1-1)-(1-10)
[7] Yuan Z, Tang G, Yu Z and Qi J 2012 *Carbon* **50** 3064-73
[8] Cho C, Wallace K, Tzeng P, Hsu J, Yu C, and Grunlan J 2016 *Adv. Energy Mater.* **6** 1502168
[9] Zong P, Liang J, Zhang P, Wan C, Wang Y and Koumoto K 2020 *ACS Appl. Energy Mater.* **3** 2224-39
[10] Ube T, Koyanagi J, Kosaki T, Fujimoto K, Yokozeki T, Ishiguro T and Nishio K 2019 *J. Mater. Sci* **54** 3904-13
[11] Khalid M, Tumelero M, Zoldan V, Pla Cid C, Franceschini D, Timm R, Kubota L, Moshkalev S and Pasa A *RSC Adv.* **64** 34168-78
[12] Liu S, Li H, Li P, Liu Y and He C 2021 *CCS Chem.* **10** 2547-60
[13] Hummers W and Offeman R 1958 *J. Am. Chem. Soc.* **80** 1339
[14] Harima Y, Fukumoto S, Zhang L, Jiang X, Yano J, Inumaru K and Imae I 2015 *RSC Adv.* **5** 86855-86860
[15] Shao Y, Wang J, Engelhard M, Wang C and Lin Y 2010 *J. Mater. Chem.* **20** 743-48
[16] Wang X, Kholmanov I, Chou H and Ruoff R 2015 *ACS Nano* **9** 8737-43
[17] Kim H, Skuza J, Park Y, King G, Choi S and Nagavalli A 2012 System to measure thermal conductivity and seebeck coefficient for thermoelectrics
[18] Fomo G, Waryo T, Feleni U, Baker P and Iwuoha E 2019 Electrochemical polymerization *Functional Polymers (Polymers and Polymeric Composites: A Reference Series)* ed Mazumder M, Sheardown H et al. (Cham: Springer International Publishing) pp 1-28
[19] Shao Y, Wang J, Engelhard M, Wang C and Lin Y 2010 *J. Mater. Chem.* **20** 743-48
[20] Zhang Y, Liu J, Zhang Y, Liu J and Duan Y 2017 *RSC Adv.* **7** 54031-38
[21] Seung-Beom Y, Yoon E and Kim K 2011 *J. Power Sources* **195** 10791-977
[22] Vaqueiro P 2022 Synthesis and property measurements of thermoelectric materials *Inorganic Thermoelectric Materials: From Fundamental Concepts to Materials Design (Inroganic Materials Series* vol 9*)* ed Powell A (Cambridge: Royal Society of Chemistry) pp 1-52
[23] Martin J, Tritt T, Uher C 2010 *J. Appl. Phys.* **108** 121101
[24] Borup K, de Boor J, Wang H, Drymiotis F, Gascoin F, Shi X, Chen L, Fedorov M, Müller E, Iversen B et al. 2015 *Energy Environ. Sci.* **2** 423-35
[25] Jaafar W, Wan M, Snyder J and Min G 2013 *Rev. Sci. Instrum.* **84** 054903
[26] Da Rosa A and Ordóñez J 2022 Thermoelectricity *Fundamentals of Renewable Energy Processes (Fourth edition)* (Oxford: Academic Press) chapter 5 pp 187-247
[27] Shao Y, Wang J, Engelhard M, Wang C and Lin Y 2010 *J. Mater. Chem.* **20** 743-48
[28] Harfouche N, Gospodinova N, Nessark B and Perrin F 2017 *J. Electroanal. Chem.* **786** 135-44
[29] Shalini V, Navaneethan M, Harish S, Archana J, Ponnusamy S, Ikeda H and Hayakawa Y 2019 *Appl. Surf. Sci.* **493** 1350-60

IOP Conf. Series: Materials Science and Engineering **1318** (2024) 012041 doi:10.1088/1757-899X/1318/1/012041

Hydrothermal extraction of microcrystalline cellulose from post-consumer indigo denim fabrics and evaluation of its UV blocking property

F M Sarmiento[1], C E Aquino[2] and M L Balela[1]

[1]Sustainable Electronic Materials Group, Department of Mining, Metallurgical and Materials Engineering, University of the Philippines Diliman, Velasquez St. cor C.P. Garcia Ave., Quezon City 1101, Philippines
[2]Department of Engineering Science, University of the Philippines Los Baños, Pili Drive, Laguna 4031, Philippines

Email: faithlaurencesarmiento815@gmail.com

Abstract. Waste cotton clothes, including post-consumer denim fabrics, are abundant biopolymer sources due to their high cellulose content. These waste materials commonly end up in landfills, posing health and environmental concerns and losing valuable cellulosic materials. To address these, this study aims to extract and characterize microcrystalline cellulose (MCC) from post-consumer indigo denim fabrics using a one-step hydrothermal method and evaluate its UV-blocking properties. FTIR and TGA analysis of raw and bleached denim confirmed the presence and purity of cotton cellulose and the indigo dye. Extracted MCC yielded 70-84%, with a degree of polymerization (DP) of 281-299. XRD analysis of the MCC showed a cellulose Type I structure. FTIR suggests the removal of amorphous phases of cellulose, leaving the crystalline structure. SEM meanwhile revealed rod-shaped and rough-surfaced MCC particles with diameters of 10-20 μm and lengths of 45-60 μm. Both FTIR and SEM indicated the retention of indigo dye on MCC surface. Increased acid hydrolysis time led to smaller particle sizes and higher degree of crystallinity (CrI). Indigo PVA-MCC films showed good transparency and effective UV blocking. These results indicate the successful conversion of denim to MCC via the hydrothermal method and the stability of indigo dye in the cellulose matrix.

1. Introduction

Pollution due to discarded textile fabric waste like denim in landfills results in loss of potential cellulose source. In 2015, 66% of textile waste was disposed in landfills in the USA, the largest textile user in the world. On the other hand, 19% are combusted, while only 15% get recycled [1]. Landfilling causes tremendous material loss and harmful environmental effects, such as anaerobic soil degradation and greenhouse emissions. A significant amount of these textile wastes is cotton, containing up to 99% cellulose [2]. Microcrystalline cellulose (MCC) can be extracted from waste cotton and used as fillers to polymers to enhance their mechanical and thermal properties. Since denim contains 95-100% cotton [2], it can be engineered for upcycling into MCC. The most common extracting method for MCC is acid hydrolysis. Since cotton cellulose is highly crystalline, acids with stronger concentration and higher temperatures are commonly employed [3]. This makes the process expensive and hazardous. Meanwhile, the hydrothermal method is a medium-temperature, high-pressure, one-step process involving an enclosed autoclave reactor. This method utilizes significantly milder acid concentrations since the main driving force of the reaction is pressure [3]. On the other hand, polymers are widely used in numerous applications due to their low cost, workability, and corrosion resistance. However, they have relatively weak covalent bonding, making them susceptible to molecular chain scission and degradation due to UV. Due to this, UV shielding is necessary for polymers to be usable in environments with UV rays, such as PVC piping and protective equipment. Recent developments in UV-blocking involve micro and nanoparticles from zinc oxide (ZnO), titanium oxide (TiO_2), etc., but the main disadvantage is their incompatibility with organic polymeric materials. Thus, viable and green UV-blocking materials must be explored. This research aims to produce MCC from indigo denim waste through a one-step hydrothermal acid hydrolysis process and analyze its UV-blocking properties.

2. Methodology

Indigo denim was obtained from post-consumer garments, washed with water, and dried. Raw fiber analysis was conducted using a ThermoFisher Scientific Nicolet iS50 Fourier Transform-Infrared spectrometer (FTIR), TA Instruments Thermogravimetric Analyzer (TGA) Q500, and JSM-IT500HR InTouchScope™ Scanning Electron Microscope (SEM). Two sets of dried samples were processed for extraction, one unbleached and the other was bleached using technical grade sodium hypochlorite (NaClO) from Just-In-One-Marketing Laboratory Supplies with a 1:25 (g/mL) solid-to-liquid ratio at 70 °C for 45 min. For MCC extraction, 2 g of fabric was mixed with 60 mL of 1.2 M hydrochloric acid (37% HCl from RCL Labscan) at a 1:30 (g/mL) solid-to-liquid ratio. Then, the mixtures were heated at 150 °C in a Teflon-lined stainless steel autoclave reactor at different durations (80, 100, 120 min). The products were filtered using a 125 mm Grade 1 Whatman filter paper, then washed with distilled water until neutral pH. Finally, the residue was dried at 80 °C for 5 h and then manually ground into powders. MCC yield was calculated from the ratio of the dried weight of extracted MCC over its initial weight, reported as a percentage. The morphology of MCC was observed using an optical microscope and SEM. Then, the particle size was analyzed using ImageJ and Gaussian fitting. FTIR spectra were also obtained to analyze functional groups on the surface of the samples. The degree of polymerization (DP) was determined via CED (copper ethylenediamine) method, adapted from the US Pharmacopeial Convention (USP) [4]. The crystallinity index (CrI) was determined via XRD deconvolution using Shimadzu Maxima XRD7000 (Cu $K\alpha$), where X-ray diffractograms were fitted in a Lorentz function using (110), $(1\bar{1}0)$, (021), and (200) crystalline phase reflections of Type I cellulose [5]. The CrI was calculated from the ratio of the cumulative area of these crystalline regions over the area of the original diffractogram (sum of amorphous and crystalline). To analyze the UV-blocking of MCC, neat PVA and PVA-MCC composite films were prepared via solution casting. For the neat PVA, 5 wt%/v solution was prepared by dissolving 2 g PVA powders from Sigma-Aldrich (87-90% hydrolyzed, average mol wt 30,000-70,000) in 40 mL distilled water, stirred at 400 rpm (25 °C for the first 30 min, 90 °C for the succeeding 6 h). The same procedure was done for PVA-MCC, except with the varying amounts of PVA and MCC (5% and 10%). MCC was added after the 30 min wetting of PVA. Afterward, the solution was poured into a 10 x 10 cm silicon mold and cast at 25 °C for 12 h, followed by oven heating at 40 °C for 24 h. The films were conditioned at 20 °C for 12 h, and thickness was measured at ten different points using a digital caliper. UV-Vis analysis was conducted using Shimadzu UV-VIS Spectrophotometer (UV-2600i) from 200-800 nm to analyze the optical properties of the films.

3. Results and Discussions

3.1. Characterization of Raw Fiber

Figure 1 shows the FTIR spectra of raw and bleached denim fabric. The peak at 3270, 3330, 896 cm^{-1} represent the intermolecular stretching of H-bonds, intramolecular stretching of H-bonds in -OH, and β-glycosidic linkages between cellulose [3][6][7][8]. As observed, these peaks have no significant change after bleaching, indicating no considerable scission of cellulose monomers. The peaks at 2900, 1360, 1026-1100 cm^{-1}, which represent C-H symmetrical and asymmetrical stretching of $-CH_2$, C-H bending, and stretching absorptions of C-O-C also match with literature, indicating that the raw denim fabric is composed of cellulose cotton [7]. The absence of peaks at 1700-1740 and 1510-1590 cm^{-1} indicates the lack of hemicellulose and lignin, indicating that the fabric is composed of almost pure cellulose [6][9][10]. The peak at 1650 cm^{-1} in both spectra is due to the absorption of water molecules [2][3], attributed to the -OH groups in cellulose [11]. The bands in 1630, 1609, and 1482 cm^{-1} in the denim fabric could refer to the stretching vibration of C=O and C=C in the aromatic ring in indigo, and the C-H rocking vibration of indigo molecules [2][12]. This characteristic indigo peak, particularly the C=O, is not present in the bleached fabric due to the reduction of indigo into leuco-indigo, which is the absence of indigo color and conversion to the water-soluble, white to yellowish color of the dye. Aside from this, the FTIR graphs are almost identical, indicating no significant chemical change in the cellulose after bleaching.

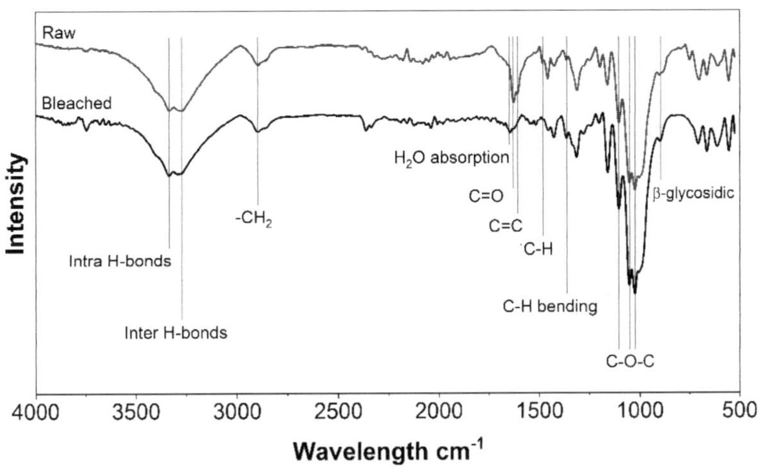

Figure 1. FTIR-ATR of raw and bleached denim.

Figure 2 shows the TGA curves of raw and bleached denim fabric. The initial mass loss at around 150 °C in both denim is due to the dehydration of absorbed water in the fibers. In comparison, the second stage at 250-500 °C represents the decomposition of the cotton cellulose due to the yielding of aliphatic char and volatile products [6]. The raw denim has an earlier onset of degradation temperature at 315 °C compared to the bleached denim with 329 °C, showing that raw denim exhibits less thermal stability at the beginning of degradation due to the presence of chemicals and indigo dyes that might have catalyzed its degradation [2]. These substances could have been removed during bleaching, resulting in the higher thermal stability. Moreover, the single decomposition observed in the TG plots of the two fibers exhibits the thermal behavior of pure cotton without any other fiber blends [3][7]. The weight loss in the 220-310 °C range attributed to hemicellulose and lignin, was also not observed [7]. This indicates negligible amount of hemicellulose, confirming the purity of cellulose in cotton fabrics as observed in the FTIR.

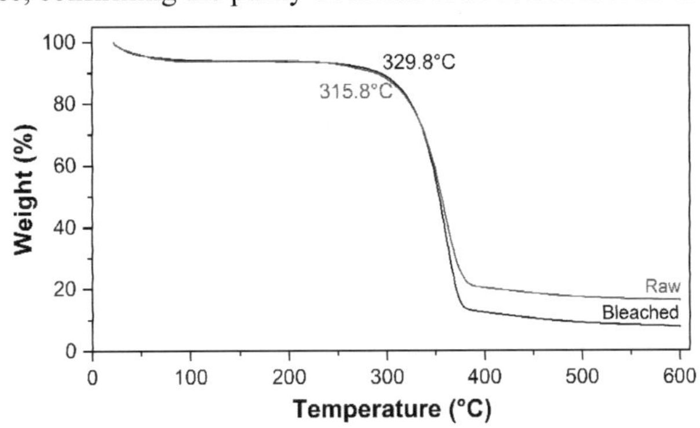

Figure 2. TGA of raw and bleached denim.

3.2. Extraction of MCC – Effect of Reaction Time

Figure 3 shows the FTIR spectra of MCC produced at different acid hydrolysis times. The spectra show the retention of the characteristic indigo peaks in the MCC with increasing reaction time, indicating that indigo dye is stable even in prolonged acidic, pressurized, and high-temperature environments. Moreover, the broadening of the -OH peak indicates the formation of more hydroxyl groups attributed to the cleaving of MCC molecules. However, the retention of the peak at 894 cm^{-1} attributed to the β-glycosidic bonds in MCC means that not all linkages are broken, suggesting the presence of oligomers [3][6][7]. The peak at 3330 cm^{-1} in raw denim due to the intramolecular stretching vibrations of H-bonds in -OH groups within the cellulose was retained in the extracted MCCs and became more apparent as the reaction time progressed [8]. Meanwhile, the 3270 cm^{-1} peak in raw denim representing the intermolecular stretching vibrations of H-bonds attraction between the cellulose molecules weakened [8] and merged with the broadening of -OH peak. This means that as reaction time proceeds, more hydroxyl groups are formed, and the H bonds between molecules weaken but not the internal H-bonds, implying the scission of molecules.

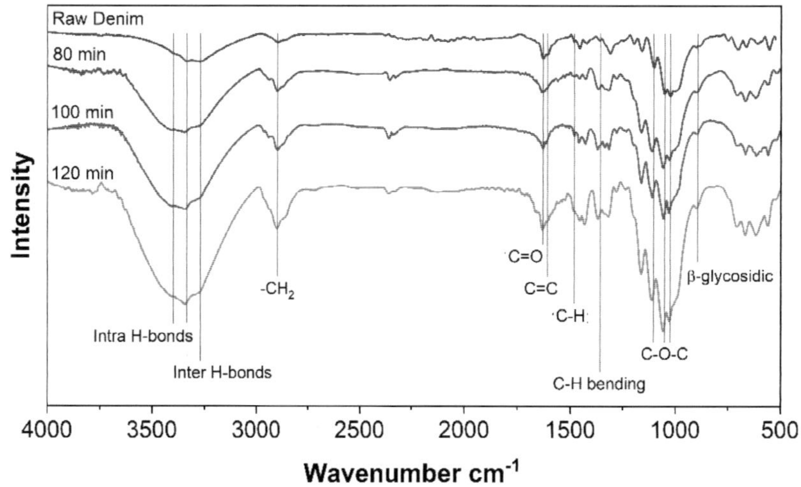

Figure 3. FT-IR spectra of raw denim cellulose, and MCC extracted at 80, 100, and 120 min.

The XRD patterns of extracted MCC at 80 and 120 min are shown in figure 4. The peaks at 2θ = 14.7°, 16.5°, 20° (shoulder), and 22.8° in both samples indicate that the formed MCC has a cellulose Type I monoclinic crystal structure [3][13]. The absence of a doublet peak in 2θ = 22.8° for both samples confirms that no other cellulose types are present. The calculated crystallinity index (CrI) shows that MCC extracted after 80 and 120 min has 48.5% and 52.9% CrI, respectively. This indicates that a longer reaction time hydrolyses more amorphous phases and leaves the crystalline parts intact, which agrees with the broken intermolecular H-bonds identified in the FTIR spectra.

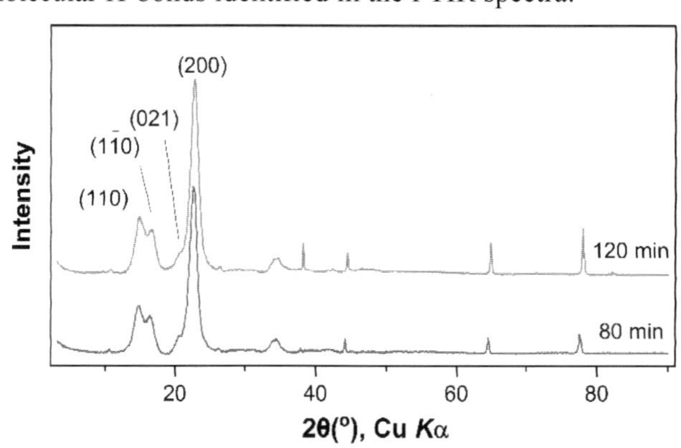

Figure 4. XRD patterns of MCC extracted at 80 and 120 min.

IOP Conf. Series: Materials Science and Engineering **1318** (2024) 012041 doi:10.1088/1757-899X/1318/1/012041

The broadening of -OH bonds and weakening of intermolecular H-bonds in FTIR and the calculated CrI from XRD can be related to the decrease in yield and degree of polymerization (DP) of MCC as reaction time increases (figure 5).

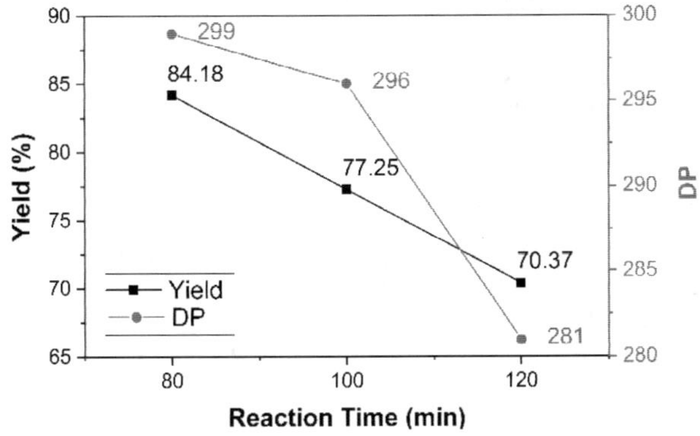

Figure 5. Effect of reaction time on the yield and degree of polymerization (DP) of MCC.

This is attributed to the hydrolysis of cellulose, which selectively breaks down the more exposed amorphous phases to retain the closely-packed crystalline phases. This reaction is illustrated in figure 6 where the presence of H^+ ions catalyzes the cleavage of β-glycosidic bonds of the amorphous phases in cellulose, forming glucose and other reducing sugars (oligosaccharides) [6]. Moreover, as seen in figure 5, there is a visible decrease in the DP. This is due to the fast removal of amorphous phases, which are readily hydrolyzable in acids, supporting the results of the XRD. However, only the initial rapid decrease of DP is observed and not the leveling of the degree of polymerization (LODP)[14]. Thus, the absence of LODP until 120 min of reaction time suggests that amorphous cellulose phases from denim are still not completely hydrolyzed.

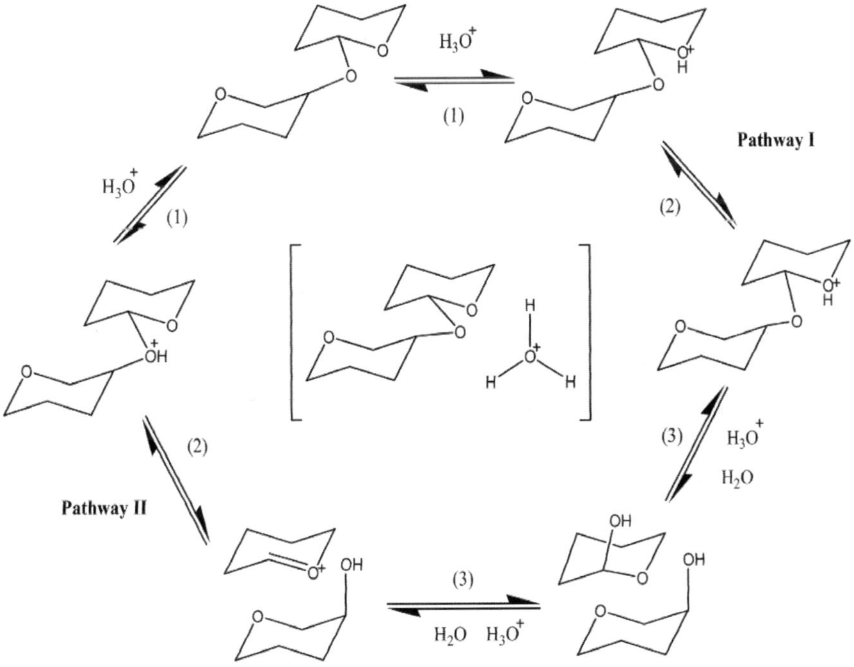

Figure 6. Acid hydrolysis mechanism of molecular scission of cellulose to microcrystalline cellulose [6].

The morphology of the extracted MCC can be observed in figure 7. After hydrothermal treatment, the smooth, elongated fibrillar structure of raw denim in figure 7 (a-b) was transformed into flaky, almost flattened, rod-like microparticles in figure 7 (c). The retention of blue indigo is also observed in figure 7 (d). Surface roughening and uneven, sharp ends are also observed from the start of the acid attack at

80 min. In addition, cracks at the surface of the particles are more apparent in the MCC extracted at 120 min, indicating sites of chain scission, as shown in figure 7 (f). At longer acid hydolysis time, the particle diameters of 10-20 μm were unchanged while particle length decreased from 45.5 μm in 80 min, 55.5 μm in 100 min, and 59.0 μm in 120 min due to increased molecular scission from prolonged acid attack.

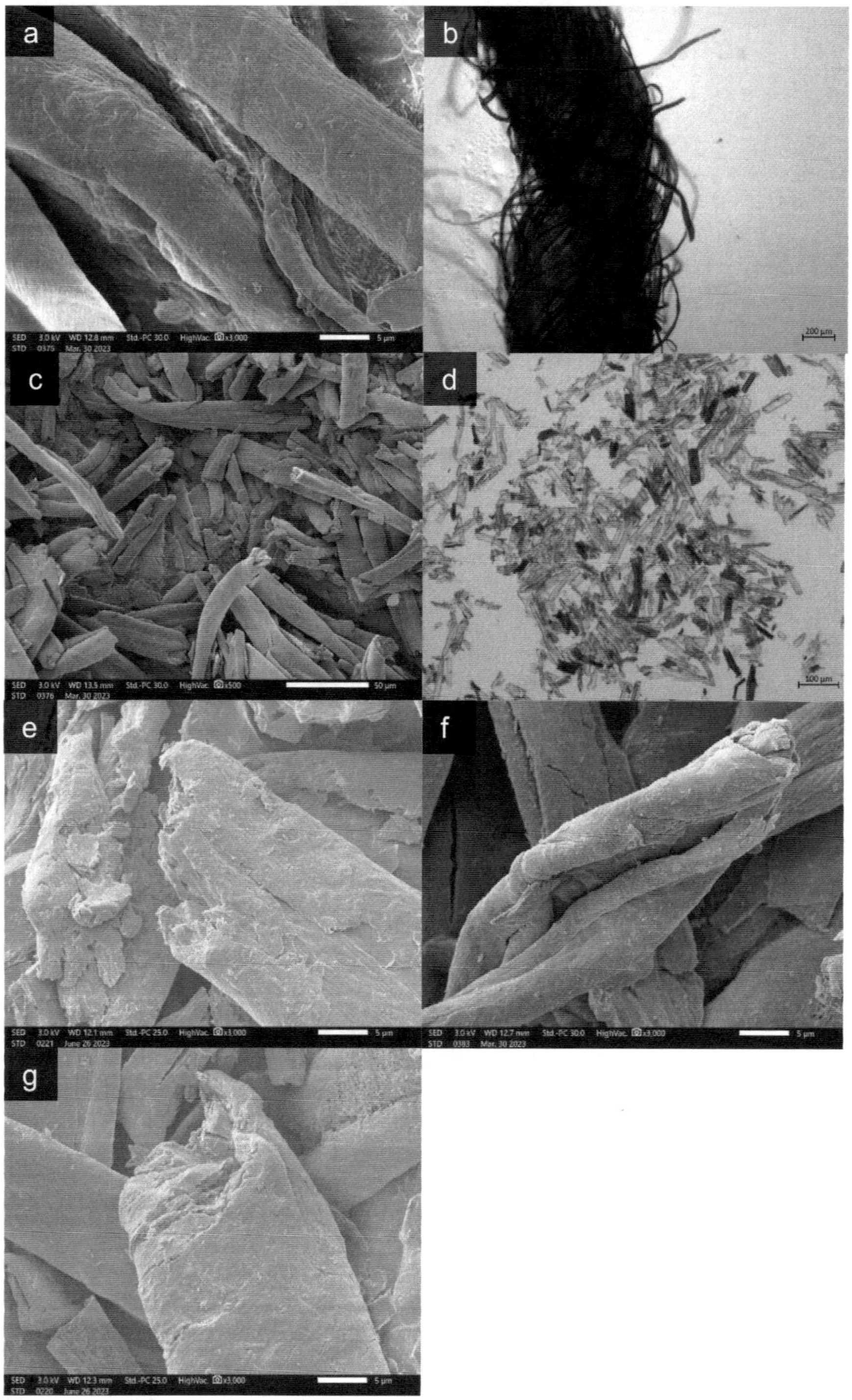

Figure 7. Raw denim under (a) SEM; (b) optical microscope; MCC from denim at 100 min under (c) SEM; (d) optical microscope. MCC from denim under SEM at (e) 80 min; (f) 100 min; (g) 120 min.

3.3. UV Blocking Property of PVA-MCC Films

Figure 8 shows the transmittance of neat PVA and PVA-MCC films. At 200-400 nm (UV Region), the film with bleached MCC blocked 27% more UV light than neat PVA, indicating that MCC has natural UV blocking properties. When indigo MCC was incorporated, a peak at 680 nm was identified, as shown in the dip in the visible light region. This peak is attributed to the absorbance of red, orange, and yellow light, a characteristic of indigo dye [2]. The 5% indigo MCC blocked 24% more UV light than bleached MCC. When the loading of indigo MCC was increased to 10%, the films blocked up to 97% UV light. This indicates that indigo dye, in addition to MCC, effectively blocks UV light. The resonance of the aromatic ring and the high photostability of indigo dye contributed to the overall light blocking of indigo-MCC [2].

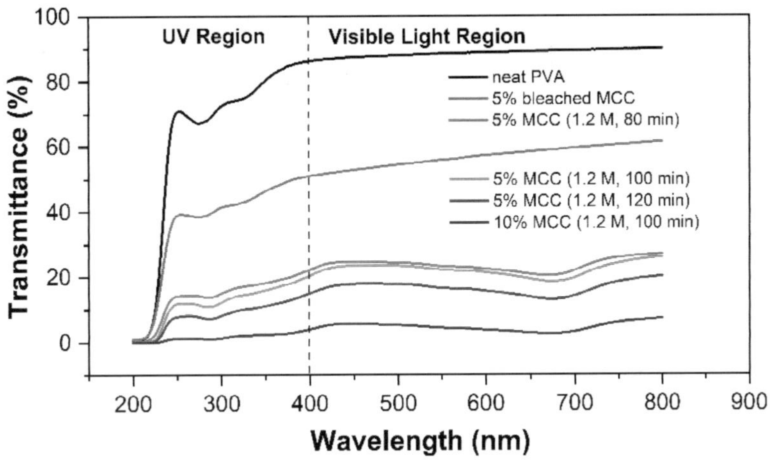

Figure 8. UV-Vis spectra of PVA-MCC films.

The thickness of the films (123 μm in neat PVA, 133-177 μm in 5% MCC, and 199 μm in 10% MCC) shows no trend with the factors varied in the MCC production. However, higher-loading MCC films have relatively higher thickness, which could mean a lower density of MCC than PVA [15]. As observed in figure 9, the films with MCC from raw denim produced uniformly blue-colored films, indicating homogeneous dispersion of MCC in the matrix. Although the transparency of the films decreases as MCC loading is increased, the background images are still visible, implying good transparency on all films. Due to its enhanced UV-blocking performance and transparency, indigo-MCC could be utilized for UV-C disinfection, piping, photochemical equipment, packaging, and manufacturing applications.

Figure 9. Photos of films with 5% MCC extracted at (a) 80 min; (b) 100 min; (c) 120 min. Films with (d) neat PVA; (e) 5% bleached MCC; (f) 10% indigo MCC.

IOP Conf. Series: Materials Science and Engineering **1318** (2024) 012041 doi:10.1088/1757-899X/1318/1/012041

4. Conclusions

MCC was successfully extracted from indigo denim by hydrothermal acid hydrolysis at varying reaction times. Indigo was found stable in the MCC even after the process.The amorphous cellulose phases were hydrolyzed, leaving a Type I crystalline cellulose phase. The elongated fibrillar structure of raw denim was transformed into rod-like microparticles with diameters of 10-20 μm and lengths of 45-60 μm. With increased reaction time, the MCC length decreases, and CrI increases. In all conditions, the DP of MCC was about 281-299, within the standard range of <350 set by USP. The LODP was not observed at the set reaction times, indicating that the reaction could still proceed beyond 120 min. Lastly, all PVA-MCC films showed good transparency. Films with bleached MCC could block UV light, indicating that MCC from cotton can effectively block UV. Indigo PVA-MCC was more effective than bleached PVA-MCC. Increasing the concentration of indigo MCC also blocked more UV light, reaching up to 97% blocking at 225 nm wavelength, indicating that indigo was an effective UV light blocker attributed to its photostability.

Acknowledgments

The authors would like to thank the Department of Science and Technology – Science Education Institute Undergraduate Scholarship and the Webfocus Thesis Grant under the UP Engineering Research and Development Foundation, Inc.

References

[1] Shirvanimoghaddam K, Motamed B, Ramakrishna S and Naebe M 2020 *Science of The Total Environment* **718** 137317
[2] Zhong T, Dhandapani R, Liang D, Wang J, Wolcott M P, Fossen V and Liu H 2020 *Carbohydr Polym.* **240** 116283
[3] Shi S, Zhang M, Ling C, Hou W and Yan Z 2018 *Waste Management* **82** 139–46
[4] United States Pharmacopoeia Convention 2018 [cited 2023 Apr 12] Microcrystalline Cellulose Monograph Available from: https://www.usp.org/harmonization-standards/pdg/excipients/microcrystalline-cellulose
[5] Park S, Baker J O, Himmel M E, Parilla P A and Johnson D K 2010 *Biotechnol Biofuels* **3**(1) 10
[6] Trache D, Hussin M H, Hui Chuin C T, Sabar S, Fazita M R N, Taiwo O F A, Hassan T M and Haafiz M K M 2016 *Int J Biol Macromol* **93** 789–804
[7] Portella E H, Romanzini D, Angrizani C C, Amico S C and Zattera A J 2016 *Materials Research* **19**(3) 542–7
[8] Leppänen K, Andersson S, Torkkeli M, Knaapila M, Kotelnikova N and Serimaa R 2009 *Cellulose* **16**(6) 999–1015
[9] Yulina R, Gustiani RrS, Kasipah C and Danny Sukardan M 2020 *E3S Web of Conferences* **148** 02004
[10] Chaiwutthinan P, Pimpan V, Chuayjuljit S and Leejarkpai T 2015 *J Polym Environ* **23**(1) 114–25
[11] Han S et al 2022 *Atmos Chem Phys* **22**(6) 3985–4004
[12] Chakraborty J N and Mazumdar P 2020 *AATCC Journal of Research* **7**(1_suppl) 1–8
[13] Naduparambath S and Purushothaman E 2016 *Cellulose* **23**(3) 1803–12
[14] Battista O A, Coppick S, Howsmon J A, Morehead F F and Sisson W A 1956 *Ind Eng Chem* **48**(2) 333–5
[15] Pan L, Li P and Tao Y 2020 *Materials* **13**(19) 4370

Preparation and Characterization of Water Hyacinth-Derived Cellulose Nanofibers for Fuel Cell Membrane application

Alyssa Marie U. Sansolis[1,2], Ronaldo P. Parreño Jr.[2], and Arnel B. Beltran[1,3*]
[1]Department of Chemical Engineering, De La Salle University, Manila, Philippines
[2]Industrial Technology Development Institute, Department of Science and Technology, Taguig City, Philippines
[3]Center for Engineering and Sustainable Development Research, De La Salle University, Manila, Philippines

*Corresponding author (arnel.beltran@dlsu.edu.ph)

Abstract. Cellulose nanofibers (CNF) derived from biomass materials were used for various applications such as medicine, packaging, wastewater treatment, and fuel cells. In this study, CNF was synthesized from water hyacinth plants obtained from Laguna de Bay. Collected water hyacinth stems were cleaned, washed, air-dried, and oven-dried for fiber extraction. Chemical and mechanical treatments were used to extract the CNF from the water hyacinth. The fibers were treated with 4 % potassium hydroxide solution and homogenized at 25,000 rpm for 3 mins. The resulting CNF was characterized using Scanning Electron Microscopy (SEM) to investigate the microstructures and morphology. Attenuated Total Reflectance-Fourier Transform Infrared Spectroscopy (AR-FTIR) was used to determine the chemical composition of the CNF. Lastly, crystallinity was examined by using an X-Ray Diffraction (XRD) to ensure that the extraction of CNF was successful. The resulting properties of CNF ensure its potential when used as a pure material or added as a composite for the membrane of fuel cells.

1. Introduction

Chemical energy is an alternative energy source that is being researched and developed due to its low carbon emissions as well as to promote green development and sustainability [1]. This form of energy is mostly used in a fuel cell device to convert it directly into electricity. Fuel cells are used for industrial buildings, trains, and electric vehicles. Ion-exchange membrane fuel cell (IEMFC) technology with polymer electrolyte membrane fuel cell (PEMFC) is the widely used fuel cell. A PEMFC is made up of two electrodes, an anode and a cathode[2]. A membrane is placed in between these electrodes to prevent the crossover of chemicals. [3]. Perfluorosulfonic acid (PFSA) polymer commonly called Du Pont's Nafion is the most popular kind of membrane which is the current standard used for PEMFC due to its good proton conductivity. However, due to its poor conductivity in very high temperatures it suffers from membrane swelling due and is very costly [4]. Due to these limitations, alternative polymer electrolyte membranes (PEM) are being studied that can be in comparison, are sustainable, and has a lesser cost than Nafion. In recent years, the use of alternative membranes for fuel cells has been developed an example of which are cellulose-based membranes. Researchers are developing this alternative membrane to lower the cost of membrane and to address sustainability by using biodegradable materials.

Cellulose-based membranes are prepared by the extraction of cellulose from plants and animals. Alkali or acid treatments are used to extract cellulose from these materials. The treatment breaks down the fiber from the source to convert it to CNF [5]. Cellulose nanofibers then undergo different functionalization techniques before to synthesis to cellulose-based membrane. The preferred material in the extraction of cellulose for cellulose-based membrane is one with low lignin and high cellulose content. This is to use less chemicals for a greener approach and avoid the delignification process [6]. Water Hyacinth is one example of a plant that can be used to extract cellulose due to its high cellulose content.

Eichhornia crassipes or commonly known as water hyacinth, is an aquatic plant that can be found in freshwater bodies. This plant has long stems, purple flowers, and wide leaves that grow in freshwater that prevents sunlight from reaching the water due to its structure. Due to this, water hyacinth is considered a pest in some countries due to its high growth rate [7]. Water hyacinths can be found in Pasig River and Laguna de Bay in the Philippines. When the water hyacinths grow exponentially, it disrupts the ecosystem and the fisherfolks that traverse into the water to make a living. Government agencies and non-government organizations are making a way to clear and remove these plants. However, these plants are just removed and dumped into landfills. Some homemakers dry the water hyacinth and use them in weaving handicrafts such as trays and baskets. Moreover, the high cellulose content of the water hyacinths can be extracted to make cellulose nanofibers (CNFs) for further synthesis into cellulose-based membrane. Water hyacinth contains 50.38% cellulose [3] and less than 10 % lignin [6]. Cellulose-based membranes can be used in different technologies such as water purification, gas purifications, moisture-electricity generation and humidity sensing, and for fuel cells [8] [9] [10].

In this paper, the extraction of cellulose nanofiber from water hyacinth and its characterization were discussed. The characterization of the raw material and the cellulose nanofiber were undertaken for the different properties by using Fourier Transformed Infrared Spectroscopy (FTIR) for its structural composition, Scanning Electron Microscope (SEM) for its morphology, and X-Ray Diffraction (XRD) for its crystallinity. This is to determine the functionalities of the CNF that will make it suitable as cellulose -based membrane for fuel cell application.

2. Methodology

2.1. Materials

Nine kilograms of Water Hyacinth was collected from Laguna de Bay in Brgy. Malaya, Pilillia Rizal last May 2023. Potassium Hydroxide (Analytical grade 85%) was purchased from Loba Chemie PVT. Ltd. and was used as received.

2.2. Synthesis of Cellulose Nanofiber

The collected stems were separated from the roots and flowers and the stems were washed with tap water to remove any dirt and impurities. The washed stems were cut into 2-inch lengths, air-dried for 48 hours, and oven-dried for at least 24 hours at 60°C or until the texture of the stems feels brittle to the touch. Wiley Hammer mill was used to reduce the particle size of the water hyacinth fibers. A 4 wt% potassium hydroxide (KOH) solution was prepared, and this was used as a soaking solution for the fibers. A 1: 33 water hyacinth fiber to KOH solution ratio was used for the alkali treatment. The water hyacinth fibers and 4% KOH solution were mixed for 4 hours at 90°C using a hotplate and stirring bar.

This solution was cooled at room temperature before mixing using a homogenizer for 3 minutes at 25000 rpm. The homogenized CNF solution was filtered using a Büchner funnel and was neutralized using deionized water until the filtrate reads at pH 7. After the neutralization of the CNF, this was dried at 60°C for at least 18 hours. After drying the extracted cellulose nanofibers were set aside for characterization.

2.3. Characterization of Cellulose Nanofiber

The synthesized cellulose nanofibers were characterized for their crystallinity, morphology, and microstructure, and structural composition.

2.3.1. Crystallinity

The crystallinity of the water hyacinth and water hyacinth CNF was investigated using a Shimadzu LabX XRD-6000. The sample was placed in a sample holder and scanned at a wavelength of 5-50 operating at 40kV and 30 mA.

2.3.2. Microstructure and Morphology

The microstructure and morphology of the water hyacinth CNF was analyzed using the Thermo Scientific Phenom XL SEM. 15 kV acceleration voltage and 6.8-6.9 mm working distance was used in the analysis of its morphology. Gold sputter was used to coat the sample to avoid the ion sputter coater to charge the sample **[6]**. The degree of crystallinity will be computed using the following equation.

$$Degree\ of\ Crystallinity = \frac{I_{002}}{I_{002}+I_{am}} x\ 100 \qquad [1]$$

I_{002} refers to the area under the peaks and I_{am} refers to the area under the amorphous regions [11].

2.3.3. Structural Composition

The structural composition of the water hyacinth and water hyacinth CNF was analyzed using a Shimadzu FTIR in attenuated total reflectance (ATR) mode. Background monitoring was performed at the beginning of the measurement and the samples were measured at a wavelength range of 40-4500 cm^{-1}.

3. Results and Discussion

3.1. Crystallinity

The crystallinity of the water hyacinth CNF was measured using XRD to investigate the effect of the alkaline treatment on the structure of CNF. Shown on Figure 1is the XRD of the water hyacinth fiber and the extracted Water Hyacinth CNF. There was an increase in crystallinity after the alkaline treatment of the Water Hyacinth fibers. The Water Hyacinth fibers has 32% degree of crystallinity while the Water Hyacinth CNF showed 54% crystallinity. The increase in crystallinity is due to the removal of hemicellulose and lignin in the sample [6]. The 54% crystallinity of the Water Hyacinth CNF is much lower than the crystallinity of Oil Palm empty fruit bunches at 72% [12]. The resulting WH showed the removal primarily of the hemicellulose and lignin contents and to a certain degree of the amorphous content that added to the high crystallinity of the extracted CNF.

IOP Conf. Series: Materials Science and Engineering **1318** (2024) 012042 doi:10.1088/1757-899X/1318/1/012042

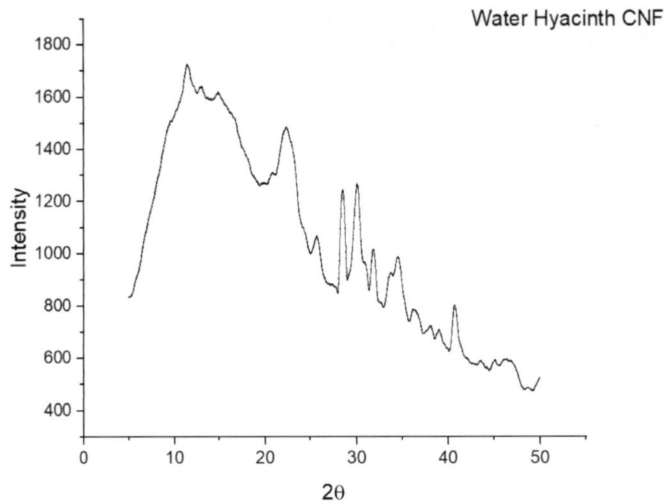

Figure 1. XRD of Water Hyacinth CNF

3.2. Microstructure and Morphology

Untreated Water Hyacinth fibers usually have long fibers that are porous in nature to make them float in water [11] [6]. The microstructure and morphology of the extracted Water Hyacinth CNF is shown in Figure 2. The CNF was analyzed using SEM using 1000x, 2500x, and 500x magnification with 80, 30, and 10 µm scale. The SEM showed that the alkali-treated Water Hyacinth CNF have long fibers with round-shaped, thick, and flat fibers. The SEM showed that the Water Hyacinth CNF are amorphous in nature due to the voids and spaces that are visible in the image. The diameter of the Water Hyacinth CNF was measured using Image J software and it ranges from 1.06 to 9 µm with an average of 3.97 µm. Only the diameter of the surface of the CNF was measure and not its cross-sectional area.

3.3. Structural Composition

The chemical structure of the Water Hyacinth CNF prepared by alkaline treatment is shown on Figure 3. The peak at 3312 cm^{-1} is from the OH stretching of the carboxylic acid and the 1622 cm^{-1} is from the C=O stretching of the cellulose in the Water Hyacinth. The peaks from 3312 to 3746 cm^{-1}, 1622 cm^{-1} and 1392 cm^{-1}. corresponds to the increase in cellulose content and removal of its lignin component [12]. The C=C bending and C-O stretching of Water Hyacinth CNF corresponds to the weak lignin and content and higher cellulose content of the CNF [6]. The peak at 2917 is associated with the C-H stretching of hemicellulose and cellulose which was increased due to the removal of lignin [13]. However, the peak at 702 cm^{-1} indicates that there is still a presence of an aromatic compound that is from the lignin found in the Water Hyacinth CNF.

IOP Conf. Series: Materials Science and Engineering **1318** (2024) 012042 doi:10.1088/1757-899X/1318/1/012042

Figure 2. SEM image of Water Hyacinth CNF under alkali treatment (a) 80 μm at 1000x magnification (b) 30 μm at 2500x magnification (c) 10 μm at 5000x magnification.

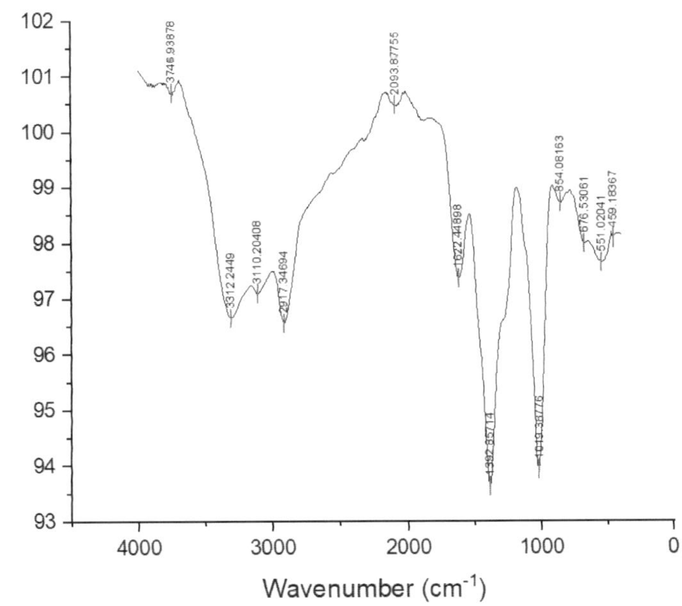

Figure 3. FTIR Spectra of Water Hyacinth CNF

IOP Conf. Series: Materials Science and Engineering **1318** (2024) 012042 doi:10.1088/1757-899X/1318/1/012042

4. Conclusion and Recommendations

The present study showed that CNF can be derived from Water Hyacinth fibers. In this study, the use of 4% (w/w) alkaline treatment was shown to be an effective way to extract the CNF. The CNF extracted had a 54% degree of crystallinity and showed the removal of lignin and the increase in cellulose content due to the C-H stretching shown by FTIR analysis. However, the CNF showed a large diameter range of 1.06 to 9 μm with an average of 3.97 μm with the presence of porous microstructure. It is recommended to explore alkaline treatment concentrations and longer homogenization time to achieve a smaller CNF diameter. It is also recommended to get the SEM image of the cross-sectional area of the CNF to get a more intensive view of the CNF. Furthermore, other characterizations such as contact angle, thermal stability, and tensile strength can be done to analyze the Water Hyacinth CNF. Overall, the extracted Water Hyacinth CNF can be used to develop a cellulose-based membrane for fuel cell applications.

References

[1] Q. Wang and Z. Fan, "Green finance and investment behavior of renewable energy enterprises: A case study of China," *International review of Financial analysis,* 2023.

[2] A. J. Samaniego and R. Espiritu, "Prospects on utilization of biopolymer materials for ion exchange membrane in fuel cells," *Green Chemistry Letters and Reviews,* 2022.

[3] A. Pambudi, A. Priyangga, D. Hartanto and L. Atmaja, "Fabrication and characterization of modified microcrystalline cellulose membrane as proton exchange membrane for direct metahnol fuel cells.," *Materials Today: Proceedings,* 2021.

[4] Q. Zhao, W. Y., C. Ni, L. Wang, B. Liu, J. Liu, M. Zhang, Y. Men, Z. Sun, H. Xie, W. Hu and Y. Lu, "Effect of aminated nanocrystal cellulose on proton conductivity and dimensional stability of proton exchange membranes," *Applied Surface Science,* pp. 691-702, 2019.

[5] H. Zhang, X. An, L. Liu, L. Z., H. Liu and Y. Ni, "Preparation of cellulose-based lithium ion battery membrane enhanced with alkali-treated polysulfonamide fibers and cellulose nanofibers," *Journal of Membrane Science,* vol. 591, 2019.

[6] S. Tanpichai, S. Biswas, S. Witayakran and H. Yano, "Water Hyacinth: A Sustainable Lignin-Poor Cellulose Source for the production of Cellulose Nanofibers," *ACS Sustainable Chemistry & Engineering ,* 2019.

[7] N. Rumjit, P. Thomas, C. Lai, Y. Wong, V. George, P. Basilraj and M. Johan, "Recent Advancements of Supercapacitor Electrode Materials Derived Fom Agriculture Waste Biomass," *Encyclopedia of Energy Storage,* vol. 4, pp. 382-397, 2022.

[8] S. Li, X. Wang, Y. Guo, J. Hu, S. Lin, Y. Tu, C. L., Y. Ni and L. Huang, "Recent advances on cellulose-based nanofiltration membranes and their applications in drinking water purification: A review," *Journal of Cleaner Production,* vol. 333, 2022.

[9] T. Araujo, A. Parnell and G. M. A. Bernardo, "Cellulose-based carbon membranes for gas separations - Unraveling structural parameters and surface chemistry for superior separation performance," *Carbon,* vol. 204, pp. 398-410, 2023.

[10] T. Chen, D. Zhang, T. X., S. Qiang, C. Sun, L. Dai, M. Zhang, Y. Ni and X. Jiang, "Highly ordered asymmetric cellulose-based honeycomb membrane for moisture-electricity generation and humidity sensing," *Carbohydrate Polymers,* vol. 294, 2022.

[11] T. Yasim Anuar, H. Ariffin and M. Hassan, "Characterization of cellulose nanofiber from oil palm mesocarp fiber produced by ultrasonication," *IOP Conference Series: Materials Science and Engineering,* 2018.

[12] S. Gea, A. Siregar, E. Zaidar, M. Harahap and D. P.-A. Indrawan, "Isolation and Characterisation of Cellulose Nanofibre and Lignin from Oil Palm Empty Fruit Bunches," *PubMed Central,* 2020.

[13] J. Singh, A. Dubey, Chaurasiam B and A. Daneite, "Biological Characterization and Instrumental Analytical Comparison of Two Biorefining Pretreatments for Water Hyacinth (Eichhnornia crassipes) Biomass Hydrolysis," *Sustainability,* 2020.

[14] S. a. T. A. Ventura-Cruz, "Extraction and characterization of cellulose nanofibers from Rose stems (Rosa spp.)," *Carbohydrate Polymers,* vol. 220, pp. 53-59, 2019.

[15] C. Hernandez, F. Ferreira and R. D.S, "X-ray powder diffraction and other analyses of cellulose nanocrystals obtained from corn straw by chemical treatments," *Carbohydrate Polymers,* pp. 39-44, 2018.

Synthesis of iron-loaded MFI-type (Fe/ZSM-5) zeolite using Philippine natural zeolites (PNZ) for potential adsorption and catalytic applications

Apraile Hope P. Dumrigue[1,4], Hannah Shamina O. Cosiñero[1,3,4], Don Nelson C. Potato[1,2,4] and Ramuel John I. Tamargo[1,2,4,*]

[1] Department of Chemical Engineering, College of Engineering, University of the Philippines Diliman, Quezon City, Philippines, 1101
[2] Environmental Engineering Program, College of Engineering, University of the Philippines, Diliman, Quezon City, Philippines, 1101
[3] Materials Science and Engineering Program, College of Science, University of the Philippines Diliman, Quezon City, Philippines, 1101
[4] AirSAVER DOST-NICER Center for Environmental Technology and Compliance, University of the Philippines Diliman, Diliman, Quezon City, Philippines, 1101
[*]ritamargo@up.edu.ph

Abstract. Philippine natural zeolites (PNZ) are locally abundant aluminosilicates used in different applications such as filtration and purification. However, the presence of impurities limits their usability. Structure modification of PNZ into a new framework was conducted in the study to address this issue. This study used PNZ as a precursor in the sustainable synthesis of iron-loaded MFI-type (Fe/ZSM-5) zeolite, a synthetic zeolite extensively studied for its excellent catalytic activity and high adsorption capacity. PNZ was depolymerized into active silica and alumina, then recrystallized via hydrothermal synthesis. Ammonium iron citrate (AIC) and TPAOH were used as the iron source and structure-directing agent, respectively. The disappearance and appearance of FTIR peaks in the depolymerized PNZ spectra deduced the successful destruction of the initial zeolite framework of the PNZ. The synthesized material was characterized using FTIR, XRD, SEM-EDS, TGA, and BET analyses. FTIR showed the crystallinity peak of MFI-type zeolite, while XRD confirmed its successful synthesis. SEM images show hierarchical aggregated particles suggesting the presence of intercrystalline and intracrystalline mesopores. The BET surface area of the synthesized zeolite increased by 168% compared to the raw PNZ, and TGA results show its high thermal stability. EDS analysis confirmed the presence of iron in the sample.

1. Introduction

Over the past decades, zeolites have become an area of interest for material research and development due to their unique properties such as high surface area, thermal stability, well-defined pores and structure, shape selectivity, hydrothermal stability, mobility of their cations to act as a catalyst, and their ability to tailor their properties such as wettability and auxiliary mesopore generation in the crystals [1-4]. The unique properties of zeolites make them suitable to perform functions in different processes, including catalysis, adsorption, and ion exchange [3,5–7]. The most common natural zeolite frameworks in mineral deposits used in commercial applications are mordenite, clinoptilolite, heulandite, chabazite, and erionite [8].

Interestingly, the Philippines has an enormous potential to produce natural zeolites due to its geographic location within the Pacific Ring of Fire. Mainly, natural zeolites exist in rocks, volcanoes, and mines in many parts of the world. In the Philippines, zeolites come from a mineral mined in West Central Luzon. Accordingly, the site of the bed of zeolite layers consists primarily of heulandite with mordenite, clinoptilolite, and minor quartz minerals [9]. Due to the properties and their potential applications, PNZ has gained more attention from the public, market, and research areas. In recent years, most PNZ-based researches had focused on improving its properties, such as alkali and acid pretreatment using sodium hydroxide and acetic acid. This effectively modified its structure for adsorption efficiency [10]. Cationic surfactant modification using cetyltrimethylammonium bromide (CTAB) enhanced its affinity towards anionic heavy metals [11]. Esterification functionalized PNZ with a hydrophobic surface [12]. Metal cation modification improved its electrochemical properties [13], thermal properties [14,15], and antibacterial properties [16,17]. Lastly, gamma-ray irradiation pretreatment increased its crystallinity [18]. These studies focusing on the improvement of the properties of PNZ have led to different applications, including removal of heavy metals in water such as copper (Cu^{2+}), nickel (Ni^{2+}), lead (Pb^{2+}), arsenic (As^{3+}), and zinc (Zn^{2+}), and as a catalyst for the reduction of air pollutants such as carbon monoxide (CO), hydrocarbon (HC), and nitrogen monoxide (NO) emissions.

However, with the ever-growing demands of PNZ in the research and the market, its potential has yet to be fully explored, and its applications are still minimal compared to the application range of synthetically manufactured zeolites. Modifying not only the surface structure of PNZ but also its entire crystal structure and framework will widen its scope of application and maximize its potential. The International Zeolite Association (IZA) has identified over 200 different zeolite frameworks. Each has a unique and well-defined pore structure, different Si/Al ratios, extremely complex framework topologies, and different physical and chemical properties, potentially leading to various applications [19,20,3,21,22]. Significant efforts have been made to pursue well-engineered zeolites with desired framework structures, such as synthetic zeolite types W, ZSM-5, MCM-22, and other frameworks [23,24]. Moreover, synthetic zeolites address the disadvantages of natural zeolites in terms of impurities, intercrystalline diffusion limitations, relatively smaller micropores, and other limitations [25]. Thus, exploring the feasibility of PNZ to be transformed into another framework, novel to PNZ's initial framework, makes it highly valuable in function.

Therefore, to utilize and explore the high potential of PNZ, this study aims to transform the locally sourced PNZ into a synthetic zeolite framework. Specifically, this study aims to synthesize hierarchically porous MFI-type (ZSM-5) zeolite loaded with iron for potential adsorption and possible catalytic applications. Also, the structural characterization data are discussed alongside the effects of the depolymerization technique on the PNZ material.

2. Methodology

The study used PNZ, a naturally mined aluminosilicate from Pangasinan, Philippines, as raw material. The initial composition of the reaction mixture was set to be 1 Al_2O_3: 50.4 SiO_2: 1.5 TPAOH: 5.2 Fe_2O_3: 1248 H_2O, and the synthesis was done in two steps, depolymerization and crystallization.

2.1. Depolymerization

In a PTFE-lined autoclave, the PNZ was mixed with 5.0 M NaOH solution (3 g PNZ, 10.6 mL of 5 M NaOH solution). The autoclave was heated for two (2) hours at 180°C. After heating, the sample was washed, filtered, and dried in an oven set at 100°C.

2.2. Crystallization

The depolymerized PNZ (DPNZ) was mixed with NaOH solution, colloidal silica, ammonium iron citrate (AIC), and tertapropylammonium hydroxide (TPAOH) as the mineralizing agent, silica source, iron source, and structure-directing agent, respectively. The mixture was stirred in a beaker with the following parameters: 400 rpm, 50°C, and 31 hours. Crystallization was done inside a PTFE-lined autoclave heated to 180°C for 72 hours. The synthesized sample was washed, filtered, and dried, ready for calcination inside an air-flowing furnace at 550°C for 6 hours.

2.3. Characterization
The PNZ, DPNZ, and crystallized sample were subjected to Fourier Transform Infrared Spectroscopy (FTIR) (Instrument: Thermo Nicolet 6700) to account for changes in the functional groups of the material. The crystalline framework was identified using X-ray Diffraction Spectroscopy (XRD) (Instrument: Lab X XRD-6000 by Shimadzu). The morphology and composition of the samples were observed using Scanning Electron Microscopy (SEM) coupled with Energy Dispersive X-ray Spectroscopy (EDS) (Instrument: Hitachi S-3400N Type II). The thermal and textural properties of the PNZ and the synthesized zeolite were examined using thermogravimetric analysis (TGA) (Instrument: TGA Q500 V20.13 Build 39) and Brunauer-Emmett-Teller (BET) analysis (Instrument: Quantachrome Instruments NOVA 2200e Surface Area and Pore Analyzer).

3. Results and Discussion

3.1. Depolymerization of Raw Material
The effect of the depolymerization process was investigated by comparing the FTIR spectra of PNZ and DPNZ shown in figures 1a and 1b. There are disappearances of peaks for the free and bound waters initially found in the PNZ spectrum at bands around 1630 cm^{-1} and 3620 cm^{-1} followed by a broad band at 3380 cm^{-1} in the DPNZ spectrum [26]. Removal of the bound water or crystallization water can be achieved by heating at the expense of losing the crystal structure of the sample. This implies the destruction of the initial zeolitic framework of the PNZ, which can be further supported by the disappearance of the peak of PNZ at 510 cm^{-1} in the spectrum of the DPNZ. Rohayati et al. reported that bands between 500 cm^{-1} and 650 cm^{-1} are considered the zeolite framework's fingerprint [27]. Additionally, a decrease in the vibrational intensity of the 1010 cm^{-1} band, assigned to the asymmetric stretching of the TO_4 tetrahedra, and a slight shifting to a lower wavenumber of 965 cm^{-1} in the DPNZ can be observed. The shifting of the band may have resulted from the collapse of the aluminum matrix [25], while the decrease in intensity might be due to the solubility of silica to basic solutions, which

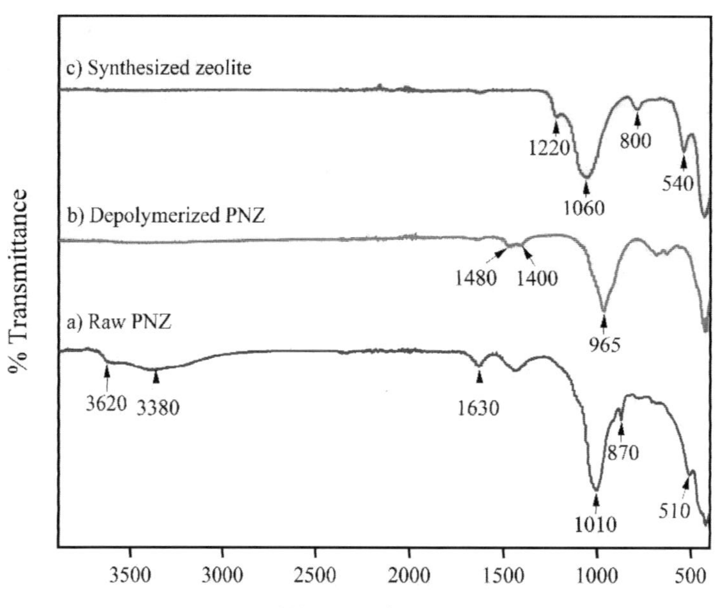

Figure 1. FTIR spectra of (a) raw PNZ, (b) DPNZ and (c) synthesized zeolite

caused the desilication of the depolymerized samples, thus decreasing the amount of silica. Peaks similar to $NaSiO_2$ and $NaAlO_3$, the starting materials in zeolite synthesis at 1400 cm^{-1} and 1480 cm^{-1}, appeared in the DPNZ spectrum. All these confirm the successful depolymerization of the original zeolite framework and the formation of aluminate and silicate anions [27]. Lastly, the peak at 870 cm^{-1}, which originated from the PNZ's impurity, probably calcite, disappeared after depolymerization [28]. This may imply that the process destroyed the initial structure of PNZ and caused impurities to be removed.

3.2. Crystallization

Figure 1c shows the FTIR spectrum of the synthesized zeolite sample. The characteristic absorption bands for the synthesized zeolite were found at around 1220 cm^{-1}, 1060 cm^{-1}, 800 cm^{-1}, and 540 cm^{-1}. The bands at 1060 cm^{-1} and 800 cm^{-1} are due to the internal asymmetric stretching vibration and external symmetric stretching of Si-O-T linkages, respectively [29]. The shift to a higher wavenumber from 1010 cm^{-1} for PNZ to 1060 cm^{-1} for the synthesized sample may be due to dealumination or the process of releasing aluminum in the zeolite structure to become aluminum outside the structure [30]. This might result in a lower framework aluminum, thus producing a zeolite with a higher Si/Al ratio compared to the initial value of PNZ. The absorption band at 1220 cm^{-1} and 540 cm^{-1} appeared due to the presence of double five-membered ring systems in the framework, suggesting that the synthesized sample belongs to zeolites with MFI type framework [29].

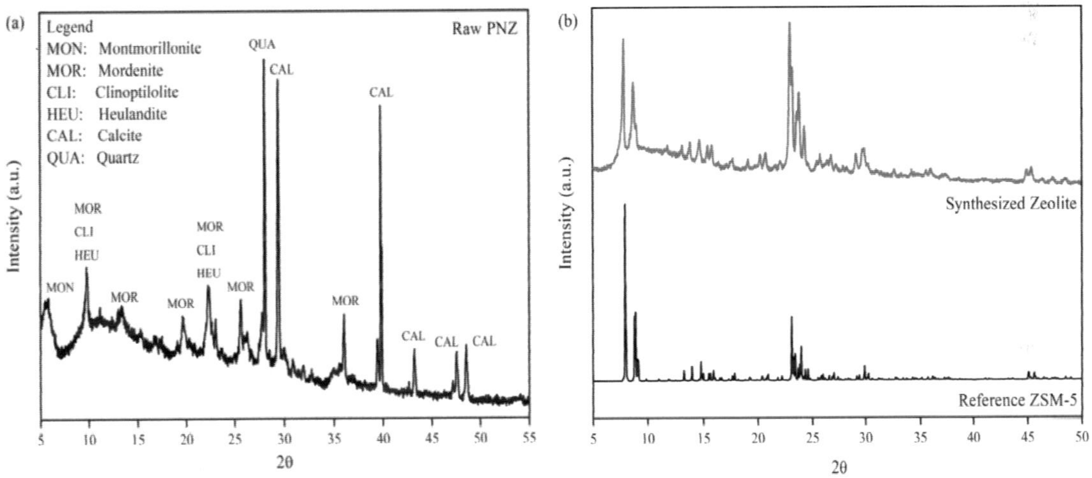

Figure 2. XRD pattern of the (a) raw PNZ and (b) synthesized zeolite material superimposed with the reference XRD pattern for ZSM-5 [31]

X-ray diffraction (XRD) analysis was used to confirm and identify the structure of the PNZ and the synthesized material in this study, as shown in figure 2. High-intensity diffraction peaks for minerals such as quartz, calcite, and montmorillonite; and peaks that correspond to zeolite frameworks such as mordenite, clinoptilolite, and heulandite were observed in the raw PNZ, as shown in figure 2a. There is evidence of amorphous characteristics in the raw material as seen in $2\theta = 5°- 40°$. Meanwhile, as shown in figure 2b, the synthesized sample shows similar peaks and d-spacing to a reference ZSM-5 pattern [31]. At the synthesis parameters used in this study, crystallization duration of 72 h and crystallization temperature of 180°C, there is a visible amorphous region between $2\theta = 5°- 15°$. A previous related study by Nada and coworkers in 2017 used sodium aluminate and silica sol as precursors and examined the effect of crystallization temperature and crystallization time on the crystallinity of synthesized ZSM-5 zeolites. In their study, it was observed that crystallization temperatures below 165°C show evidence of amorphous material. Meanwhile, an increase in crystallization time from 14 h to 24 h shows a positive effect on the crystallinity of the samples [32]. Therefore, studying the effect of crystallization time and temperature using PNZ as a precursor is essential. The absence of peaks other than those for ZSM-5 shows the high purity of the synthesized sample.

3.3. Composition of the Synthesized Zeolite

Elemental composition analysis of the synthesized sample was done using EDS analysis. Table 1 confirms the successful incorporation of iron elements in the synthesized ZSM-5 particles. Iron-loaded ZSM-5 zeolites (Fe/ZSM-5) have potential applications in different photocatalytic degradation of VOCs due to the added ability to undergo Fenton reactions [33,34]. Furthermore, the synthesized zeolite material has a high Si/Al ratio of 20.37. This ratio can significantly affect the adsorption property, catalytic activity, and lifetime of ZSM-5 zeolites because it influences the zeolite's acidic properties and hydrophobicity. An increase in the Si/Al ratio leads to a decrease in the aluminum content and thus, a decrease in the acidity of zeolites. In general, more and stronger acid sites increase the catalytic activity of zeolites. However, these acid sites also produce high amounts of coke, which can cause the fast deactivation of the catalyst. Therefore, it is critical to find the optimum Si/Al ratio that would give the highest catalytic activity and provide the longest lifespan for the zeolite [35]. The Si/Al ratio of ZSM-5 zeolite also significantly influences its hydrophobicity. Higher Si/Al ratios are associated with increased hydrophobicity in ZSM-5, suitable for organic reaction applications [36].

Table 1. Composition of the synthesized material

Element	% weight
Sodium (Na)	4.92
Silicon (Si)	85.47
Aluminum (Al)	4.03
Iron (Fe)	5.57

3.4. Morphology and Textural Properties

SEM was used to gain insights on the surface morphology of the PNZ and the synthesized zeolite sample. PNZ exhibits a large, irregularly shaped crystal with rough surfaces, as shown in figure 3a. Crystallization via the hydrothermal synthesis method transformed the PNZ into a material with aggregated spheroidal particles that are narrowly distributed between 3 and 7 μm and are formed from the stacking of plate-like nanocrystals, as seen in figure 3b. A similar observation was reported in past studies, which described that the synthesized particles are hierarchical in structure, suggesting the presence of intercrystallite mesopores [33,37]. The existence of mesopores increases the total surface area and provides additional sorption sites giving hierarchical ZSM-5 zeolites their excellent sorption capacities over the conventional ZSM-5 zeolites [38].

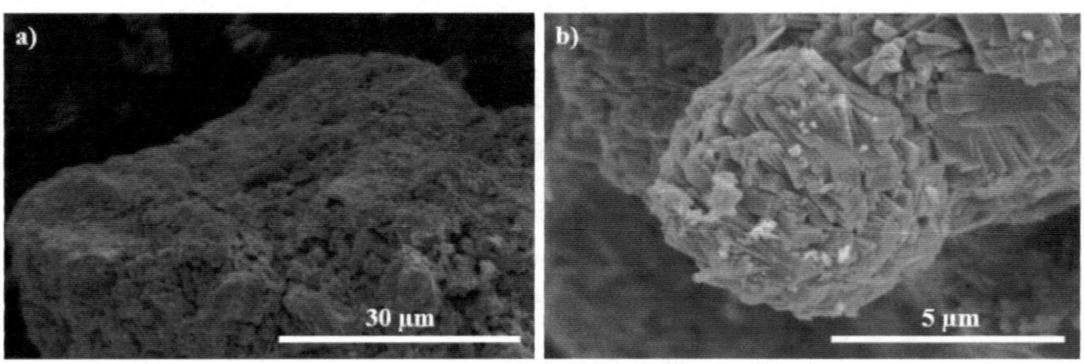

Figure 3. SEM micrographs of (a) raw PNZ at 1,900X magnification and (b) synthesized zeolite sample at 10,000X magnification

IOP Conf. Series: Materials Science and Engineering **1318** (2024) 012043 doi:10.1088/1757-899X/1318/1/012043

The BET surface area of the raw material is 90.894 m^2/g, and the synthesized material is 243.67 m^2/g, listed in table 2. This value is close to the obtained surface area with a value of 250 m^2/g by Aziz et al. in their synthesized Fe/ZSM-5 zeolites using pure chemicals as precursors [33]. A previous study by Yaripour and company on the synthesis of ZSM-5 via hydrothermal method using TPAOH as a structure-directing agent produced a ZSM-5 sample with a BET surface area of 369.4 m^2/g, [39]. In this study, the surface area is lower by 34%, and a probable cause of this decrease is the presence of Fe particles on the surface that may have clogged the pores of the synthesized sample. Additionally, a study by Shen in 2017 observed the effects of TPAOH on the surface properties of ZSM-5 synthesized samples. Specifically, their study showed that the surface area has a negative correspondence with the TPA^+/OH^- molar ratio [40]. In this study, only one TPA^+/OH^- molar ratio was observed, and potential future studies to increase the surface area can be done by optimizing TPAOH molar ratio with the crystallization parameters.

Table 2. Surface area of the raw material and the synthesized sample calculated using the BET analysis method

Material	BET Surface Area (m^2/g)
Raw PNZ	90.894
Synthesized zeolite	243.67

3.5. Thermal Stability

A possible application for the synthesized zeolite is its use as an adsorbent and catalyst inside high-temperature equipment and exhaust systems, such as plastic recycling industries. The thermal property of the raw PNZ and synthesized zeolite was observed using thermogravimetric analysis by heating in the presence of air up to 900°C, as shown in figure 4. The raw PNZ has lost 3.5% of the initial weight at 160°C and 5% at 700°C. Meanwhile, the synthesized zeolite lost 4.5% of the initial weight at 300°C and the weight was stable from 300°C to 900°C. The results show high thermal stability of the synthesized Fe/ZSM-5 zeolite prepared with TPAOH, heated at 180°C for 72 h, and calcined at 550°C for 6 h, relevant for high-temperature applications.

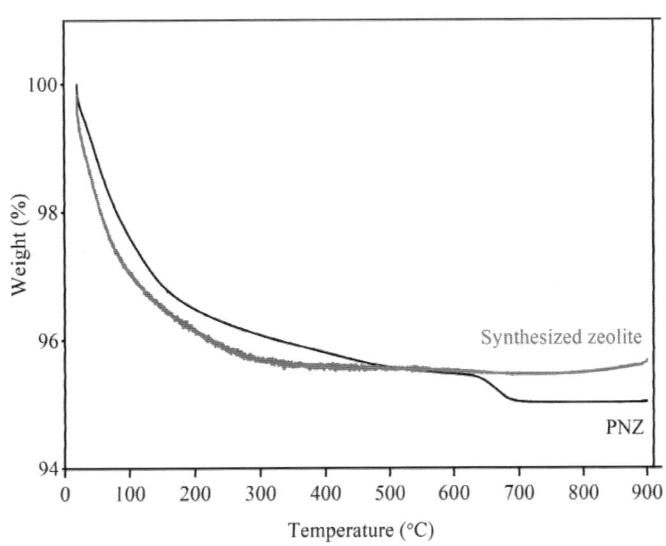

Figure 4. Thermogravimetric plot of the raw material and the synthesized sample shows thermal stability of up to 900°C.

4. Conclusion

MFI-type zeolite loaded with iron metal (Fe/ZSM-5) from PNZ via the hydrothermal synthesis method has been successfully carried out as supported by the characterization studies on the synthesized sample. The depolymerization process destroyed the initial structure of the PNZ thus, creating active silica and alumina species for zeolite synthesis. Recrystallization of the obtained precursors into the MFI framework significantly improved the properties of the PNZ. SEM images showed the transformation of the irregularly shaped PNZ crystals into aggregated spherical crystals, which are hierarchical. This significantly increased the BET surface area of the synthesized material by 168%, from 90.894 m^2/g to 243.67 m^2/g. This implies the potential of the synthesized material in different adsorption applications. It also has high thermal stability with no observable degradation up to 900°C, making it suitable for applications that require high-temperature operating conditions. The composition of the synthesized product indicates a zeolite with a high Si/Al ratio, which suggests the hydrophobicity of the product, making it suitable for reactions involving organics. Lastly, the successful incorporation of iron in the synthesized zeolite sample widened its scope of application as it may impart photocatalytic degradation properties to the synthesized zeolite sample.

Acknowledgments

This work is under the AirSAVER Project (Project no. 10107) of the Center for Environmental Technologies and Compliance funded by the Department of Science and Technology-Grants-In-Aid (DOST-GIA) through the Niche Centers in the Regions for R&D (NICER) Program and monitored by the Department of Science and Technology - Philippine Council for Industry, Energy and Emerging Technology Research and Development (DOST-PCIEERD). The authors also acknowledge the Office of the Chancellor of the University of the Philippines Diliman, through the Office of the Vice Chancellor for Research and Development, for the funding support through the Source of Solutions Grant. Lastly, the authors also express their sincerest gratitude to the Department of Science and Technology—Science Education Institute (DOST-SEI) through the Engineering Research and Development for Technology (ERDT) for additional funding support.

References

[1] Kianfar E 2020 *Zeolites: Advances in Research and Applications* (New York: Nova Science Publishers, Inc.)
[2] Kouser S, Hezam A, Khadri M and Khanum S 2022 *J. Porous Mater.* **29** 683-81
[3] Mgbemere H, Ekpe I and Lawal G 2017 *Int. Res. J. Environmental Sci.* **6** 45-59
[4] Veerapandian S, De Geyter N, Giraudon J-M, Lamonier J-F and Morent R 2019 *Catalyst* **9** 98
[5] Medina M, Platero-Prats A, Snejko N, Rojas A, Monge A, Gándara F, Gutiérrez-Puebla E and Camblor M 2011 *Adv. Mater.* **23** 5283-92
[6] Rhodes C 2010 *Science Progress* **93** 223-84
[7] Tosheva L, Garbev K, Miller G and Mihailova B 2023 *Crystal Growth & Design* **23** 3834-44
[8] Marantos I, Christidis G and Ulmanu M 2011 *Handbook of Natural Zeolites* (Bentham Science Publishers Ltd.) pp. 19-36
[9] Gili M, Olegario-Sanchez L and Conato M 2019 *Int. Symposium on Frontier of Applied Physics (1-2 November 2018, Indonesia)* vol 1191 (IOP Publishing Ltd) 012042
[10] Dignos E, Gabejan K, Olegario-Sanchez E and Mendoza H 2019 *The 2nd Mineral Processing and Technology Int. Conf. (1 November 2018, Indonesia)* vol 478 (IOP Publishing Ltd) 012030
[11] Barola C, Dusaban I, Olegario-Sanchez E and Mendoza H 2019 *The 2nd Mineral Processing and Technology Int. Conf. (1 November 2018, Indonesia)* vol 478 (IOP Publishing Ltd) 012039
[12] Osonio A and Olegario-Sanchez E 2017 *Proc. of the 3rd Advanced Materials Conf. 2016 (28-29 November, 2016, Malaysia)* vol 1901 (AIP Publishing) 080003
[13] Olegario E, Pelicano M, Cosiñero H, Sayson L, Chanlek N, Nakajima H and Santos G 2021 *Materials Letters* **294** 129799

[14] Guerrero L, Mendoza J, Ong K, Olegario Sanchez E and Ferrer E 2019 *Arab. J. Sci. Eng.* **44** 5581-8
[15] Olegario E, Pelicano M, Felizco J and Mendoza H 2019 *Mater. Res. Express* **6** 085204
[16] Olegario-Sanchez E, Tan M, Mendoza H and Balela M 2017 *Materials Science Forum* **890** 150-4
[17] Taaca K, Olegario M and Vasquez M 2017 *Proc. of the 3rd Advanced Materials Conf. 2016 (28-29 November, 2016, Malaysia)* vol 1901 (AIP Publishing) 030015
[18] Gili M, Pares F, Nery A, Guillermo N, Marquez E and Olegario E 2020 *Mater. Res. Express* **6** 125552
[19] Kerstens D, Smeyers B, Van Waeyenberg J, Zhang Q, Yu J and Sels B 2020 *Adv. Mater.* **44** 2004690
[20] Chen L-H, Li X-Y, Rooke J, Zhang Y-H, Yang X-Y, Tang Y, Xiao F-S and Su B-L 2012 *J. Mater. Chem.* **22** 17381-403
[21] Nabavi M, Zhou M, Mouzon J, Grahn M and Hedlund J 2019 *Microporous and Mesoporous Materials* **278** 167-74
[22] Narayanan S, Tamizhdurai P, Mangesh V, Ragupathi C, Krishnan S, Ramesh A 2020 *RSC Adv.* **11** 250–67
[23] Petrov I and Michalev T 2012 Synthesis of zeolite A: a review **51** 30-5
[24] Tang L *et al.*, 2008 *Nature Mater* **7** 381-5
[25] Liu H, Shen T, Wang W, Li T, Yue Y and Bao X 2015 *Applied Clay Science* **115** 201-11
[26] Reddy R, Kaneko S, Endo T and Reddy L 2017 *J. Laser Opt. Photonics* **04** 1000171
[27] Rohayati, Krisnandi Y and Sihombing R 2017 *Proc. of the 2nd Int. Symp. on Current Progress in Mathematics and Sciences 2016 (1–2 November 2016, Indonesia)* vol 1862 (American Institute of Physics) 030094
[28] Hajji S, Turki T, Boubakri A, Ben Amor M and Mzoughi N 2017 *Desalination and Water Treatment* **83** 222-33
[29] Dey K, Ghosh S and Naskar M 2013 *Ceramics International* **39** 2153-7
[30] Mazlan, Lestari A, Istiqomah A, Krisdiyanto D, Pramono E and Afrianto A 2020 *4th Int Conf. on Advanced Material for Better Future 2019 (7-8 October 2019, Indonesia)* vol 858 (IOP Publishing Ltd) 012015
[31] Baerlocher Ch and McCusker L 2001 *Collection of Simulated XRD Powder Patterns for Zeolites* (Amsterdam, London, New York, Oxford, Paris, Shannon, Tokyo: Elsevier)
[32] Nada M and Larsen S 2017 *Microporous and Mesoporous Materials* **239** 444-52
[33] Aziz A, Kim S and Kim K 2016 *Journal of Environmental Chemical Engineering* **4** 3033-40
[34] Bokare A and Choi W 2014 *Journal of Hazardous Materials* **275** 121-35
[35] Sharifi K, Halladj R and Royaee S 2020 *Rev. Adv. Mater. Sci.* **59** 188-206
[36] Al-Jubouri S 2020 *Microporous and Mesoporous Materials* **303** 110296
[37] Yue Y, Liu H, Yuan P, Li T, Yu C, Bi H, and Bao X 2014 *Journal of Catalysis* **319** 200-10
[38] Tang Q, Deng W, Chen D, Liu D and Guo L 2021 *Dalton Trans.* **50** 16694-702
[39] Yaripour F, Shariatinia Z, Sahebdelfar S and Irandoukht A 2015 *Journal of Natural Gas Science and Engineering* **22** 260-9
[40] Shen Q, Zhang L, Wu M, He C, Wei W, Sun N and Sun Y 2017 *J. Porous Mater.* **24** 759-67

IOP Conf. Series: Materials Science and Engineering **1318** (2024) 012044 doi:10.1088/1757-899X/1318/1/012044

Titanium Dioxide-Decorated and Silanized Electrospun Recycled Polyethylene Terephthalate (PET) Fibers for Filtration

L Macaraig*, J Zingapan, K V N Alvarez, and E P Enriquez

Department of Chemistry and Program in Materials Science and Engineering, School of Science and Engineering, Ateneo de Manila University, Loyola Heights, Quezon City, Philippines

lmacaraig@ateneo.edu

Abstract. Electrospinning methods have produced polymer-based non-woven fibers that have excellent filtration for use as air filter media, however most work have been focused on producing thinner fibers by reducing polymer loading. In this work, polyethylene terephthalate (PET) plastic from used soda plastic bottles was electrospun and decorated with nano-TiO_2 to produce nonwoven media with potential antimicrobial properties. Electrospinning parameters were optimized to maintain higher loading percentage of PET to as much as 30% (w/v) with resulting fibers having an average diameter of 1.8 μm. Electrospun fiber mats were then treated with a silane-based binder for surface-decoration with nano-TiO_2 coating to impart antimicrobial properties to these fibers. XRF mapping showed uniform coating of the nanoparticles. Atomic ratios from XRF of Ti to Si did not significantly change before and after ultrasonication washing signifying the covalent coating of the nanoparticles onto the fiber matrix.

1. Introduction

Through the years, several polymer spinning methods have been tested to produce polymer-based non-woven fibers that can have great filtration properties and at the same time produce filters that would experience lower pressure drops making them suitable for air circulators or air-conditioning units. One of those methods is electrospinning. This method has been utilized because of its versatile and straightforward technique in creating nanofibers as it involves the utilization of an electrical field to release charged threads of polymer solution that forms into non-woven fibers upon contact with a conductive collector [1]. Electrospinning is most suitable in creating the polymer-based non-woven fibers for air filters because the fiber diameter can be manipulated in the sub-micron range [2–4]. Furthermore, the morphology can be highly porous, and the fiber is charged, which are both excellent properties for air filters for particulates [3,5,6]. Among the polymers suitable for electrospinning is polyethylene terephthalate (PET) and its resulting fiber mat has been utilized especially for filtration. One notable work is that of Strain et. al. [7] wherein they used electrospun fibers from recycled PET bottles for smoke filters.

One of the limitations of typical air filters is they only entrap microbial particulates but lack the ability to deactivate them. One advantage of using polymer-based non-woven fibers is that they can be integrated with metal oxide nanoparticles to give them antimicrobial properties [5,8]. One of the commonly used metal oxides for this is nano-TiO_2. Several studies have already shown and proven that introducing TiO_2 to materials can help deactivate viruses by photocatalysis with UV [9–11]. Several studies have already shown that silane precursors such as tetraethyl orthosilicate (TEOS) can be used as a binder between polymeric materials and nano-TiO_2 [12,13]. Veronovski et al. [14] have facilitated hydrolysis of tetraethyl orthosilicate (TEOS) to act as a binder in the adhesion of TiO_2 onto cellulose fibers. In this work, we report the use of TEOS to bind 18-nm anatase nano-TiO_2 onto electrospun fiber mats from recycled PET to render them anti-microbial properties for possible use in air filtration systems.

2. Methodology

2.1. Materials

PET bottles (500 mL soda bottles) were washed in water with detergent and then dried with isopropanol. It was then manually shredded to less than 1 cm square size chips. Then the prepared PET chips were dissolved in mixtures of dichloromethane (DCM, 99.5%, Scharlau), trifluoroacetic acid (TFA, 99.8%, Loba Chemie Pvt Ltd.) and acetic acid (AA, 99.8%) using a magnetic stirrer set at 600 rpm for 24 hours. Different solvents were tested, and electrospinning parameters were optimized to produce uniform and thin fibers.

2.2. Electrospinning

The dissolved PET preparations were then loaded into the electrospinning apparatus (Inovenso NS1). The injection pump speed was set at 0.5 mL per hour and the rotating drum collector speed was set at 500 rpm. The needle voltage and working distance were controlled. The electrospinning was carried out at temperature 28 ± 3 °C and a relative humidity of 60 ± 10%. After collection, the fibers were cut into 5 cm x 5 cm square. Each fiber sample was cleaned via sonication with acetone (99.5%, RCI Labscan), ethanol (99.7%, RCI Labscan), and deionized water, then dried with a stream of nitrogen gas.

2.3. Silanization of Polyethylene Terephthalate (PET) Fibers and Deposition of nano-TiO_2

The cleaned electrospun fiber was treated using a UV-ozone cleaner (Model 18, Jelight Company). The fibers were then sprayed with a silane solution of TEOS (99%, Sigma Aldrich), deionized water, ethanol, and glacial acetic acid with a mole ratio of 1.00:10.0:12.3:0.158. The silanized fibers were then partially cured in an already pre-heated oven set at 80 °C for 5 minutes. Nano-TiO_2 suspension was then sprayed onto the silanized side of the fiber mat. The TiO_2 suspension was prepared using the protocol of San Esteban and Enriquez [15]. Anatase, 18-nm TiO_2 nanoparticles (0.30 g, US Nano) were dispersed in 10 mL of acidic ethanol (400:5:1 v/v of absolute ethanol (99.9%), deionized water, and 12 M HClvia sonication. After the deposition of the TiO_2 suspension, the fibers were then fully cured at 80 °C in the ove for about 1 hour. After curing, the nano-coated fiber was washed with ethanol using a sonicator for 15 min to remove non-adhered,excess TiO_2 nanoparticles. Lastly, the fibers were dried in the oven at 80 °C for about 15 min.

2.4. Characterization

The nano-TiO_2 -decorated fibers were also characterized similarly as the untreated, and silanized-only fibers as control. The fibers were imaged using scanning electron microscopy (Hitachi SU1510) to determine the fiber diameter and morphology. The fibers were also analyzed for crystallinity using a Rigaku Ultima IV X-ray Diffractometer and thermal properties using a Shimadzu DSC-60 Differential Scanning Calorimeter. The bound nano-TiO_2 on the PET fibers was also confirmed using a Horiba XGT 7200 X-ray fluorometer with the probe size set of 1.2 mm, and by Raman microspectrometry (Horiba Xplora).

IOP Conf. Series: Materials Science and Engineering **1318** (2024) 012044 doi:10.1088/1757-899X/1318/1/012044

3. Results and Discussion

3.1. PET solution optimization

The PET precursor mixtures to be used for electrospinning was formulated with a suitable solvent system that can produce a high load of PET at room temperature (ca. 30 °C). Here, different mixtures of solvents (70:30 and 60:40 DCM:TFA) and (80:10:10 and 60:20:20 DCM:TFA:AA) were tested across different percentage loads (% w/v) of PET samples (15%, 20%, 25%, 30% and 35%). The different mixtures of solvents were first tested at the lowest PET load (15%). Table 1 shows that the 60:40 and 70:30 DCM:TFA solvent systems are able to completely dissolve the PET samples up to 30% PET load.

3.2. Optimization of electrospinning of PET solutions

Electrospinning the precursor mixtures with 15 and 20% PET load was not successful as these dripped down from the tip even at the highest electrospinning voltage (30 kV) and at the closest working distance of 10 cm. Increasing the concentration increased the viscosity, while increasing the voltage and reducing the working distance should increase the electrostatic force the between tip and collector to produce stable jets[16]. Thus, the PET load for the subsequent runs was increased to 25% and 30% PET and using the highest electrospinning needle voltage of 30 kV to still allow varying working distances.

Table 1. Dissolution test of PET with different type of solvent mixtures and PET loads.

Solvent System	PET Loading Percentage (w/v)				
	15%	20%	25%	30%	35%
60:40 DCM:TFA	Completely dissolved	Completely dissolved	Completely dissolved	Completely dissolved	Partially dissolved
70:30 DCM:TFA	Completely dissolved	Completely dissolved	Completely dissolved	Completely dissolved	--
60:20:20 DCM:TFA:AA	Partially dissolved	--	--	--	--
80:10:10 DCM:TFA:AA	Partially dissolved	--	--	--	--

The optimization test for electrospinning conditions was done for the 25 and 30% PET loading. Besides testing the type of solvent system, electrospinning parameters such as the working distance (10, 15, and 20 cm), and the type of collector (flat and rotating drum) were adjusted. Meanwhile, other electrospinning parameters like the needle voltage (30 kV) and injection pump speed (0.5 ml/h) were kept constant during the optimization tests. Each type of polymer solution with their respective parameters was elecrospun for about 10 min or until enough fibers were produced for characterization. Diameters of the electrospun fibers were estimated from SEM images using ImageJ [17] analysis. Table 2 summarizes the diameters obtained from the different parameters tested. Sample images are also shown in Figure 1.

Among all the parameters tested, optimal fibers—thin and uniform in morphology—were electrospun from 60:40 DCM:TFA with either of these conditions: 25% PET, working distance of 15 cm or from 30% PET loading and 20 cm working distance. The longer working distances suggest that there was longer time for the fibers to splay and thin-out before hitting the collector plate. These fibers had an average diameter of 1.6 (s. d. 1.0) and 1.8 (s. d.1.2), respectively (Figure 1). Achieving thinner fibers could also usually lead to increased tensile strength due to a high degree of molecular alignment induced by the electric field during the electrospinning process [7]. Higher loading of PET in the precursor solutions also has two additional advantages: aside from saving on solvent use, it is less prone to the effect of humidity (which affects volatilization). Thus, the highest load possible (30% PET, 60:40 DCM:TFA) and 20 cm working distance was chosen as the electrospinning parameters for the subsequent batch production of fibers.

In the SEM images shown in Figure 1, there were still minimal beads present in the fiber mat produced from the 25% PET compared with that from the 30% PET. These beads on the fiber are due to non-volatilized jet drippings during electrospinning. The dripping can largely be attributed to a humid environmental condition (RH >70%). In electrospinning, the liquid coming off the needle is jetted, and it splays (or spreads out into thinner jets) forming fibers before hitting the collector. At high humidity, spray elongation happens—this is when the volatility of the solvent is lower, and so the jetted solution does not completely form the splay before it hits the collector, in turn, some of the jetted liquid forms drippings on the collected fiber[18]. Without humidity control during electrospinning, the only parameter that can be controlled would be polymer concentration. An increase in polymer concentration would allow for the decrease in spray elongation, allowing for more uniform fibers.

3.3. Electrospinning of PET fiber mat

In order to produce fiber mats, electrospun fibers were collected from the rotating drum for a total of 47 hours of collection. The fiber mat was divided into 5 cm strips: left, right, and center. It was observed that deposition rate averaged to about 38 μm/h for the left and right sides of the fiber mat, and about 47 μm/h for the center of the mat. Several 5x5 cm2 were sampled (Figure 2) from the strips with average mass of 0.62-0.64 g which corresponds to about 248 -256 gsm.

Table 2. Average diameter of electrospun fibers with different electrospinning parameters at 30kV and 0.5 mL/h flow rate

DCM/TFA Ratio	PET % Load	Working Distance	Type of Collector	Diameter (μm)	Standard Deviation
60:40	25	10	Flat	1.8	0.7
			Rotating Drum	2.9	2.0
		15	Flat	2.0	1.1
			Rotating Drum	1.6	1.0
		20	Flat	2.2	1.7
			Rotating Drum	2.7	1.7
	30	10	Flat	3.7	2.8
			Rotating Drum	3.5	2.9
		15	Flat	2.1	1.4
			Rotating Drum	2.5	1.7
		20	Flat	2.9	2.0
			Rotating Drum	1.8	1.2
70:30	25	10	Flat	5.2	3.1
			Rotating Drum	2.6	1.0
		15	Flat	3.1	1.2
			Rotating Drum	2.4	0.6
		20	Flat	3.4	2.6
			Rotating Drum	-	-
	30	10	Flat	4.3	2.5
			Rotating Drum	3.0	2.3
		15	Flat	3.4	1.6
			Rotating Drum	3.3	1.3
		20	Flat	3.6	1.7
			Rotating Drum	3.6	2.5

IOP Conf. Series: Materials Science and Engineering **1318** (2024) 012044 doi:10.1088/1757-899X/1318/1/012044

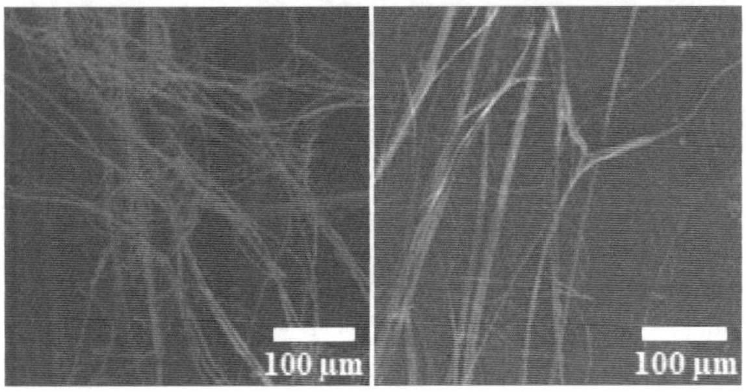

Figure 1. SEM images of electrospun fibers from 60:40 DCM:TFA electrospun onto a rotating drum with a PET load of 25% (left) with a 15 cm working distance, and a PET load of 30% (right) with a 20 cm working distance.

3.4. Characterization of the electrospun PET

XRD patterns of the electrospun PET (Figure 3) are similar to those reported for microplastic PET which are mostly from textile fibers [19,20]. The XRD of the PET of the soda bottle has a single peak at 25.6° ascribed to biaxial molecular orientation of semicrystalline PET (ICDD 00-061-1413). PET bottles are formed through the blow-molding process which results in a biaxial orientation [21]. Randomized PET has three main peaks at 2Θ values of 17.5° (010), 23.0° (110) and 25.6° (100). The broad peak at 23.0° for the nanofibers indicate its mainly amorphous nature, which is ascribed to the rapid solidification process in electrospinning [22]. The 17.5° and 25.6° peaks change their relative intensities depending on the material [23].

The DSC results of the PET bottle and nanofibers (Figure 3) show significant differences on the first heating scan given the difference in thermomechanical histories of the samples. For instance, the glass transition temperature (Tg) for the electrospun PET nanofibers is significantly lower at 52.5 °C compared to that of the soda bottle, at 78.2 °C. The curve for the electrospun fiber showed cold crystallization (an exothermic peak indicated by the red arrow) around 126 °C which is not present for the soda bottle curve, which indicates a difference in percent crystallinity. The smaller endothermic peak for the melting at 245.0 °C for the electrospun fiber compared with that for the soda bottle (246.9 °C) also indicates lower crystallinity of the electrospun fiber. This agrees with the XRD results discussed above. On second heating, the thermomechanical history would have been erased for both samples, and therefore they show very similar DSC profiles as shown in Figure 3.

The fact that there is more amorphous component in the electrospun fiber implies that this can be subject to molecular alignment due to the electric field in the electrospinning process. This is attributed to a constrained mesomorphic phase formed during their rapid solidification in the electric field due to the electrospinning process, an internal morphology known to enhance the tensile strength of fiber mats [7].

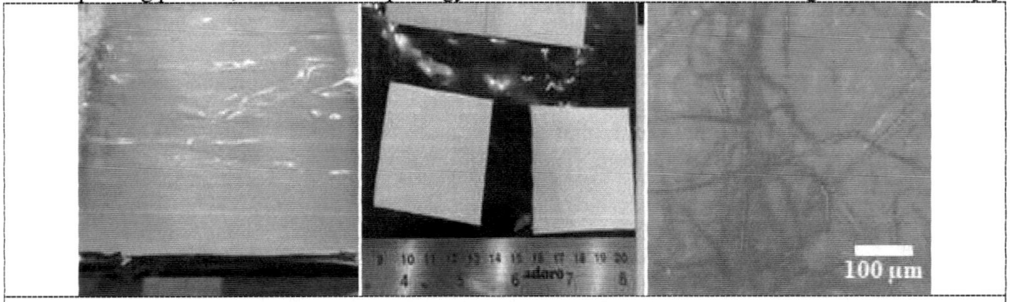

Figure 2. Pictures of collected fiber mat on aluminum foil (left) and PET square samples (center), 100x image of the fiber mat visualized through the ViewSharp™ imaging of the Raman Microscope.

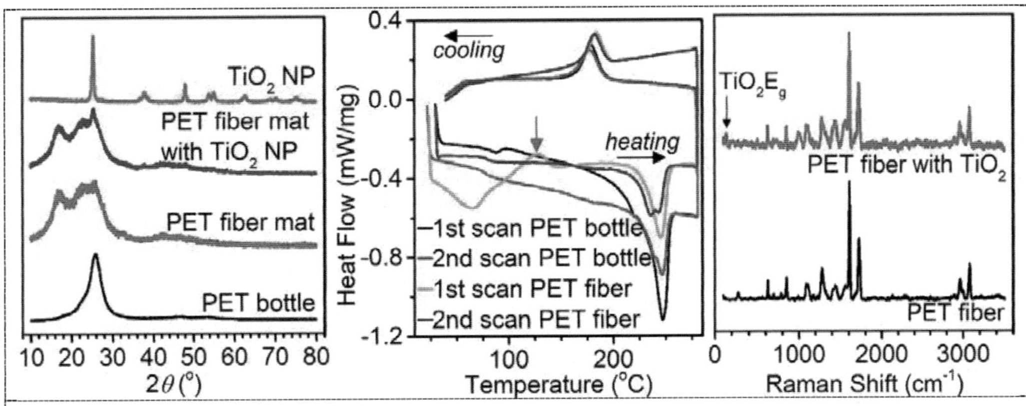

Figure 3. XRD patterns of the PET bottle, electrospun PET fiber mat, washed TiO_2 nanoparticle-decorated electrospun PET fiber mat and TiO_2 nanoparticles (left). DSC heating and cooling curve of the the electrospun PET fibers and bulk PET (Cola Bottle) (center). Raman spectra of the PET nanofiber mat with (washed) and without titanium dioxide nanoparticles (right).

3.5. TiO_2 nanoparticle decoration of the PET fibers

The electrospun nanofiber mats were then decorated with nano-TiO_2 in conjunction with a silanization process. Table 3 shows the relative amounts of the silane and nano-TiO_2 components, expressed in terms of Ti and Si mean atomic percent via XRF analysis. The results show that the fibers have both the silane layer and nano-TiO_2 coating. The presence of Si across all front side of the samples is important since it confirms that the silane is present to bond with the PET fiber. Meanwhile, the relatively high Ti:Si ratio also implies sufficient loading of nano-TiO_2 on the silanized PET fiber. Elemental maps were also conducted using the XRF (Figure 4). The elemental maps show that both Si and Ti were deposited uniformly across the sample showing good coverage of the spray coating method that was used. It is also noted that even the back side of the PET sample the presence of both Si and Ti was observed as shown in Table 3 and Figure 4, possibly due to residual spray. It is important to note though that XRF can only measure relative percentages of Ti and Si, and so the absolute amount of Ti and Si on the back side of the samples may still be significantly less than that on the front side of the sample.

To observe the durability of the adhesion of the nano-TiO_2 coating to the PET fiber, the prototype nano-decorated filter was sonicated in water for 15 minutes. The mean atomic % of Ti decreased by a small margin; however, the values were still within the standard deviation of that before sonication indicating minimal to no significant change. This shows that the nano-TiO_2 decoration of the PET fibers were durable and could withstand washings with water, allowing for the prototype fiber filter to be reusable. The presence of nano-TiO2 in the washed PET fiber mat was also confirmed in its Raman spectrum and XRD pattern (Figure 3).

Table 3. XRF data on average Ti and Si contents for the two sides of the PET sample before and after sonication

Treatment	Fiber Side	Site	Mean Atomic %				Ti/Si	SD
			Ti	SD	Si	SD		
Before Sonication	Front	1	39.0	9.5	61.0	9.5	0.7	0.3
		2	46.8	3.2	53.2	3.2	0.9	0.1
	Back	3	35.0	1.1	65.0	1.1	0.5	0.02
		4	37.0	11.1	63.0	11.1	0.6	0.3
After Sonication	Front	5	30.0	5.8	70.0	5.8	0.4	0.1
		6	41.8	1.6	58.2	1.6	0.7	0.04
	Back	7	42.0	1.5	58.0	1.5	0.6	0.3
		8	30.0	6.2	70.0	6.2	0.4	0.2

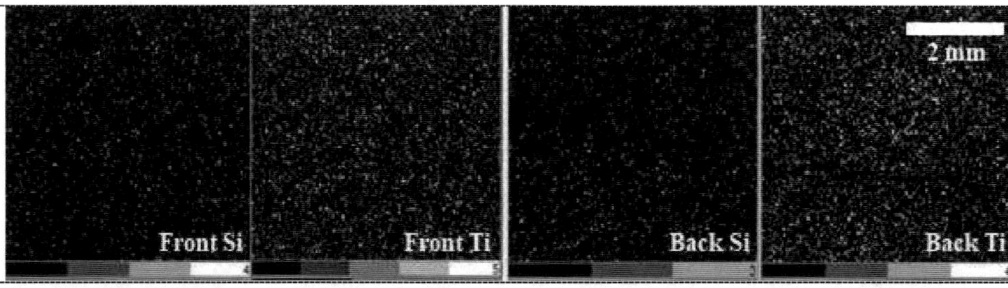

Figure 4. Elemental maps for Si and Ti present in the spray-coated (front) and uncoated (back) side of the PET fiber sample.

4. Conclusion

The study was able to create a prototype recycled PET fiber mat by electrospinning PET from soda bottles, which was subsequently decorated with nano-TiO_2. Results showed that fibers with an average diameter of 1.8 (s. d.1.2) μm can be electrospun from a polymer solution with high PET loading (30%) dissolved using the solvent system of 60:40 DCM:TFA. The fibers were shown to have high amorphous content via XRD and DSC analysis, which is characteristic of the electrospinning process. The fibers were successfully coated with a silane-based binder for better chemical adhesion with the nano-TiO_2 coating, which would impart photo-induced antimicrobial properties. The coated PET fiber mat had a durable and even coating of nano-TiO_2 which would be good for reusability.

Acknowledgments
The authors would like to acknowledge support from Science and Technology for a Resilient Community against the Pandemic (STRAP) of the Department of Science and Technology (DOST) through project number 09345 of 2020 and the Ateneo Research institute for Science and Engineering (ARISE). LM also acknowledges the AdMU University Research Council for an Early Career Grant.

References

[1] Li D and Xia Y 2004 Electrospinning of Nanofibers: Reinventing the Wheel? *Advanced Materials* **16** 1151–70
[2] Leung W W F and Sun Q 2020 Electrostatic charged nanofiber filter for filtering airborne novel coronavirus (COVID-19) and nano-aerosols *Sep Purif Technol*
[3] Leung W W F and Sun Q 2020 Charged PVDF multilayer nanofiber filter in filtering simulated airborne novel coronavirus (COVID-19) using ambient nano-aerosols *Sep Purif Technol*

[4] Bonfim D P F, Cruz F G S, Guerra V G and Aguiar M L 2021 Development of filter media by electrospinning for air filtration of nanoparticles from pet bottles *Membranes (Basel)*

[5] Covaliu-Mierlă C I, Matei E, Stoian O, Covaliu L, Constandache A C, Iovu H and Paraschiv G 2022 TiO2–Based Nanofibrous Membranes for Environmental Protection *Membranes (Basel)* **12**

[6] Borojeni I A, Gajewski G and Riahi R A 2022 Application of Electrospun Nonwoven Fibers in Air Filters *Fibers 2022, Vol. 10, Page 15* **10** 15

[7] Strain I N, Wu Q, Pourrahimi A M, Hedenqvist M S, Olsson R T and Andersson R L 2015 Electrospinning of recycled PET to generate tough mesomorphic fibre membranes for smoke filtration *J Mater Chem A Mater*

[8] Karuppannan S K, Ramalingam R, Mohamed Khalith S B, Musthafa S A, Dowlath M J H, Munuswamy-Ramanujam G and Arunachalam K D 2021 Copper oxide nanoparticles infused electrospun polycaprolactone/gelatin scaffold as an antibacterial wound dressing *Mater Lett* **294** 129787

[9] Fahimirad S, Fahimirad Z and Sillanpää M 2021 Efficient removal of water bacteria and viruses using electrospun nanofibers *Science of The Total Environment* **751** 141673

[10] Monmaturapoj N, Sri-on A, Klinsukhon W, Boonnak K and Prahsarn C 2018 Antiviral activity of multifunctional composite based on TiO2-modified hydroxyapatite *Materials Science and Engineering: C* **92** 96–102

[11] Kumaravel V, Nair K M, Mathew S, Bartlett J, Kennedy J E, Manning H G, Whelan B J, Leyland N S and Pillai S C 2021 Antimicrobial TiO2 nanocomposite coatings for surfaces, dental and orthopaedic implants *Chemical Engineering Journal* **416** 129071

[12] Scalarone D, Lazzari M and Chiantore O 2012 Acrylic protective coatings modified with titanium dioxide nanoparticles: Comparative study of stability under irradiation *Polym Degrad Stab* **97** 2136–42

[13] Labuto G, Sanches S, Crespo J G, Pereira V J and Huertas R M 2022 Stability of polymeric membranes to UV exposure before and after coating with TiO2 nanoparticles *Polymers (Basel)* **14** 124

[14] Veronovski N, Sfiligoj-Smole M and Viota J L 2009 Characterization of TiO2/TiO2—SiO2 Coated Cellulose Textiles *Textile Research Journal* **80** 55–62

[15] San Esteban A C M and Enriquez E P 2013 Graphene–anthocyanin mixture as photosensitizer for dye-sensitized solar cell *Solar Energy* **98** 392–9

[16] Leach M K, Feng Z-Q, Tuck S J and Corey J M 2011 Electrospinning Fundamentals: Optimizing Solution and Apparatus Parameters *JoVE* e2494

[17] Schneider C A, Rasband W S and Eliceiri K W 2012 NIH Image to ImageJ: 25 years of image analysis *Nat Methods* **9** 671–5

[18] Nezarati R M, Eifert M B and Cosgriff-Hernandez E 2013 Effects of Humidity and Solution Viscosity on Electrospun Fiber Morphology *Tissue Eng Part C Methods* **19** 810–9

[19] Billon N 2023 Strain induced crystallization of PET under biaxial conditions. From laboratory tests to injection stretch-blow molding *Polymer (Guildf)* **277** 125953

[20] Hadjizadeh A, Ajji A and Bureau M N 2011 Nano/micro electro-spun polyethylene terephthalate fibrous mat preparation and characterization *J Mech Behav Biomed Mater* **4** 340–51

[21] Bhat N and Deshmukh R 2002 X-ray crystallographic studies of polymeric materials *Indian Journal of Pure and Applied Physics* **40** 361–6

IOP Conf. Series: Materials Science and Engineering **1318** (2024) 012045 doi:10.1088/1757-899X/1318/1/012045

Structural and electrical properties of Ni-doped MnZn ferrite synthesized via solid state reaction

JC Mendoza[a], M Arcangel[b], JA Nucum[c] and RB Cervera[d]

Department of Mining, Metallurgical, and Materials Engineering, College of Engineering, University of the Philippines Diliman, Quezon City, 1101 Philippines

[a]jjmendoza2@up.edu.ph, [b]mrarcangel@up.edu.ph, [c]jvnucum@up.edu.ph, [d]rmcervera@up.edu.ph

Abstract. MnZn ferrite is commonly studied due to its exceptional magnetic, electric, and catalytic properties, making it a promising material for hyperthermia applications, magnetic fluid, memory storage devices, drug delivery, virus detection, and photocatalysis. It was identified that divalent nickel cation substitution increases the ferrite conductivity and dielectric constant. This study targets to synthesize Ni-doped MnZn ferrite by a simpler, more convenient, and economic solid state reaction, and to investigate its influence on the structural and electrical properties of MnZn ferrite. $Mn_{0.5-x}Ni_xZn_{0.5}Fe_2O_4$ (x=0.1 and 0.2) was synthesized via solid state reaction with calcination and sintering temperature at 1000 °C and 1200 °C, respectively. The structural and electrical properties of the resulting pellet were characterized by x-ray diffraction (XRD), scanning electron microscopy (SEM), and electrochemical impedance spectroscopy (EIS). The XRD profile, indexed as cubic spinel, indicated good crystallinity and no impurity peaks were detected. As Ni^{2+} dopant concentration increases, a decrease in lattice parameter and an increase in theoretical and apparent densities were observed. This is attributed to the smaller ionic radius and greater mass of Ni^{2+} relative to Mn^{2+}. Varying Ni^{2+} concentration significantly modified the morphology of the ferrite. At higher Ni^{2+}, less uniformity in shape and size was evident in the SEM micrograph since Ni promotes aggregation at the surface. An increase in dielectric constant was also observed with increasing Ni^{2+} molar concentration. Since Ni^{2+} presents a high tendency to occupy B sites, its substitution promotes Fe^{2+} migration to A sites, augmenting Fe^{2+}/Fe^{3+} hopping resulting in an increase in polarization and dielectric constant.

1. Introduction

Ferrite particles are commonly studied due to their exceptional magnetic, electric, and catalytic properties that can cater to a wide range of applications like in electronics, information technology, biomedicine, and wastewater treatment [1, 2]. This ferrimagnetic ceramic comprises iron oxide (Fe2O4) in large proportions and divalent metals like Mn, Ni, Zn, Co, and Cu in smaller amounts [3]. MnZn ferrite with spinel cubic structure is a soft type of ferrite with low coercivity and magnetization that can be easily modified. It has gained attention due to its high resistivity, low production cost, high permeability, low power loss, and high magnetic induction making it a promising material for hyperthermia applications, magnetic fluid, memory storage devices, drug delivery, virus detection, and photocatalysis [3, 4]. The synthesis procedure and the cation concentration and distribution cause variation in MnZn ferrite properties suited for specific types of application. Sol-gel method, combustion, solid state reaction, and hydrothermal precipitation are some of the several techniques to synthesize MnZn ferrites [3].

The structural, electrical, and magnetic properties of MnZn ferrite were modified by substituting divalent nickel cation through sol-gel method and an increase in conductivity and dielectric constant was recorded [5]. This improvement is highly favored in high frequency applications. Mn-doped Ni-Zn ferrite was prepared via sol-gel combustion method and a highly crystalline single-phase spinel structure was successfully fabricated [2]. Limited synthesis procedure in preparing Ni-MnZn ferrite was available in literature. Hence, this study targets to synthesize Ni-doped MnZn ferrite via a facile, more convenient, and economic route — the solid state reaction; and to investigate the influence of Ni-doping on the structural and electrical properties of MnZn ferrite.

2. Methods

Stoichiometric amounts of NiO (Sigma-Aldrich, 99.99%), MnO2 (Techno PharmChem, 85%), ZnO (Dalkem, 99.99%), and FeO (Sigma-Aldrich, 99.9%) were ground using an agate mortar and pestle for 2 hours while being moistened with ethanol. The well-combined powders were calcined in air at 1200 ^{o}C for 2 hours with a heating and cooling ramp rate of 5 ^{o}C/min, pelletized in 14.3 mm diameter die (MTI Corp) at 100 bar by a hydraulic press with 5 minutes dwell time, and finally sintered at 1200 ^{o}C for 3 hours.

The phase structure and crystallinity of the synthesized $Mn_{0.5-x}Ni_xZn_{0.5}Fe_2O_4$ (x=0.1 and 0.2) were analyzed using the Rigaku SmartLab X-ray Diffractometer. Cu Kα (λ = 1.5406 Å) x-ray beam at a scan rate of 5°/ min in the 2**θ** range between 10° and 90o was employed during the analysis. From the indexed peaks, the lattice parameter was calculated using the CellCalc software. Apparent density measurement was performed through the Archimedes principle in ethanol while the theoretical and the relative densities of the prepared pellets were calculated. The microstructure of the pellets was characterized using a scanning electron microscope (Hitachi S3400N) and the capacitance of the pellets was determined through impedance spectroscopy from 20 Hz to 1 MHz at room temperature using Biologic VSP300 potentiostat.

3. Results and discussions

3.1. Structural property

$Mn_{0.5-x}Ni_xZn_{0.5}Fe_2O_4$ (x=0.1, 0.2) was indexed (JCPDS No. 74-2403) as cubic spinel. The patterns of the synthesized ferrites are in good agreement with the diffraction pattern of MnZn ferrite synthesized via solid state reaction [7]. No other characteristic phase was observed at x=0.2 and the peaks suggest good crystallinity. The presence of extra peak (♦) in x=0.1 matches γ-Fe_2O_3. The formation of hematite is attributed to the oxidation of unreacted precursor, FeO, brought by the exposure in air at high temperature during calcination and sintering [8, 9]. Since no other traces of unreacted oxides are detected, it can be deduced that the doping of Ni in $Mn_{0.5-x}Ni_xZn_{0.5}Fe_2O_4$ with x=0.1 and 0.2 can be fully realized by solid state reaction.

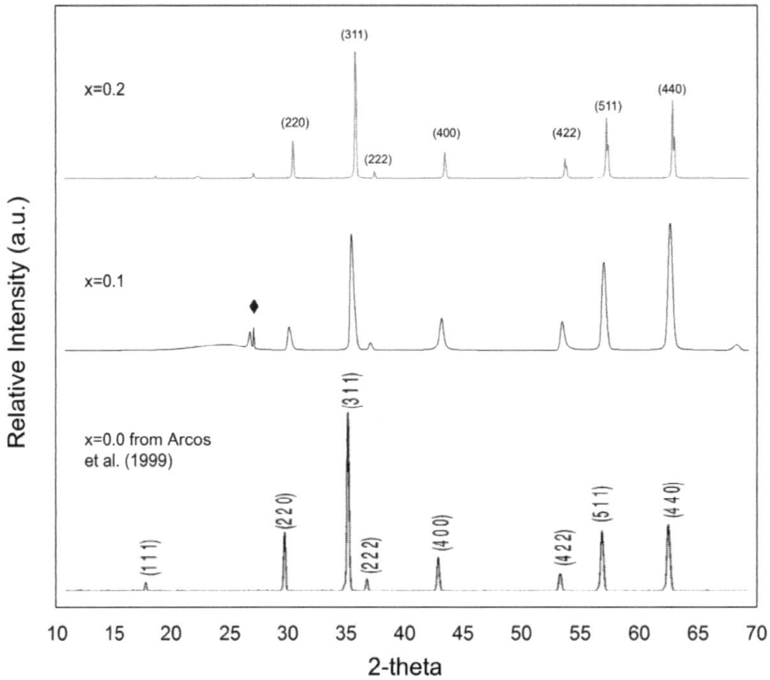

Figure 1. XRD patterns of synthesized Ni-doped MnZn ferrite.

From [3] and [8], the lattice parameter of un-doped MnZn ferrite prepared via solid state reaction was around 8.5 Å. When doped with Ni^{2+}, the lattice parameter of MnZn ferrite decreases as observed from the results of this study presented in Table 1. This change is caused by the difference in the ionic radius of Mn^{2+} (0.8 Å) and of Ni^{2+} (0.69 Å) [5, 6]. Since Ni^{2+} is relatively smaller than Mn^{2+}, its substitution in the spinel structure reduces the lattice parameter as the cubic symmetry is being preserved [6]. It was also noted in the density measurement results that both apparent and theoretical density increase with increasing Ni concentration. This is attributed to the decrease in volume brought about by the decrease in lattice parameter [6] and an increase in mass since Ni^{2+} (58.69 amu) is essentially heavier than Mn^{2+} (54.94 amu). Since density is the ratio of mass to volume, the increase in mass and the decrease in volume will increase the density as Ni is substituted in MnZn ferrite.

Table 1. Lattice parameter and density measurements of synthesized Ni-doped MnZn ferrite.

Chemical Composition	**Lattice parameter, *a***	$\rho_{apparent}$, g/cm^3	$\rho_{theoretical}$, g/cm^3	**Relative Density**
$Mn_{0.4}Ni_{0.1}Zn_{0.5}Fe_2O_4$	8.384 Å	4.2985	5.3244	80.73%
$Mn_{0.3}Ni_{0.2}Zn_{0.5}Fe_2O_4$	8.345 Å	4.6435	5.4079	85.87%

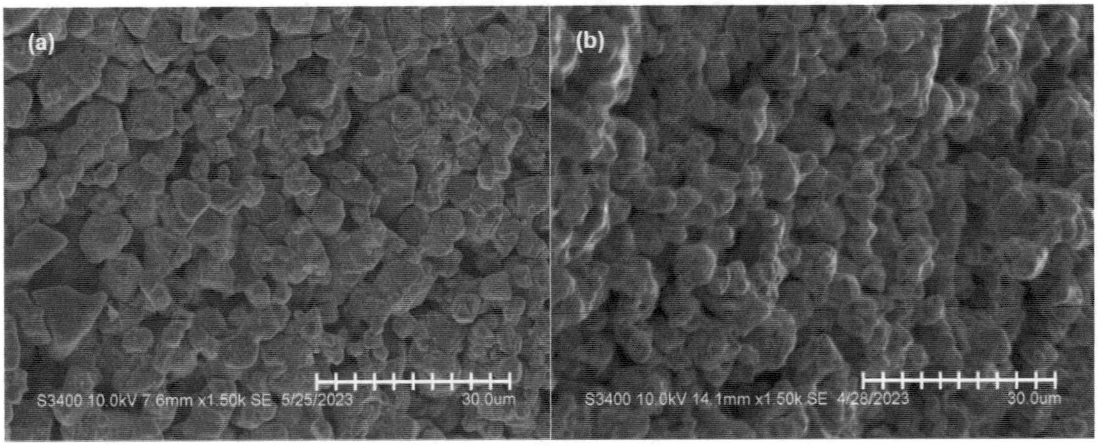

Figure 2. SEM micrographs of $Mn_{0.5-x}Ni_xZn_{0.5}Fe_2O_4$ with (a) x=0.1 and (b) x=0.2.

The observed agglomeration of particles in the micrographs is due to the inherent magnetic property of ferrite [11]. At higher Ni-doping, there is less uniformity in shape and size because Ni promotes aggregation at the surface [5, 12]. From the micrographs, a porous structure is observed which is also reflected in the calculated relative density.

3.2. Electrical property

The decreasing dielectric constant with increasing frequency is a normal dielectric behavior of ferrites. Dielectric constants are higher at lower frequency, lower at higher frequency, and become saturated as frequency further increases. At lower frequency, the dielectric constant is higher in Ni-doped MnZn ferrite with higher Ni molar concentration. From the findings of Jalaiah & Babu (2017), at 40 kHz the dielectric constant of $Mn_{0.85}Ni_{0.09}Zn_{0.06}Fe_2O_4$ and $Mn_{0.85}Ni_{0.15}Fe_{2\ 4}$ are around 4.5 and 5, respectively, whereas in the synthesized $Mn_{0.4}Ni_{0.1}Zn_{0.5}Fe_2O_4$ and $Mn_{0.3}Ni_{0.2}Zn_{0.5}Fe_2O_4$ in this study, the dielectric constants are 2.93 and 5.08, correspondingly. Mn^{2+} occupies both A and B sites but with preference in A site while Ni^{2+} shows a high tendency to occupy the B site [6]. Now with the Ni^{2+} substitution of Mn^{2+}, Mn^{2+} concentration in A site is reduced while Ni^{2+} occupies the B sites. This promotes the migration of Fe^{2+} to A site and augments the hopping of Fe^{3+}/Fe^{2+} in B sites. Hence, polarization increases resulting in an increase in dielectric constant.

IOP Conf. Series: Materials Science and Engineering **1318** (2024) 012045 doi:10.1088/1757-899X/1318/1/012045

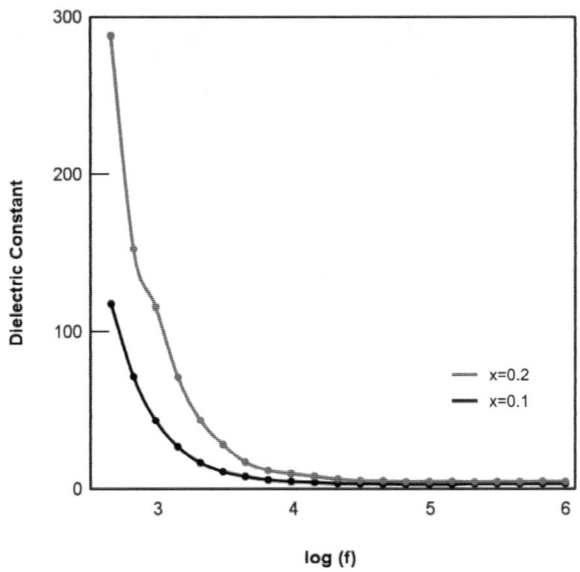

Figure 3. Dielectric constant vs log(f) of synthesized $Mn_{0.5-x}Ni_xZn_{0.5}Fe_2O_4$ (x=0.1 and 0.2).

4. Summary and conclusion

$Mn_{0.5-x}Ni_xZn_{0.5}Fe_2O_4$ (x=0.1 and 0.2) was synthesized via solid state reaction with calcination and sintering temperature at 1000 °C and 1200 °C, respectively. The lattice parameter and the apparent and theoretical densities increase with increasing Ni dopant concentration due to the difference in ionic radius of Ni^{2+} and Mn^{2+} with the former being smaller. Varying Ni^{2+} molar concentration modifies significantly the morphology as observed in the SEM micrographs. For the electric property, dielectric constant increases with increasing Ni^{2+} concentration because its substitution promotes Fe^{2+} migration to A site resulting in an increase in Fe^{2+}/Fe^{3+} hopping that leads to an increase in polarization and dielectric constant. Thus, Ni-doped MnZn ferrite can be synthesized through solid state reaction and its structural and electrical properties can be modified by varying Ni dopant concentration.

References

[1] Pham TN, Huy TQ, Le AT 2020 *RSC Adv.* **10** 31622–31661
[2] Ramesh S, Dhanalakshmi B, Sekhar BC, Rao PS, Rao BB 2016 *Appl. Phys. A* **122** 984
[3] Thakur P, Chahar D, Taneja SP, Bhalla N, Thakur A 2020 *Ceram. Int.* **46** 15740–15763
[4] Kim HJ, Hyun SW, Kim S, Choi H 2021 *J. Radioanal. Nucl. Chem.* **330** 445–454
[5] Jalaiah K, Babu KV 2017 *J. Magn. Magn. Mater.* **423** 275–280
[6] Mane, Birajdar DD, Shirsath SE, Telugu RA, Kadam RH 2010 *Phys. Status Solidi (A) Appl. Mater.* **207** 2355–2363
[7] Arcos D, Vázquez M, Valenzuela R, Vallet-Regí M 1999 *J. Mater. Res.* **14** 861–865
[8] Hamed MH, Mueller DN, Müller M 2020 *J. Mater. Chem. C* **8** 1335-1343
[9] Satalkar M, Kane SN, Ghosh A, Ghodke N, Barrera G, Celegato F, et al. 2014 *J. Alloys Compd.* **615** S313–316
[10] Thakur P, Sharma R, Sharma V, Sharma P 2017 *Mater. Chem. Phys.* **193** 285-289
[11] Xiao L, Zhou T, Meng J 2009 *Particuology* 7 491–495
[12] Singh LN, Ahmed F 2020 *International Conference on Physics of Materials and Nanotechnology*

IOP Conf. Series: Materials Science and Engineering **1318** (2024) 012046 doi:10.1088/1757-899X/1318/1/012046

Adaptive Neuro-Fuzzy Interference System (ANFIS) Control of a Dynamic, Two-Region Nuclear Power Plant Pressurizer Model

D A Bautista[1], **J A Mendoza**[1]

[1] School of Chemical, Biological, and Materials Engineering and Sciences, Mapúa University, Intramuros, Manila, 1002, Metro Manila, Philippines

jammendoza@mapua.edu.ph

Abstract. Designing modern control systems is a critical factor in producing rigid control and effective performance of non-linear and complex systems—particularly employing intelligence-based controllers, derived from soft computing algorithms. It is known to be well capable of exploiting tolerances for uncertainty and nonlinearity. Fuzzy logic, which utilizes soft computing techniques, deals with vague information and approximate reasoning, yet it lacks effective learning capability, while ANN is known for exceptional learning and adaptation of training data. Adaptive neuro-fuzzy interference system (ANFIS), a hybrid of these two algorithms, is a fuzzy interference system optimized by neural networks. Most nuclear generating stations still use PID controllers, which does not show good controlling capabilities, resulting in large coolant level overshoot and longer settling time. This paper bridges this gap by testing an ANFIS approach that yields high predictive and control capability. In validating the configured control loop model, three loss-of-load transients from a Shippingport reactor were tested, trained and demonstrated using the simulations in the MATLAB/Simulink environment. The minimal training RMSE for the 51 MW(e), 74 MW(e), and 105 MW(e) loss-of-load transients are 18.85%, 19.37%, and 24.36%. The pressure response showed ideal output, with all three steady-state pressures close at setpoint (14 MPa), proving that the ANFIS controller is well capable of rejecting disturbances with lesser overshoot and faster settling times in maintaining level setpoint. Validating solely the pressurizer model was executed in a turbine leading power trip from a 100% to 75% reduction range in PCTRAN, a pressurized water reactor (PWR) simulator made available by IAEA. The comparative performance exhibited a good fit between the reported simulation data and model data. Both results against simulation and experimental transients captured the effective predictive and control capabilities of the pressurizer unit and control system.

1. Introduction

The instrumentation and control framework of a nuclear power plant (NPP) is considered its' central nervous system, regulating plant process parameters within design limits. In order to efficiently produce power, process and plant parameters such as temperature, pressure, water level, power output and flow rates are to be kept within the operating set points—which is the responsibility of the control system. Essentially, these systems measure certain process parameters, calculate how much it has deviated from the design set point and/or control demands, and implement corrective actuation commands to its' corresponding field devices. Its' primary objectives are to reduce construction and life cycle costs, enable optimized operations, and ultimately provide investment protection [1]. The majority of the control systems used to monitor NPPs today are based on process outdated technology from 1950s and 1960s. Following the advancement of I&C technology for the past decades, NPPs are only recently adapting in aspects of operation and maintenance. The highest priority key component in a NPP is the pressure vessel, as it requires great reliability to control deviations in temperature and pressure-making it the most critical pressure boundary in the NPP [2]. Therefore, one of the critical process parameters observed is the operating pressure of the pressurizer, which is an integral part of the primary heat transport (PHT) of a NPP. This makes the pressurizer function crucial to the entire loop's security. The pressure is controlled by acting on the water temperature inside the pressurizer. In order to avoid boiling in the primary loop of the system, equilibrium is maintained between water and steam. This is to avoid coolant malfunction in the reactor, which is a crucial safety concern for the entirety of the primary loop.

Advanced control algorithms have emerged in the recent developments of modern digital I&C systems. Soft computing technology including fuzzy logic and artificial neural network (ANN) are most suited for highly non-linear processes that conventional controllers such as PID cannot effectively control. ANFIS controllers have been proven to control more accurately and safely than conventional controllers such as PI, PID and even fuzzy PID in terms of settling time, peak overshoot and steady state error [3–5]. Hybrid models like these are most suitable for industrial processes such as heating and cooling [6]. This study will resolve validating of pressurizer model in PCTRAN, modifying the adopted pressurizer control loop, training the reported loss of load transients from Shippingport Nuclear Station Pressurized Water Reactor (PWR) and testing the proposed ANFIS controller on the PWR model using MATLAB Simulink software, and create and report an overall ideally damped response which contains minimal overshoot and lower settling time and ultimately test the comparative performance between the modified control model and transient data. This research shall prove if a hybrid adaptive neuro-fuzzy interference system control algorithm can effectively replace a PID controller for pressure control, at the same time ultimately propagate nuclear safety research, which is a growing environmental and societal concern. This is in pursuit of greater research for better control in accordance with strict control requirements of NPPs since simulations of control systems such as this paper allows optimization and evaluation of the reliability and efficiency of the control of an ANFIS model in a pressurizer system in presence of specific set disturbances.

2. Modelling

2.1. Process description

A NPP pressurizer, which is part of the reactant cooling system (RCS) or the primary loop, is a fundamental unit component. It is used to compensate for the fluctuations caused by load changes in the NPP by maintaining water and steam saturation in balance. In this study, a pressurizer unit is modeled in MATLAB which was adopted from literature. It was based on a reference simulation paper submitted to Proceeding of International Conference on Electrical Engineering, Computer Science and Informatics (EECSI) in the 2015 issue [7]. As a result of temperature changes, coolant surges in from either the hot or cold leg line. When pressure inside the tank deviates from set point, various components will actuate to revert back pressure to the normal operating point. The usual cause of this occurrence is associated with a change in temperature of the primary loop. Since the pressurizer is connected to the primary loop through surge leg, the coolant will tend to expand into the pressurizer, inducing steam at the top of the pressurizer to be compressed, therefore increasing the pressure inside the vessel. This will trigger spray line cold water to be sprayed into steam space or pressure relief and safety valves will open. On the other hand, when the temperature of the RCS decreases, the pressure level decreases and electrical heaters will operate and boil water to transform into steam, increasing pressure. From reference paper, the model is composed of step blocks, one PID controller, a state-space block and integrators and gains. The state-space block contains the equations formed from mathematical modeling of the pressurizer unit. The integrator blocks allow for integration of input signal with respect to time, while the gain blocks are simply a multiplier, both are utilized in order to produce better steady state response from fed step disturbances. However, the neuro-fuzzy designer is a MATLAB design tool designated for multiple-input, single-output controllers. The MATLAB ANFIS tool tunes Sugeno-type FIS which is used to design a multiple-input, single-output (MISO) controller. Therefore, arrangements in the feedback loop were made in order to avoid dimension mismatch when ANFIS controller is employed. The modified MISO control model is assigned one controller for each input and increasing gain multiplier to 10. This arrangement allows for ANFIS controllers to work and to obtain an optimal response.

2.2. Modelling equations

Simulation and modelling of a pressurizer control loop is based on a reference simulation paper submitted to Proceeding of International Conference on Electrical Engineering, Computer Science and Informatics (EECSI) in the 2015 issue with the following approximations:

1) Pressure is uniform throughout the pressurizer all the time;
2) The mass exchanges between the liquid and steam interface occur in instant;
3) The spray water reaches saturation temperature before leaving the steam phase and going into the liquid phase;
4) The in-surge flow of coolant water mixes perfectly with the hot water in the bottom of surge tank;
5) The liquid phase is either saturation or overcooled, and the steam phase could be either saturation or overheating.

The pressurizer model adapted is a dynamic, two-phase mechanism model which is subdivided into two regions, that is two volumes of homogeneous fluid sections: steam region and water region, not necessarily in equilibrium with each other. The model governing equations are used to form the state space representation. The dynamic model of the pressurizer system is described by the conservation of mass and energy equations.

2.3. ANFIS architecture

For each input, one ANFIS controller is assigned as shown in Fig. 1. Membership functions (m) corresponds to each one input (n), therefore, the total number of rules needed is m^n. The assigned number of membership function is 3, therefore there were also 3 established rules and output membership functions as stated by the formula.

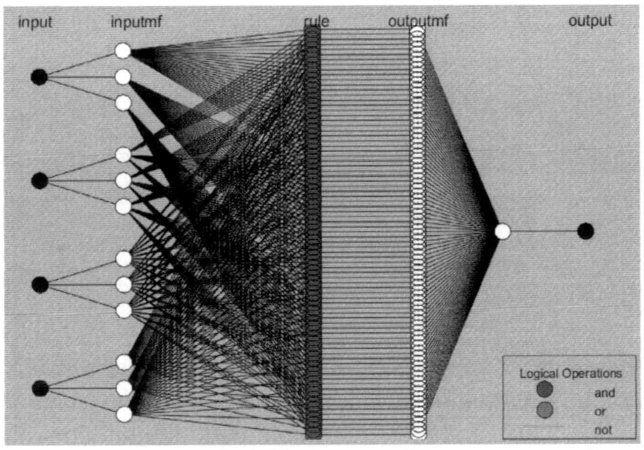

Figure 1. Proposed ANFIS model structure for every step disturbance input

In generating the membership functions, the number of fuzzy labels to be used is determined by the user. In this study, for each feature of data, default three membership functions were used. Since there is only one input and one output, there are two features, each with three equally spaced fuzzy membership functions. Fig. 2, which was taken as a screenshot from the Neuro-Fuzzy Designer app in MATLAB, individually shows each curve of the triangular membership function, while Fig. 3 shows them in a single plot. The x-axis shows the range of input values. If for example, the input value is around 4, this feature falls roughly between 90% of the first function and 30% of the second.

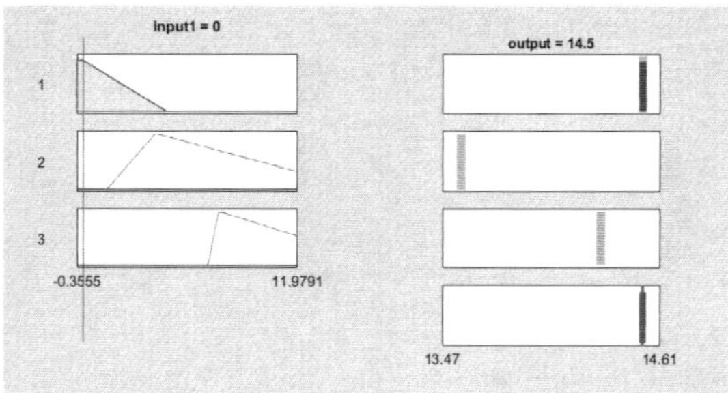

Figure 2. Input and output membership functions for one input

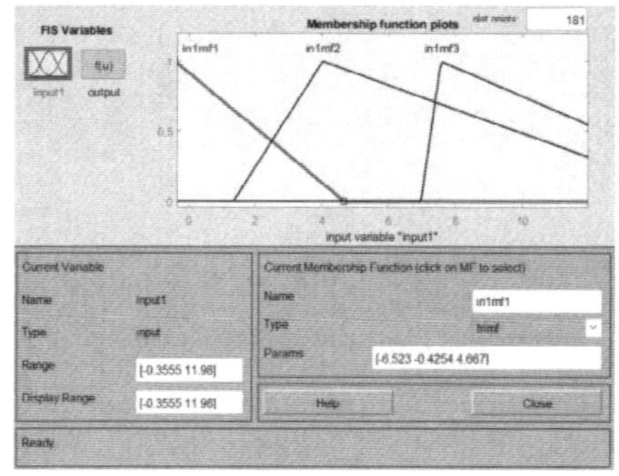

Figure 3. Membership function plot and properties

3. Simulation

Unit modelling requires validation through simulated softwares or comparison to actual measured output of an operating plant. It is an assessment on whether or not the quantity of interest (QOI), which is in this study, pressure, is within a reasonable tolerance. The adopted pressurizer model is validated through comparing predictions to PCTRAN simulation data and its' control loop though reported loss-of-load transients from Shippingport PWR.

3.1. Pressurizer model validation

On startup, a set of initial conditions that indicate various power, flow, reactor coolant system pressure, etc. is provided. In this study, the first initial condition (IC1), or 100% Powe EOC is when the plant operates at full power and fuel is at the end of cycle. This is where power demand is at 100%, the reactant pressure setpoint is 155 bar, the average coolant temperature is 306.9°C, and the steam generator pressure is 70 bar. The load power of the pressurizer unit is decreased from 100% to 75%, also known as turbine leading mode or reactor following mode is the standard preferred mode of operation in a pressurized water reactor. The adopted model parameters, as summarized in Table 1, were fed to the software. The initial pressure level (amount of fraction full) was changed to 50.92%. Both the reference paper and PCTRAN incorporated the 1995 formulation of the thermodynamic properties of ordinary water substance intended for general and scientific use which was issued by The International Association for the Properties of Water and Steam (IAPWS). For PCTRAN, the water properties based on IAWPS were embedded inside PCTRAN codes in a compiled executable program. The first step in operating the software is by running the software to 5 seconds in order to reach steady state. Fig. 4 shows that the simulation is frozen at exactly 5 seconds to initiate power reduction via turbine load decrease. The turbine power demand and ramp rate of power demand were decreased to 75% and 5%/min respectively. The simulation was resumed and allowed to finish at 1000s. The variables listed in Table 1 summarize the steady state values. Fig. 5 shows interface of plotting transient variables such as power outputs of core reactor, nuclear flux, and turbine load during a run to show instantaneous performance.

Table 1. Steady state variables at t=1000s

Variable	Value	Variable	Value
Rod position	90%	Pressurizer level	48.3%
Reactor power	1351 MWt	Generator load	450 MWe
T_{avg}	296.8°C	Steam generator pressure	57 bar
Pressurizer pressure	155 bar	Steam generator level	50%

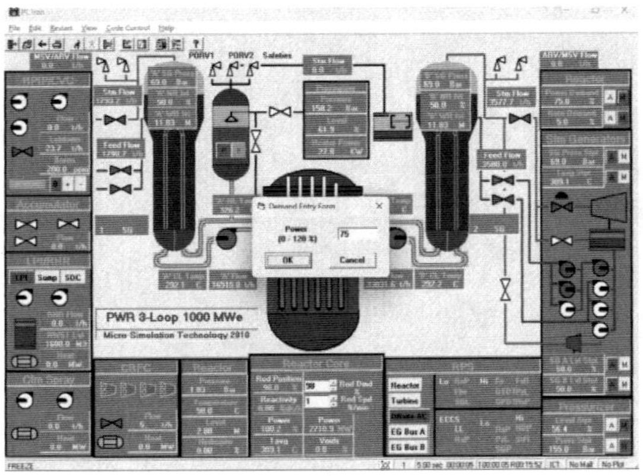

Figure 4. Power reduction step

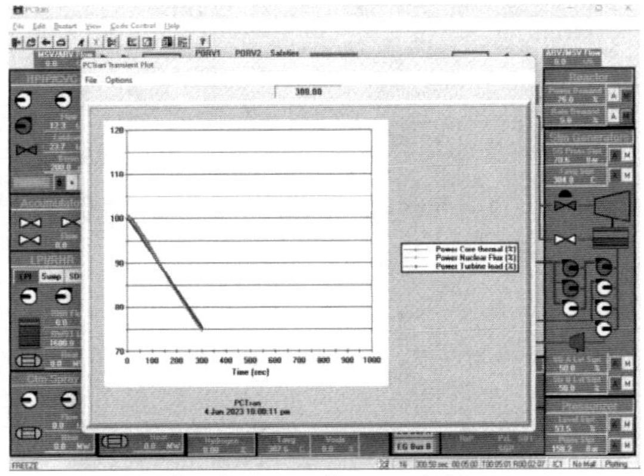

Figure 5. Plotting transient variables step

3.2. ANFIS controller training

The data used in training each ANFIS controller is from the Shippingport PWR experiment. It is then fed as training data in the ANFIS function of MATLAB. Usually there is a train-test split in evaluating machine learning (ML) algorithms. The training dataset is utilized to fit the model while the testing dataset is used to assess this trained model. Common split percentages for training and testing are 80-20, 67-33, and 50-50. However, there are instances where splitting the whole dataset is no longer needed. In this study, we have 284 data points which is relatively small. Also when an ANFIS or ANN model is loaded to a controller instead of classification purposes, it is proper to use the entire dataset so that the ML model is able to adapt to the data and provide more efficient control.

Three constant, triangular membership functions for each member of data were employed. Since the dataset is composed of one input for each controller, there are three equally spaced membership functions for each one of the features. In generating the fuzzy interference system (FIS), grid partitioning was used, wherein essentially, the data space is divided into rectangular sub spaces identified as "grids" based on the amount of membership functions and its' types. In this study, there are three membership functions (m) for each one input (n), therefore, the total number of rules needed is m^n which is 3. This is why the amount of fuzzy rules tend to increase exponentially along with the increase of input variables. For this reason, grid partition is only suitable for systems with inputs less than 6, to keep an optimal number of rules. In training the FIS, the hybrid optimization method was used. It is a combined method of least-squares and back propagation gradient descent. For this ANFIS system, the training epochs set at 1000 and training error tolerance is set to 0.

4. Results and Discussion

Pressurizer model validation was performed during a load change from 100% to 75% in order to evaluate the developed reference pressurizer model. The predictions of the reference pressurizer model were compared to reported simulation data of PCTRAN and is shown in Fig. 6 which displays great compatibility between simulation and model data. RCS pressure decreased until 300 seconds which is the initial cool down from insertion of control rods. Pressure insurges and is restored to around 157 bar from the effect of the electric heater and its' associated control system. Pressure insurges and is restored to around 157 bar from the effect of the electric heater and its' associated control system. In this turbine leading mode of operation, the control rods are inserted which induces reactor power following the set turbine load. At around 300s, as shown in Fig. 7, turbine load reaches steady state and at around 340s, the rod system reduces the rate of rod insertion, which results in thermal power undershooting wherein the power core thermal output dipped at around 73%.

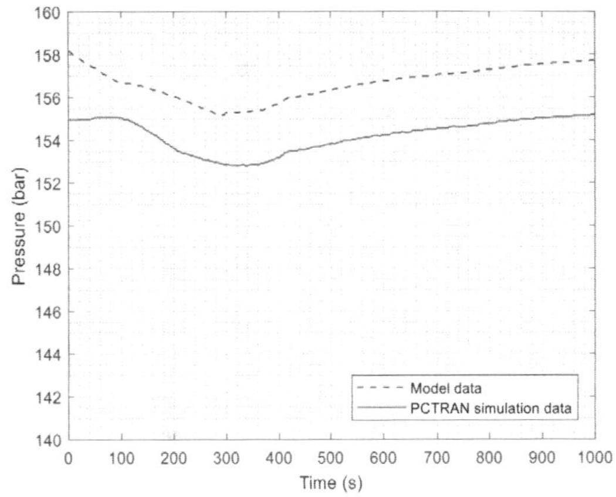

Figure 6. Pressure response of developed model and PCTRAN pressurizer unit

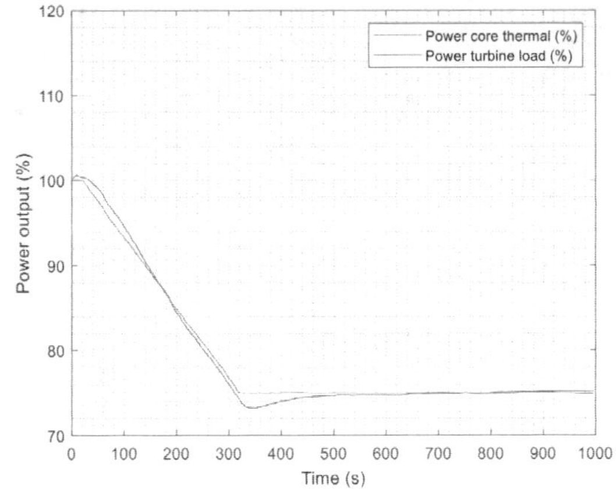

Figure 7. Turbine load and reactor core power output

In order to establish a comparative performance of the predictive capability of ANFIS, the transient data is compared with the trained data using the Neuro Fuzzy Designer App in MATLAB. Since it is essentially a fuzzy control based on a competitive algorithm provided by neural network, the ability to classify and identify patterns in a dataset allows for excellent predictive and control capability. The number of membership functions set is generally decided by a professional or someone who is familiar with the system to be modeled. In this study, the ANFIS parameters assigned were chosen empirically, that is by trial and error, to an output which produces the lowest training error value. Loss-of-load transients, or the system response to sudden load excitation from a steady state, especially from full or high-power levels generate the most rapid insurges that a pressurizer encounters. Shippingport transients are considered the benchmark since it demonstrates all insurge, outsurge, spray actuation, and heater cycling. The first load-drop test from Shippingport was initiated from 51 to 0 MW(e), whose pressure response curve is shown in Fig. 8a. The second load-drop test was initiated from 74 to 10 to 0 MW(e), whose pressure response curve is shown in Fig. 8b. The last load-drop test was initiated from 105 to 35 to 10 and then to 0 MW(e), whose pressure response curve is shown in Fig. 8c.

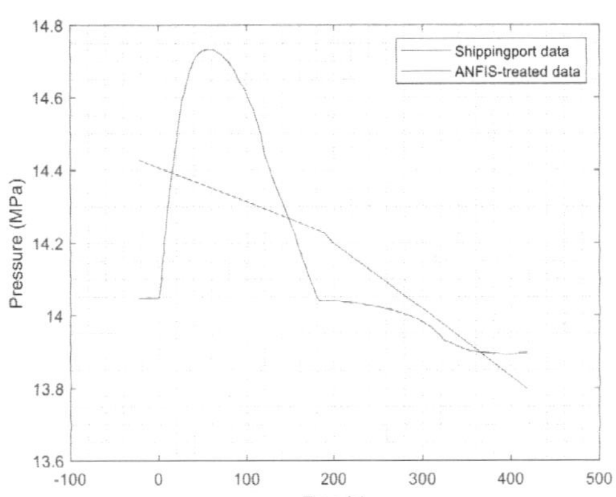

(a)

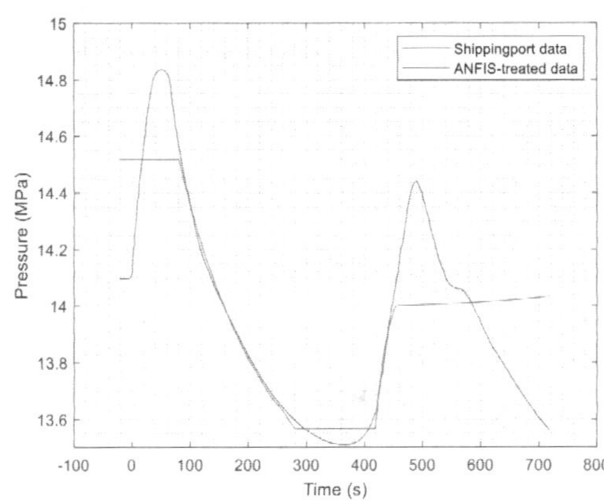

(b)

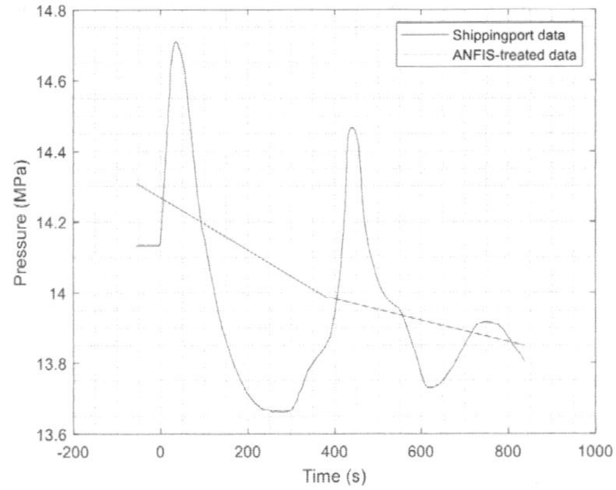

(c)

Figure 8. Time series plot of actual pressurizer transient loss of load from (a) 51 MWe, (b) 74 MWe, and (c) 105 MWe Shippingport PWR pressurizer unit vs. associated ANFIS-treated data

During steady state full power operation, the pressure is maintained at 14.2 MPa by actuating three banks of electrical heaters, a liquid spray system, and a relief valve. In training the first loss-of-load data, the designated epoch number was reached and ANFIS training was completed at epoch 2 with a minimal training RMSE of 0.188477. In training the second and third loss-of-load data, the minimal training RMSE is 0.193746 and 0.243594. The root mean square error (RMSE) is the standard method in measuring model error when it comes to predicting qualitative data as it provides a good estimate of the standard deviation of the observed value from the model's prediction. A RMSE lower than 0.75 reflects the efficiency of the model in predicting data which our system agrees to. The FIS model generated from training the Shippingport transient data is fed into each of the fuzzy logic controller. The response for the 51 MW(e), 74 MW(e), and 105 MW(e) loss-of-load transients is shown in Fig. 9 wherein the final values of pressure are peaked at 14.4193117, 14.17285483, and 14.27768721 MPa respectively, with no overshoot and nonzero initial slope. The rise time, or the time required for the system response to progress from 10% to 90% of its final steady-state value, is 24.94669235 seconds, 22.53800761 seconds, and 21.28534042 seconds. Peak time, or the time required to reach the peak value for the first time is at 418.85, 361.95, and 224.63 seconds respectively. Settling time is the time required for the system to enter ±2% of final steady-state value, which is not applicable to the response since there is no overshoot and there are no present oscillations within tolerance bands.

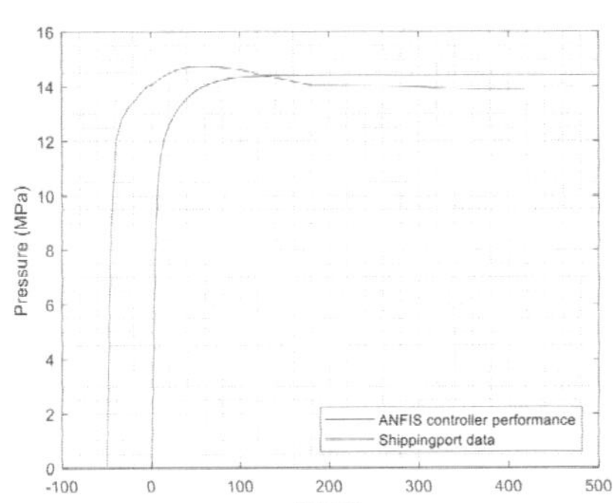

(a)

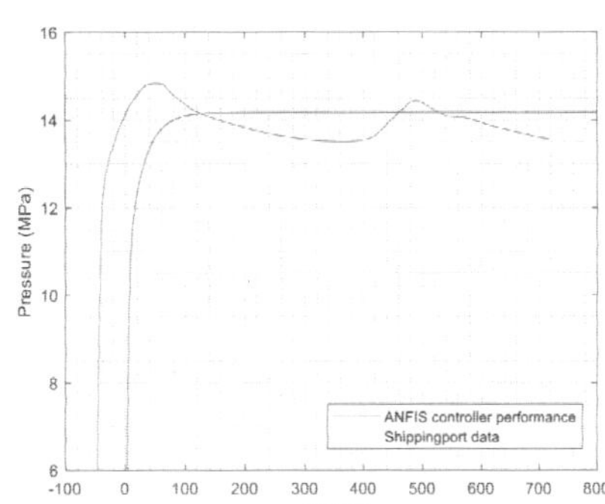

(b)

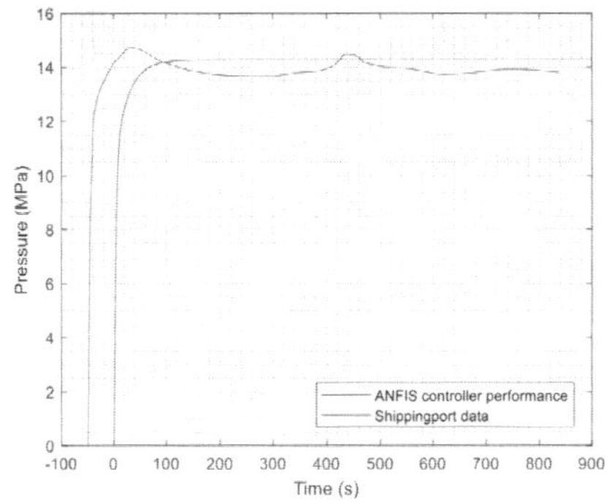

(c)

Figure 9. Control loop pressure response curve from (a) 51 MWe, (b) 74 MWe, and (c) 105 MWe loss of load operations

5. Conclusion

Daily variations in power demand challenge the load following control of NPPs. The pressurizer unit in a PWR is regulated to compensate for load fluctuations. Studying a pressurizer's dynamic behaviour requires transient data and/or analysis codes or software to aid in predictions in the presence of arbitrary perturbations such as load power in the primary reactor cooling circuit. PCTRAN, which is a globally renowned training simulation software in nuclear power plants and research institutions, is used to validate the pressurizer unit, while Shippingport experiment was used to validate the whole control loop and to train the ANFIS Controller. The performance of the reference pressurizer model agreed well to the reported simulation data of PCTRAN. Comparison between the simulation data and model response showed great fit. On the other hand, there are a few publications that has executed pressurizer experiments but the most notable is the Shippingport pressurizer test, which was the only one conducted on an actual, large-scale, plant pressurizer unit. Its' data includes a set of loss-of-load transients to test relief system malfunction. The three load-drop transients from Shippingport pressurized water reactor were trained in an ANFIS environment in MATLAB and generated as a FIS model that was fed to the controller. The control response agreed well with the trained transients by producing a damped curve with a final, steady-state response at 14 MPa. This proves that the ANFIS controller is able to reject disturbances with lesser pressure overshoot and maintain level set-point.

References

[1] IAEA Core Knowledge on Instrumentation and Control Systems in Nuclear Power Plants *Nuclear Energy Series*

[2] Anon Mitsubishi Power | Reactor Pressure Vessels

[3] Wang X K, Yang X H, Liu G and Qian H 2009 Adaptive neuro-fuzzy inference system pid controller for SG water level of nuclear power plant *Proceedings of the 2009 International Conference on Machine Learning and Cybernetics* **1** 567–72

[4] Gu J, Reddy K H, Ramasamy S and Ramanathan P 2017 Hybrid Adaptive Neuro Fuzzy based speed Controller for Brushless DC Motor *Sci* **30** 93–110

[5] Hosham Salim A, Khalid Faisal S and Malak M 2018 Comparative between Neuro–fuzzy and PI Controller Temperature of Condenser in Thermal Power Plant (160 MW) *Indian J Sci Technol* **11** 1–9

[6] Psichogios D C and Ungar L H 1992 A hybrid neural network-first principles approach to process modeling *AIChE Journal* **38** 1499–511

[7] Suryabrata A G, Mulyana T and Witarsyah D 2015 Pressurizer simulator *International Conference on Electrical Engineering, Computer Science and Informatics (EECSI)* 6

IOP Conf. Series: Materials Science and Engineering **1318** (2024) 012047 doi:10.1088/1757-899X/1318/1/012047

Implementation of Skogestad's Internal Model Control Rule in Tuning Proportional-Integral Controllers for Two-Input and Two-Output Systems Considering Individual Controller Performance Trade-off

J M Gacuan[1*], J C Muñoz[1]

[1] Process Systems Engineering Laboratory, Department of Chemical Engineering, University of the Philippines, Diliman, 1101, Quezon City, Philippines

* jakegacuan@gmail.com, jmgacuan@up.edu.ph

Abstract. In industrial plants, tuning of decentralized Proportional-Integral-Derivative feedback (PID) controllers in Multiple-Input and Multiple-Output (MIMO) systems is a complex problem, due to input and output interactions, that must be done to ensure individual control objectives are met, while system being sufficiently robust. In this study, a two-input and two-output (TITO) PI controller tuning algorithm was developed considering controller performance prioritization given performance trade-offs, expressed as the novel Individual Controller Performance Trade-off Coefficient (ICPTC), and system robustness, expressed as sensitivity peaks from the generalized Nyquist stability plot. Simplified effective open-loop transfer functions (EOTFs) of the two input-output channels were derived and used along with Skogestad's Internal Model Control (SIMC) to calculate PI controller gains. Extensive simulations using Wood-Berry (WB) and Vinante-Luyben (VL) distillation column models were made to observe how the sensitivity peaks behave across the simulations, which was then used as the basis of the developed tuning algorithm. A simplified tuning algorithm, which requires less information compared to the main algorithm, was also derived. A mapping algorithm was also created, allowing specific controller gains from other studies to be mapped to an equivalent set of controller gains in the algorithm's search field. The effectiveness of the three algorithms were tested using Wardle-Wood distillation column (WW) and Xiong and Cai (XC) process models. It was demonstrated using the main algorithm that there is a trade-off between controller performances when ICPTC is varied, at different values of design sensitivity peaks between 1.96 to 4. The usefulness of the simplified tuning and mapping algorithms were also illustrated in the study. This study is potentially useful in tuning TITO PI controllers in industrial plants, where meeting individual control objectives while maintaining robustness is more important than optimizing overall control performance.

1. Introduction

In industrial plants, feedback control is important in meeting production rates at optimal specifications while ensuring safety and environmental compliance. Proportional-Integral-Derivative feedback control algorithm or PID control is the most common feedback algorithm used in the industry due to its availability and simplicity, with 97% of control variables are still PI- or PID-governed [1]. When not working as intended, PID controller tuning has to be done and in single-input and single-output systems, this is a multi-objective problem considering controller performance and system robustness.

IOP Conf. Series: Materials Science and Engineering **1318** (2024) 012047 doi:10.1088/1757-899X/1318/1/012047

PID tuning algorithms either fix system robustness and optimize for controller performance such as the Approximate M Integral Gain Optimization (AMIGO) algorithm [2] or use one adjustable parameter which takes care of both objectives, such as the closed-loop response time in Skogestad's Internal Model Control (SIMC) or lambda tuning [3].

For multiple-input and multiple-output (MIMO) systems, PID controller tuning is a more complex problem due to interactions between input and output variables. To solve this, studies commonly frame the problem as optimizing overall system performance i.e., sum of individual controller performance metrics such as integral of absolute errors (IAE), while meeting certain robustness metrics i.e., robustness index or SISO-based Nyquist-related metrics [2, 4-6]. While these studies yield controller performances with optimized sum of IAEs, looking at overall system performance may not be the best approach to take in certain situations. In industrial plants in particular, tuning MIMO systems to meet individual control objectives is the common practice. For example, if a temperature controller dictates the specifications of an important product stream, control engineers will prioritize its performance even if the other performance of the controllers in the system suffer to an extent. Also, given that industrial plants are broken down into multiple interconnected MIMO systems, dictating how an individual controller performs is important to account how the variable it controls affects the next MIMO system where it is connected to.

With this in mind, a simple tuning algorithm for two-input and two-output (TITO) systems, the most common type of MIMO systems, was developed considering individual controller performance and its trade-offs, and overall system robustness. The basic idea is that the algorithm's user can set the desired system robustness and the prioritization of the controller performances. The former using the sensitivity peak derived from the generalized Nyquist plot for MIMO systems, and the latter using the novel individual controller performance trade-off coefficient (IPCTC) or η_k. Afterwards, the tuning algorithm was simplified to require less information when tuning and a mapping algorithm was derived from the main tuning algorithm to map existing PI(D) controllers into equivalent PI controllers on the algorithm's search field, to compare the performances of the existing and mapped controllers.

The rest of the paper is organized as follows. Section 2 discusses the novel concept of ICPTC and the other components required to derive the main tuning algorithm. Section 3 shows the results of the extensive training simulations, emphasizes findings used to derive the tuning algorithm and illustrates its effectiveness using test systems. Sections 4 and 5 discuss, and illustrate the effectiveness, of the simplified tuning and mapping algorithms, respectively. Section 6 lists the conclusions of the study.

2. Components of the Algorithm

The following requirements are needed to perform training simulations and develop the algorithm – (1) a coefficient to define controller performance prioritization, (2) a metric to measure overall system robustness, (3) a way to independently tune PI controllers in a TITO system without using iterations and (4) training and test TITO systems. The importance of each is discussed in the subsections.

2.1. Controller prioritization – individual controller performance trade-off coefficient (ICPTC)

If overall system robustness is a constraint, then improving the performance of one input-output channel in a TITO system will degrade the performance of the other. Thus, defining the extent of control performance prioritization is important, and this will be expressed as ICPTC or η_k for a kth controller. The η_k's should add to unity and for TITO systems, that is $\eta_1 + \eta_2 = 1$. At $\eta_1 = 1$, the first controller is given priority while the second controller is made to be as slow as reasonably necessary. The reverse is true for $\eta_1 = 0$, with the second controller given priority, and as an example, $\eta_1 = 0.5$ represents balanced tuning for both controllers. The value of η_1 will be defined by the algorithm user and should depend on control objectives, process-related constraints and other concerns.

2.2. Overall system robustness – sensitivity peak from generalized Nyquist plot

To measure overall system robustness, the sensitivity peak (M_S) from the generalized Nyquist plot is used because it accounts for the worst-case scenario of perturbation that can lead to system instability

[7]. It is the reciprocal of the shortest distance between (-1, 0) and the Nyquist plot is generated using the generalized Nyquist expression $W = -1 + \det\left[\mathbf{I} + \mathbf{G}(s)\mathbf{C}(s)\right]$ which for TITO systems simplify to

$$W = C_1 G_{11} + C_2 G_{22} + C_1 C_2 \left(G_{11} G_{22} - G_{12} G_{21}\right) \tag{1}$$

The less robust it is, the higher M_S is. The range of M_S explored in this study is between 1.96 to 4, with 1.96 and 4 representing conservative and aggressive control, respectively. It was derived by squaring the recommended range of 1.4 to 2, based on the heuristics derived by Luyben [8].

2.3. Individual controller tuning – Effective Open-loop Transfer Function (EOTF) and Skogestad's Internal Model Control (SIMC)

Because of interactions, SISO tuning techniques cannot be readily extended to TITO systems. However, the interaction terms in a process transfer function matrix can be accounted for by deriving the effective open-loop transfer function (EOTF) of a particular input-output channel. This is done by assuming that the input-output channel of interest is open while the other loops are closed [5]. Assuming perfect control at lower frequencies, the EOTFs become independent of initial controller settings, and given the process transfer function matrix $\mathbf{G} = \left[G_{11} \quad G_{12};\ G_{21} \quad G_{22}\right]$, the EOTFs become

$$G_1^{\text{eff}} = \frac{y_1}{u_1} = G_{11} - \frac{G_{12}G_{21}}{G_{22}} \qquad G_2^{\text{eff}} = \frac{y_2}{u_2} = G_{22} - \frac{G_{21}G_{12}}{G_{11}} \tag{2}$$

Equation (2) still yields complex transfer functions, thus, Bode plot fitting with time-delay assumptions are done to come up with a first-order process with time delay (FOPTD) model. With the TITO system mathematically simplified to two SISO systems, Skogestad's Internal Model Control (SIMC) rule [3], which was proven to yield optimal controls in terms of performance and robustness, can now be used to independently tune the two TITO system controllers. Given an FOPTD model $G_{\text{n}}^{\text{eff}} = K_{\text{n}}^{\text{eff}} e^{-\theta_{\text{n}}^{\text{eff}} s} / \left(\tau_{\text{n}}^{\text{eff}} s + 1\right)$, derived from the simplified EOTFs, the controller gains are

$$K_P = \frac{1}{K_{\text{n}}^{\text{eff}}} \frac{\tau_{\text{n}}^{\text{eff}} + \theta_{\text{n}}^{\text{eff}}/3}{\tau_C + \theta_{\text{n}}^{\text{eff}}}; \qquad \tau_I = \min\left[\tau_{\text{n}}^{\text{eff}} + \frac{\theta_{\text{n}}^{\text{eff}}}{3}, 4\left(\tau_{\text{C,n}} + \theta_{\text{n}}^{\text{eff}}\right)\right] \tag{3}$$

In the training simulations, $m_{\text{n}} = \tau_{\text{C,n}} / \theta_{\text{n}}^{\text{eff}}$ representing the closed-loop response time for an input-output channel was used to get the controller gains, with range of m_{n}'s in the runs being $m_{\text{n}} \in [-0.5, 5]$.

2.4. Training and test TITO systems

The Wood-Berry (WB) and Vinante-Luyben (VL) distillation column systems [5] were used for training simulations while the Wardle-Wood distillation (WW) [9] and Xiong-Cai model (XC) [10] systems were used for testing algorithm effectiveness. The process transfer function matrices of these systems are available in the cited references.

3. Main Tuning Algorithm and Test Results

From the process transfer function matrices of the four TITO systems, the simplified EOTFs expressed as FOPTDs were derived and are shown in table 1. These were used in the simulations.

Table 1. FOPTD values for simplified EOTFs of training and test TITO systems

	K_1^{eff}	τ_1^{eff}	θ_1^{eff}	K_2^{eff}	τ_2^{eff}	θ_2^{eff}
WB	5.9503	7.2057	1.0431	-9.1666	5.7656	3.0300
VL	-1.3326	6.0549	1.0431	2.6074	8.2195	0.3477
WW	0.0625	24.7352	6.0000	-0.0725	18.9680	8.0000
XC	14.5021	23.2693	3.0300	2.9516	25.1084	4.0235

3.1. Extensive Training Simulation Results and Major Observations

The derived EOTFs for the training systems were used to calculate 201 sets of PI controller gains, depending on m_n, for the first and second input-output channels. To account for all pairings, 40,401 simulations were made for each system, with the stability and robustness of each closed loop system in the simulation calculated using the generalized Nyquist stability criterion and the derived sensitivity peak from the Nyquist plot, respectively. The detailed results of the simulations for the VL system are shown in Figure 1 while the condensed results for the WB system containing curves of certain sensitivity peaks while also highlighting the major observations are shown in Figure 2.

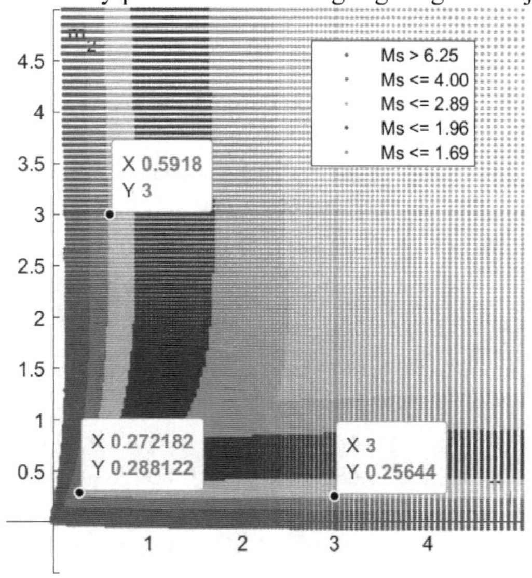

Figure 1. Detailed results for VL system, with each region containing points within a certain M_S value and illustration points A, B and C

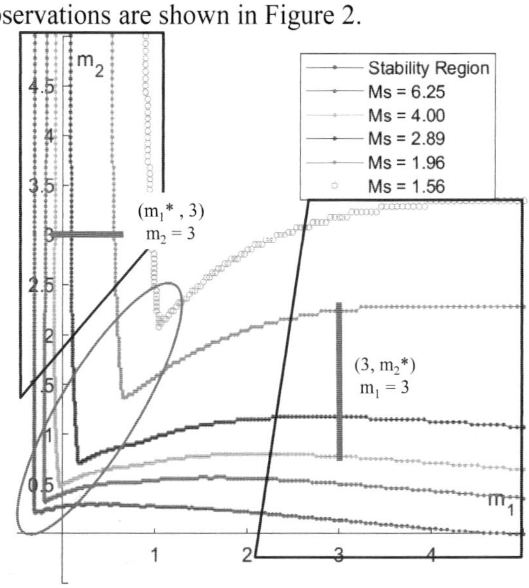

Figure 2. Condensed results for WB system, showing the stability region, curves of certain M_S and highlighting the major observations.

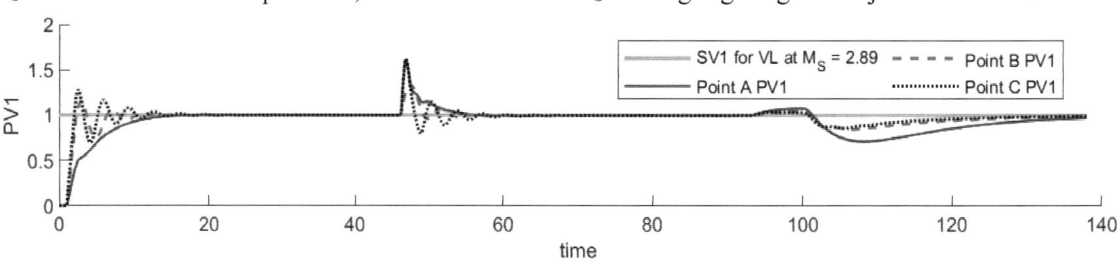

Figure 3. Illustration of false optimal solutions for VL system at $M_S = 2.89$ curve

There are two major observations in the training simulations which are the foundations of the main tuning algorithm – (1) the flattening of the M_S curves at sufficiently high values of either m_1 or m_2 as shown in the black trapezoids in Figure 2 and (2) the presence of false optimal solutions in the middle portions of the graph i.e., where m_1 and m_2 both significantly contributing to the system robustness, as shown in the blue ellipse in Figure 2. The first observation is reasonable, since when m_1 or m_2 is high, the controller derived becomes very slow like manual control. Therefore, the M_S at these segments of the curve become a function of solely of the other m, i.e., m_1 or m_2, respectively. The value of m where the M_S curves start to flatten varies, but a critical value of $m^* = 3$ was chosen, given that the M_S curves for the training systems at $1.96 \le M_S \le 4$ are already flat while not being unreasonably high that the slow controller will be performing badly at extreme values of η_1. The second observation shows that while the M_S curves protrude in the middle, indicating improved performance for both controllers, the response plot shows that control performance degraded for both controllers, as illustrated in Figure 3, using controllers represented by points A and B, at the flat segments of $M_S = 2.89$ curve, and point C, at the middle of the same curve. With this, the trade-off equation was set to be conservative by being as far from the middle part of the M_S curve as possible.

3.2. Trade-off Equation and Main Tuning Algorithm

With the first observation, the two-dimensional problem can be reduced into two one-dimensional problems, reducing the run time to ~2% of the original extensive simulations since instead of doing all simulations exhaustively, m_1 (then m_2) can be fixed at 3 to get all M_S as m_2 (then m_1) is varied. At a desired M_S, the points from these two sets of simulations represent the extreme cases of $\eta_1 = 0$ and $\eta_1 = 1$, respectively, or the end points the trade-off curve. Several families of curves were considered and experiments were then conducted to ensure that the trade-off curve generated from the equation is close to flat near the end points, that points at different values of η_1 are as equally spaced on the curve as possible, and that it is sufficiently conservative in avoiding false optimal solutions while ensuring individual controller performance trade-offs. With these, the trade-off equation used in the main tuning algorithm is shown in equation (4), with $\varepsilon = 0.1$ as the parameter selected from experiments.

$$m_1 = \left(3 - m_1^*\right)\left[\left(1+\varepsilon\right)^{\eta_2} \varepsilon^{\eta_1} - \varepsilon\right] + m_1^* \qquad m_2 = \left(3 - m_2^*\right)\left[\left(1+\varepsilon\right)^{\eta_1} \varepsilon^{\eta_2} - \varepsilon\right] + m_2^* \tag{4}$$

In equation (4), m_1* and m_2* represents the m_n's which yields the design M_S at $\eta_1 = 1$ and 0, respectively. For the training systems, the sensitivity peaks using the calculated controller gains from the trade-off equation is less than the design M_S. Now, the main tuning algorithm is shown in Figure 4.

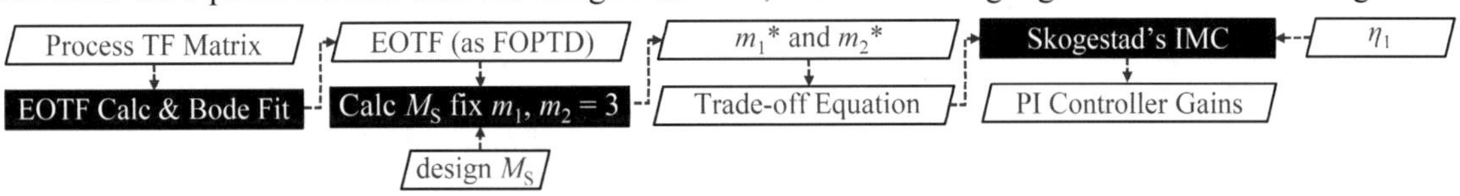

Figure 4. Main TITO PI tuning algorithm considering M_S and ICPTC

3.3. Main Tuning Algorithm Tests and Results

The algorithm was then tested by calculating the controller gains for the two test systems, WW and XC systems, at three robustness levels, namely, $M_S = 1.96$, 2.89 and 4, and at varying the values of η_1 at 0, 0.25, 0.5, 0.75 and 1. The response of the first and second control variables to set point changes and disturbances were plotted and the presence of trade-offs in terms of overshoots, settling times and disturbance rejections were observed as η_1 is varied. The results all paint the same picture, that the algorithm yields controller gains that has fast first controller performance at $\eta_1 = 1$, then it slows down as η_1 is decreased, up to 0. The reverse is true for the second controller performance. Sensitivity peaks from the calculated controller gains is at worse, only 4% less robust than the design M_S across all runs while most of the time, it is more robust than the design M_S. To illustrate these results, two response plots at varying values of η_1 are shown below, the WW at design $M_S = 4$ and XC at design $M_S = 2.89$.

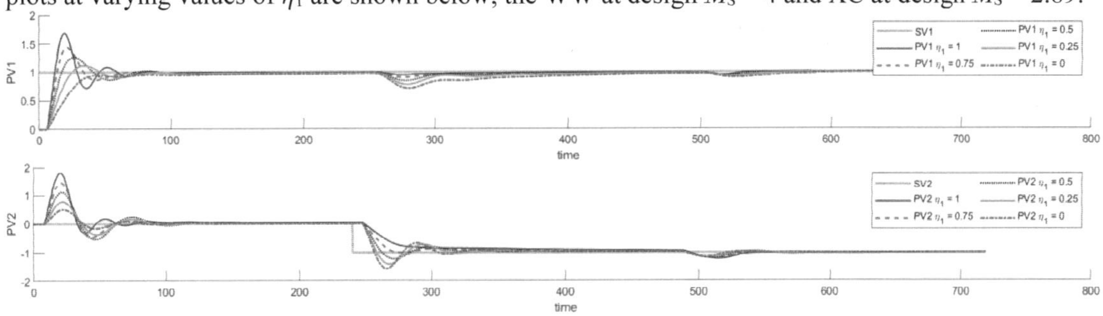

Figure 5. Response plots of PV1 and PV2 for WW at $M_S = 4$

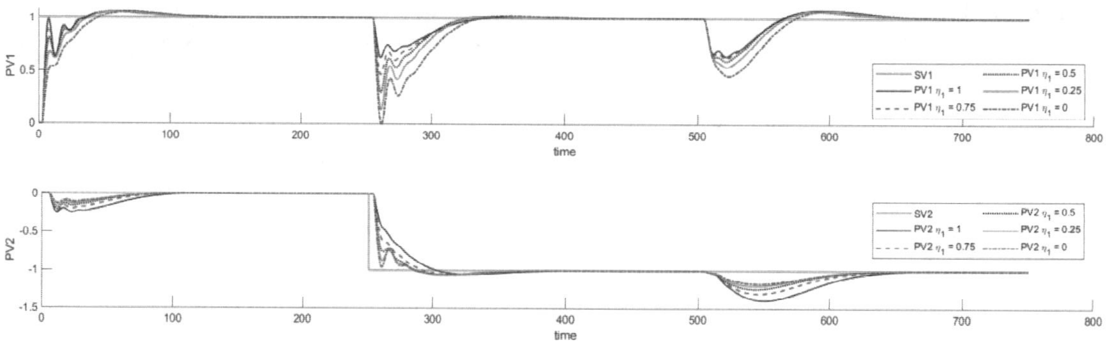

Figure 6. Response plots of PV1 and PV2 for XC at M_S = 2.89

Notice how the PV1 of both systems are relatively fast when $\eta_1 = 1$, and these become slower until $\eta_1 = 0$, with PV1 slow to catch up to the set point change. On the other hand, the PV2 of both systems have improved disturbance rejection and relatively faster set point response as η_1 increases. Overall, this illustrates that the main tuning algorithm yields controller gains that prioritizes individual controller performances given η_1 while being more or almost as robust as the design M_S.

4. Simplified Tuning Algorithm

The main tuning algorithm requires the complete process transfer function matrix and the calculations of sensitivity peaks at multiple values of m_1 and m_2, albeit at reduced number of runs. With the former requiring complex system identification or modelling while the latter requiring calculation of robustness measures that are usually abstract in industry settings, identifying a simpler way to implement the algorithm will improve its usefulness in industrial plants.

First, simplifying assumptions on equation (1) led to the conclusion that the EOTFs when m_1 or m_2 is fixed at 3 is close to the transfer functions approximated when step test experiments are designed for the first (or second) input-output channel while the second (or first) input-output channel is closed with the controller tuned to be very slow. Since SISO experiments are simpler, approximating process transfer function models is simpler too. Then, instead of getting m_1* and m_2* using M_S calculations, simple simulations to determine values of m at borderline stability are instead done, then the following equation is implemented to calculate the approximate values of m_1* and m_2* at different settings:

$$\left(m^*_{i,Ms} + 1\right) = F_{Ms}\left(m^*_{i,stable} + 1\right) \tag{5}$$

with F_{Ms} equal to 1.45, 1.70 and 2.45, for aggressive, intermediate and conservative tuning settings, respectively. These values were derived by getting the relationship between the values of m at borderline stability and the values of m at M_S = 4, 2.89 and 1.96, through linear regression. Afterwards, the equation (4) is used. The simplified algorithm is summarized in Figure 7.

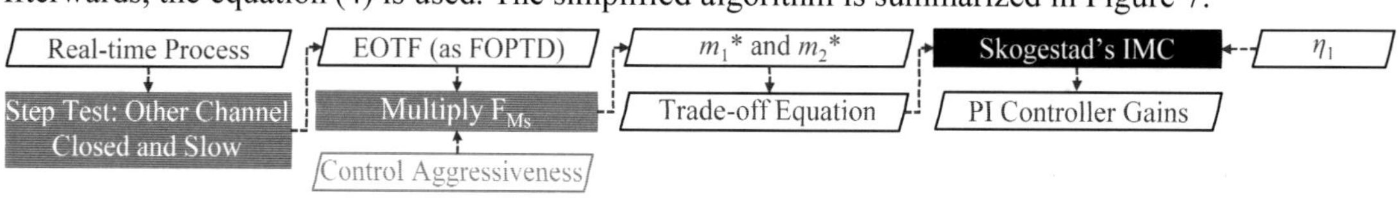

Figure 7. Simplified TITO PI tuning algorithm considering controller aggressiveness and ICPTC

To show its effectiveness, the response plots for the WW system using the controller gains calculated from the main tuning algorithm is compared to the response of the same system using the controller gains calculated from the simplified tuning algorithm. Results were compared at three values of design M_S, namely 4, 2.89 and 1.96 and at $\eta_1 = 0.5$. All three performance comparisons show that the main and simplified tuning algorithm results are comparable. The response plot for the WW system at M_S = 2.89 and $\eta_1 = 0.5$ is shown in Figure 8 to illustrate the said performance comparability.

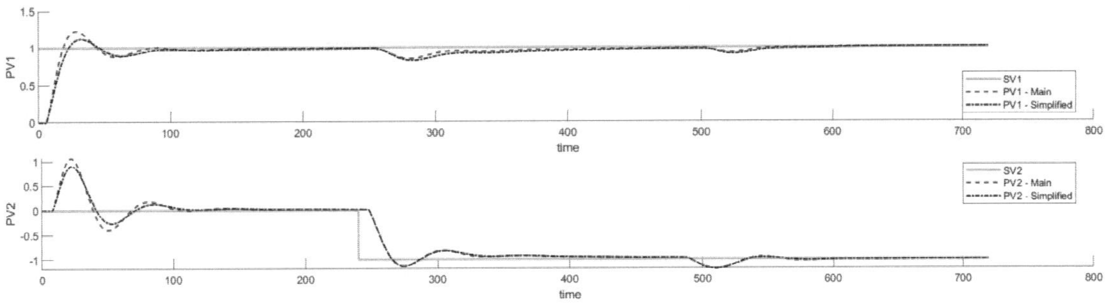

Figure 8. Response plot comparison for WW system using the main and simplified tuning algorithms

5. Mapping Algorithm

With the main and simplified algorithms, the user just needs η_1 to set the controller prioritization and either the design M_S or the controller tuning aggressiveness to come up with the controller gains. The premise though is that the user would perform tuning as if it is done from scratch. However, a better way to do this is to tune directionally, by knowing how robust the current system is and knowing which controller should be prioritized. If the starting point when using the proposed algorithm is known, then there is an idea on how to adjust the controller gains using the main tuning algorithm. Noting that the existing controllers are likely derived from a different algorithm or were derived using an old process model, it is likely that the existing controllers are not on the defined search field of the main tuning algorithm. With this, a mapping algorithm that calculates a set of controller gains on the defined search field that is equivalent to the existing controllers was developed. Also, since the main algorithm considers controller prioritization instead of overall controller performance, the results of the main algorithm are more general compared to the results of algorithms considering overall controller performance. Thus, the mapping algorithm is a way for comparisons to be made between the performances of the main tuning algorithm and existing algorithms.

The gist of the mapping algorithm is to map the existing controller to an equivalent controller in the search field one at a time while the other controller is assumed to be sufficiently slow. This is done by pairing the first existing controller C_1 to the slow controller represented by $m_2 = 3$ and then finding the sensitivity peak of the closed-loop system with this controller set. Then, the equivalent controller represented by m_1** is obtained, which when paired to $m_2 = 3$ yields the same sensitivity peak as the C_1–m_2 controller pairing. The same is done to get the second mapped controller, by pairing the second existing controller C_2 to the slow controller represented by $m_1 = 3$. The results will be a mapped controller set represented by (m_1**, m_2**). Equivalent design M_S and η_1 can be calculated as supplemental information as well. The mapping algorithm is summarized in Figure 9.

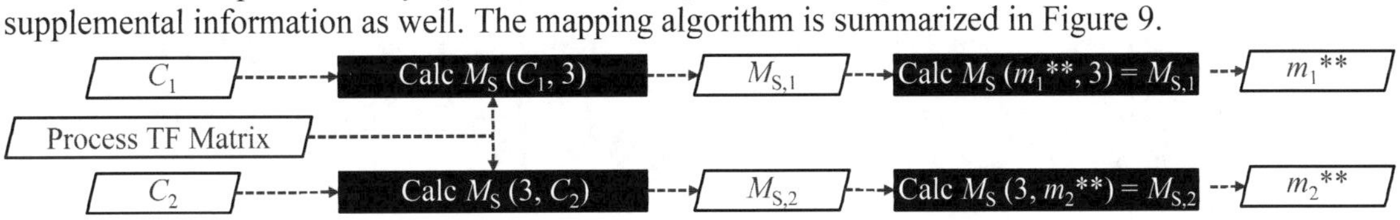

Figure 9. Mapping algorithm for existing controllers outside algorithm search field

To show the algorithm effectiveness, it was tested using two results from different algorithms, which are the PI controller gains obtained by Garrido et al in 2021 [4] for a WB system and the gains obtained by Vu et al in 2007 [11] for a WW system. The existing and mapped gains are shown in table 2. The plots when the existing and mapped gains are used in the said systems are shown in Figure 10.

Table 2. Existing and mapped controller gains calculated from the mapping algorithm

	Existing Controller Gains				**Mapped Controller Gains**			
	K_1	$\tau_{I,1}$	K_2	$\tau_{I,2}$	K_1	$\tau_{I,1}$	K_2	$\tau_{I,2}$
WB [4]	0.71	13.95	-0.042	2.60	0.64	7.55	-0.078	6.78
WW [11]	48.10	18.99	-25.40	26.30	58.06	26.74	-23.03	21.63

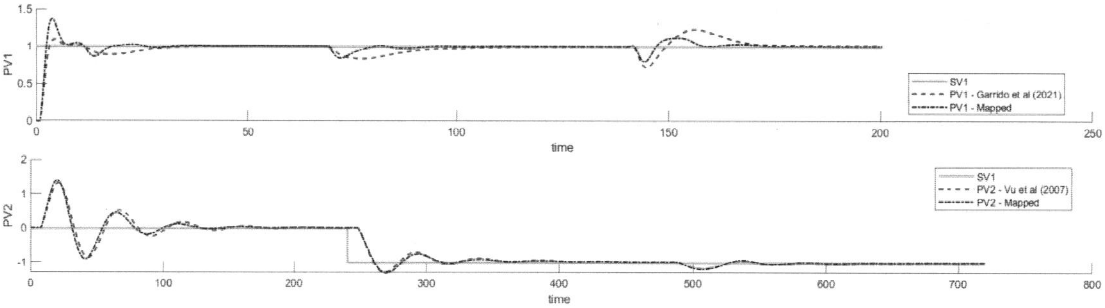

Figure 10. PV1 of WB and PV2 of WW systems plots when using existing/mapped controllers

Both response plots show that performances are comparable when using the existing and mapped controller gains, with the WB system showing that the mapped controllers have better disturbance rejection. This shows that the mapping algorithm is effective in calculating equivalent controllers.

6. Conclusions

In this study, the MIMO PID controller tuning problem was viewed by considering the individual controller performances and the overall system robustness when tuning a set of controllers. EOTFs and SIMC were combined to perform training runs, which was used to develop the main tuning algorithm considering design sensitivity peak and ICPTC. A simplified tuning algorithm was designed to require less information for tuning while a mapping algorithm was derived to compare existing controllers with mapped controllers in the search field. The effectiveness of the three algorithms were illustrated. These algorithms were developed with industrial plants in mind, knowing that individual control objectives are the main considerations in tuning PI controllers in TITO systems. More in-depth evaluation of the interactions between variables is recommended to further improve the work.

References

[1] Desborough L and Miller R 2002 Increasing customer value of industrial control performance monitoring – Honeywell's experience *AICHE Symp. Ser.* **98** 173-91

[2] Pérez J A R and Herrero P B 2012 Extending the AMIGO PID tuning method to MIMO systems. *IFAC Proc. Vol.* **45** (3) 211–6

[3] Grimholt C and Skogestad S 2012 Optimal PI-Control and verification of the SIMC tuning rule. *IFAC Proc. Vol.* **45** (3) 11–22

[4] Garrido J, Ruz M L, Morilla F and Vázquez F 2021 Iterative method for tuning multiloop PID controllers based on single loop robustness specifications in the frequency domain *Processes* **9** (1)

[5] Vu T N L and Lee M 2010 Independent design of multi-loop PI/PID controllers for interacting multivariable processes *J. Process Control* **20** (8) 922–33

[6] Boyd S, Hast M and Åström K J 2016 MIMO PID tuning via iterated LMI restriction. *Int. J. Robust Nonlinear Control* **26** (8) 1718–31

[7] Skogestad S and Postlethwaite I 2005 *Multivariable Feedback Control: Analysis and Design* (West Sussex: John Wiley & Sons Ltd) p 152

[8] Luyben W L 1986 Simple method for tuning SISO controllers in multivariable systems *Ind. Eng. Chem. Process Des. Dev.* **25** (3) 654-60

[9] Monica T J 1987 *An exploration of solutions to the multivariable control problem* (Doctoral dissertation, Lehigh University)

[10] Xiong Q and Cai W J 2006 Effective transfer function method for decentralized control system design of multi-input multi-output processes *J. Process Control* **16** (8) 773–84

[11] Vu T N L, Lee J and Lee M 2007 Design of multi-loop PID controllers based on the generalized IMC-PID Method with Mp criterion *Int. J. Control Autom. Syst.* **2** (2) 212–7

IOP Conf. Series: Materials Science and Engineering **1318** (2024) 012048 doi:10.1088/1757-899X/1318/1/012048

Application of Dynamic Simulation for the Design of the LNG Inlet Gas Compressor Anti-Surge Protection

K M Soriano[1] and J C Muñoz[1]

[1] Process Systems Engineering Laboratory, Department of Chemical Engineering, University of the Philippines, Diliman, Quezon City, 1011, Philippines

kmsoriano1@up.edu.ph and jcmunoz@up.edu.ph

Abstract. An inlet gas compressor system is simulated in Aspen HYSYS Dynamics v12.1. The emphasis of the simulation is to design the system to prevent or mitigate surges during an emergency shutdown (ESD). The dynamic simulation aims to assess the performance of the anti-surge control valve (ASCV), to add and size a cold gas bypass valve (CGBV) or hot gas bypass valve (HGBV) if the ASCV is insufficient, to determine the corresponding bypass valve size, and to select which bypass valve configuration is appropriate. An ASCV accompanied by an 8" HGBV is shown to be the recommended anti-surge protection configuration to mitigate surge due to its faster response and is more appropriate for motor-driven compressors which slow down more quickly during ESD. The ASCV provided by the compressor vendor with the recommended 8" HGBV can mitigate the effects of compressor surges during an emergency shutdown scenario.

1. Introduction

Natural gas has nonetheless emerged as a crucial alternative fuel in both home and industrial contexts due to its light composition, which contributes less to carbon emissions when burned in comparison to other fossil fuels. The Inlet Facilities are installed in Liquified Natural Gas (LNG) plants to receive natural gas from the offshore pipeline to separate unwanted materials such as solids, oil, and water. Inlet facilities are designed to run at steady-state well-head pressure, and this proves to be a problem down the line in a case when the natural gas operating pressure is insufficient for processing, such as when the production reservoir pressure decreases as the gas is extracted over time. Inlet Gas Compressors (IGCs) are installed in the inlet facilities to raise the gas pressure to a sufficient level that is suitable for efficient processing. These compressors are highly specialized and designed specifically for the process in which they are employed. Consequently, they are mostly custom-made, expensive, and challenging to replace. Designing an efficient compressor system is a necessity for the current LNG plants to prevent or mitigate surges.

Compressor systems are often designed based on set operating points in steady-state conditions. However, during actual operations, the compressor undergoes transient phenomena that can cause a disturbance known as a surge. A surge occurs when a compressor loses its ability to develop a pressure equal to its discharge causing rapid flow reversals. This damages the compressor and leads to significant operational and capital losses [1]. Figure 1 illustrates the surge cycle in the compressor performance map [1-2].

Typically, the surge is avoided through an anti-surge line where part of the compressor discharge is sent back to the compressor suction. The recycled gas is taken downstream of the compressor aftercooler which helps in reducing the temperature of the returning gas to safe levels, as illustrated in figure 2. The recycle loop configuration is equipped with an anti-surge controller that protects the compressor from surge by opening the Anti-Surge Control Valve (ASCV), bringing the compressor's operating point above the surge line. If the ASCV is not enough, a surge protection device such as a Cold Gas Bypass Valve (CGBV) or Hot Gas Bypass Valve (HGBV) is added to aid the ASCV to prevent or mitigate compressor surge, as illustrated in figure 2. The CGBV is installed parallel to the ASCV and helps to increase the flow rate of the cold gas in the anti-surge line or the recycle loop eventually increasing the flow rate at the suction of the compressor. The HGBV is installed in a recycle pipe, where the uncooled recycled gas is taken upstream of the compressor aftercooler. The HGBV helps the ASCV by increasing the flow rate of the gas at the suction of the compressor.

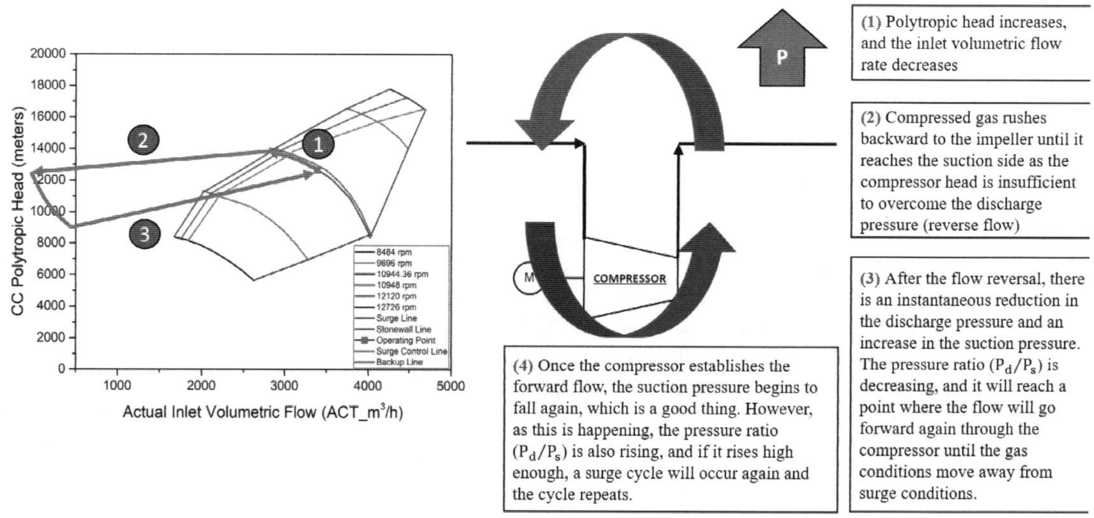

Figure 1. Surge cycle in the compressor performance map.

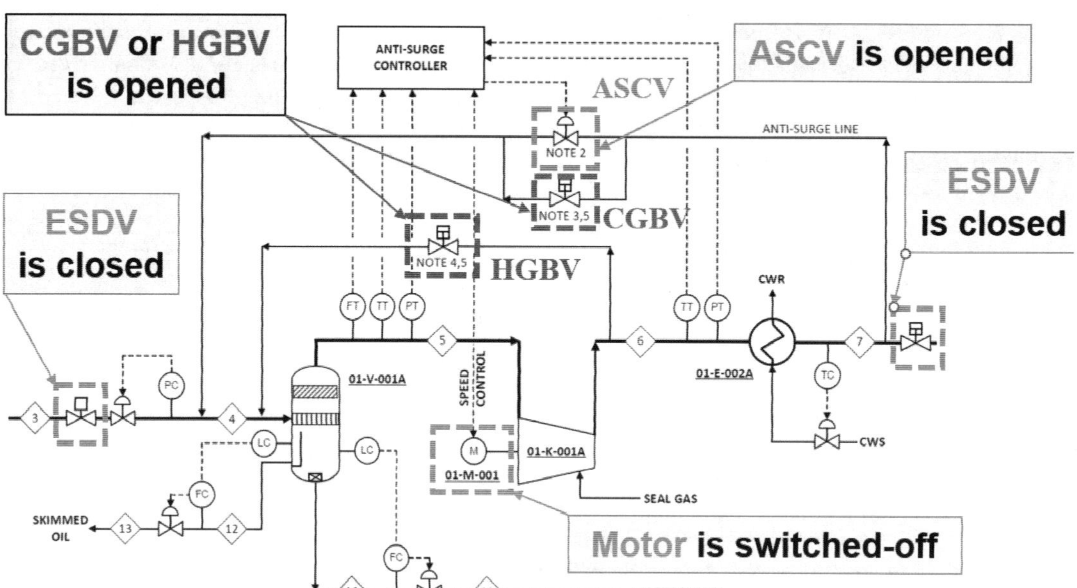

Figure 2. Process Flow Diagram for the Inlet Gas Compressor System.

If the surge is mitigated and not prevented, the maximum accepted surge duration after the compressor trip is 3 seconds for speeds above 70% as recommended by the compressor vendor. A typical plant operation that can cause significant disturbances and challenges for a compressor system is emergency shutdown (ESD).

ESD scenario is triggered when alarm trips are activated or the trip is initiated by the plant operator, leading to the initialization of safety control systems, which subsequently initiate the shutdown process. In this situation, the electric motors that drive the compressor are switched off, and the anti-surge (ASCV) and bypass (CGBV or HGBV) valves are immediately and completely opened. Additionally, the system is isolated from upstream and downstream operations by closing the Emergency Shutdown Valves (ESDVs). Figure 2 illustrates the shutdown process.

The dynamic simulation predicts how a process, and its associated control system will respond to various disturbances as a function of time. Compressor surge and emergency shutdown (ESD) are inherently dynamic and cannot be modeled in a steady state. The transition between operating conditions is best studied using dynamic simulation. Dynamic simulation can determine the surging time of the compressor during upset scenarios and allows better sizing of ASCV and CGBV or HGBV.

The findings of this study can serve as a basis for future modifications to the IGCs installed in the LNG Inlet Facilities, particularly when operational parameters approach critical values and instances of surge exceed the current working conditions. These results provide valuable insights that can guide the necessary adjustments and improvements to ensure the long-term stability and reliability of the facility. Furthermore, based on the current literature available [1-10], either CGBV or HGBV has been selected in combination with the ASCV as compressor anti-surge protection. There are no clear guidelines or procedures on how to select the bypass valve configuration. The results of this study can provide a rationale for anti-surge protection selection in inlet gas compressor systems.

The objectives of the study are the following: (1) to determine the response of the ASCV protection configuration measured in terms of the shortest time duration in the surge region (maximum acceptable surge duration after the compressor trip is 3 seconds for speeds above 70%) during an emergency shutdown, (2) to determine the response of the surge protection devices such as CGBV and HGBV in combination with the ASCV measured in terms of the shortest time duration in the surge region during an emergency shutdown, and (3) to determine the corresponding bypass valve size and select which bypass valve configuration is appropriate.

2. Methodology

The inlet stream data in the Inlet Gas Compressor System simulation input including the operating conditions, flow rates, and physical properties are tabulated in Table 1. The natural gas components and respective compositions are shown in Table 2.

Table 1. Inlet Process Conditions in the Inlet Gas Compressor System.

Process Variable	Unit	Value
Phase	-	Two-Phase
Operating Temperature	°C	34
Operating Pressure	kPag	2930
Molecular Weight	-	20.25
Mass Flow Rate	kg/h	89016
Density	kg/m^3	26.45

IOP Conf. Series: Materials Science and Engineering **1318** (2024) 012048 doi:10.1088/1757-899X/1318/1/012048

Table 2. Natural Gas Composition.

Component	Unit	Value
Methane	mole%	0.8478
Ethane	mole%	0.0527
Propane	mole%	0.0264
n-Butane	mole%	0.0129
n-Pentane	mole%	0.0062
n-Hexane	mole%	0.0172
Benzene	mole%	0.0024
Nitrogen	mole%	0.0266
Carbon Dioxide	mole%	0.0055
Water	mole%	0.0023

The compressor maps of both Polytropic Head (m) vs. Volumetric Flow rate (m^3/hr) and Polytropic Efficiency (%) vs. Volumetric Flow rate, as shown in figures 3 and 4, respectively. The surge control line defines the primary set point of the surge controller. The backup line is located between the surge line and the surge control line. The backup line is added closer to the surge line to enable more aggressive action if needed. The surge control line is 10% above the guaranteed surge flow, and the backup line is 5% above the guaranteed surge flow.

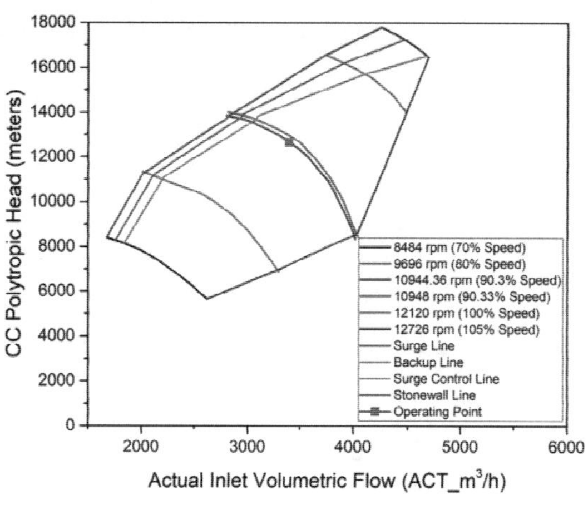

Figure 3. Compressor map for the Polytropic Head vs. Volumetric Flow rate at various speeds.

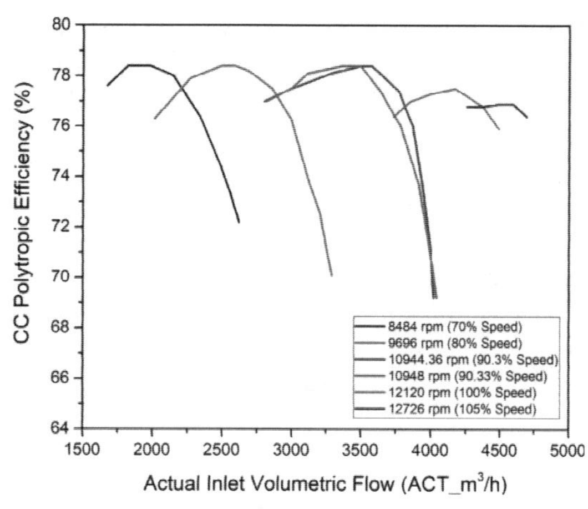

Figure 4. Compressor map for the Polytropic Efficiency vs. Volumetric Flow rate at various speeds.

The steady-state and dynamic models are prepared using Aspen HYSYS V12.1. The Peng-Robinson Equation of State thermodynamic package is used. The compressor, vessel, heat exchanger, piping, valve, pressure relief valve, and control systems are modelled using HYSYS standard module.

Figure 2 shows the Process Flow Diagram (PFD) for the inlet gas compressor system installed in the inlet facilities. The three-phase (oil/gas/water) mixture of natural gas at 2928 kPag and 34°C is controlled by a Pressure Control Valve (PCV) to maintain the compressor inlet pressure when there is a sudden increase in the upstream pressure. Normally, when the upstream pressure is expected to be low, the PCV is set for a minimum pressure drop and in a fully open position. The natural gas mixture is then routed to the Inlet Gas Compressor Knockout Drum (01-V-001A) to separate the gas and liquid (oil and

IOP Conf. Series: Materials Science and Engineering **1318** (2024) 012048 doi:10.1088/1757-899X/1318/1/012048

condensate) streams. The oil and condensate are level-controlled and sent to downstream facilities for further processing. The separated natural gas is routed to the Inlet Gas Compressor (01-K-001A) to increase the pressure up to 7452 kPag. The high-pressure natural gas is then cooled at 50°C in the Inlet Gas Compressor Aftercooler (01-E-002A), this is provided upstream of the anti-surge line. Since compressed gas undergoes an increase in temperature, the gas should be cooled before returning to the suction side of the compressor. The pressurized cooled natural gas is sent to the downstream facilities for further processing.

Dynamic simulations begin with a steady-state simulation with the desired operating conditions. The results from the steady-state case are used as the base point for transitioning to dynamic simulation. The steady-state simulation model depicts the uninterrupted operation of the inlet gas compressor. The steady-state model is created in Aspen HYSYS with the process flow diagram as the basis. Only one train of the actual three trains is simulated since the main objective of the study is to determine the anti-surge protection during an ESD scenario. The three inlet gas compressor trains are identical. Thus, modeling a single train is sufficient to study the dynamic behavior of the compressor during disturbances. The dynamic model is developed to match the equipment characteristics as much as possible. Figure 5 shows the Aspen HYSYS dynamic model.

Aspen HYSYS Dynamics function called Event Scheduler is used to simulate the case study scenario. This will automate the desired action sequence or task during dynamic simulation runs. The event schedule with tasks for the ESD scenario is shown in table 3. The tasks in this scenario considered time delay in opening and closing of valves. There is a 0.2-second time delay in the opening of the recycle and bypass valves (ASCV, CGBV, and HGBV) and a 1.2-second time delay in the closing of the inlet and outlet ESDVs from the time of the compressor trip. The simulation will be stopped at t = 5 minutes (300 seconds).

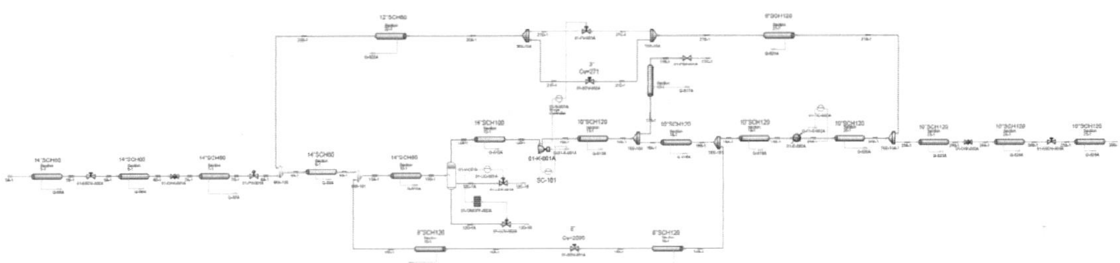

Figure 5. Inlet Gas Compressor System Aspen HYSYS Dynamic Model.

Table 3. ESD Event Schedule.

Time (seconds)	Event	Remarks
at 5	Start of Simulation	
at 5	Start of compressor trip – motor power starts decreasing to zero	
at 5.2	ASCV will open; or ASCV + CGBV will open; or ASCV + HGBV will open	Time delay of 0.2 seconds from the time of the compressor trip.
at 6.2	ESDV at suction and discharge will close	Time delay of 1.2 seconds from the time of the compressor trip.
at 300	End of simulation	

3. Results and Discussion

During an ESD where the compressor tripped instantly, the ASCV is not sufficient to mitigate compressor surge as depicted in figure 6. Even for cases where surpassing the surge line is acceptable given a 3-second margin for pushing it back to the safe operating region (above the surge control line) within the compressor map, the ASCV was not able to bring it to safe limits at any point in time. Thus, an additional bypass line and valve to assist the recycle flow of the gas is needed to further increase the suction flow rate during an emergency shutdown.

For the resulting dynamic calculation runs of the ASCV and CGBV during an emergency shutdown, this setup is still not sufficient to mitigate or prevent compressor surge as depicted in figure 7. Even for cases where a large bypass valve is used, the CGBV failed to assist the ASCV during coast down. Therefore, the installation of a hot gas bypass valve (HGBV) in combination with the ASCV is the final option to prevent or mitigate compressor surges.

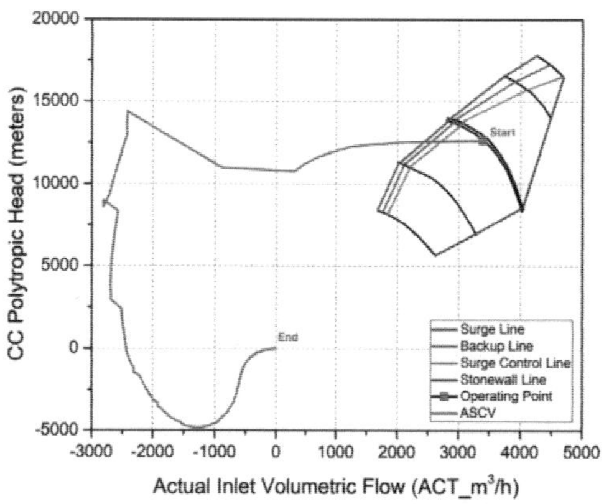

Figure 6. ASCV as anti-surge protection configuration.

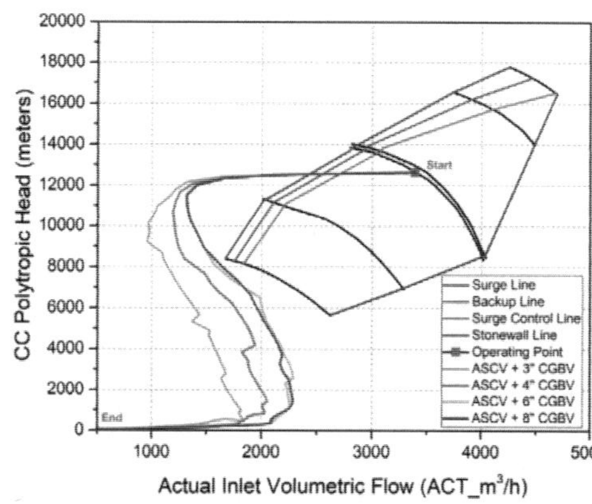

Figure 7. ASCV + CGBV as anti-surge protection configuration.

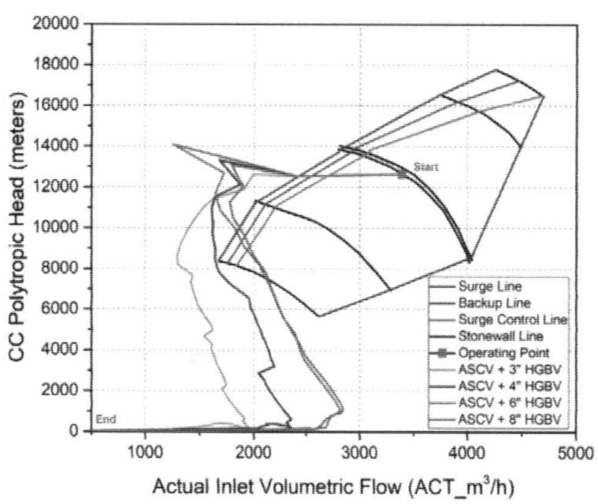

Figure 8. ASCV + HGBV as anti-surge protection configuration.

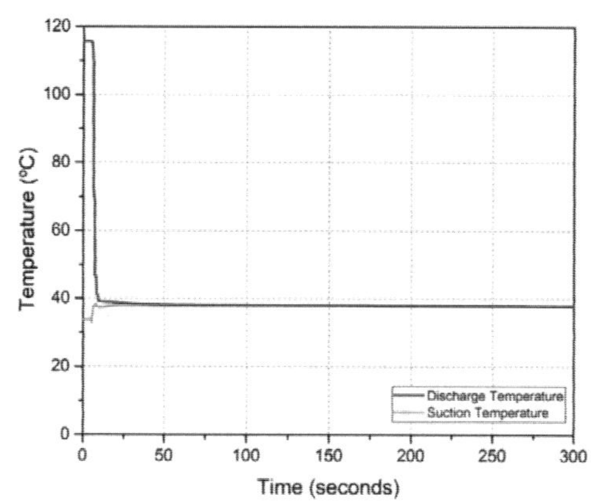

Figure 9. Compressor suction and discharge temperature profile during an emergency shutdown.

For the resulting dynamic calculation runs of the ASCV and HGBV during an emergency shutdown, the 3" and 4" valves failed to mitigate compressor surge due to smaller return volumes. For cases where large bypass valves are used, such as 6" and 8" HGBVs, the compressor experiences surge but the operating point managed to return to the area of the compressor map above the surge line, backup line, and surge control line as illustrated in figure 8. Thus, these HGBVs successfully assisted the ASCV in mitigating the surge during the coast down.

When the ASCV alone is not sufficient to prevent or mitigate surging, the ASCV in combination with the CGBV is usually the next option in anti-surge protection configuration since it keeps the system in a safe temperature range (below ~135°C) [7], by recycling cold gas to the compressor suction. Its disadvantage is that it will take a longer route and time for the gas molecules to return to the compressor suction. Hence, a slower anti-surge response from this configuration is inevitable. Moreover, the results of this study show that this setup for the inlet gas compressor system is not sufficient to aid the anti-surge control valve to mitigate the surge. Therefore, the installation of HGBV in combination with the ASCV is the final option to mitigate the compressor surge, if it will not exceed the allowable bearing temperature limit of ~135°C which is a common practice of compressor vendors. Exceeding the allowable bearing temperature will cause mechanical damage to the compressor.

Due to the slower anti-surge response of the ASCV + CGBV case, the ASCV + HGBV Combination is preferred by process designers. In the experience of the researcher in the EPC industry, the ASCV + CGBV combination is more commonly used for compressor systems that are driven by either a gas turbine (GT) or a steam turbine (ST). As these turbine types tend to have more inertia than their electric counterparts, GT/ST drivers have lesser tendencies to operate below the surge curve. This can be attributed to the delay brought about by the gradual decrease in the speed of the compressor impeller rotation.

Table 4 shows the summary of dynamic calculation results for the emergency shutdown scenario. The table includes the anti-surge protection configuration, results, surge duration, and which anti-surge protection is selected for the shutdown scenario.

Table 4. Dynamic Calculation results summary for ESD.

Anti-Surge Protection Configuration	Emergency Shutdown (ESD)		
	Results[a] (N/M/P)	Surge Duration (seconds)	Selected? (Yes/No)
ASCV	N	300	No
ASCV + 3" CGBV	N	300	No
ASCV + 4" CGBV	N	300	No
ASCV + 6" CGBV	N	300	No
ASCV + 8" CGBV	N	300	No
ASCV + 3" HGBV	N	300	No
ASCV + 4" HGBV	N	300	No
ASCV + 6" HGBV	M	0.7	No
ASCV + 8" HGBV	M	0.6	Yes

[a]N = Surge is Not Mitigated, M = Surge is Mitigated, and P = Surge is Prevented.

The ASCV + 8" HGBV anti-surge protection configuration is selected for the ESD scenario. This is due to its sufficient gas recycle volume at the suction side of the compressor to keep the operating point above the surge region. Furthermore, the HGBV/line also provides a shorter route for the gas to return to the compressor suction. Thus, a faster anti-surge response is expected in this configuration. Figure 9 shows the temperature profile of the compressor's suction and discharge during shutdown scenarios.

4. Conclusion

An inlet gas compressor system of the LNG Inlet Facilities is simulated in Aspen HYSYS Dynamics. The emphasis of the simulation is to properly design the system to mitigate compressor surge during transient operations such as ESD.

An anti-surge valve accompanied by an 8" HGBV/line is shown to be the recommended anti-surge protection configuration to mitigate surges in the inlet gas compressor system during shutdown scenarios. The type of compressor driver (motor, steam turbine, gas turbine) and its inertia have an impact on bypass valve selection. The ASCV + HGBV combination which has a faster anti-surge response is more appropriate for motor-driven compressors which slow down more quickly during ESD. The ASCV provided by the vendor with the recommended 8" HGBV can mitigate the effects of compressor surges during normal and emergency shutdown scenarios.

The use of dynamic simulation as a tool in the process design of the LNG inlet gas compressor system anti-surge protection allows multiple procedures to be tested in advance of the actual plant scenarios making the design more robust, safe, reliable, and efficient.

References

[1] Piercy, B. (2011). Objectives of Anti-Surge Control. Paper presented at the Year 2011 Spring National Meeting Ethylene Producers' Conference, IL, March 13-17, 2011. Retrieved from https://www.aiche.org/node/59319 [Accessed 19 February 2023]

[2] Saidin, S. (2019). An Overview on Transient Simulation Analysis. IChemE's Oil & Natural Gas Special Interest Group Webinar. April 29, 2019.

[3] Amin, A. A., Maqsood, M. T., & Mahmood-ul-Hasan, K. (2021). Surge protection of centrifugal compressors using advanced anti-surge control system. Measurement and Control, 54(5–6), 967–982. https://doi.org/10.1177/0020294020983372

[4] Botros, K. K., & Bakker, D. (2012). Application of Three Methods in Determining the Effectiveness of Surge Protection Systems in Gas Compressor Stations. Volume 1: Upstream Pipelines; Project Management; Design and Construction; Environment; Facilities Integrity Management; Operations and Maintenance; Pipeline Automation and Measurement, 781–793. https://doi.org/10.1115/IPC2012-90618

[5] Liedman, J., & Månsson, R. (2013). Dynamic simulation of a centrifugal compressor system, 69. Retrieved from https://publications.lib.chalmers.se/records/fulltext/179072/179072.pdf [Accessed 01 June 2023]

[6] Patel, V. (2007). Application of Dynamic Simulation in the Design, Operation, and Troubleshooting of Compressor Systems. Proceedings of the Thirty-sixth Turbomachinery Symposium, 95-106. https://doi.org/10.21423/R1TM0X

[7] Sarathy, J.V. (2021). Selected Topics in Oil and Gas Process Design for Graduate Chemical Engineers. Retrieved from https://jefersoncosta.com/vijaysarathy/ [Accessed 20 December 2021]

[8] Sodemann AA, Li YY, Lee JJ, et al. (2006). Data-Driven Surge Map Modeling for Centrifugal Air Compressors. ASME International Mechanical Engineering Congress and Exposition, Dynamic Systems and Control, Parts A and B, 611-618.

[9] Turbomachinery Magazine. (2017, May 11). Turbomachinery Magazine. Retrieved from https://www.turbomachinerymag.com/compressor-surge-and-its-detection.html. [Accessed 20 May 2017]

[10] Wolfswinkel, O. va., & Simeulers, J. P. M. (2001). Evaluation of Anti-Surge Control Concepts during ESD Simulation of dynamic process behaviour with PULSIM.

IOP Conf. Series: Materials Science and Engineering **1318** (2024) 012049 doi:10.1088/1757-899X/1318/1/012049

Removal of tetracycline from aqueous solution via Fenton process using iron-impregnated bamboo biochar

J D Calangi[1], J L Borres[1], M K P F Grande[1], D C Ong[1] and J M C Malones[2]

[1]School of Technology, University of the Philippines Visayas, Miagao, Iloilo 5023
[2]Chulalongkorn University, Bangkok, Thailand 10330

jdcalangi@up.edu.ph

Abstract. Tetracycline (TC) is a pharmaceutical that has been used extensively that its presence in water sources causes detriment to both aquatic life and human health. TC is not completely removed in wastewater treatment processes and could enter the environment through effluent and sludge. This study evaluated the potential of iron-impregnated bamboo biochar (Fe-biochar) as catalyst in the removal of tetracycline from aqueous solution via Fenton process. Characterization via Fourier transform infrared (FTIR) and energy dispersive X-ray spectroscopy (EDX) analyses verified the successful iron impregnation on biochar. TC removal was determined at 357nm absorbance through UV-Visible spectroscopy. Highest TC removal of 63.24% was achieved under conditions of 75 mg Fe-biochar, 0.45 M H_2O_2, and pH 5. Experimental data were assessed using the three factors Box-Behnken response surface experimental design where Fe-biochar dosage, H_2O_2 concentration, and solution pH were found to be significant parameters in TC removal. Furthermore, it was determined that the data fitted well in a reduced-cubic model with high coefficient of determination (R^2 = 0.9990) which indicates good regression. The optimum conditions were found to be 39.44 mg Fe-biochar dosage, 0.05 M H_2O_2, and pH 4, with TC removal efficiency of 33.90%. Overall, the results of this study suggested that Fe-impregnated bamboo biochar can be used as an effective catalyst in removing tetracycline from aqueous solution via heterogeneous Fenton process.

1. Introduction

Tetracycline (TC), with its multifaceted applications in human health, veterinary care, and agricultural practices, stands as one of the primary contributors to global antibiotic consumption with a rate of approximately 5,000 metric tons per year [1,2]. It is widely used in European Union countries and Switzerland with 2,575 tons and 1 ton yearly consumption, respectively. High consumption levels are also observed in Russia (13,579 tons/year), China (6,950 tons/year), and the USA (3,230 tons/year). Moreover, South Korea consumes 732 tons, and the UK consumes 240 tons annually [3]. However, almost 75% of TC consumed by human and livestock are excreted through feces and urine in the form of parent compounds, active metabolite, or combination of both leading to its accumulation in aquatic environments [4]. Subsequently, this accumulation of TC causes disruption in the food chain, toxicity to the microbial community, and facilitates the propagation and spread of antibiotic-resistant bacteria posing a palpable threat to the environment and human health [3,5].

Studies show that TC in wastewater presents a greater challenge for conventional treatment processes compared to other types of organic wastewater due to its complex composition and potent antibacterial attributes [6]. However, advanced oxidation processes (AOPs), particularly Fenton process, have been broadly used to degrade a variety of persistent pharmaceuticals in aqueous solutions [7]. Fenton process uses metallic catalysts, such as iron catalyst, to react with hydrogen peroxide (H2O2) to produce highly reactive OH• radicals that are responsible for TC degradation. The OH• radicals generated in this process have greater oxidation potential than conventional oxidants, making them extremely effective in degrading TC [8].

Fenton process can be categorized into two main types: homogeneous and heterogeneous Fenton process. In homogenous Fenton process, iron sludge generation occurs, hence, requiring further treatment process. On the other hand, in heterogenous Fenton process, the iron catalyst is recovered, thus, no iron sludge is generated. Furthermore, in heterogeneous Fenton process, the iron catalyst is loaded on a supporting material [7]. Biochar, a biomass-derived material produced by pyrolysis, can be used as a matrix for the iron catalyst in this process, where iron is being impregnated on the biochar [9]. To date, no study has been conducted for iron catalyst impregnation on bamboo biochar.

Thus, this study focused on the evaluation of the potential of iron-impregnated bamboo biochar (Fe-biochar) as catalyst in the removal of TC via Fenton process.

2. Materials and methods

2.1. Synthesis and characterization of iron-impregnated bamboo biochar (Fe-biochar)

Bamboo wastes, particularly the stem part, were collected and washed using distilled water to remove dirt and other impurities. Moisture was removed by sun drying for 2 days. The dried bamboo waste was then pyrolyzed inside a fine carbonizer until biochar was produced. The resulting bamboo biochar (BB) was treated with 1.0 M nitric acid for about 24 h to remove metallic properties. It was then washed with distilled water to achieve neutral pH and dried again until constant weight. The dried BB was pounded using mortar and pestle and was sieved through 250 μm wire mesh.

Iron from ferric nitrate nonahydrate ($Fe(NO_3)_3{\cdot}9H_2O$) was used as the catalyst in this study. Iron solution of 50 ppm was prepared by dissolving $Fe(NO_3)_3{\cdot}9H_2O$ to distilled water. Impregnation of iron catalyst onto the BB was done through 8 h agitation of 5.0 g BB in a 100 mL of the prepared 50 ppm iron solution using magnetic stirrer at room temperature. The resulting mixture was filtered and washed with distilled water until the filtrate became clear. The resulting solid residue was oven dried to a constant weight. The dried residue, which was the iron-impregnated bamboo biochar (Fe-biochar), was stored in a sealed container.

Samples of BB and Fe-biochar were characterized using Energy dispersive X-ray (EDX) and Fourier transform infrared (FTIR) analyses to analyze the elemental composition and surface functional groups present, respectively.

2.2. Optimization of fenton process

2.2.1. Experimental design. Box-Behnken Design, a response surface methodology, was used in the optimization of TC removal. Design Expert 7.0.0 was used to generate experimental runs and statistically analyze the data. This study explored the effect of varying parameters such as Fe-biochar dosage (25 mg, 50 mg, 75 mg), H_2O_2 concentration (0.05 M, 0.25 M, 0.45 M) and solution pH (4, 5, 6) on the TC removal efficiency(%).

2.2.2. Fenton Process. A 50 ppm TC solution was prepared by dissolving powdered tetracycline hydrochloride (TRICOR) in distilled water. Batch experimental runs were carried out in a 250 mL glass beaker with a working volume of 100 mL of 50 ppm TC solution. The Fenton process was conducted by adding certain dosage of Fe-biochar to the 100 mL 50 ppm TC solution. The pH was then adjusted to the desired value using 0.1 M HCl and/or 0.1 M NaOH. The addition of H_2O_2 solution dictated the start of reaction time. After an hour of stirring using magnetic stirrer, 5 mL of solution sample was collected and analyzed using a UV-VIS spectrophotometer. Absorbance of tetracycline at 357 nm was determined for each run. Concentration of TC was determined using the standard calibration curve. TC removal efficiency was calculated using the initial and final concentration of the TC in aqueous solution as shown in equation (1).

$$\%\,TC\ removal = \frac{initial\ TC\ concentration - final\ TC\ concentration}{initial\ TC\ concentration} \quad (1)$$

The mathematical model for the process was then generated by fitting the obtained data from the experiment on different models. Furthermore, the optimized conditions of the Fenton process using Fe-biochar in the removal of TC from aqueous solution were determined.

3. Results and discussion

3.1. Synthesis and characterization of iron-impregnated bamboo biochar (Fe-biochar)
FTIR analysis of bamboo biochar (BB), and Fe-biochar before and after Fenton process suggests a successful iron impregnation as shown in figure 1.

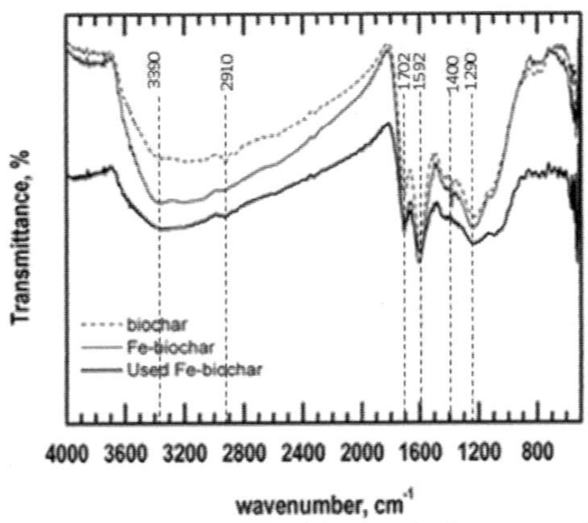

Figure 1. FTIR analysis of BB and Fe-biochar before and after Fenton process

As seen in figure 1, new broad peak was observed at 3390 cm^{-1} which implies an alteration in the surface functional groups before and after impregnation [2]. No reaction could have occurred between the Fe-biochar and the aqueous solution as there were no observed peak changes before and after Fenton process. This shows that Fe-biochar acted merely as catalyst for the hydroxyl formation and has not reacted with TC [10].

Additionally, EDX analysis of the biochar revealed the absence of iron, while carbon (66.03%), oxygen (27.37%), silicon (4.22%) and iron (2.38%) were detected in the Fe-biochar before the Fenton process, which further confirms the successful iron impregnation process.

3.2. Effect of varying parameters

3.2.1. Effect of solution pH
Figures 2 and 4 shows a parabolic trend on the effect of solution pH to TC removal. Solution pH showed little effect on TC removal, maintaining a TC removal of more than 25% between pH 4-6. Starting at pH 4, as pH was increased, removal efficiency also increased. However, after reaching maximum removal efficiency at pH 5, increasing pH leads to a decrease in the removal efficiency, reaching its minimum removal efficiency at pH 6. Comparing the removal efficiencies at different pH levels, the values obtained were not significantly far from each other. Based on the figure, adjacent removal efficiencies of TC, approximately 28%, 33%, and 25% can be obtained for pH 4, pH 5, and pH 6, respectively. These results indicate that the process works for a wide range of pH levels. At acidic conditions, •OH have higher oxidation potential, thus, TC removal is increased. Since the pH was within the acidic range, satisfactory TC removal was obtained for all levels [11].

3.2.2. Effect of Fe-biochar dosage
As shown in figures 3 and 4, Fe-biochar dosage and TC removal had a parabolic relationship. At Fe-biochar dosage of 50 mg, maximum TC removal was observed. Fe-biochar serves as the catalyst in the Fenton reaction. Therefore, its amount in the reaction could dictate the amount of OH● radicals produced. As catalyst dosage increased, generation of OH• radicals also increased, thus, higher removal efficiency is obtained. However, after reaching the maximum consumption of the catalyst, ferric ions could scavenge OH● radicals in competition with TC. Thus, further increasing catalyst dosage can significantly induce OH● scavenging and decrease TC removal. In this case, the overall Fenton removal was partially inhibited and partially favored [12].

3.2.3. Effect of H_2O_2 concentration

TC removal increased with increasing H_2O_2 concentration as shown in figures 2 and 3. This is expected since H_2O_2 is the main source of OH● radicals which can remove an electron from any substance, forming hydroxide ion and accounts for the degradation of TC. At low concentration of H_2O_2, it can be observed that removal efficiency was also low. This was due to the low generation of •OH which was not enough to reach the required amount for an effective TC removal. Higher H_2O_2 concentration leads to higher removal efficiency, for there was enough •OH to decompose the antibiotic [13].

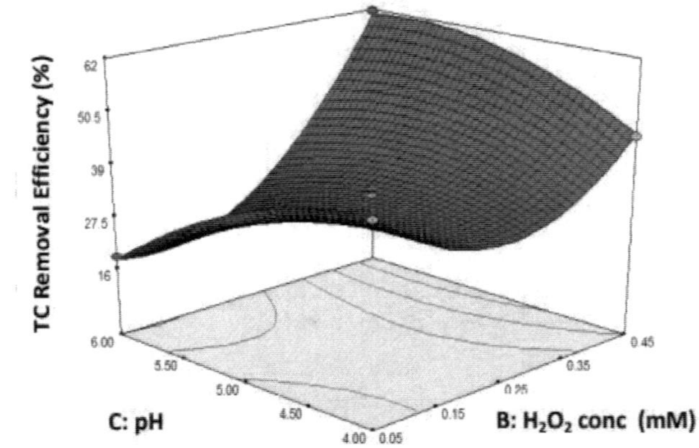

Figure 2. Solution pH and H_2O_2 concentration on TC removal

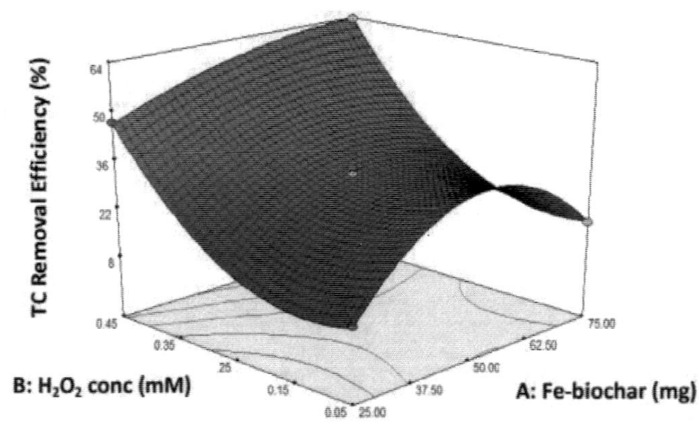

Figure 3. H_2O_2 concentration and Fe-biochar dosage on TC removal

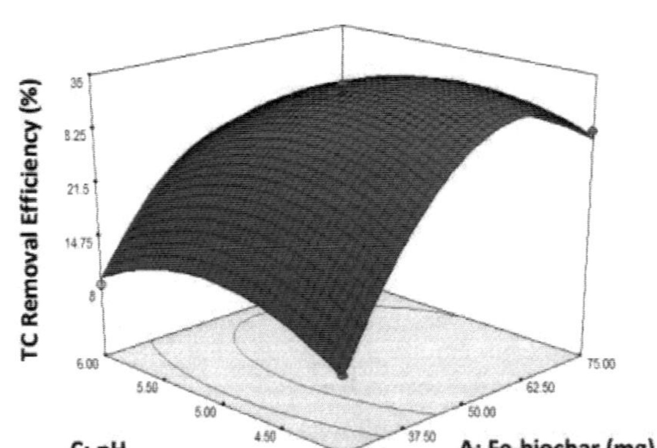

Figure 4. Solution pH and Fe-biochar dosage on TC removal

3.3. *Statistical Analysis*

Experimental data were assessed using the three factors Box-Behnken response surface experimental design to yield approximate functions for each dependent variable and to investigate the individual and interactive effects of process variables on the percentage TC removal in aqueous solution.

The highest TC removal efficiency of 63.24% was found under conditions of 75 mg catalyst, 0.45 M H_2O_2, and pH 5. The mathematical model for tetracycline removal, generated using Design Expert 7.0.0, can be expressed by the reduced-cubic model shown in equation (2).

$$\text{Removal Efficiency (\%)} = -152.35403 + 4.68059A - 220.31239B + 44.95935C - 6.91669AB - 0.13333AC + 51.31822BC - 0.038781A^2 + 363.13179B^2 - 5.38191C^2 + 0.073706A^2B \quad (2)$$

where A, B and C corresponds to Fe-biochar dosage, H_2O_2 concentration, and solution pH, respectively.

Fe-biochar dosage, H_2O_2 concentration, and solution pH were found to be significant parameters. Furthermore, model terms AB, AC, BC, A^2, B^2, C^2, A^2B were also identified to be significant, indicating that there were mutual effect and interaction between the three parameters. The Analysis of Variance (ANOVA) for tetracycline removal indicated that the model was highly significant with high coefficient of determination (R^2) value of 0.9990 which indicates good regression.

3.4. *Optimization of Fenton process*

For optimization, operational conditions were set with percent removal as maximum, amount of Fe-biochar as minimum, H_2O_2 concentration as minimum, and pH as within range. As generated by the Design Expert, the optimum conditions that would yield maximum removal of 38.69% were 41.33 mg Fe-biochar, 0.05 M H_2O_2, and at pH 4.

4. Conclusions and recommendations

In conclusion, this study demonstrated the viability of iron-impregnated bamboo biochar as a catalyst in the removal of TC via heterogeneous Fenton process. FTIR analysis characterized the biochar and iron-impregnated biochar, identifying abundant functional groups before and after the Fenton reaction. Successful iron impregnation was evidenced by alterations in the surface functional groups of the biochar. On the other hand, no alteration on the surface of the iron-impregnated biochar was observed which implies that no reaction has occurred between the catalyst and tetracycline after the Fenton process.

By exploring varying parameters including catalyst dosage, H_2O_2 concentration, and solution pH, and employing the three factors Box-Behnken response surface experimental design, it was observed that these parameters exhibited a parabolic relationship with respect to tetracycline removal. Notably, it showed that at optimum conditions of 41.33 mg Fe-biochar, 0.05 M H_2O_2, and pH 4, maximum TC removal of 38.6899% can be achieved.

Although the study has showed promising results, there are still several aspects that need further investigation. To improve this study, it is highly recommended to widen the range of the process parameters in which TC removal is investigated. Moreover, to gain a comprehensive understanding of the reaction rate and mechanism, it is advisable to study the kinetics of TC removal using Fe-biochar catalyst. Lastly, it is recommended to analyze the degraded products using High Performance Liquid Chromatography (HPLC), as well as use X-ray Diffraction (XRD) and Scanning Electron Microscopy (SEM) to characterize the biochar.

5. References

[1] Balakrishnan A, Chinthala M, Polagani R K and Vo D N 2023 *Environmental Research* **216** 114660
[2] He J, et al 2022 *Spectrochimica Acta Part A: Molecular and Biomolecular Spectroscopy* **277** 121252
[3] Amangelsin Y, Semenova Y, Dadar M, Aljofan M and Bjørklund G 2023 *Antibiotics* **12(3)** 440

IOP Conf. Series: Materials Science and Engineering **1318** (2024) 012049 doi:10.1088/1757-899X/1318/1/012049

[4] Xu L, Zhang H, Xiong P, Zhu Q, Liao C and Jiang G 2021 *Science of The Total Environment* **753** 141975
[5] Zhang Y, Zhao Y, Yang D and Zhao Y 2022 *Journal of Hazardous Materials Advances* 7 100129
[6] Wang X, Gao J, Xie J, Hu Z, Xue R, Chen S and Ge P 2023 *Journal of Hazardous Materials Advances* **10** 100302
[7] Deng J, Dong H, Zhang C, Jiang Z, Cheng Y, Hou K and Fan C 2018 *Separation and Purification Technology* **202** 130-37
[8] Homem V and Santos L 2011 *Journal of Environmental Management* **92(10)** 2304-47
[9] Boshir M, Zhou J L, Ngo H H, Guo W and Chen M 2016 *Bioresource Technology* **214** 836-51
[10] Viglašová E, Galamboš M, Danková Z, Krivosudský L, Lengauer C L, Hood-Nowotny R, Soja G, Rompel A, Matík M and Briančin J 2018 *Waste Management* **79** 385-94
[11] Park J, Wang J J, Xiao R, Tafti N, Delaune R D and Seo D 2018 *Bioresource Technology* **249** 368-76
[12] Gligorovski S, Strekowski R, Barbati S and Vione D 2015 *Chemical Reviews* **115(24)** 13051-92
[13] Mirzaei A, Chen Z, Haghighat F and Yerushalmi L 2017 *Chemosphere* **174** 665-88

IOP Conf. Series: Materials Science and Engineering **1318** (2024) 012050 doi:10.1088/1757-899X/1318/1/012050

Investigation of the influence of fluorine on metal elution from cathodic active material liberated from spent lithium-ion batteries with a focus on the underwater electrical pulse method

Yanhui Kong[1], Yutaro Takaya[2,3], Mauricio Córdova-Udaeta[4], Chiharu Tokoro[2,3*]

[1] Graduate School of Creative Science and Engineering, Waseda University, Japan

[2] Faculty of Engineering, The University of Tokyo, Japan

[3] Faculty of Science and Engineering, Waseda University, Japan

[4] Research Institute for Science and Engineering, Waseda University, Japan

*Corresponding author's e-mail address: tokoro@waseda.jp

Abstract: As a promising candidate to separate the cathode materials and aluminum foils, the underwater electrical pulse method is an environmentally friendly method that can effectively liberate cathodic active material in a very short time. However, performing the electric pulse separation in aqueous media may cause the release of fluoridated elements, which may come from fluoridated components formed during battery cycles or due to the remaining electrolyte within the cathode particles. Thus, the purpose of this work is to study the solution obtained after pulse discharge in terms of fluoridated entities, to understand their effect on the elution of active cathodic materials. Herein, to investigate the potential source of F and its possible effects on metal extraction, leaching experiments were conducted in the presence of different fluoride concentrations with a solid/liquid ratio of 0.5g/L under circumneutral pH conditions (pH range: 5.7-7.5) using a spent nickel-cobalt-manganese lithium battery sample liberated by N-Methyl pyrrolidone. The results show that the amounts of F ions in solution increased with leaching time, with a maximum value close to 2.34 ppm F obtained after 2 h. It is worth noting that the presence of fluoride-ion can promote the elution of aluminum, as a maximum concentration of 1.85 ppm Al was observed at 2 h in the presence of 40 ppm F, whereas only about 0.32 ppm Al was detected in the case of fluoride-free solutions. Consequently, our results demonstrated that the amounts of transition metals (Ni, Mn, and Co) released towards liquid changed very little irrespective of the concentrations of dissolved fluoride, suggesting that the released F ions do not affect metals elution.

1. Introduction

In the past decades, the worldwide demand for lithium-ion batteries (LiBs) has been increasing dramatically due to the growth of electric vehicle production, accompanied with an increased number of spent LiBs [1,2]. As a result, the disposal of spent LIBs has become a big issue, not only because LiBs contain various heavy metals, but also because toxic electrolytes found inside the battery would endanger human health and the environment if not handled properly [3,4]. In addition, with the increasingly unstable international situation and supply chain crisis, the recycling of used lithium batteries has become more and more important for the sustainable supply of strategic resources [5].

Roughly, the whole recycling process of spent lithium batteries can be summarized in three steps: pretreatment, recovery, and purification [6]. Among them, the pretreatment process includes sorting, discharging, dismantling and separation methods. In addition, elucidating how to effectively and economically separate cathode material bound to the aluminum foil is a critical step, but the high stability and strong adhesion of the binder – polyvinylidene fluoride (PVDF) – makes this process very difficult [7]. According to the literature, separation methods at the laboratory stage include solvent dissolution with N-Methyl pyrrolidone (NMP) [8]. However, to effectively promote the peel-off rate, a temperature-assisted ultrasonic treatment method is necessary in this process. On the other hand, some new methods were also proposed; including electrochemistry, using supercritical fluids, and so on. Recently, our laboratory introduced a novel separation method – the underwater electric pulse –, which has been developed to achieve the release of cathode powders with Al foils [9]. This method can liberate cathodic active particles instantaneously and at relatively lower temperatures (below 300°C) when compared to other typical thermal treatment methods. A simplified schematic for the separation process by underwater electrical pulse is shown in Figure 1. Nonetheless, the whole separation process via electric pulse is conducted in water which may pose the risk of lithium and transition metal losses towards the liquid. Furthermore, an analysis of the solution after discharge revealed the presence of large amounts of fluoride ions (up to 20 ppm). This behavior is expected since it is well known that fluoride ions, which can promote some cations solubility via forming soluble compounds, were eluted during underwater pulse experiments.

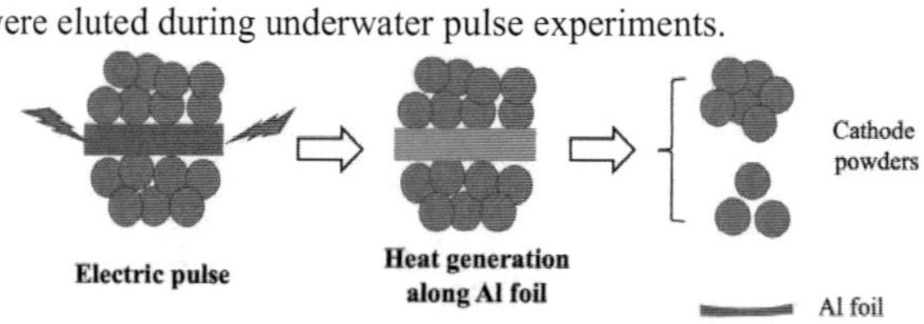

Figure 1. Schematic for the separation of active powder and aluminum current foil by underwater electric pulse

To address the abovementioned problem that may arise during the underwater electric pulse separation process of cathode materials and Al foils, it is necessary to investigate the potential source of fluoride and its possible effects on the elution of transition metals. Therefore, this research work will focus on studying the leaching behavior of cathodic active material using a solution medium with different fluoride concentrations (0, 10, 20, 40 ppm).

2. Material and Method

2.1 Sample preparation

The spent cathode electrodes used in this study were provided by a local recycling company. To obtain the spent samples from the foils, cathode electrodes were cut into small sheets (10 mm × 20 mm) using scissors and then immersed into a 1-methyl-2-pyrrolidone (NMP, purity > 99%) solvent with ultrasonic agitation for 2 h at 60°C. Then, the black powders were separated by vacuum filtration and freeze-dried. The dried positive electrode powder was further sieved to remove the residual Al foil pieces.

In addition, a fluorine stock solution (1000mg/L) was prepared by using solid NaF. A predefined volume of F-stock solution was added to a plastic conical flask to prepare the solutions with different F amounts for leaching experiments. All reagents used were of analytical grade and purchased from Fujifilm Wako Pure Chemical Corporation, Japan.

2.2 Analytical methods

To determine the composition and average content of the active materials, 0.1g of cathode sample was dissolved in aqua regia (HCl: HNO_3 = 3:1, v/v) at 80°C for 4h. After dilution, the elemental composition and their concentrations in the digestate were analyzed by an ICP-OES machine (ICAP 6000 Series, Thermo Fisher Scientific, USA).

In addition, the crystal structure of the prepared spent sample was characterized using X-ray diffractometry (XRD; Smartlab, Rigaku, Japan) with a Cu-Kα source operating at 40 kV and 30 mA. Morphologies of the cathode sample were characterized by a field emission scanning electron microscope coupled with EDS mapping (FE-SEM/EDS, JSM-6500F, JEOL, Japan). Also, X-ray photoelectron spectroscopy (XPS, JPS-9010MX, JEOL) measurement was performed on the surface of the spent cathode material.

2.3 Leaching experiments

The experiments were conducted in an PTFE beaker with a predetermined fluoride-ion concentration of 0, 10, 20, and 40 ppm, respectively. The total leaching time was set to 2h at room temperature (25±1°C) with a constant magnetic stirring speed. During the leaching process, 2 mL of liquid was withdrawn at a predetermined time interval (1, 5, 10, 15, 20, 40, 60, and 120min) and then filtered through a 0.45μm syringe filter to quantify the leached elements. The residue solid particles were collected, washed several times with ultrapure water to remove excess F, and then freeze-dried for further investigation. Meanwhile, changes in the solution pH throughout the whole leaching process were measured with a pH meter.

2.4 Measurement of fluoride concentration in the leachate

Aside from the concentration change of metals (Li, Mn, Co, Ni, and Al), the variation of fluoride concentration during the leaching process was also studied. Due to the presence of Al in the solution, liquid samples were firstly diluted with a 1:1 ratio in volume with TISAB II, which can provide a constant background ionic strength and then dissociate loosely bound fluoride ions [10]. The detailed procedure of the experiment is shown in Figure 2.

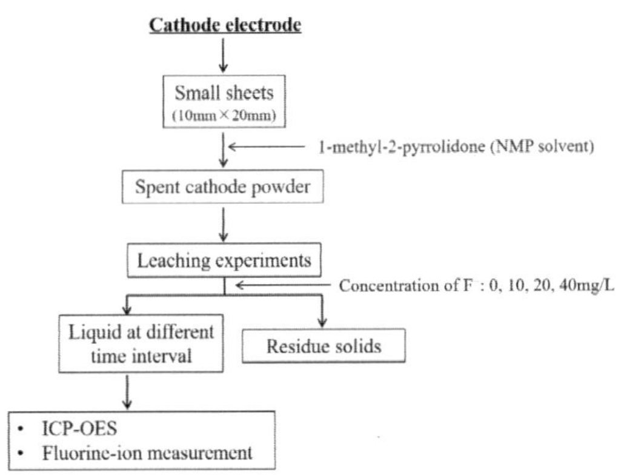

Figure 2. Flowchart of the experimental procedure in this study

3. Results and Discussion

3.1. Materials characterization before leaching

The results of concentrated aqua regia digestion are shown in Table 1. The molar ratio of Li, Ni, Co, and Mn was determined to be 1: 0.16:0.66: 0.16, respectively. The presence of trace amounts of Al in the prepared spent samples may result from the Al foil and/or its degradation products (Al_2O_3, AlF_3, and so on) caused by the corrosion reaction during the battery cycles [11]. The presence of phosphorus was also detected, which can be ascribed to residual $LiPF_6$ and/or its degradation products on the particle surface [12]. As presented in Figure 3, the XRD pattern of cathode materials was assigned to $LiNi_{0.16}Mn_{0.16}Co_{0.66}O_2$ (PDF No:00-063-0183), which was consistent with ICP-OES results.

Table 1. Main compositions of spent cathode materials

Element	Li	Mn	Co	Ni	Al	P
wt.%	8.19	10.36	44.21	10.54	1.56	0.31

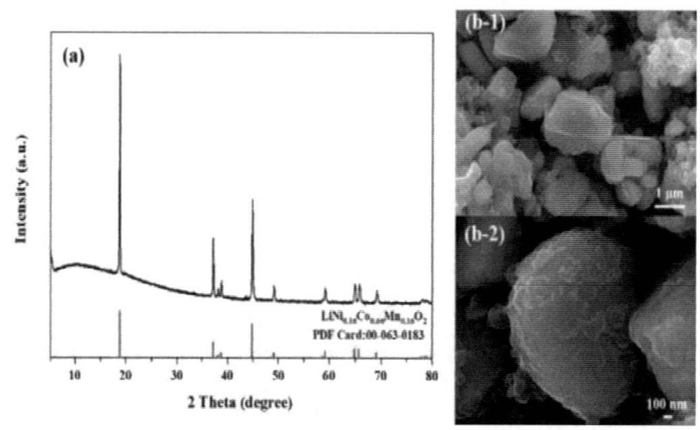

Figure 3. (a) XRD pattern and (b) FE-SEM images of spent cathode materials

The surface morphology of the spent cathode electrode was also analyzed by FE-SEM coupled with EDS to study the elemental composition and distribution on the electrode surface. Figure 4 shows the FE-SEM/EDS results of the spent cathode electrode. It can be seen that there are some white spots on the cathode sheet's surface. From the EDS mapping, besides O and F, elements such as Al, P were distributed around the corrosion spots. According to the literature, these white spots can be formed due to the chemical corrosion induced by HF generated by the decomposition of $LiPF_6$ during long-term electrochemical cycling, as described in Eqs (1) ~ (3) [13,14]. Besides, it was seen that the electrode has a porous structure, thus HF can penetrate the interior of the electrode through the pores and gaps to cause chemical corrosion of the Al_2O_3 film on the Al foil surface [15,16]. Likewise, the generated AlF_3 and other corrosion products could have been continuously deposited and eventually formed white spots on the electrode surface. In addition, when the spent cathode electrodes are exposed to air, the moisture in the ambient environment would also induce the decomposition of $LiPF_6$ to generate HF, which further aggravates the corrosion of aluminum foil. Therefore, it cannot be ruled out that HF was generated through the reaction with moisture in the ambient environment when the spent cathode electrodes were exposed to air.

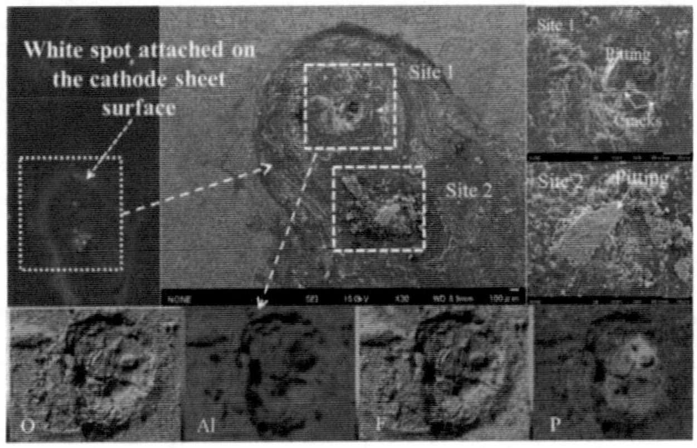

Figure 4. FE-SEM/EDS results of the spent cathode sheet

3.2 Leaching results

3.2.1 Leaching behavior of metals in the fluorine-containing solution

As shown in Figure 5, the effect of F on the eluted amount of lithium and transition metals was not evident regardless of the initial F concentration. However, as illustrated in Figure 5(v), contrary to the other metals, F concentration shows a much more pronounced influence on the elution of Al, whereby the amount of Al eluted notoriously increased with the increase of initial F concentration in the solution. For comparison purposes, the spent cathode material separated by the underwater electric pulse was also used as a sample for the experiments under the same experimental conditions. As depicted in Figure 5, a significantly lower element elution was observed for the spent cathode material separated by underwater electric pulse, when compared to spent samples separated by NMP solvent. This may be due to the fact that the spent cathode material has undergone an immediate dissolution process during the separation via underwater electric pulse, resulting in a reduction of water-soluble species. The amount of dissolved Al also showed an increasing trend, indicating that the presence of F in the solution does indeed contribute to the increase in Al elution.

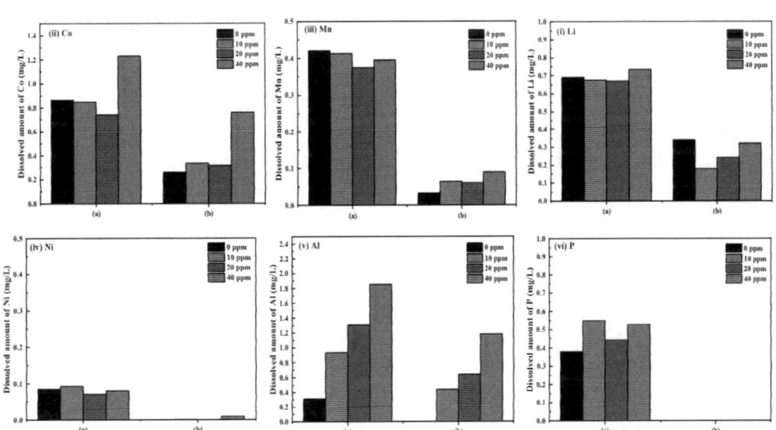

Figure 5. Effect of initial fluoride ion concentrations on the concentration of different element elution during the leaching process: (a) separated by NMP solvent and (b) separated by underwater electric pulse (0.5g/L of solid-to-liquid ratio, 2 hours)

3.2.2 The variation of F and P in the fluorine-containing solution

The changes in the F amount during the leaching process were investigated. Figure 6 shows the net variation of F concentration in solutions with different initial F concentrations. When the spent cathode material was added to ultrapure water, water-soluble components such as $LiPF_6$ remaining on the particle surface would dissolve, leading to an increase of F in the solution. Therefore, in all cases, a clear upward trend was observed in the initial stages of leaching, as described in Figure 6(a). However, as time went by, the F concentration showed a gradual declining trend. Figure 6(b) depicts the changes in solution pH under different initial F concentrations during leaching. The curves show that the solution pH increases slightly with the increase of the initial F amount.

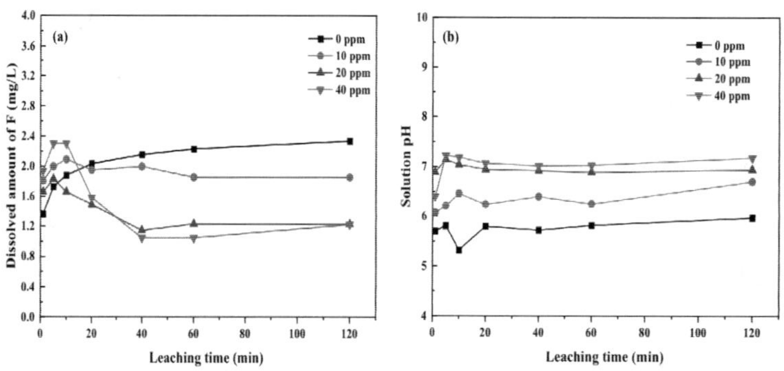

Figure 6. (a) The net variation of F concentration in solution and (b) the corresponding change in solution pH under the different initial F concentrations

3.3 XPS analysis for the evolution of surface chemical components

To study the chemical composition of the surface before and after leaching, XPS analysis was conducted. From the full scan results shown in Figure 7(a), it can be seen that C, O, F, and Al were the main components present in the spent cathode material. Conversely, P was not detected due to its low content on the spent cathode powder.

The fine spectra of O 1s can be deconvoluted into five peaks (Figure 7(b)). Here, the peak belonging to lattice oxygen was found at ~529.7eV[17], and three peaks of ~531.6eV,~533.0eV,~534eV can be assigned to surface oxygen, which contributes from the organic electrolyte remaining on the cathode surface. In addition, a higher binding energy sub-peak located at ~535.2eV was observed, which can correspond to the O-F species resulting from the degradation of $LiPF_6$. This was corroborated by the appearance of the corresponding peaks in the F 1s spectra observed at ~686.4eV which has been reported to be the result of Li_xPOF_y contribution [14]. Aside from this, another possible contribution of the O-F species is the corrosion products of Al foils caused by HF, observed in species such as Al-O-F [18,19]. This was confirmed by the fine spectra of Al 2p, as shown in Figure 7(d). For F 1s (Figure 7(c)), a relatively higher binding energy observed at ~687.5eV is considered to be associated with AlF_3, and at a lower binding energy a sub-peak at ~685.1eV indicates the formation of LiF generated by the degradation of $LiPF_6$ and the dissolution of spent cathode material caused by HF. Additionally, for the Al 2p fine spectra presented in Figure 7(d), the peaks located at ~74.2eV and ~77.3eV [15], can be ascribed to Al_2O_3 and AlF_3, respectively. Meanwhile, the peak at ~75.8eV can be assigned to fluorinated alumina (Al-O-F species), which was also one of the possible products of chemical corrosion of alumina caused by HF [20]. Furthermore, a sub-peak was observed at ~79.5eV, which to the best of our knowledge has not been reported in the previous literature.

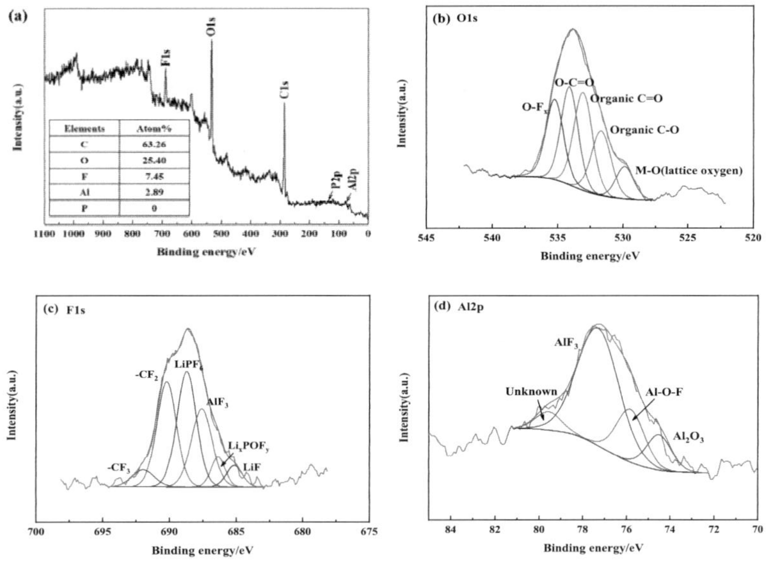

Figure 7. XPS spectra of the spent cathode powders before leaching: (a) full survey, (b) O1s, (c) F1s, and (d) Al2p

XPS results show that the surface of spent cathode particles contains Al (about 3.0 %), which mainly exists on the surface in the form of Al_2O_3, Al-O-F, and AlF_3. As discussed above, the increase of F in solution can be ascribed to the dissolution of $LiPF_6$ and/or its degradation products. Nonetheless, the decreased F amount in the solution observed with the prolongation of leaching time seems to imply that there is another reaction mechanism occurring in the leaching system. Previous studies have verified that Al_2O_3 can be used as an absorbent to remove F^- from aqueous media, forming aluminum-associated F compounds such as Al_2O_3F [21]. Meanwhile, a surface reaction might have happened between F ions and the hydroxyl groups in terms of ion exchange, resulting in the adsorption of F^- and the release of OH^- [22]. This likely occurred during leaching, which may explain the observed increase in solution pH (Figure 6(b)). On the other hand, in addition to the possible adsorption process of F leading to a decrease in the amount of F in solution, the dissolution process of Al also occurred simultaneously and resulted in an increase in the amount of Al ions. One conceivable reason for this phenomenon is the formation of water-soluble substances in the pH range of 6~7, such as AlF_4^-, AlF_2^-, $AlOHF_2$, and so on [10]. However, the exact reaction mechanism for this process is still unclear and requires further study.

5. Conclusions

This study addressed the effect of different initial F concentrations on the leaching behavior of metals. The main results indicate that F^- in solution did not significantly affect the leaching behavior of lithium and transition metals (Mn, Co, and Ni). In contrast, it was seen that fluoride-containing solutions induce a substantial removal of aluminum from the cathode active powders, whereby an increase of initial fluoride concentrations also increased the leached amount of Al. On the other hand, the eluted fluoride demonstrates a clear declining trend in the later stages of leaching, suggesting that F^- was removed from the aqueous solution rather quickly, but the specific reasons for the dissolution of Al and removal of F^- are still unknown. Finally, although the presence of F^- does not have a promoting effect on the elution of lithium and transition metals from the cathode, the wastewater generated after cathode liberation processes such as the underwater electric pulse must be properly treated to avoid potential harm to groundwater.

References

[1] Kim S, Bang J, et al., 2021 A comprehensive review on the pretreatment process in lithium-ion battery recycling. *Journal of Cleaner Production,* **294**.

[2] Jin S, Mu D, et al., 2022 A comprehensive review on the recycling of spent lithium-ion batteries: Urgent status and technology advances. *Journal of Cleaner Production*, **304** 130535.

[3] He H, Feng J, et al., 2022 Selective separation and recovery of lithium, nickel, MnO2, and Co2O3 from LiNi0.5Mn0.3Co0.2O2 in spent battery. *Chemosphere*, **286(Pt 3)**: p.131897.

[4] Huang Z, Liu F, et al., 2022 Metal Reclamation from Spent Lithium-Ion Battery Cathode Materials: Directional Conversion of Metals Based on Hydrogen Reduction. *ACS Sustainable Chemistry & Engineering*, **10(2)**: p756-765.

[5] Mathieux F, Ardente F, et al., 2017 Critical Raw Materials and the Circular Economy – Background report.

[6] Chan KH, et al.,2021 Closed-Loop Recycling of Lithium, Cobalt, Nickel, and Manganese from Waste Lithium-Ion Batteries of Electric Vehicles. *ACS Sustainable Chemistry & Engineering,* **9(12)**: p. 4398-4410.

[7] He K, Zhang ZY, et al., 2019 A green process for exfoliating electrode materials and simultaneously extracting electrolyte from spent lithium-ion batteries. *J Hazard Mater,* **375**: p. 43-51.

[8] Wang M, Liu K, et al.,2023 Challenges in Recycling Spent Lithium-Ion Batteries: Spotlight on Polyvinylidene Fluoride Removal. *Glob Chall,***7(3)**: p.2200237.

[9] Tokoro C, Lim S, et al., 2021 Separation of cathode particles and aluminum current foil in Lithium-Ion battery by high-voltage pulsed discharge Part I: Experimental investigation. *Waste Manag*,**125**: p. 58-66.

[10] Pitre D, Boullemant A, and Fortin C, 2014 Uptake and sorption of aluminium and fluoride by four green algal species. *Chem Cent J,***8(1)**: p. 8.

[11] Yoon E, Lee Jihye, et al, 2022 Passivation Failure of Al Current Collector in LiPF6-Based Electrolytes for Lithium-Ion Batteries. *Advanced Functional Materials*, **32** 2200026.

[12] Eshetu G.G, Diemant T, Grugeon S, et al., 2016 In-Depth Interfacial Chemistry and Reactivity Focused Investigation of Lithium-Imide- and Lithium-Imidazole-Based Electrolytes. *ACS Appl Mater Interfaces,***8(25)**: p. 16087-100.

[13] Gauthier M, Karayaylali P, Giordano L, et al.,2018 Probing Surface Chemistry Changes Using LiCoO2-only Electrodes in Li-Ion Batteries. *Journal of The Electrochemical Society,***165(7)**: p. A1377-A1387.

[14] Hekmatfar M, Kazzazi A, et al.,2019 Understanding the Electrode/Electrolyte Interface Layer on the Li-Rich Nickel Manganese Cobalt Layered Oxide Cathode by XPS. *ACS Appl Mater Interfaces*, **11(46)**: p. 43166-43179.

[15] Ma T, Xu G.L, Li Y, et al., 2017 Revisiting the Corrosion of the Aluminum Current Collector in Lithium-Ion Batteries. *J Phys Chem Lett*,**8(5)**: p. 1072-1077.

[16] Gao H, Ma T, Duong T, et al., 2018 Protecting Al foils for high-voltage lithium-ion chemistries. *Materials Today Energy*, **7**: p. 18-26.

[17] Billy E, Joulie M, et al.,2018 Dissolution Mechanisms of LiNi1/3Mn1/3Co1/3O2 Positive Electrode Material from Lithium-Ion Batteries in Acid Solution. *ACS Appl Mater Interfaces,***10(19)**: p. 16424-16435.

[18] Myung S.-T, Hitoshi Y, and Sun Y.-K,2011 Electrochemical behavior and passivation of current collectors in lithium-ion batteries. *Journal of Materials Chemistry*,**21(27)**.

[19] Hofmann A, Schulz M, Winkler V, et al., 2014 Anodic Aluminum Dissolution in Conducting Salt Containing Electrolytes for Lithium-Ion Batteries. *Journal of The Electrochemical Society*,**161(3)**: p. A431-A438.

[20] Ramos R, Cunge G, Pelissier B, et al., 2007 Cleaning aluminum fluoride coatings from plasma reactor walls in SiCl4/Cl2plasmas. Plasma Sources Science and Technology, **16(4)**: p. 711-715.

[21] Wu X, Zhang Y, Dou X, et al., 2013 Fluoride adsorption on an Fe–Al–Ce trimetal hydrous oxide: Characterization of adsorption sites and adsorbed fluorine complex species. *Chemical Engineering Journal*, **223**: p. 364-370.

[22] Xu C, Li J, He F, et al., 2016 Al2O3–Fe3O4–expanded graphite nano-sandwich structure for fluoride removal from aqueous solution. *RSC Advances*,**6(99)**: p. 97376-97384.

Ionic Liquids and its Potential as a New Green Solvent for REE Extraction in Philippine Bauxite Residues

Ivanna Raena V. Campos[1], Jillian Aira S. Gabo-Ratio[2], Betchaida D. Payot[2], Mary Donnabelle L. Balela[1]

[1]Sustainable Electronic Materials Group, Department of Mining, Metallurgical, and Materials Engineering, University of the Philippines Diliman, Quezon City 1101, Philippines
[2]National Institute of Geological Sciences, University of the Philippines Diliman, Quezon City 1101, Philippines

ivcampos@up.edu.ph; jgratio@nigs.upd.edu.ph; bdpayot@up.edu.ph; mlbalela1@up.edu.ph

Abstract. Rare earth elements (REEs) have become necessary for high-technology applications. This has prompted an increase in demand for REE exploration, production, and extraction. Despite its name, REEs are fairly rich in the Earth's crust. REEs are estimated to have a crustal abundance of 9.2 ppm relative to other elements in our crust. In the Philippines, deposits are found to give adequately high concentrations of REE, ranging from approximately 200 ppm to 600 ppm. Samar bauxite deposits especially had around 500 ppm REE. In line with the planned bauxite production in the Philippines, it is of interest to further uncover ways to maximize its recovery. One of the possible optimizations in its extraction is to utilize its residue. There have been a number of methods to extract REE in bauxite residues. The use of ionic liquids (IL) to recover REE in bauxite offers a possibility of an effective and environmentally safe recovery process. ILs are molten salts that have melting points below 100degC. They exhibit high electrochemical stability, non-flammability, non-volatility, and can be tailored to the desired application. Hence, they have been the topic of study for different research regarding REE extraction in bauxite residues. Herein, the potential of ILs as a green solvent for REE extraction in bauxite is reviewed for future use in the extractive metallurgy of bauxites in the Philippines.

1. Rare Earth Elements (REE)

Elements under the lanthanide group (La, Ce, Pr, Nd, Pm, Sm, Eu, Gd, Tb, Dy, Ho, Er, Tm, Yb, Lu) along with Sc and Y are classified as rare earth. According to the International Union of Pure and Applied Chemistry (IUPAC), rare earth elements (REE) can be divided into two groups: light REEs and heavy REEs. Elements lanthanum (La) to samarium (Sm) fall under the light REEs. Meanwhile, elements europium (Eu) to lutetium (Lu) are considered heavy REEs. "Medium" or "mid" REEs are sometimes used to refer to those in the middle of light and heavy REEs, such as elements samarium (Sm) to dysprosium (Dy).[1]

Rare earth elements share similar geochemical properties, hence they commonly occur together as oxides instead of individual elements. This causes difficulty in their recovery as pure metals, thus the term "rare" earth is coined. Their properties can be traced back to the inner electrons in their atoms. The filling of the "f" electron shell in the lanthanide series induces magnetic and spectroscopic properties in REEs.[1] Such characteristics make rare earth elements highly suitable for modern technology applications. For example, REEs aretypically used in ceramics and glass due to their optical properties. Cerium oxide is a common material added to glass for precision polishes. REEs lanthanum (La) and lutetium (Lu) are used to alter the refractive index of glass products. La can be used in camera lenses, while Lu is used for materials requiring a higher refractive index.[2] REEs can also be found in magnets and catalytic converters. Neodymium magnet, alloyed with iron and boron, is one of the most common rare earth magnets. Its strength is comparably 5 to 7 times stronger than regular ferrite magnets. This makes them suitable for high technology purposes such as imaging devices in the medical field and the muscular tonus of astronauts in NASA.[3-4]

Despite its name, REEs are relatively rich in the Earth's crust. They are estimated to have a crustal abundance of 9.2 ppm relative to other elements in our crust. Cerium, the most abundant element among rare earths, has a level of crustal abundance similar to copper. Incidentally, precious metals such as gold, silver, and platinum are less common in the Earth's crust compared to all REEs except promethium (Pr).[5] Worldwide rare earth oxide (REO) reserves amount to approximately 130,000,000 MT, with China accounting for 33.85% of these resources. Consequently, the world mine REE production is also dominated by China, which produced 210,000 MT of rare earth in 2022 as listed in Table 1. United States is second in REE production, followed by Australia and Burma.[6]

Table 1. A table of REE production and reserve values by country in 2022[6]

Country	Production (MT)	Reserves (MT)
China	210,000	44,000,000
Vietnam	4,300	22,000,000
Brazil	80	21,000,000
Russia	2,600	21,000,000
India	2,900	6,900,000
Australia	18,000	4,200,000
United States	43,000	2,300,000
Greenland	-	1,500,000
Tanzania	-	890,000
Canada	-	830,000
South Africa	-	790,000
Other Countries	80	280,000
Thailand	7,100	-
Madagascar	960	-
Burma	12,000	-

Rare earth deposits can be categorized into primary and secondary deposits. Primary deposits are carbonatite-associated alkaline igneous rock-associated, iron-REE, and hydrothermal occurrences. Meanwhile, secondary deposits include marine placers, alluvial placers, palaeoplacers, lateritic, and ion-adsorption occurrences. Carbonatite-related deposits appear to be the main source of REE production worldwide. The largest REE carbonatite deposit is in Bayan Obo, Inner Mongolia, China. It is estimated to contain over 10,000 KT of REOs, garnering a grade of around 4%. The production of REEs in Bayan Obo started in the late 1980s and is still currently the world's largest source of rare earths.[1]

IOP Conf. Series: Materials Science and Engineering **1318** (2024) 012051 doi:10.1088/1757-899X/1318/1/012051

2. Rare Earth Elements in Philippine Bauxite

The Philippines currently has 55 operating mines in the entire country as of 2022. None of them, however, have explored any economic REE deposits. In fact, there have been minimal studies regarding rare earth elements in the Philippines. The early research on Philippine REE deposits mostly focuses on the beach placer deposit in Ombo, San Vicente, Palawan. Its results show that the beach area contained approximately 19,000 MT of sand deposits with an average of 22.19% REE (Ce and La), 0.85% Th, and 0.55% U.[7]

Recently, a study has established several potential economic REE deposits in the Philippines. It investigated 94 different samples of various mineral deposits in the Philippines, which included Samar bauxites, ion adsorption clay in El Nido, a granite in San Vicente, Pulang Daga limonites, laterites in Surigao and Zambales, coal deposit in Isabela, Panganiban and Mt. Labo iron skarn deposits, epithermal Au deposit in Paracale, the Santo Tomas II porphyry Cu deposit, the Rapu-Rapu VMS deposit, and the La Union beach placer deposit. Its locations are annotated on the Philippine map in Figure 1 below. The highest concentrations of REE were around 300 ppm to 600 ppm, led by Erawan beach placer deposit (~600 ppm) and Samar bauxites (~500 ppm). Other deposits contained an estimated 100 ppm to 200 ppm. The average results can be compared to REE deposits in other countries, such as the karstic bauxites in China, Spain, Turkey, and Montenegro, and the ion adsorption clay in southern China.[8] This implies that the Philippine REE deposits have a high potential to become economic sources of rare earth.

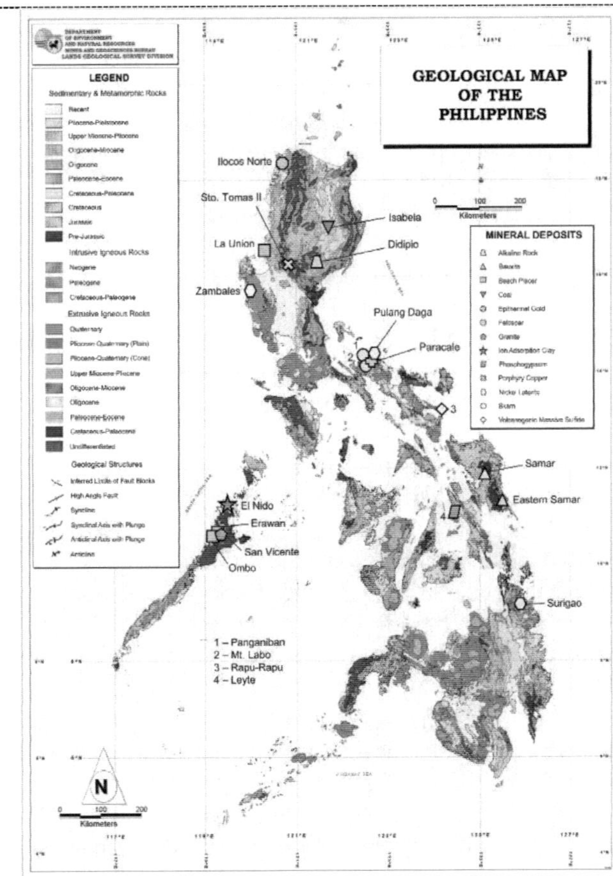

Figure 1. Geological map of the Philippines with annotations of places where samples were taken from[9]

Recovery of REE from secondary sources is another viable way for rare earth production in the Philippines. Approximately 1.3 to 2.4 million tons of phosphogypsum (PG) is produced annually by fertilizer plants in the country. The PG is stacked in a tailings pond yearly and has accumulated to around 10.1 million tons. An estimated 2,678 tons of REO has been determined in the stack of PG. This amounts to 266 mg kg^{-1} of the accumulated phosphogypsum and can fulfill 10% of the country's demand for rare earth.[9] However, as with any other potential secondary REE resources, its extraction and processing would need further study.

3. Ionic Liquids (ILs)

Ionic liquids (ILs) are molten salts containing an organic cation and varying anions. They usually are in a liquid state at temperatures below 100ºC and are often described as non-volatile/flammable with high thermal and electrochemical stability. Altering the structural composition of the IL's cations and anions can change its physiochemical properties per the desired application. This ability is what brands them as "designer solvents."[10]

Throughout the years, new classifications of ionic liquids have been developed: room-temperature ILs (RTILs), task-specific ILs (TSILs), polyionic liquids, and supported IL membranes (SILMs) that include composites of ILs supported onmetal–organic frameworks (MOFs).[11] Additionally, they can be defined based on their interaction with water. The inherent properties of the given cation and anion dictate this. The hydrophobicity of the IL can be modified by varying the cation, whereas the anion is held responsible for the solution's water miscibility.[12] The dilution ratio of an IL solution has adverse effects on its viscosity, density, diffusivity, and electrolytic conductivity. Hence, its efficiency as a solvent is indirectly affected.[13]

The flexible properties of ILs allow them to be applicable to various fields such as electrochemistry, gas capturing, extraction processes, and other industrial applications. For instance, ionic liquid systems have become preferable in the electrodeposition of water-sensitive metals due to their nonaqueous nature. Studies regarding the electrodeposition of aluminum, zinc, and their alloys using ILs emphasize their potential as an environmentally friendly and corrosion-resistant coating. The electrochemical properties of ionic liquids enable them to control metal redox potentials and the deposition morphology.[14-16] Consequently, such properties also prompt them as a subject of interest in energy application research. The low flammability and low volatility of ILs give them the potential to be a stable electrolyte for advanced battery technologies such as lithium-ion, sodium-ion, and magnesium-ion batteries. This allows the procedures to be performed at lower temperatures, making way for a low cost/high energy system.[17]

Storage and transportation of toxic gases require adsorption at sub-atmospheric pressures. Altering the configuration of ions in the ILs modifies their reactivity with regard to these gases which enables the formation of a complex between Lewis acidic ILs and Lewis basic gases and vice versa. The simplicity of its transportation and higher heat transfer rates causes ionic liquids to be favorable compared to solid adsorbents.[18] Results from a study regarding a carbon dioxide (CO_2) extraction simulation show that greater than 90% of the CO_2 was recovered from the flue gas using ionic liquids and monoethanolamine (MEA). The IL-based process provided a 16% reduction in energy loss and a 12% reduction in equipment footprint compared to the MEA process. Additionally, investments for engineering design are lower by 11% in the ionic liquid process than in the MEA.[19]

Dissolution of transition metals using chlorouranium systems were discussed in a report in the early 1980s. This served as one of the earlier studies that tackled the extraction of metal oxides using ionic liquids. Currently, multiple research has explored various combinations of ILs for metal extraction. Most of which discuss heavy metal recovery.[20] However, REEhas become a point of interest for ionic liquid processes. The potential of ionic liquids as a new green solvent for metal dissolution and extraction owes itself to its physiochemical properties. Optimizing these features, nonetheless, would require further investigation of the appropriate operating conditions for higher recovery, solubility, and selectivity of desired metals.

IOP Conf. Series: Materials Science and Engineering **1318** (2024) 012051 doi:10.1088/1757-899X/1318/1/012051

4. Use of Ionic Liquids in REE Extraction of Bauxite Residues

In line with the growing demand for REE, the need for an efficient and environmentally friendly recovery process is increasing simultaneously. Previous extraction procedures for REE in bauxite residues (BR) mainly utilize acids as their lixiviant. Commonly used acids for REE leaching are sulfuric acid (H_2SO_4), hydrochloric acid (HCl), and nitric acid (HNO_3). Amongst the three, HNO_3 garnered the highest recovery values with an inclination towards the heavy lanthanides. The increase of ionic radii in lanthanides with higher atomic numbers was speculated to increase their solubility hence the high recoveries for heavy lanthanides. The optimal conditions for most REEs were at 0.5M acid concentration with a 1:20 pulp density ratio with the highest recovery values of 96% for Y, 72% for Yb, and 75% for Sc.[21] Despite its efficiency, acid lixiviants threaten the safety of the environment and people handling these reagents, therefore calling for a green REE leaching process.

Ionic liquids offer stable electrochemical properties and a wide range of combinations, adjustable based on the desired application. This nominates them as a suitable candidate for an eco-friendly lixiviant for REE extractions. In fact, recent studies have discussed the use of IL systems in the recovery of REE. Table 2 below lists the most common ionic liquids investigated for their rare earth extraction efficiency in BRs.

Table 2. A table of ionic liquids used for studies about REE extraction in bauxite residues

Ionic Liquid	REEs Studied	Reference
$HBetTf_2N$	La, Nd, Y, Ce, Sc, Dy	[22], [23], [24]
$EmimHSO_4$	La, Nd, Y, Ce, Sc	[23], [25], [26]
Hbet-STFSI-PS-DVB	Nd, Y, Sc, Dy	[27], [28]

The solubilizing ability of the betaine bis(trifluoromethylsulfonyl)imide [$HBetTf_2N$] ionic liquid is brought about by the acidity of the protonated betainium cation after its dissolution. It presents an acidic value higher than acetic and propionic acids, enabling improved metal dissolution. The anion Tf_2N^-, on the other hand, induces the hydrophobicity of the IL while it exists as a spectator ion chemically. $HBetTf_2N$ exhibits a behavior called thermomorphism which causes it to become miscible with water at temperatures higher than the Upper Critical Solution Temperature (UCST) and immiscible below the UCST.[22] On the contrary, ethyl-methyl-imidazolium hydrogen sulphate [$EmimHSO_4$] demonstrates hydrophilic behavior. A comparative study between these two ILs shows that $HBetTf_2N$ is more selective against Fe, Si, and Ti than $EmimHSO_4$. $HBetTf_2N$, on the other hand, has lower Sc yields (39%) compared to $EmimHSO_4$ (79%). In terms of operating conditions, the hydrophobic system requires the addition of water for metal dissolution to occur. Meanwhile, the hydrophilic process can proceed water-free. It can also be noted that $EmimHSO_4$ is less expensive than $HBetTf_2N$.[23] Equations 1 and 2 below show the chemical reactions of $HBetTf_2N$ and $EmimHSO_4$ wuth regards to REE dissolution.

Dissolution of REE with $HBetTf_2N$:

$$6HbetTf_2N + REE_2O_3 \rightarrow 2REE(bet)_3(Tf2N)_3 + 3H_2O \quad (1)$$

Dissolution of REE with $EmimHSO_4$:

$$REE_2O_n + n[Emim][HSO_4] \rightarrow REE_2(SO_4)_n + nOH^- + n[Emim]^+ \quad (2)$$

Support ionic liquid phase (SILPs) are ILs attached to solid support through physisorption or covalent bonds. Adding a support mechanism provides several advantages to the IL system, such as a reduction in the amount of IL used, enhanced reusability and recovery of ILs, and prevention of IL leaching. All these benefits are achieved while still maintaining the extraction properties of ionic liquids.[27] Initially, the SILP betainium sulfonyl(trifluoromethanesulfonylimide) poly(styrene-co-divinylbenzene) [Hbet-STFSI-PS-DVB] was investigated to observe its adsorption of Sc in acidic solutions for a recovery process from bauxite residues (BR). Sc was extracted by the $HBetTf_2N$ in the SILP while the polystyrene–divinylbenzene (PS–DVB) served as its support. The positive results from this study prompted the use of SILP in REE recovery from BR. The latter research reported high recoveries of REE, with Y and Sc achieving 100% yield. It also extracted 71.8% Nd, but negligible values were recorded for Dy. The SILP also exhibited stability and reusability without affecting its efficiency as a lixiviant.[28]

Despite its reputation as a 'green' solvent, ionic liquids cannot be tagged as 100% environmentally safe. ILs can have increasing toxicity based on their structural composition. Some cations used in ionic liquids may have higher toxicity values than others. For instance, imidazolium-based ILs have been reported to be less toxic than pyridinium-based ILs. Cholinium-based ILs, on the other hand, are less harmful than the former mentioned ionic liquids. Toxicity also depends on the length of the alkyl chain on the cation. The increase in alkyl chain length often indicates higher toxicity in ILs. Depending on its solubility with water, ionic liquids may cause pollution of water sources if not disposed of properly. This can lead to hazardous consequences for the ecosystems and living organisms near the contaminated water. As such, future studies regarding IL extraction systems should consider sustainable designs without limiting their properties as a leaching reagent.[29-30]

5. Conclusion

The rapid emergence of advanced technologies has instigated the exploration of suitable materials, one of which is rare earth elements. Their exceptional properties make them favorable in different applications, which causes a surge in their demand. This, in turn, triggers the call for environmentally friendly REE recovery processes. Accordingly, a potential sustainable path for REE extraction can be rendered by leaching secondary sources, such as bauxite residues and phosphogypsum, using green lixiviants.

Ionic liquids, also known as designer solvents, are generally liquid salts at temperatures below 100°C. They are described as having high chemical and thermal stability with the capacity to have modifiable chemical structures. This allows them to cater to different applications such as energy production and storage, gas capturing, and extraction processes. Free ILs, like $HBetTf_2N$ and $EmimHSO_4$, have been studied for their efficiency as a lixiviant for REEs in bauxite residue. A comparative report presented a high Sc yield for $EmimHSO_4$ while REEs, such as La, Y, Ce, and Nd, were recovered more using $HBetTf_2N$. Operation-wise, $EmimHSO_4$ is cheaper than $HBetTf_2N$ and is capable of metal dissolution without the presence of water. In another study, the ionic liquid $HBetTf_2N$ was deposited on a PS-DVB matrix, forming a support ionic liquid phase. Its results show complete recoveries of Y and Sc, and 71.8% recovery of Nd while demonstrating high stability and reusability. Compared to the free ILs, the SILP produces higher recovery values while having economic advantages. However, $HBetTf_2N$ displays selectivity against Fe, Ti, and Si, which is deemed desirable for selective leaching.

The promising outcomes of these studies solidify the potential application of ILs for REE extraction in the Philippines. With the recent discovery of multiple REE deposits in the country, the interest in their recovery comes to light especially for ores with high concentrations like the Erawan beach placer deposit (~600 ppm) and Samar bauxites (~500 ppm). Using the previous references for REE extraction in BR, Philippine bauxite residues can be processed. Nonetheless, this would require further characterization and observation of the interaction between ILs and Philippine BRs. Thus, if ILs were to be the future of the green REE extraction systems in the Philippines, research should consider optimizing their operating conditions and the sustainability of their design.

References

[1] Wall F 2014 Rare earth elements *Critical Metals Handbook* ed Gunn G (Nottingham: John Wiley & Sons, Ltd) pp 312–339

[2] Gosen B S V, Verplanck P L, Seal R R II, Long K R and Gambogi J 2017 Rare-earth elements *Critical mineral resources of the United States—Eco. and env. geology and prospects for future supply: U.S. Geological Survey Professional Paper 1802* ed Schulz K J et. al. (Reston: U.S. Geological Survey) chapter O pp O1–O31

[3] Shewane P G, Singh A, Gite M and Narkhede A 2014 An Overview of Neodymium Magnets over Normal Magnets for the Generation of Energy *Int. J. on Recent and Innov. Trends in Computing and Communication* vol 2 pp 4056–4059

[4] Yuksel C, Ankarali S and Yuksel N A 2018 The use of neodymium magnets in healthcare and their effects on health *North Clin Istanb* 2018;5(3):268–273

[5] Long K R, Van Gosen B S, Foley N K and Cordier D 2012 The Principal Rare Earth Elements Deposits of the United States: A Summary of Domestic Deposits and a Global Perspective *Non-Renewable Resource Issues (Int. Yr. of Planet Earth)* ed Sinding-Larsen R and Wellmer F W (Dordrecht: Springer) pp 131–155

[6] U.S. Geological Survey 2023 Rare Earths *Mineral commodity summaries 2023* (Reston: U.S. Geological Survey) pp 142–143

[7] Ramos A F, Santos G Jr, Magsambol W N, Castillo M K and Tabora E U 2001 Preliminary geological assessment for rare earths at Ombo Area, San Vicente, Northern Palawan (Quezon: Philippine Nuclear Research Institute)

[8] Gibaga C R L, Samaniego J O, Tanciongco A M, Quierrez R N M, Montano M O, Gervasio J H C, Reyes R C G and M J V Peralta 2022 The rare earth element (REE) potential of the Philippines *J. Geochem. Exploration* **242** 107082

[9] Gaustad G, Williams E and Leader A 2021 Rare earth metals from secondary sources: Review of potential supply from waste and byproducts *Resources, Conservation and Recycling* **167** 105213

[10] Pei Y, Zhang Y, Ma J, Fan M, Zhang S and Wang J 2022 Ionic liquids for advanced materials *Materials Today Nano* **17** 100159

[11] Lei Z, Chen B, Koo Y-M and MacFarlane D R 2017 Introduction: Ionic Liquids *Chem. Rev.* **117** 10

[12] Huddleston J, Visser A, Reichert W, Willauer H, Broker G and Rogers R 2001 Characterization and Comparison of Hydrophilic and Hydrophobic Room Temperature Ionic Liquids Incorporating the Imidazolium Cation *Green Chem.* **3** pp 156–164

[13] Yee P, Shah J K and Maginn E J 2013 State of Hydrophobic and Hydrophilic Ionic Liquids in Aqueous Solutions: Are the Ions Fully Dissociated? *J. Phys. Chem. B* **117** 41

[14] Abbott A P, Frisch G and Ryder K S 2013 Electroplating Using Ionic Liquids Annual Review of Materials Research **43** pp 335-358

[15] Caporali S, Fossati A, Lavacchi A, Perissi I, Tolstogouzov A and Bardi U 2008 Aluminium electroplated from ionic liquids as protective coating against steel corrosion *Corrosion Sci.* **50** 2 pp 534–539

[16] Maniam K K and Paul S 2020 Progress in Electrodeposition of Zinc and Zinc Nickel Alloys Using Ionic Liquids *Applied Sci.* **10**(15) 5321

[17] MacFarlane D R, Tachikawa N, Forsyth M, Pringle J M, Howlett P C, Elliott G D, Davis J H Jr., Watanabe M, Simon P and Angell C A 2014 Energy applications of ionic liquids *Energy Envi. Sci.* **7** pp 232–250

[18] Greer A J, Jacquemin J and Hardacre C 2020 Industrial Applications of Ionic Liquids *Molecules* **25**(21) 5207

[19] Shiflett M B, Drew D W, Cantini R A and Yokozeki A 2010 Carbon Dioxide Capture Using Ionic Liquid 1-Butyl-3-methylimidazolium Acetate *Energy Fuels* **24** 10 pp 5781–5789

[20] Kim B-K, Lee E J, Kang Y and Lee J-J 2018 Application of ionic liquids for metal dissolution and extraction *J. Indus. Engg. Chem.* **61** pp 388–397

[21] Ochsenkühn-Petropulu M, Lyberopulu T, Ochsenkühn K M and Parissakis G 1996 Recovery of lanthanides and yttrium from red mud by selective leaching *Analytica Chimica Acta* 319(1-2)

pp 249–254
[22] Davris P, Balomenos E, Panias D and Paspaliaris I 2016 Selective leaching of rare earth elements from bauxite residue (red mud) using a functionalized hydrophobic ionic liquid *Hydrometallurgy* 164 pp 125–135
[23] Bonomi C, Davris P, Balomenos E, Giannopoulou I and Panias D 2017 Ionometallurgical Leaching Process of Bauxite Residue: a Comparison between Hydrophilic and Hydrophobic Ionic Liquids *Proceedings of the 35th International ICSOBA Conference, Hamburg, Germany* pp 2–5
[24] Mikeli E, Balomenos E and Panias D 2021 Utilizing Recyclable Task-Specific Ionic Liquid for Selective Leaching and Refining of Scandium from Bauxite Residue *Molecules* **26**(4) 818
[25] Davris P, Balomenos E, Panias D and Paspaliaris I 2014 Leaching of rare earths from bauxite residues using imidazolium based ionic liquids *Proceedings of the 1st European Rare Earth Resources Conference (ERES 2014), Milos, Greece* pp 4–7
[26] Bonomi C, Alexandri A, Vind J, Panagiotopoulou A, Tsakiridis P and Panias D 2018 Scandium and Titanium Recovery from Bauxite Residue by Direct Leaching with a Brønsted Acidic Ionic Liquid *Metals* **8**(10) 834
[27] Avdibegović D, Regadío M and Binneman K 2017 Recovery of scandium(III) from diluted aqueous solutions by a supported ionic liquid phase (SILP) *RSC Adv.* **7** pp 49664–49674
[28] Avdibegović D, Regadío M and Binnemans K 2018 Efficient separation of rare earths recovered by a supported ionic liquid from bauxite residue leachate *RSC Adv.* **8** pp 11886–11893
[29] Made M, Liu J-F and Pang L 2015 Environmental Application, Fate, Effects, and Concerns of Ionic Liquids: A Review *Environ. Sci. Technol.* 49(21) pp 12611–12627
[30] Magina S, Barros-Timmons A, Ventura S P M and Evtuguin D V 2021 Evaluating the hazardous impact of ionic liquids – Challenges and opportunities *J. Hazardous Materials* **412** 125215

IOP Conf. Series: Materials Science and Engineering **1318** (2024) 012052 doi:10.1088/1757-899X/1318/1/012052

Selection of Solar-Based Water-Splitting Hydrogen Production and Hydrogen Storage Technologies Using Neutrosophic Analytic Hierarchy Process and Neutrosophic Complex Proportional Assessment

R Oseña, J Merto, S Tayona and G Rustia*

Department of Chemical Engineering, College of Engineering and Technology, University of the City of Manila, NCR, City of Manila 1002, Philippines

*Corresponding author: gsrustia2018@plm.edu.ph

Abstract. The production of hydrogen through water-splitting using solar energy has the potential to meet the world's future energy demand while also addressing the issue of greenhouse gas emissions. To compare and select the best technology among solar thermochemical cycles, photoelectrochemical, and photocatalytic water-splitting in terms of solar-to-hydrogen efficiency, production cost, safety, life expectancy, maintenance cost, and detrimental impact on the environment, this study used a multi-criteria decision analysis hybrid called Neutrosophic Analytic Hierarchy Process (NAHP) and Neutrosophic Complex Proportional Assessment (NCOPRAS). The same method was used to determine the best technology for automobiles among compressed hydrogen storage, metal hydrides, metal-organic frameworks, and chemical storage in terms of gravimetric system capacity, volumetric system capacity, safety, system cost, cycle life, energy efficiency, detrimental impact on the environment, and refueling time. There were 4 experts each in hydrogen storage and hydrogen, specifically from the Philippines, Australia, India, Romania, and Italy. All experts' judgments have a consistency ratio (CR) < 0.1. The priority criterion for hydrogen production was life expectancy (weighted value (WV) = 0.189), and gravimetric system capacity for hydrogen storage (WV = 0.137). The result showed that among all alternatives for hydrogen production, the solar thermochemical cycle was the best (relative significance value (RSV) = 0.3432) as for hydrogen storage, the best technology among the alternatives was metal hydrides (RSV = 0.2575). A sensitivity analysis was used to check the variability and robustness of the solution.

1. Introduction

Across the years, as the need for energy arise, significant attention was given to develop hydrogen technologies. Because of its high energy density, low environmental impact, and long-term energy storage, hydrogen is the most promising energy carrier known today [1]. However, the variety of hydrogen production and storage methods being developed had limited its progress given the division of resources.

Some of the promising hydrogen production methods that work as mentioned are photocatalysis, photoelectrochemical, and thermochemical water-splitting. These technologies for hydrogen production are still in the developmental or RnD stage [2]. Still, many researchers are continuously investigating and experimenting with these processes. This is especially so because they are long-term solutions for

IOP Conf. Series: Materials Science and Engineering **1318** (2024) 012052 doi:10.1088/1757-899X/1318/1/012052

environmental issues (e.g., global warming, climate change, etc.) caused by the abundance of greenhouse gases.

Hydrogen production necessitates storage options. Researchers primarily focus on developing storage systems for mobile applications, particularly for automobiles, as they require extensive testing and development. Compressed hydrogen storage is commonly used for stationary applications [3]. Various storage methods, such as compressed vessels, cryogenic vessels, metal hydrides, metal-organic frameworks (MOFs), and chemical storage, are being developed, specifically for automobile use [3, 4]. To determine the best technology for further development, this study used a Multi-Criteria Decision Analysis (MCDA) method to rank the criteria according to their importance and later select the best technology. The Analytic Hierarchy Process (AHP) was a widely used method that involved consistency checking while Complex Proportional Assessment (COPRAS) only required low calculation time and was easy to use [5, 6]. Both methods were taken under a neutrosophic environment to address the uncertainties and provide more accurate results [7]. The Neutrosophic Analytic Hierarchy Process (NAHP) and Neutrosophic Complex Proportional Assessment (NCOPRAS) are used within a neutrosophic environment to account for uncertainties and improve accuracy. These methods are utilized to compare and select the best water-splitting technology (photocatalysis, photoelectrochemical systems, and solar thermochemical cycles) and hydrogen storage systems (compressed hydrogen, MOFs, metal hydrides, and chemical storage) for automobiles.

2. Methodology

2.1. General procedure

The steps the authors had taken to meet the objectives are shown in figure 1. It was divided into two stages: preparation and NAHP-NCOPRAS stage. The whole process started from the selection of alternatives and criteria, defining the criteria and linguistic variables, constructing the questionnaire and sampling design, gathering data, performing the procedure for NAHP followed by NCOPRAS, constructing tables and charts based on the results, discussing the findings, and finally made conclusions from them.

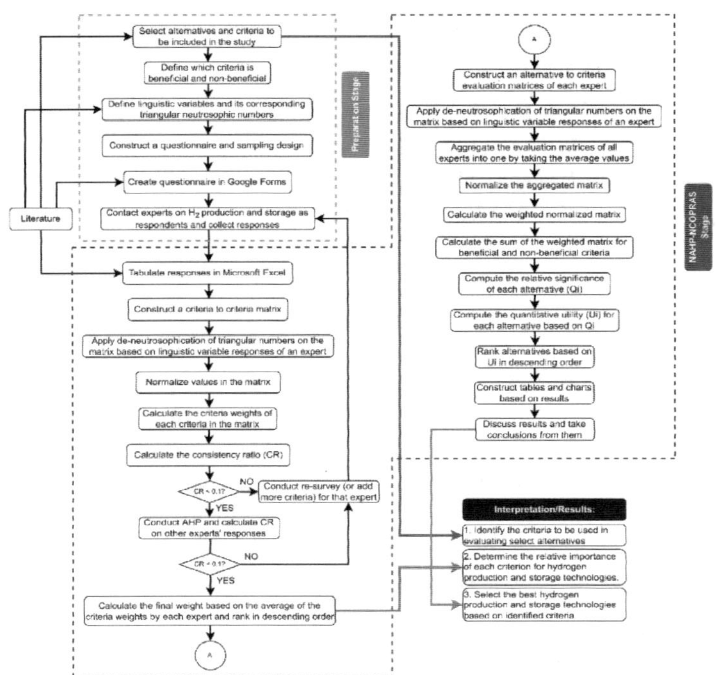

Figure 1. Process Flowchart.

2.2. Utilization of MCDA

The Analytic Hierarchy Process is simple and easy to use and did not compromise its purpose [5]. However, more accurate outcomes may result from AHP when used along with a neutrosophic environment [7]. The utilization of NAHP for the ranking of the criteria is discussed below [7]:

Step 1: Define the linguistic scale and the triangular neutrosophic scales for use in evaluating the criteria. The scales and variables were adopted based on the study of Abdel-Basset et al. [7], and are defined as follows:

Table 1. Linguistic variable and triangular neutrosophic scale for ranking criteria [7].

Linguistic variables	Triangular neutrosophic scale (l, m, u); α, β, θ
Low importance (LM)	(0.08,0.28,0.28); 0.50,0.10,0.30
Simple importance (SM)	(0.35,0.40,0.50); 0.80,0.20,0.20
Primary importance (PM)	(0.50,0.65,0.80); 0.90,0.20,0.10
Highly robust importance (HRM)	(0.80,0.82,0.88); 0.90,0.10,0.10
Extremely prioritized (EL)	(0.85,0.90,1.00); 0.90,0.10,0.00

Step 2: Construct a criteria-to-criteria matrix that is based on experts' judgement in terms of linguistic variables using the matrix formula. The number of matrices created will depend on the number of experts surveyed.

$$C_1 \quad C_2 \quad \dots \quad C_i A = \begin{matrix} C_1 \\ C_2 \\ \vdots \\ C_j \end{matrix} \begin{bmatrix} - & a_{12} & \cdots & a_{1i} \\ 1/a_{21} & - & \cdots & a_{2i} \\ \cdots & \cdots & - & \cdots \\ 1/a_{j1} & 1/a_{j2} & \cdots & - \end{bmatrix} \tag{1}$$

Step 3: Convert the linguistic variables into their respective triangular neutrosophic scales. After, conduct the de-neutrosophication process using the score function to obtain usable values for the AHP method.

$$A = \begin{bmatrix} - & \langle(l,m,u);\alpha,\beta,\theta\rangle_{12} & \cdots & \langle(l,m,u);\alpha,\beta,\theta\rangle_{1i} \\ 1/\langle(l,m,u);\alpha,\beta,\theta\rangle_{21} & - & \cdots & \langle(l,m,u);\alpha,\beta,\theta\rangle_{2i} \\ \cdots & \cdots & - & \cdots \\ 1/\langle(l,m,u);\alpha,\beta,\theta\rangle_{j1} & 1/\langle(l,m,u);\alpha,\beta,\theta\rangle_{j2} & \cdots & - \end{bmatrix} \tag{2}$$

$$\text{Score Function (A)}=\frac{(l+m+u)\times(2+\alpha-\theta-\beta)}{8} \tag{3}$$

Step 4: Calculate each criterion's weight using the AHP method.

$$\sum_{i=1}^{g} A_{ij} \tag{4}$$

From the computed sum of each column, normalize the matrix:

$$Z_{ij}=\frac{A_{ij}}{\sum_{i=1}^{g} A_{ij}} \tag{5}$$

Then compute for the criteria weights as:

$$W_j=\frac{\sum_{j=1}^{h} Z_{ij}}{h} \tag{6}$$

Once done, calculate the consistency ratio (CR) of the matrix where readers are advised to refer to the steps for its computation on literature by Mu and Pereyra-Rojas [8].

Step 5: If the calculated CR < 0.1, then the degree of consistency is acceptable, and the authors would proceed to the next step. If not, the authors will conduct either a resurvey with the expert or increase the number of criteria to satisfy the acceptable consistency ratio for the matrix in question [6, 7].

Step 6: Finally, compute for the final weights by taking the average of the criteria weights by each expert and ranking the criteria in descending order based on the computed final weights.

COPRAS is the simplest method compared to VIKOR, EVAMIX, TOPSIS, and AHP and is one of the methods with low calculation time yet has the best-ranking performance, especially when it comes to material selection [6]. However, one of the weaknesses of the conventional method is that it does not consider the uncertainties and imprecise information of experts' judgement. Thus, a neutrosophic environment is introduced to improve the accuracy in selecting alternatives [7]. The utilization of NCOPRAS for the ranking of the alternatives is discussed below [7]:

Step 1: Define the linguistic scale and the triangular neutrosophic scales for use in evaluating the criteria. The scales and variables were adopted based on the study of Abdel-Basset et al. [7], and are defined as follows:

Table 2. Linguistic variable and triangular neutrosophic scale for ranking alternatives [7].

Linguistic variables	Triangular neutrosophic scale (l, m, u); α, β, θ
Highly low (HL)	(0.10,0.30,0.35); 0.10,0.20,0.15
Low (LW)	(0.20,0.25,0.25); 0.60,0.20,0.30
Intermediate low (IW)	(0.35,0.40,0.50); 0.60,0.10,0.20
Intermediate (IE)	(0.60,0.65,0.70); 0.80,0.10,0.10
Intermediate strong (IG)	(0.65,0.70,0.80); 0.90,0.20,0.10
Robust (RT)	(0.85,0.90,0.90); 0.70,0.20,0.20
Highly robust (HT)	(0.90,0.95,0.95); 0.90,0.10,0.10

Step 2: Define which criteria were beneficial and non-beneficial. The criterion was labeled beneficial if it was desired to be higher (the greater the better) and non-beneficial if it was desired to be lower (the lower the better). Shown in table 3 were the criteria used in the study with their respective labels:

Table 3. Beneficial and non-beneficial criteria.

Hydrogen production criteria	Type
Solar-to-Hydrogen efficiency	Beneficial
Detrimental impact on the environment	Non-beneficial
Hydrogen production cost	Non-beneficial
Safety	Beneficial
Maintenance cost	Non-beneficial
Life expectancy	Beneficial
Hydrogen storage criteria	**Type**
Detrimental impact on the environment	Non-beneficial
Energy efficiency	Beneficial
Safety	Beneficial
System cost	Non-beneficial
Refueling time	Non-beneficial
Cycle life	Beneficial
Volumetric system capacity	Beneficial
Gravimetric system capacity	Beneficial

Step 3: Construct an alternatives-to-criteria matrix based on the responses of the experts in terms of linguistic variables. The number of matrices created will depend on the number of experts surveyed.

$$X=\left[x_{ij}\right]_{m\times n}=\begin{bmatrix} x_{11} & x_{12} & \cdots & x_{1n} \\ x_{21} & x_{22} & \cdots & x_{2n} \\ \cdots & \cdots & \cdots & \cdots \\ x_{m1} & x_{m2} & \cdots & x_{mn} \end{bmatrix} \tag{7}$$

Step 4: Convert the linguistic variables into their respective triangular neutrosophic scales. After, conduct the de-neutrosophication process to obtain usable values for the COPRAS method.

$$A=\begin{bmatrix} - & \langle(l,m,u);\alpha,\beta,\theta\rangle_{12} & \cdots & \langle(l,m,u);\alpha,\beta,\theta\rangle_{1i} \\ 1/\langle(l,m,u);\alpha,\beta,\theta\rangle_{21} & - & \cdots & \langle(l,m,u);\alpha,\beta,\theta\rangle_{2i} \\ \cdots & \cdots & - & \cdots \\ 1/\langle(l,m,u);\alpha,\beta,\theta\rangle_{j1} & 1/\langle(l,m,u);\alpha,\beta,\theta\rangle_{j2} & \cdots & - \end{bmatrix} \tag{8}$$

$$\text{Score Function }(A)=\frac{(l+m+u)\times(2+\alpha-\theta-\beta)}{8} \tag{9}$$

Step 5: Aggregate the individual matrices into one single matrix by computing the average of the values in each matrix.

Step 6: Normalize the aggregated matrix, and additionally, construct a weighted normalized matrix through the use of the dedicated equation for its computation.

$$R=\left(r_{ij}\right)_{m\times n}=\frac{x_{ij}}{\sum_{i=1}^{m} x_{ij}} \tag{10}$$

$$D=\left[y_{ij}\right]_{m\times n}=r_{ij}\times w_j \tag{11}$$

Step 7: Calculate the sum of the beneficial and non-beneficial criteria from the weighted normalized matrix, obtain the relative significance of each alternative Q_i and quantitative utility U_i from the computed Q_i.

$$S_{+i}=\sum_{j=1}^{n} y_{+ij} \tag{12}$$

$$S_{-i}=\sum_{j=1}^{n} y_{-ij} \tag{13}$$

$$Q_i=S_{+i}+\frac{S_{-min}\sum_{i=1}^{m} S_{-i}}{S_{-i}\sum_{i=1}^{m}(S_{-min}/S_{-i})} \tag{14}$$

$$U_i=\left(\frac{Q_i}{Q_{max}}\right)\times 100\% \tag{15}$$

Step 8: Finally, rank the alternatives based on U_i in descending order to obtain the best alternative based on expert opinion.

2.3. Sensitivity analysis

Defined as the study of how distinct sources of uncertainty in the model output (numerical or otherwise) might be assigned to different sources of uncertainty in the model input [9], sensitivity analysis is a tool that assesses the effects of changes in input data created by MCDA methods. This is primarily done to determine the MCDA method's robustness; to highlight any differences in the main criterion weights that could affect the proposed MCDA method's selection of alternatives as well as to eliminate bias that may arise during the collection of data [10, 11].

Several methods for sensitivity analysis exist, where the study will adopt the sensitivity analysis conducted by Büyüközkan & Göçer [10] due to it being utilized in a similar MCDA method as the study.

3. Results and Discussion

3.1. Identification of the criteria

Shown in table 4 were the criteria used and the related studies that used them. These criteria were chosen because they were commonly used compared to other criteria like the prospective market, labor influence, and predictability. Other criteria were given an umbrella term, such as emission of contaminants and waste generation, which were classified under the detrimental impact on the environment. Material and labor costs, on the other hand, were covered by the hydrogen production cost.

Table 4. Related studies which used similar criteria.

Criteria	Description	Sources
Hydrogen Production Cost	The money needed to produce a kilo of hydrogen. It includes the material and labor cost.	[7, 12-15]
Detrimental Impact on The Environment	How the selected method has an unfavorable effect to the environment towards the sustainable use of resources.	[7, 12, 14, 16, 17]
Maintenance Cost	Costs incurred by the system to preserve its good working condition. It includes the cost needed for the maintenance of the equipment.	[7, 13, 14]
Life Expectancy	Amount of time or period a given technology is expected to remain operational from the period beginning its formal operations until the time that the facility should properly be closed or shut down.	[7, 13]
STH Efficiency	Total energy that is procured divided by the total energy that enters.	[15, 18]
Safety	A state wherein the possible risks and injuries that can come from a system are being controlled.	[13, 17, 19, 20]
System Cost	Cost that includes all the materials and components used in the development and design of the storage system.	[16, 20, 21]
Gravimetric System Capacity	Amount of hydrogen stored over the mass of the storage system.	[19-21]
Volumetric System Capacity	Amount of hydrogen stored over the total volume of the storage system.	[19-21]
Cycle Life	The number of cycles that can be done before the capacity falls to 80% and before a noticeable reduction in the performance of the storage system took place.	[19, 20]
Refueling Time	Time it takes for fuel, or in this case, hydrogen, to fill the storage, and it is also measured in minutes.	[19, 20]
Energy Efficiency	Defined as the energy necessary to intake or release hydrogen.	[17, 19, 20]

3.2. Relative importance of each hydrogen production criterion, and ranking of hydrogen production alternatives

Table 5 presents the consistency ratios of the four experts with their corresponding values. All CR calculated were less than 0.1.

Table 5. Consistency ratios of four experts in hydrogen production criteria.

Expert	1	2	3	4
Consistency ratio	0.0793	0.0667	0.0768	0.0273

The results of the values of each hydrogen production criterion weights from individual expert opinions were averaged across the four experts which resulted in the following final weight criteria values and their corresponding ranking shown in figure 2. From these findings, it was determined that the overall relative importance of each hydrogen production criterion showed that the life expectancy of hydrogen production alternatives is the most important criterion to consider when determining what alternative to select from other options.

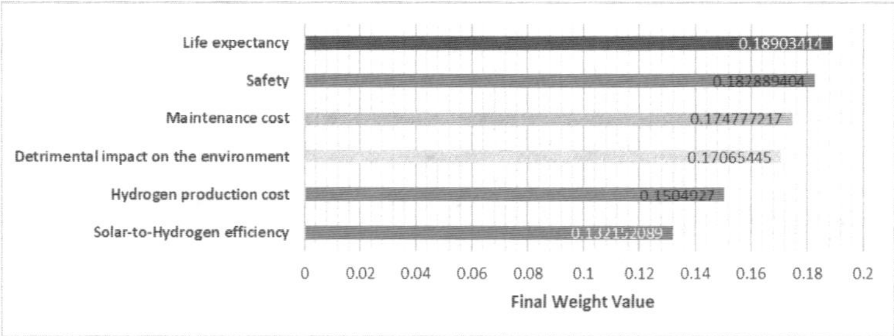

Figure 2. Ranking of hydrogen production final criteria weights by all experts.

Shown in figure 3 is the ranking of alternatives for solar-based hydrogen production technologies. Solar thermochemical cycle obtained a value of 0.3432. Thus, the best solar-based hydrogen technology is the solar thermochemical cycle. Although, solar thermochemical cycle is the best technology, both photocatalytic and photoelectric are still good alternatives since their relative significance values are close to the value of solar thermochemical.

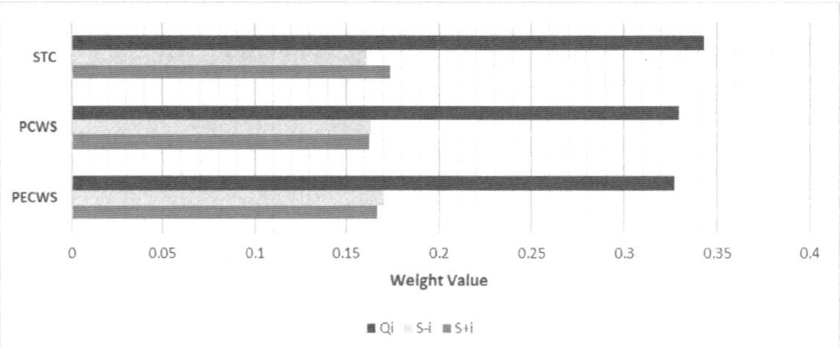

Figure 3. Ranking and weights of three hydrogen production alternatives based on experts' opinion.

3.3. Relative importance of each hydrogen storage criterion, and ranking of hydrogen storage alternatives

According to table 6, the experts who answered the survey had consistency ratios of less than 0.1.

Table 6. Consistency ratios of four experts in hydrogen storage criteria.

Expert	1	2	3	4
Consistency ratio	0.0467	0.0467	0.0480	0.0505

Figure 4 summarizes the final weights of the experts which accordingly determined the order of the criteria with regards to importance. The most important criterion was gravimetric system capacity. Therefore, this meant that the gravimetric system capacity held the most weight when evaluating the storage technologies for vehicles.

Hydrogen storage technologies were ranked based on the results obtained from experts' opinions. It revealed that the relative significance value or Q_i closely relates to the given alternatives as shown in figure 5. This means that there is tight competition in selecting the best hydrogen storage technology. Accumulated answers from experts worldwide revealed metal hydrides as the highest-ranking among all the alternatives.

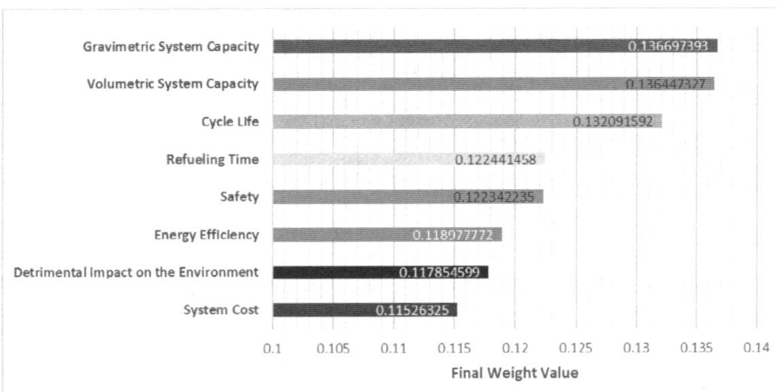

Figure 4. Ranking of hydrogen storage final criteria weights by all experts.

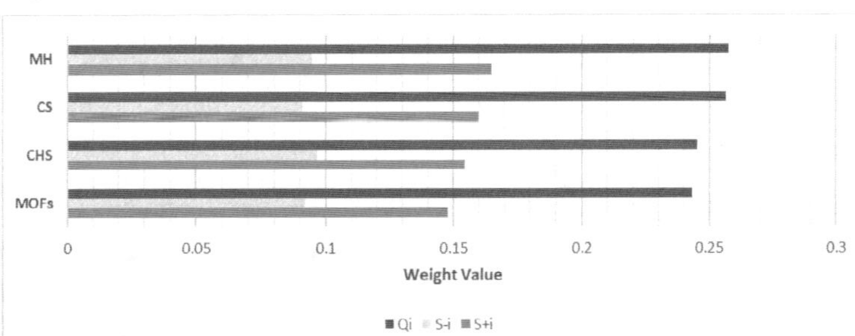

Figure 5. Ranking and weights of four hydrogen storage alternatives based on experts' opinion.

3.4. Sensitivity analysis

Figure 6 and figure 7 present the results of the sensitivity analysis conducted to assess the acquired rankings of hydrogen production and storage alternatives using NAHP-NCOPRAS. It was determined that solar thermochemical cycles were still the best hydrogen production alternative as they acquired the top rank in six out of seven cases. This was also the case for metal hydrides which acquired the top rank four out of nine times, compared to chemical storage which acquired the top rank three out of nine cases, and compressed hydrogen storage and MOFs both for one out of nine cases. Thus, metal hydrides were the best alternative out of the four hydrogen storage alternatives.

IOP Conf. Series: Materials Science and Engineering **1318** (2024) 012052 doi:10.1088/1757-899X/1318/1/012052

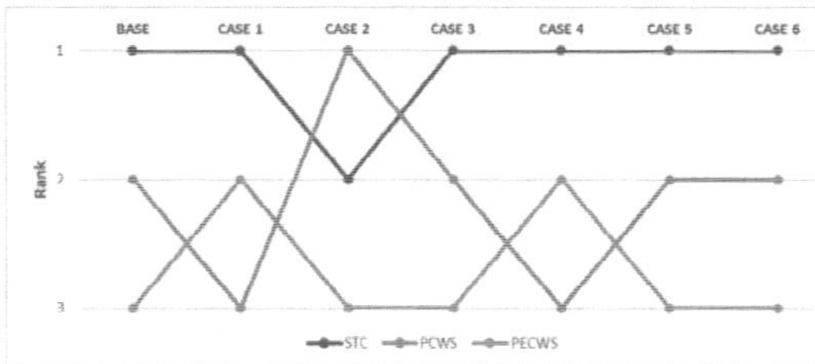

Figure 6. Sensitivity analysis on the ranking of hydrogen production alternatives.

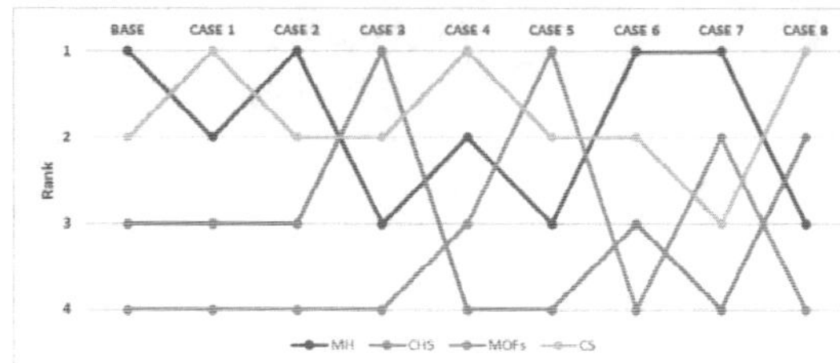

Figure 7. Sensitivity analysis on the ranking of hydrogen storage alternatives.

4. Conclusion

The criteria having the most importance in hydrogen production and storage systems were life expectancy, and gravimetric system capacity respectively. As for the alternatives, the first in rank for hydrogen production was Solar Thermochemical Cycles, followed by Photocatalytic water-splitting. For hydrogen storage, the first in rank was Metal Hydrides, followed by Chemical Storage. According to the sensitivity analysis, the ranks of the alternatives were often the same. Solar thermochemical cycles were still the best hydrogen generation option, and this was also true for metal hydrides.

References

[1] Bhandari R and Shah R 2021 *Renew. Energy* **177** 915–931

[2] International Renewable Energy Agency 2018 *Hydrogen from renewable power: Technology outlook for the energy transition* (Abu Dhabi: International Renewable Energy Agency)

[3] Hassan I A, Ramadan H S, Saleh M A and Hissel D 2021 *Renewable and Sustainable Energy Reviews* **149** 111311

[4] Barthelemy H, Weber M and Barbier F 2017 *Int. J. Hydrog. Energy* **42** 7254–7262

[5] Velasquez M and Hester P 2013 *Int. J. Oper. Res.* **10** 56-66

[6] Chatterjee P, Athawale V, and Chakraborty S 2011 *Mater. Des.* **32** 851–860

[7] Abdel-Basset M, Gamal A, Chakrabortty R and Ryan M 2020 *Int. J. Hydrog. Energy* **46** 4567–4591

[8] Mu E and Pereyra-Rojas M 2017 *Practical Decision Making* (New York: Springer Cham)

[9] Ustinovichius L and Simanaviciene R 2010 *(Cooperative Design, Visualization, and Engineering. Lecture Notes in Computer Science* vol 6240*)* (Berlin: Springer) 89–96

[10] Büyüközkan G and Göçer F 2021 *IEEE Trans Eng Manag* **68** 1486–1503

[11] Gupta H and Barua M K 2018 *Science of the Total Environment* **633** 122–139

[12] Li W, Ren X, Ding S and Dong L 2020 *Int. J. Hydrog. Energy* **45**(59) 34385–34395

[13] Xu D, Li W, Ren X, Shen W and Dong L 2019 *Int. J. Hydrog. Energy* **45** 34396-34411

[14] Pilavachi P A, Chatzipanagi A I and Spyropoulou A I 2009 *Int. J. Hydrog. Energy* **34**(13) 5294–5303

[15] Department of Energy (US-DOE) DOE Technical Targets for Hydrogen Production from Photoelectrochemical Water Splitting [Internet] 2022 [cited 2022 Jun 16] Available from: https://www.energy.gov/eere/fuelcells/doe-technical-targets-hydrogen-production-photoelectrochemical-water-splitting

[16] Deveci M 2018 *Int. J. Hydrog. Energy* **43**(19) 9353–9368

[17] Acar C, Beskese A and Temur T 2018 *Int. J. Hydrog. Energy* **43**(39) 18059–18076

[18] Ouagued M, Khellaf A and Loukarfi L 2018 *Int. J. Hydrog. Energy* **43**(6) 3451–3465

[19] Gim B and Wook J 2014 *Int. J. Hydrog. Energy* **39**(15) 7852–7858

[20] Department of Energy (US-DOE) DOE Technical Targets for Onboard Hydrogen Storage for Light-Duty Vehicles [Internet] 2022 [cited 2022 Jun 16] Available from: https://www.energy.gov/eere/fuelcells/doe-technical-targets-onboard-hydrogen-storage-light-duty-vehicles

[21] Gumus A T, Yayla A Y, Çelik E and Yildiz A 2013 *Energies* **6**(6) 3017–3032

Conversion of polypropylene (PP) plastic waste to liquid oil through catalytic pyrolysis using Philippine natural zeolite

Angelica S Bautista[1], Karl Nikolai O Rivera[1*], Trixie Anne Kimberly M Suratos[1] and Maria Natalia R Dimaano[2]

[1] Chemical Engineering Department, Faculty of Engineering, University of Santo Tomas, España, Manila, Philippines 1008
[2] Research Center for the Natural and Applied Sciences and Graduate School, University of Santo Tomas, España, Manila, Philippines 1008

karlnikolai.rivera.eng@ust.edu.ph

Abstract. The accumulation of plastic waste compelled the need to develop energy recovery methods, such as pyrolysis, which could convert plastics into valuable energy sources. Pyrolysis requires high operating temperatures; thus, a catalyst is often utilized to speed up the process. In this study, the viability of Philippine Natural Zeolite (PNZ) as a catalyst was investigated through the conversion of polypropylene (PP) waste into liquid oil using catalytic pyrolysis. The PP waste feedstocks were pre-mixed with the PNZ in a 1:10 ratio. Pyrolysis was carried out in a heating mantle for three trials of non-catalytic and catalytic pyrolysis. SEM and XRD were conducted to characterize the PNZ, while FTIR was carried out for the pyrolysis oil. SEM analysis showed the voids formed from the pores of the catalyst that influenced the pyrolysis reaction. Results from the XRD pattern also revealed clinoptilolite, mordenite, and heulandite as the main minerals found in the PNZ. Non-catalytic pyrolysis of PP yielded an average of 75.03% liquid oil, 13.66% residue, and 11.31% uncondensed gas. Meanwhile, catalytic pyrolysis produced higher yields with 86.39% liquid oil, 12.49% residue, and 1.12% uncondensed gas. FTIR results exhibited clear peaks for both processes, with similar components but different intensities.

1. Background

Plastic production and consumption continue to increase as it remains a valuable commodity due to its multiple applications. However, its adverse environmental impact must be addressed as the plastic waste produced annually increases while plastic waste management becomes insufficient. Around 1.28 billion US dollars' worth of plastics were generated in the Philippines in 2016, which grew by 6.11% per year from 2018 to 2023 due to the increasing demand. Most of the plastics produced locally are plastic packaging for consumer products, with around 59.7 billion sachets and single-use plastics used and discarded annually due to their low cost and convenience [1]. About 0.28 to 0.75 metric tons of plastic wastes accumulated critically imposes a severe problem in the Philippines [2].

Among the seven types of plastics, polypropylene (PP) is one of the most common and essential consumer plastics, generally due to its initial low density and lightweight property as an amorphous resin [3]. The country produces approximately Php 40 M worth of PP plastics as free-on-board (FOB) exports, of which 28% are recycled and a majority of which are disposed of locally or shipped internationally [4].

Pyrolysis is one of the energy recovery and cogeneration techniques for different organic feeds. It utilizes high temperatures under an inert atmosphere to promote the thermal decomposition of organic feedstocks and produce simpler hydrocarbon compounds. It is more favorable than other waste-to-energy technologies, such as open-air combustion and incineration. This is due to the minimal refuse and high recovery rates. Moreover, pyrolysis requires high energy rates to decompose the plastic waste feedstock completely [5].

Using a viable catalyst to lower the energy required in performing thermal degradation can make pyrolysis more efficient. Alumina-silica catalysts, which include zeolite minerals, are usually employed due to their high thermal, chemical, and physical stability, depending on their crystal lattice formation. Synthetic zeolites are more effective as catalysts but more expensive than naturally occurring zeolites. One of the most abundant natural zeolites in the Philippines is the Philippine Natural Zeolite (PNZ) which may be used as a catalyst [6]. Compared to other common catalysts, PNZ is cheaper and readily available but does not alter the components of the product and can maintain percent recovery relative to the residue amount.

The lack of studies on catalytic pyrolysis using PNZ and environmental concerns on plastic waste prompted the researchers to test the viability of PNZ as a catalyst for pyrolysis of PP waste by characterizing the zeolite and comparing the product yields from non-catalytic and catalytic pyrolysis. This study also aims to investigate the functional groups found in the liquid product obtained from both processes to identify whether the catalyst would affect the components present in the pyrolyzed oil.

2. Materials and Methodology

2.1. Sample preparation

The collected PP waste was cleaned and shredded into uniform smaller pieces using a plastic shredder. For non-catalytic pyrolysis, 120 g of shredded PP waste served as the feedstock for the process. The PNZ (1mm by 3mm) used for catalytic pyrolysis was procured from SAILE Industries Inc. that trades industrial minerals directly from Mangatarem, Pangasinan, Philippines. For catalytic pyrolysis, the feedstock contained PNZ mixed with the shredded PP waste in a 1:10 ratio. With this, 12 g of PNZ was used for every 120 g of shredded PP waste.

2.2. Preliminary setup

The pyrolysis setup in Figure 1 utilized an ISOPAD GmbH Heidelberg heating mantle as the reactor, and a 500 mL Quickfit Pyrex round bottom flask to contain the feedstock. Attached to the neck are glass adapters where the condenser, nitrogen gas (N_2), and thermometer are connected. The non-catalytic and catalytic pyrolysis samples were fed separately in the round-bottom flask. Before heating, the set-up was purged with N_2 at 50 kPa for 5 minutes to attain an inert atmosphere.

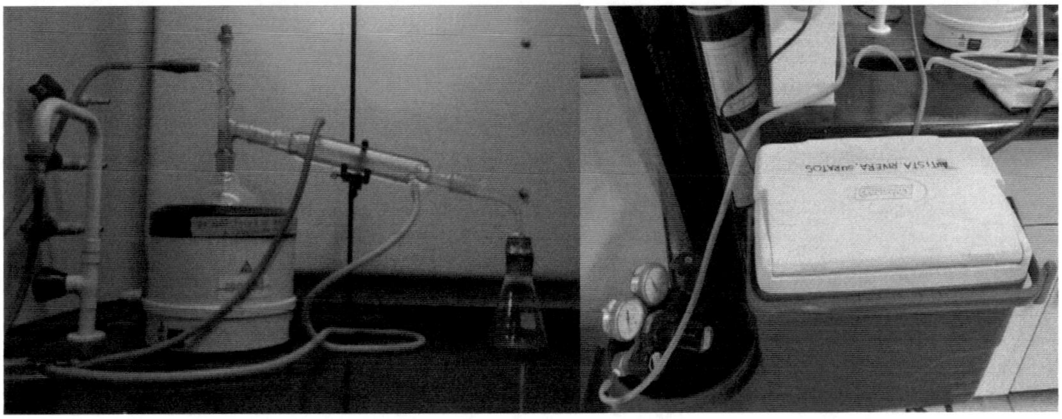

Figure 1. Experimental set-up for non-catalytic and catalytic pyrolysis.

2.3. Experimental procedure
Three trials were conducted each for the non-catalytic and catalytic pyrolysis. The heating mantle was switched on to a temperature of 320°C. The cooling water at 26°C was allowed to flow through the condenser using a pump at a rate of 4.487 L min^{-1}. Temperature was monitored throughout the process, and the PP waste was allowed to vaporize and condense for 1 hour. The temperature in the flask reached 320°C after an average time of 54 min. The pyrolyzed oil was collected from the Erlenmeyer flask attached at the end of the condenser. The liquid oil and residue were measured for the weight and volume, respectively. The catalytic pyrolysis applied the same procedure but with the PNZ-mixed PP waste as feedstock.

3. Results and Discussion

3.1. Scanning Electron Microscopy (SEM) analysis of Philippine natural zeolite
Uneven surfaces from the PNZ were observed with voids, as shown in Figure 2 (a), which became more apparent at closer magnification. The SEM image exhibited microporous surfaces with elongated ridges as magnification was increased to 2500x-5000x. Having microporous surfaces, the ridges and voids were suspected to increase the rate of pyrolysis due to an increased surface area [7].

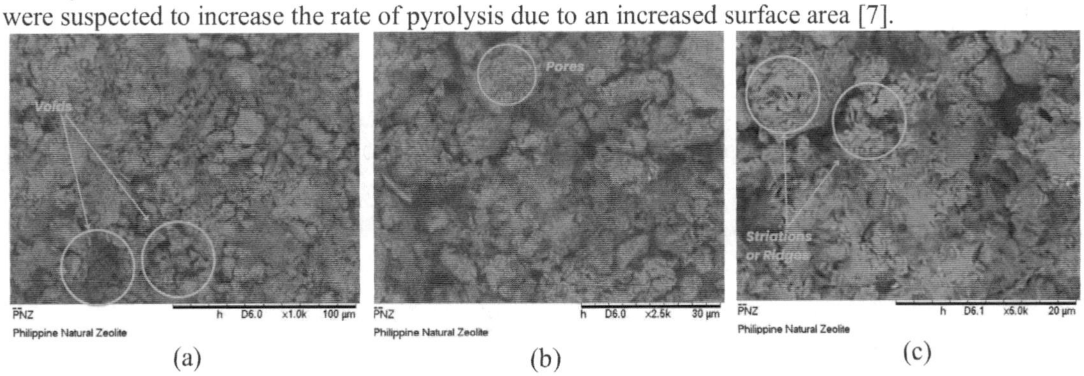

Figure 2. (a) 1000x, (b) 2500x, and (c) 5000x magnification of a PNZ granule.

3.2. X-Ray Diffraction (XRD) analysis of Philippine natural zeolite
The characterization of PNZ using XRD was carried out in continuous scan with a speed of 1.00° min^{-1} from a range of $2\theta = 2.00°$ to 70.00°. Figure 3 presents the XRD pattern of the PNZ, showing sharp and intense peaks, indicating that the material has a crystalline structure [8]. The diffraction peaks were identified through matching via Match software based on the composition of PNZ, as stated in the chemical analyses of previous works [9,10]. Major and minor characteristic peaks were mainly

attributed to clinoptilolite, mordenite, and heulandite, which are common natural zeolite minerals that primarily contain silicon and aluminum compounds. Major peaks were located around 2θ values of 9.89°, 22.36°, 25.72°, and 29.50° which have intensity values that are relatively greater than 50%.

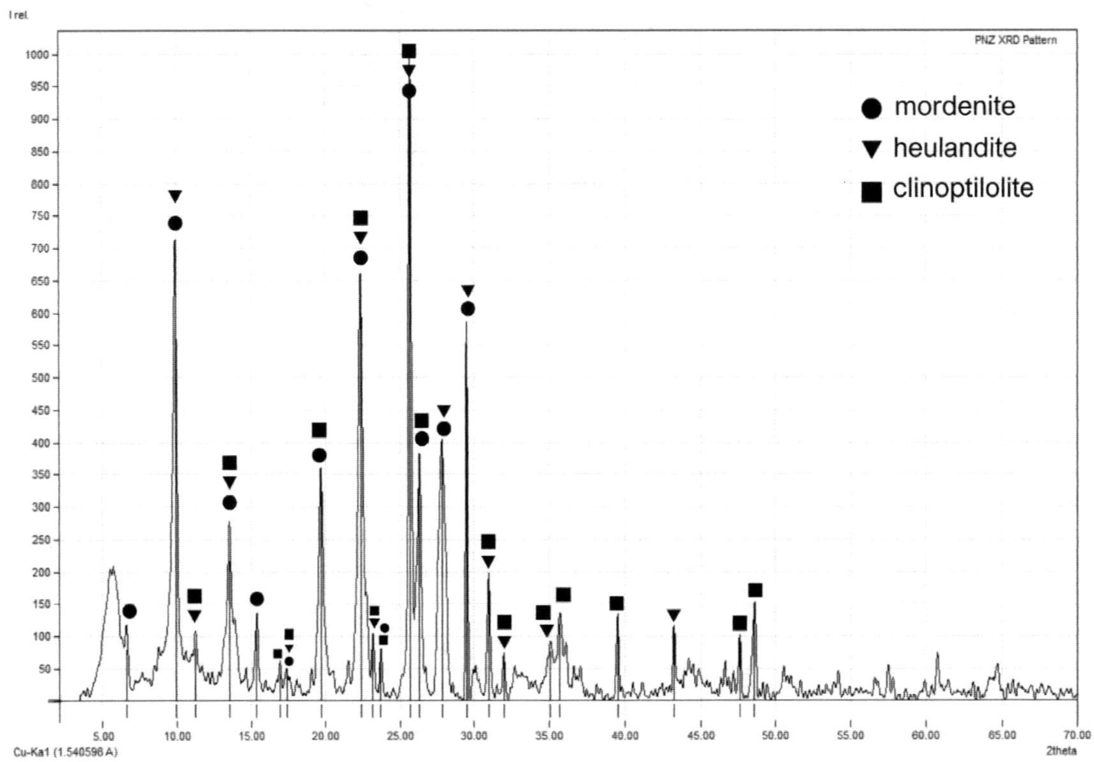

Figure 3. XRD pattern of the Philippine natural zeolite (PNZ) with identified peaks.

3.3. Yields and appearance of the products from non-catalytic and catalytic pyrolysis

The amount of liquid oil and residue collected from the non-catalytic and catalytic pyrolysis of PP waste are presented in Table 1 and Table 2, respectively. Results showed that the first drop of condensate from catalytic pyrolysis occurred at a time 9.63 min less than that of the non-catalytic pyrolysis. In addition, the temperature at which the PP waste vaporized is lower for catalytic pyrolysis, signifying that the PNZ acted as a catalyst that can speed up the degradation of the plastic.

Table 1. Measured amounts of liquid oil and residue produced from non-catalytic pyrolysis of PP waste for each trial, and the recorded temperature and time for the first drop of condensate.

	Trial 1	Trial 2	Trial 3	Average
Temperature inside the flask during the first drop of condensate (°C)	220.00	225.00	222.00	222.33
Time it took for the first drop of condensate (min)	29.00	29.36	28.50	28.95
Weight of liquid oil (g)	92.14	89.29	88.68	90.04
Volume of liquid oil (mL)	120.50	119.00	116.50	118.67
Weight of residue (g)	16.50	17.58	15.10	16.39

IOP Conf. Series: Materials Science and Engineering **1318** (2024) 012053 doi:10.1088/1757-899X/1318/1/012053

Table 2. Measured amounts of liquid oil and residue produced from catalytic pyrolysis of PP waste for each trial, and the recorded temperature and time for the first drop of condensate.

	Trial 1	Trial 2	Trial 3	Average
Temperature inside the flask during the first drop of condensate (°C)	152.00	150.00	148.00	150.00
Time it took for the first drop of condensate (min)	20.00	18.48	19.50	19.33
Weight of liquid oil (g)	103.38	102.28	105.33	103.66
Volume of liquid oil (mL)	136.60	134.50	138.50	136.53
Weight of residue with PNZ (g)	16.41	18.59	14.46	16.49

The percentage yield of the liquid oil and residue for both non-catalytic and catalytic pyrolysis were calculated using equation (1). Meanwhile, the percentage yield of the uncondensed gas was obtained by difference.

$$percentage\ yield = \frac{weight\ of\ product}{initial\ weight\ of\ sample} \times 100\% \quad (1)$$

Based on the results in Figure 4, a higher percentage of liquid oil was collected from catalytic pyrolysis, with an average of 86.39%, compared to non-catalytic pyrolysis, with an average of 75.03%. Low amounts of residue were also obtained for both processes. However, the residue from the catalytic pyrolysis includes the zeolite retained in the flask, indicating that a higher amount of the PP waste vaporized and condensed within 1 hour. This signifies that the PNZ increased the conversion rate of plastic into liquid oil. The microporous structure of PNZ creates a large surface area that provides catalytic sites, resulting in a lower activation energy, thereby speeding up the process [11].

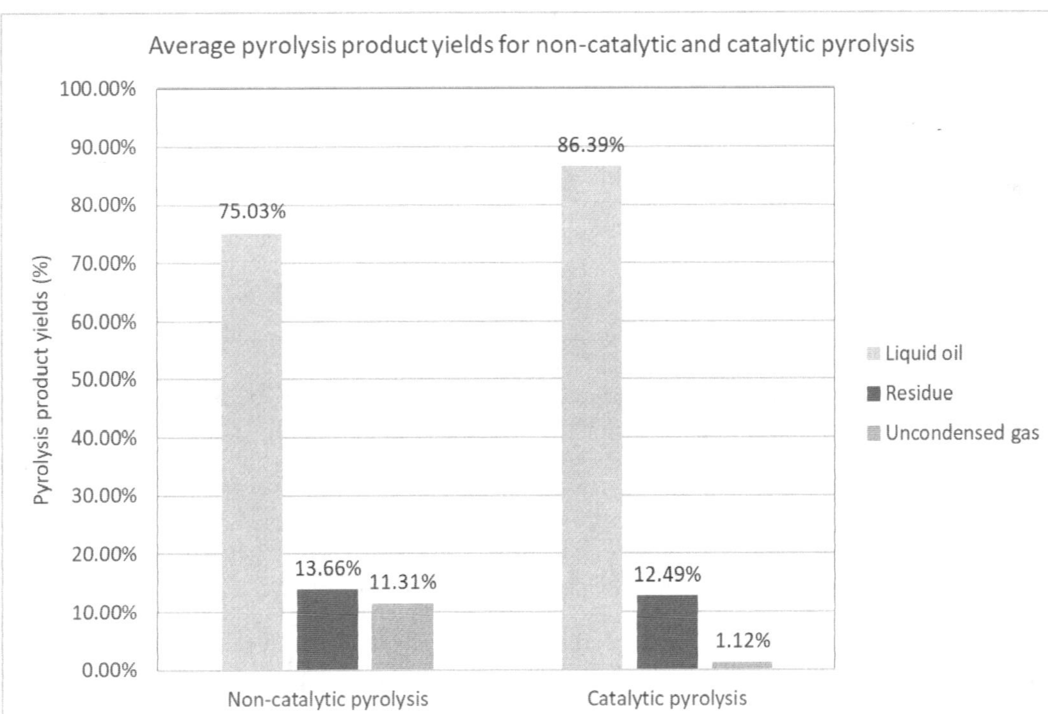

Figure 4. Average pyrolysis product yields, including liquid oil, residue, and uncondensed gas for non-catalytic and catalytic pyrolysis.

The liquid oil produced from non-catalytic and catalytic pyrolysis displayed a clear yellow to dark yellow appearance, signifying the aqueous phase of the pyrolyzed oil. Pyrolysis of plastics into liquid oil at elevated temperatures usually yields aqueous and organic phases, wherein the former fraction contains water-soluble hydrocarbons, acids, and phenols. On the other hand, the latter is rich in organic compounds that are heavier and are indicated by the black appearance of oil [12]. Significant amounts of the organic phase may be obtained at higher temperatures, as exhibited in the study of Jamiltun, *et al.* [13] when the operating temperature increased from 300 to 550°C.

(a) (b)

Figure 5. (a) Liquid oil products and (b) residue from three trials of non-catalytic and catalytic pyrolysis of PP waste.

3.4. Fourier Transform Infrared (FTIR) analysis of pyrolysis oil from PP waste

FTIR was employed to characterize the obtained liquid sample from the experiment. This method was used to identify the composition of certain mixtures by producing signals specific to the type of bonds or functional groups [14].

Figure 6 (a) shows the combined resulting infrared spectra for all three trials of the non-catalytic pyrolysis. Based on the graph, the fingerprint region lies on a wavelength of less than 1500, a critical and complex region where reading the functional groups present becomes difficult due to the existence of multiple peaks occurring at the specified range. The functional groups on this side of the spectrum may or may not be present in the sample.

Six evident peaks are present, excluding the peaks found in the fingerprint region. A peak at 3,075, or the first peak, could be an indicator of a double-bonded C-H or alkene. Meanwhile, the second peak, between 2,950 and 3,000, indicates a single-bonded methyl group or alkane. The third peak indicates a double-bonded methylene group. The fourth signifies a single-bonded methyl group. The fifth is a single-bonded methoxy, and the last peak is an aldehyde.

Figure 6 (b) exhibits the infrared spectrum for all three trials of the catalytic pyrolysis. Five evident peaks are observable for this spectrum, excluding the peaks found in the fingerprint region. The first peak indicates a double-bonded alkyl group. Meanwhile, the second peak indicates a single-bonded methyl group. The third peak indicates a double-bonded methylene group. The fourth indicates a single-bonded methyl group, the fifth shows a very weak intensity of the methoxy group, and the last peak is an aldehyde.

The graph shown in Figure 6 (c) presents the plotted spectra of all the catalytic and non-catalytic pyrolysis trials. No significant difference between the presence of functional groups in the liquid sample produced from the non-catalytic and catalytic pyrolysis using PNZ can be seen from the graph. However, the catalytic pyrolysis spectra shown in Figure 6 (b) produced six peaks with one very weak peak, compared to the non-catalytic pyrolysis, which generated six evident peaks. The weak peak for the catalytic process signifies the presence of the methoxy group [15].

IOP Conf. Series: Materials Science and Engineering **1318** (2024) 012053 doi:10.1088/1757-899X/1318/1/012053

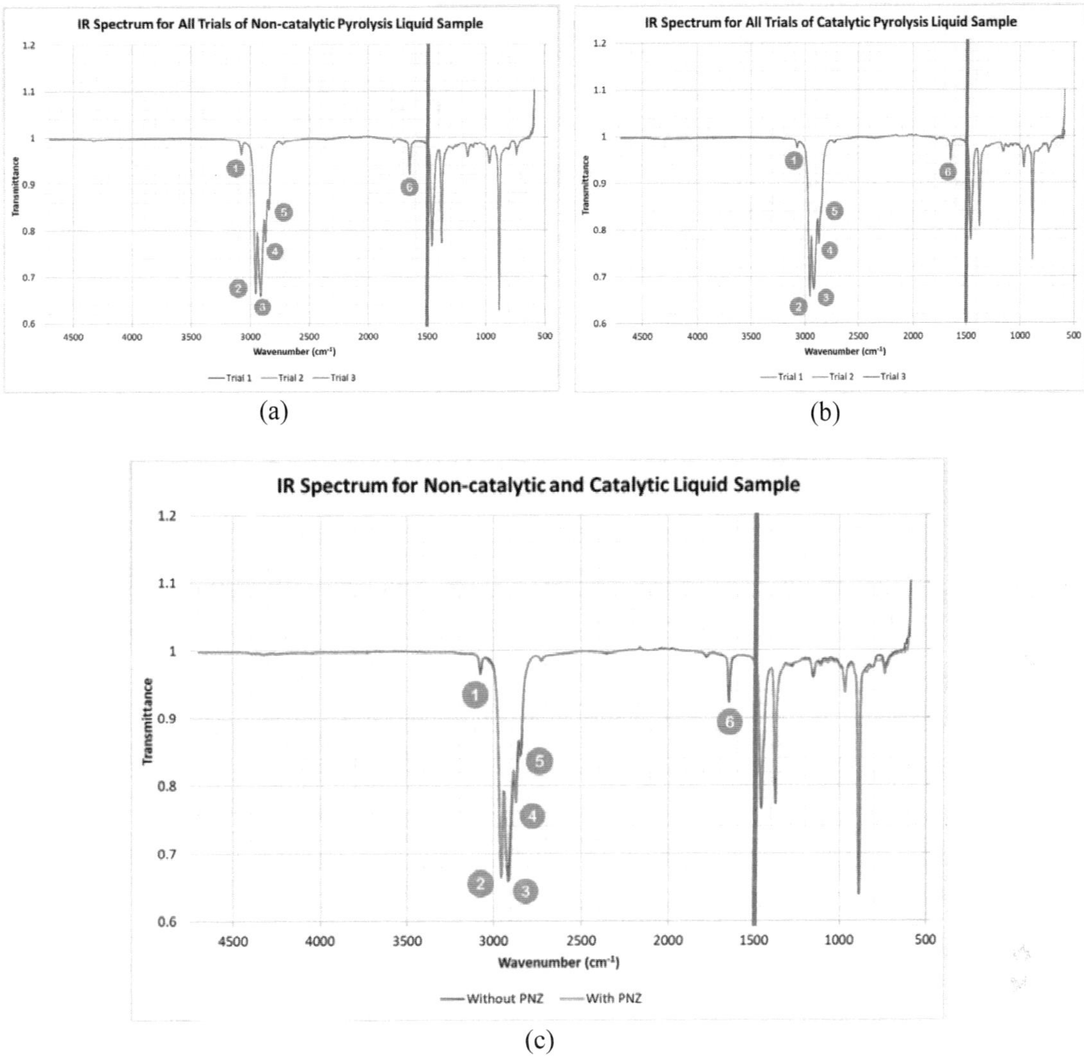

Figure 6. IR spectrum for all three trials of (a) non-catalytic pyrolysis, (b) catalytic pyrolysis, and (c) non-catalytic and catalytic pyrolysis.

4. Conclusion

Three trials were conducted each for both the non-catalytic and catalytic pyrolysis. Multiple trials were done to ensure the consistency of the obtained data from the experiment, thus, increasing the reliability and accuracy of the study. The amorphous crystalline structure from the surface of PNZ from SEM analysis exhibited an uneven surface providing additional reaction sites, similar to natural zeolites used as catalysts for pyrolysis. The XRD data also exhibited sharp peaks that correspond to clinoptilolite, mordenite, and heulandite minerals which comprise mainly of silicon and aluminum compounds, implying the suitability of the zeolite as a catalyst for pyrolysis. The catalytic pyrolysis of polypropylene plastic waste resulted in an increase in the liquid oil yield with an average of 86.39% compared to non-catalytic pyrolysis with an average of 75.03%. Both catalytic and non-catalytic pyrolysis had extremely similar functional groups. In conclusion, PNZ was effective for a catalyzed pyrolysis reaction due to its increased yield, and there were no significant differences in the functional groups present between the catalyzed and non-catalyzed liquid oil products.

Acknowledgement
The proponents would like to express their utmost gratitude to the following individuals who played an essential role in the success of this study: Mr. Louie Alberth J. Leyba, R.ChT., Engr. Ma. Cristina Joyce Manalo, M.Sc., and Engr. Divine Angela G. Sumalinog, Ph.D., for their valuable insights and guidance throughout the experimentation and completion of the study. The proponents would also like to acknowledge the efforts of the laboratory technicians at the UST Chemical Engineering Laboratory 5 and 11 for their immediate response to the requests and queries regarding the equipment needed for the study; the Department of Science and Technology: SEI for providing financial assistance; and ITDI-MSD division for assisting in the XRD analysis.

References
[1] Socio-economic profile: sea circular report in the Philippines [Internet]. SEA Circular. 2020. Available from: https://www.sea-circular.org/wp-content/uploads/2020/04/SEA-circular-Country-Briefing_THE-PHILIPPINES.pdf
[2] Onda D. The extent of plastics pollution in PH. The Manila Times [Internet]. 2021 Jun 15; Available from: https://www.manilatimes.net/2021/06/15/opinion/the-extent-of-plastics-pollution-in-ph/1803201
[3] Comprehensive guide on polypropylene (PP) [Internet]. Omnexus. Available from: https://omnexus.specialchem.com/selection-guide/polypropylene-pp-plastic
[4] Mapa D. 2020 Foreign trade statistics of the Philippines 2021. Available from: https://psa.gov.ph/sites/default/files/2020%20FTS%20Publication_signed-compressed.pdf
[5] Erdogan S 2020 Recycling of waste plastics into pyrolytic fuels and their use in IC engines *Sustainable Mobility* doi:10.5772/intechopen.90639
[6] Rollinson A and Oladejo J 2019 'Patented blunderings', efficiency awareness, and self-sustainability claims in the pyrolysis energy from waste sector *Resources, Conservation and Recycling* **141** 233–42
[7] Taylor H S 1925 A theory of the catalytic surface *Proc. of the Royal Society of London Series A, Containing Papers of a Mathematical and Physical Character* **108** 105–111
[8] Achilles C, Downs G, Downs R, Morris R, Rampe E, Ming D, et al. Lunar and Planetary Science Conference. In: NASA [Internet]. 2018. Available from: https://ntrs.nasa.gov/api/citations/20180002942/downloads/20180002942.pdf
[9] Gili M, Pares F, Nery A, Guillermo N, Marquez E and Olegario E 2020 Changes in the structure, crystallinity, morphology and adsorption property of gamma-irradiated Philippine natural zeolites. *Materials Research Express* **6** 125552 doi:10.1088/2053-1591/ab6c8b
[10] Gili M, Olegario-Sanchez L and Conato M 2019 Adsorption uptake of Philippine natural zeolite for Zn^{2+} ions in aqueous solution. *Journal of Physics: Conf. Series* 1191:012042 doi:10.1088/1742-6596/1191/1/012042
[11] Olegario-Sanchez E and Pelicano C 2017 Comparative study of As (III) and Zn (II) removal from aqueous solutions using Philippine natural zeolite and alumina *AIP Conf. and Proc.* https://doi.org/10.1063/1.5010513
[12] Abnisa F, Wan Daud W, Arami-Niya A, Ali B and Sahu J 2014 Recovery of liquid fuel from the aqueous phase of pyrolysis oil using catalytic conversion *Energy and Fuels* **28** 3074–85
[13] Jamilatun S, Pitoyo J, Amelia S, Ma'arif A, Hakika D and Mufandi I 2022 Experimental study on the characterization of pyrolysis products from bagasse (saccharum officinarum L.): bio-oil, biochar, and gas products *Indonesian J. Science and Technology* **7** 565–582
[14] Introduction to fourier transform infrared spectrometry [Internet]. Thermo Nicolet Corporation; 2001. Available from: https://www.chem.uci.edu/~dmitryf/manuals/Fundamentals/FTIR%20principles.pdf
[15] Reusch W. Infrared spectroscopy [Internet]. 2013. Available from: https://www2.chemistry.msu.edu/faculty/reusch/virttxtjml/spectrpy/infrared/infrared.htm

IOP Conf. Series: Materials Science and Engineering **1318** (2024) 012054 doi:10.1088/1757-899X/1318/1/012054

On the Viability of Hydrofluoroolefin-R245fa Mixtures as Organic Working Fluids in a Low-Temperature Solar-Powered Organic Rankine Cycle (ORC)

J Icaranom[1], W F Bolo[1], K J Yap[1] and R De Leon[1]

[1] Fuels, Energy & Thermal Systems Laboratory, Department of Chemical Engineering, College of Engineering, University of the Philippines Diliman, Quezon City, Philippines

jsicaranom@up.edu.ph, wtbolo@up.edu.ph, kryap@up.edu.ph, and rldeleon@up.edu.ph

Abstract. To address the intermittent nature of solar and other renewable energy (RE) sources, an organic Rankine cycle (ORC) can be employed to facilitate steady production of power. Organic working fluid (OWF) mixtures in ORCs are of interest due to the possibility of reducing irreversible losses in the system and having a better sense of control over its environmental impact while not compromising on the performance benefits when paired with conventional fluids such as R245fa. Hydrofluoroolefins (HFOs) are considered as one of the components due to their zero ozone depleting potential (ODP) and very low global warming potential (GWP). This study evaluates the performance of binary zeotropic mixtures of R245fa and HFOs with available thermodynamic properties (R1234yf, R1234zeE, and R1234zeZ) as OWFs in a simple low-temperature solar-powered ORC. Specifically, the optimal compositions of each pair of HFO-R245fa mixtures within the specified operating conditions are identified via the performance indicators (1 kW net power output; thermal efficiencies; and GWP); and are compared against their pure components and known conventional OWFs. The turbine power outputs and efficiencies of the mixtures, determined from ASPEN Plus simulations, were observed to be competitive as alternatives for the pure components, and conventional fluids. At optimal compositions, R245fa/R1234zeZ (0.2619 mass fraction R245fa) was shown to be the most viable OWF blend with peak efficiency at 7.5545% and GWP of 274.2, accomplishing the target 1kW turbine power output, followed closely by R245fa/R1234zeE (0.2678 R245fa; 1kW power output; 7.4824% peak efficiency; GWP = 281), and R245fa/R1234yf (0.3589 R245fa; 0.9595 kW power output; 7.4338% peak efficiency; GWP = 372.2). It is recommended that more HFO-R245fa mixtures, as well as other complex blends, be explored and be subjected to further studies.

1. Introduction

To avoid overdependence on finite resources (e.g., fossil and coal), a push for renewable energy, particularly through the harnessing of solar energy, must be initiated. In the past four decades, renewable energy sources have gained recognition in being able to supply the needed energy since the oil crisis in 1970[1]. One major source of renewable energy is the sun; however, it is still not utilized well with the uneven distribution of solar irradiation around the globe and its intermittence[2]. The high capital costs and low power density of solar technologies also remain a challenge of solar energy utilization[3]. Despite

IOP Conf. Series: Materials Science and Engineering **1318** (2024) 012054 doi:10.1088/1757-899X/1318/1/012054

this, an optimistic outlook of 11% of total electricity production globally would be via solar thermal energy by 2050[4].

Rankine cycles are the most suitable thermodynamic cycles to utilize in converting low-temperature heat sources (<300°C), such as solar thermal energy, to electricity to overcome its intermittent nature[5]. Similar to a typical Rankine cycle, an ORC also makes use of the four basic unit operations: a pump, an evaporator, a turbine and a condenser. The working fluid is pumped through the evaporator and goes through the turbine as a saturated vapor. It is then condensed to a saturated liquid before being pumped again to complete the cycle. Instead of water, an ORC utilizes an organic working fluid (OWF) to allow the utilization of the low temperature heat source.

Solar energy is still rare as an ORC's heat source, with the dominant ones being industrial waste heat and geothermal energy[15]. This might be due to the lower costs of solar PVs, which further hinders the push for solar ORCs to be one of the primary options for power generation[13]. Despite the claims of the ORC technologies' effectiveness to utilize low-grade heat sources and renewable energy sources, operating at lower temperatures continues to be a problem as it is prone to perform at lower efficiencies, especially in solar ORC systems that use non-concentrating types of solar collectors[1]. It was suggested that future works involve the more efficient and more cost-effective designs of solar ORCs[13]. The selection of the appropriate OWF is also a critical aspect of different ORC systems. A poor choice of OWF can result in expensive and low efficient operation[1].

International regulations on the preservation of the ozone layer, such as the Montreal Protocol, require nationwide enforcement of policies to monitor the ODP of substances when trying to implement organic working fluids (OWFs) and/or introducing fluids (pure or mixtures) for commercial, industrial, and even residential use[6][7]. With the current innovations, zero-ODP substances are already present in the industry, but the high GWP of some still remains a concern.

R245fa is one of the most widely used and considered pure OWF in an ORC[8][9][10]. Like isopentane and R134a, it is also considered a conventional OWF but is now considered to be replaced with alternatives due to its high GWP[8]. It was established that hydrofluoroolefins (HFOs), such as R1366mzz(Z), R1233zd(E), and R1234ze(Z), are promising alternatives due to their zero ODPs, low GWPs, and mostly are non-flammable[8]. Despite most of them performing at a lower level compared to most available working fluids, they have lower GWP. The review showed that there have been findings of R1234yf matching the performance of R134a, and R1336mzz(Z) providing higher thermal efficiency than R245fa[8].

The introduction of zeotropic mixtures can be explored as well as they reduce the challenges of system irreversibility with the temperature glide of the fluids and improve system efficiency[1][8][9][10][11]. The temperature glide of zeotropic mixtures, which refers to the difference between the bubble point temperature and dew point temperature, does not equate to zero during phase changes[9]. This, in effect, constantly changes the vapor phase and liquid phase during phase changes. The temperature glide also allows for better matching between the working fluid and the heat transfer fluids. This could reduce the irreversible loss during the heat transfer processes[9].

However, these zeotropic mixtures present a new set of challenges which include higher additional cost from the possible optimization, a larger required heat exchange area leading to additional costs, tedious calculations to estimate thermodynamic properties that are not readily available at arbitrary concentrations, and additional research regarding composition shift[9][10][11]. A composition shift is a change in the composition of the circulating fluid compared to the composition of the charged fluid that can be caused by varying velocity between the liquid and vapor phase of the working fluids, the leakage in the system, and the distinct solubility of the lubricating oil for each component[9]. It was determined that this composition shift affects the performance of an ORC system in terms of a lower net output work and thermal efficiency, and higher power consumption in the pump[16].

For zeotropic mixtures, the temperature glide remains as one of the recognized consensuses when selecting the optimal composition and fluid. However, this method is unable to explain why a selected fluid is the optimal one[14]. This can be due to various boundary conditions, assumptions, functions, and fluids identified in the reviews. Due to this lack of a unified conclusion, another step for fluid selection can be added to verify the choice of composition. The identified optimal composition may also not always result to an optimal performance due to other irreversibility sources, such as the evaporator, pump, and turbine[17].

Zeotropic mixtures present potential in overcoming the shortcomings of its individual components, however this field still needs to be studied due to its lack of experimental studies, calculations requiring estimation of thermodynamic properties, and the lack of studies tackling the variation of the circulating composition from the charged composition[9][10][11]. Despite these, mixtures still pose advantages in terms of generally better performance, and a sense of control on its impact on the environment[8]. Because of these emerging technologies, opportunities for research, and improvement of nationwide transition to clean energy, the researchers were motivated to discover binary zeotropic mixtures that can reduce negative environmental impacts without compromising performance in an ORC for low-temperature application.

Conventional fluids such as R245fa are still widely selected as the best OWF in the literature despite its high GWP. Fourth generation fluids such as HFOs are on the rise due to their practicability, low to zero ODP, and low GWP. Mixing a low GWP HFO with a R245fa is to be proposed as an OWF in an ORC. This study aimed to evaluate the performance of zeotropic mixtures of HFO (R1234yf, R1234zeE, and R1234zeZ) and R245fa as OWFs. The specific objectives of the study are to identify the optimal compositions of each pair of mixtures within the specified operating conditions via the objective functions of a target net turbine power output of 1 kW; and to compare the performance indicators [net power output, thermal efficiency, and mixture GWP] of the optimal mixtures against their pure components and known conventional OWFs.

2. Methodology

This study aimed to examine the viability of the R245fa/R1234yf, R245fa/R1234zeE, and R245fa/R1234zeZ zeotropic mixtures by assessing their performance and determining their optimal composition. These were then compared with the pure components and conventional OWFs. The compositions of each mixture have increment value for the mass fraction set to 0.1 as this was the value used by literature in the analysis of the bubble point temperatures of different zeotropic mixture compositions[17].

Working fluid calculations for optimal composition determination for mixtures only encompassed the HFO-R245fa pairs. The HFOs that were studied were R1234yf, R1234zeE, and R1234zeZ, while the conventional fluids included were R245fa, R134a, and isopentane. The limitations on the number of HFOs that were considered stemmed from the lack of readily available data in the ASPEN Databanks, subjected only to previously studied operating conditions[12]. System assumptions include steady-state operation, negligible heat loss and pressure drops across the ORC, negligible kinetic and potential energy of the fluids, and the constant mixture composition of the mixtures across the cycle. To represent the low-temperature application of the ORC, saturated steam with temperature of 200 °C was utilized as the heat transfer fluid for the evaporator[12]. ASPEN Plus was used as the simulation program due to the availability of the target fluids. The design of the equipment used were not included in the study.

A simple ORC composed of an evaporator, a turbine, a condenser, and a pump were utilized in the study. At steady state, the governing equation for the evaporator is defined by the equation below:

$$\dot{Q}_{evap} = \dot{m}_{wf}\left(h_{wf}^{evap,\ out} - h_{wf}^{evap,\ in}\right) = \dot{m}_{hf}\left(h_{hf}^{evap,\ in} - h_{hf}^{evap,\ out}\right) \tag{1}$$

Where $\dot{m}_{wf}$ is the mass flowrate of the organic working fluid in the ORC, and the enthalpy terms h signify whether they are from the working (*wf*) or the heating fluid (*hf*) and whether they are at the inlet (*in*) or outlet (*out*) of the evaporator.

The isentropic efficiency of the turbine can then be defined as the ratio of the actual work over the isentropic work of the turbine (the difference in the enthalpies of the outlet and inlet streams).

$$n_{tur,\ isen} = W_{tur}\ [\dot{m}_{wf}(h_{wf}^{tur,\ in} - h_{wf}^{tur,\ out})]^{-1} \quad (2)$$

Similar to the evaporator, the governing equation for the condenser is shown below:

$$\dot{Q}_{cond} = \dot{m}_{cf}(h_{cf}^{cond,\ out} - h_{cf}^{cond,\ in}) = \dot{m}_{wf}(h_{wf}^{cond,\ in} - h_{wf}^{cond,\ out}) \quad (3)$$

where the subscript cf indicates the cooling fluid.

The isentropic efficiency of the pump is defined by the ratio of the isentropic work required and the actual work required.

$$n_{pump,\ isen} = [\dot{m}_{wf}(h_{wf}^{pump,\ out} - h_{wf}^{pump,\ in})](W_{pump})^{-1} \quad (4)$$

For the analysis of the performance of the ORC and the system, the respective performance indicators net power output (the difference between the work output of the turbine and the required work of the pump, $\dot{W}_{tur} - \dot{W}_{pump}$) and thermal efficiency, η_I were calculated.

$$\eta_I = (\dot{W}_{tur} - \dot{W}_{pump})\left(\dot{Q}_{evap}\right)^{-1} \quad (5)$$

The pump model was utilized with a specified efficiency of 75% and a discharge pressure of 12 bar. The turbine model on the other hand was specified as an isentropic type with a mechanical efficiency of 80% and a discharge pressure of 5 bar. These initial specifications (mechanical efficiency and pressure) were utilized as the minimum cycle values in previous studies[12][18].

The evaporator and condenser were both modeled as shortcut model fidelity with countercurrent flow direction. The calculation methods were in design mode. For the evaporator, its specification was set to "Cold stream outlet vapor fraction" with a value of 1 to indicate that its outlet process stream will be a saturated vapor. On the other hand, the specification of the condenser was set to "Hot stream outlet vapor fraction" with a value of 0 to make sure that the inlet process stream of the pump is a saturated liquid. The heat transfer fluid was at 200°C and 15.5 bar with a steam flow rate of 3000 kg/hr, while the cooling fluid was at 15°C and 1 bar with a water flow rate of 2000 kg/hr[12].

The same models and database were used in the simulation to produce the performance indicators of all the proposed fluids and conventional fluids. The simulation flowsheet utilized throughout the study is seen in Figure 1.

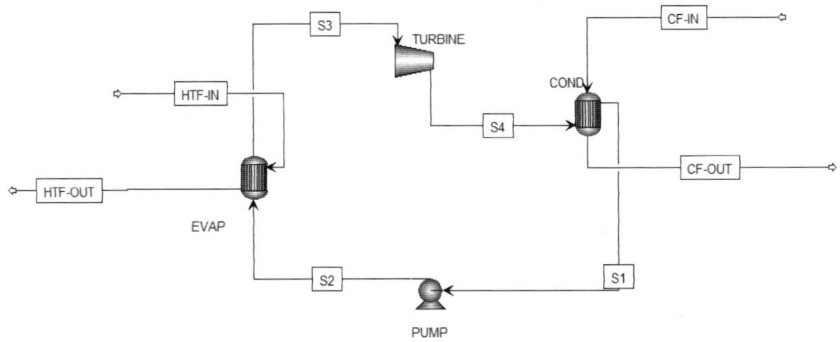

Figure 1. Simulation Flowsheet

In terms of its environmental impact, the GWP of the mixtures were also calculated using the equation below.

$$GWP_{mix} = \sum_i w_i(GWP_i) \quad (6)$$

where w_i and GWP_i are the mass fraction and the GWP of the component *"i"*.

3. Results and Discussion

Sensitivity analysis of fluid mixtures was performed and plots of varying pump outlet pressures on net turbine power output and cycle thermal efficiency at varying compositions and mass flowrates were obtained. Scatter plots and surface plots were procured to visualize differences of each fluid mixture for corresponding plots.

Preliminary simulations were performed to determine valid mass flowrate ranges for each fluid mixture through varying mass compositions expressed in mass fraction of R245fa. The criteria for determining valid mass flowrates are accomplished in simulation by taking the range where there are no substantial errors. It was observed that for the same fluid mixture, the boundaries of valid mass flowrates will vary depending on the mixture compositions.

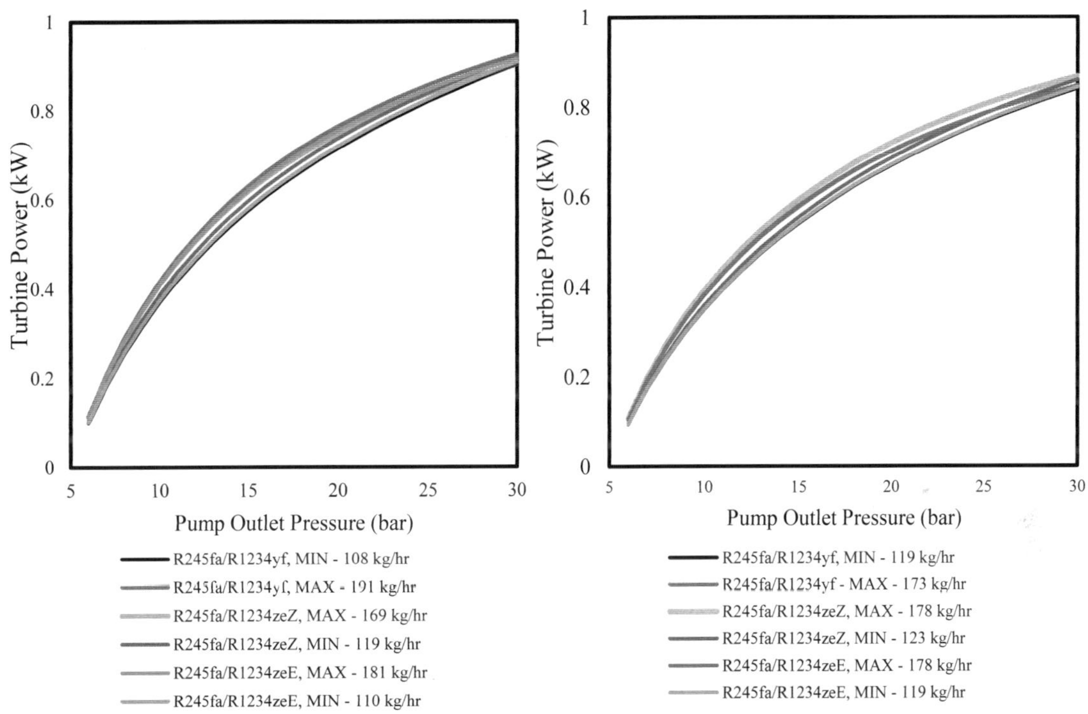

(a) 0.1 Mass Fraction R245fa (b) 0.9 Mass Fraction R245fa

Figure 2. Turbine Power (kW) vs Pump Outlet Pressure (bar) of R245fa-HFO Mixtures at (a) 0.1 and (b) 0.9 Mass Fraction R245fa thru Minimum-Maximum Valid Flowrates

IOP Conf. Series: Materials Science and Engineering **1318** (2024) 012054 doi:10.1088/1757-899X/1318/1/012054

From Figure 2, the differences between the OWF mixtures from minimum and maximum valid flowrates are within 0.1 kW. Trends for both Fig 2a and Fig 2b are consistent, with increasing turbine power output at higher pump outlet pressures for a constant turbine outlet pressure of 5 bar. This can also be attributed to a greater pressure ratio between the pump outlet pressure and turbine outlet pressure, ergo, the main driver of increasing turbine outlet power. Another consistent trend is that at 0.1 mass fraction R245fa, the turbine power output of the mixture at minimum and maximum valid flowrates is higher (>0.9 kW) than that of the 0.9 mass fraction R245fa plots (<0.9 kW).

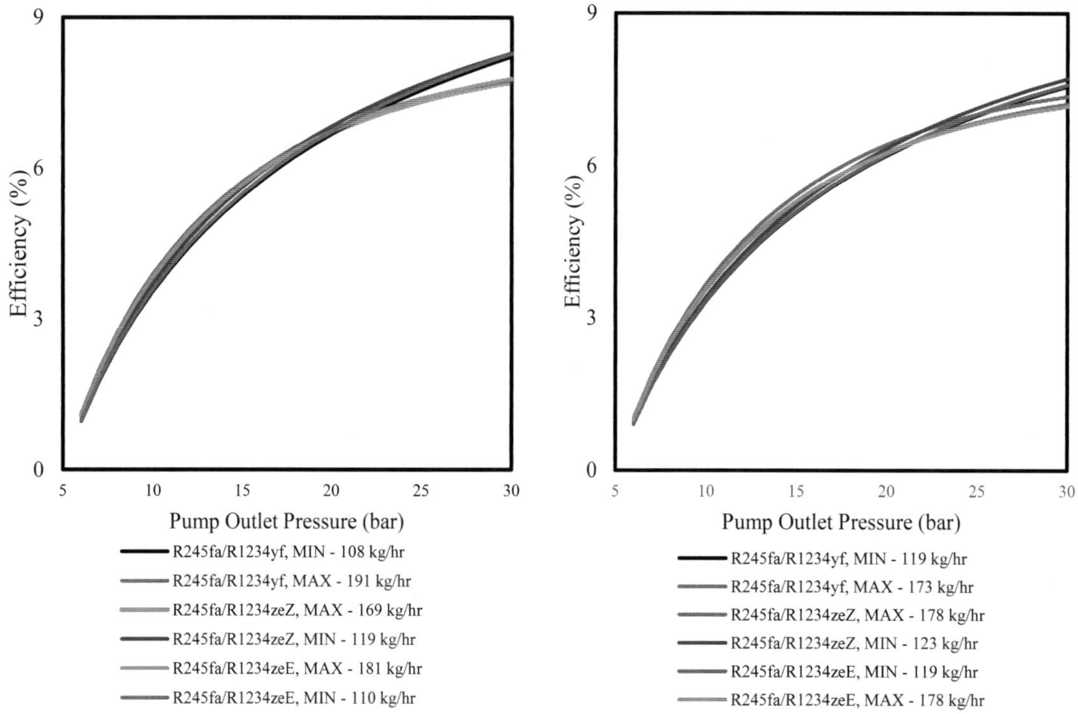

(a) 0.1 Mass Fraction R245fa (b) 0.9 Mass Fraction R245fa

Figure 3. Efficiency (%) vs Pump Outlet Pressure (bar) of HFO-R245fa Mixtures at (a) 0.1 and (b) 0.9 Mass Fraction R245fa thru Minimum-Maximum Valid Flowrates

For Figure 3, at higher pump outlet pressures (or pressure ratios), the thermal ORC efficiency at 0.1 and 0.9 mass fractions R245fa at minimum and maximum valid flowrates increases. Particularly, at Fig 3a, the efficiencies of the maximum valid flowrates for the OWF mixtures split from that of the minimum valid flowrates from ~23 bar pump outlet pressure. Beyond this pressure, the plots of the maximum valid flowrates do not increase in efficiency as much as that of the minimum valid flowrates. At low mass compositions of R245fa, a higher mass flowrate does not result to a higher cycle efficiency. A similar trend can be observed for Fig 3b, where a 0.9 mass fraction of R245fa yields varying efficiencies at higher pump outlet pressure, although not as pronounced as that for the 0.1 mass fraction of R245fa. Note that the peak efficiencies for 0.1 mass fraction of R245fa for the OWF mixtures is higher than that of the 0.9 mass fraction R245fa. Specifically, the peak efficiencies of the minimum valid mass flowrates for Fig 3a are >8% while yielding <8% for Fig 3b.

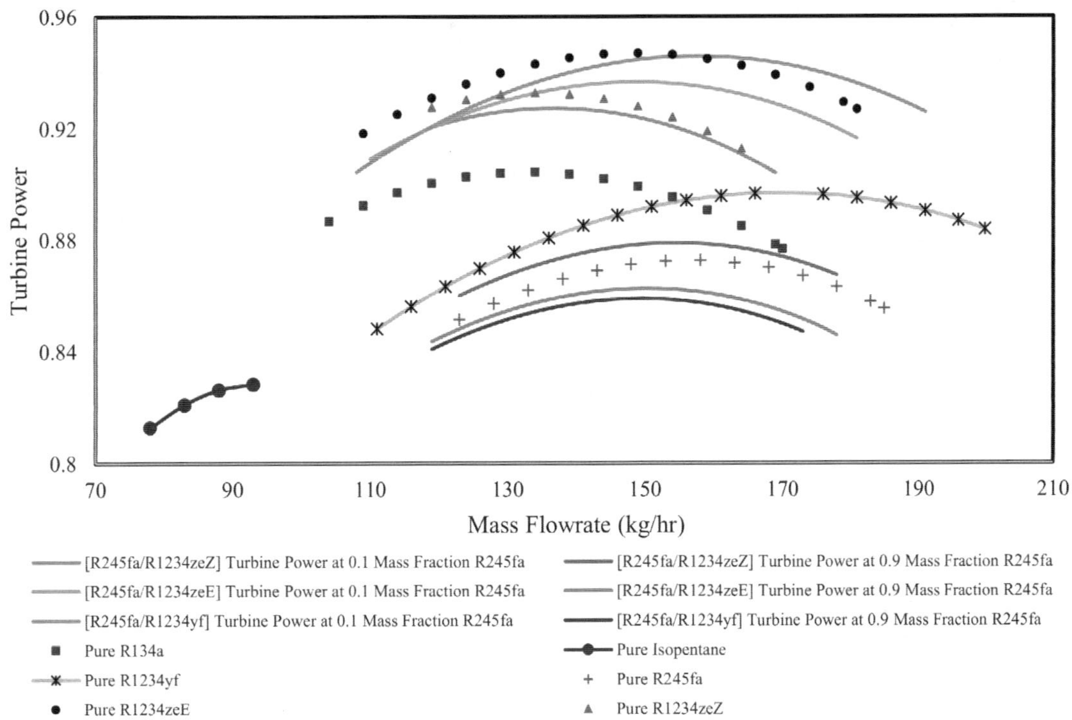

Figure 4. Turbine Power (kW) vs Mass Flowrate of HFO-R245fa Mixtures, Pure and Conventional Fluids

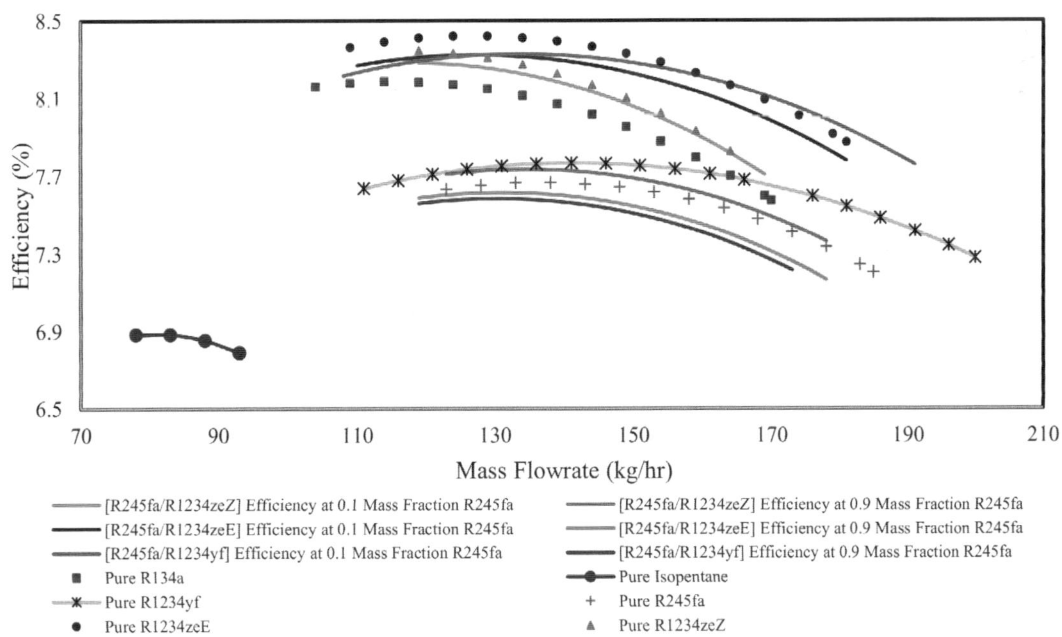

Figure 5. Efficiency (%) vs Mass Flowrate of HFO-R245fa Mixtures, Pure and Conventional Fluids

For Figure 4, the turbine power of the OWF mixtures through their valid flowrates at the cycle's operating conditions and system parameters at 0.1 and 0.9 mass fractions of R245fa prove to be viable for use in low-temperature applications as substitutes to conventional fluids and pure OWFs. Particularly, the OWF mixtures at 0.1 mass fraction R245fa are competitive against pure R134a, with the formers being less environmentally harmful. Moreover, Figure 4 shows that the HFO-R245fa mixtures at lower R245fa composition yield more power output than their pure components. Additionally, isopentane, a conventional fluid used in low-temperature ORC systems, performs worse than the R245fa/R1234yf mixture at 0.9 mass fraction R245fa, showing the effectiveness of these new generation HFOs and their blends in terms of power generation.

A similar trend is shown for Figure 5 regarding the efficiencies against the valid mass flowrates of the fluid mixtures, pure components, and conventional fluids. The fluid mixtures at lower R245fa composition are the most competitive in terms of peak efficiencies. For both the efficiency and the turbine power output vs mass flowrate plots, pure R1234zeE yields higher values at lower values of the valid mass flowrates, which is also higher than that of pure R245fa.

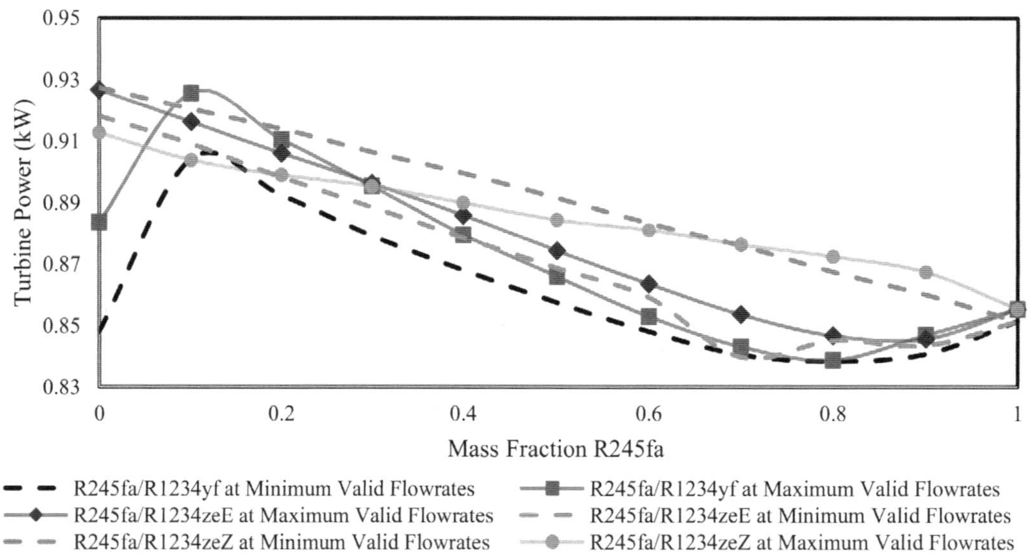

Figure 62. Turbine Power (kW) vs Mass Fraction R245fa

Figure 6 shows the trends of turbine power output versus the mass fractions of R245fa for the OWF mixtures at minimum and maximum valid flowrates. The plots for the minimum and maximum valid flowrates for R245fa/R1234yf yield an initial increase in turbine power output at 0.1 mass fraction R245fa, decreasing further before converging at an intersection point around 0.8 mass fraction R245fa. For minimum and maximum valid flowrates for R245fa/R1234zeE, a pinch point between the turbine power values can be observed at >0.8 mass fraction R245fa, after the differences yielded a sudden increase at 0.7 mass fraction R245fa. For R245fa/R1234zeZ, the turbine power plot at maximum valid flowrates is generally below the plot for the minimum valid flowrate, before overtaking it beyond 0.7 mass fraction R245fa. This contrasts with R245fa/R1234yf and R245fa/R1234zeE, where the power output at maximum valid mass flowrate with increasing mass fractions of R245fa is consistently above that of the power output plot at minimum valid mass flowrate.

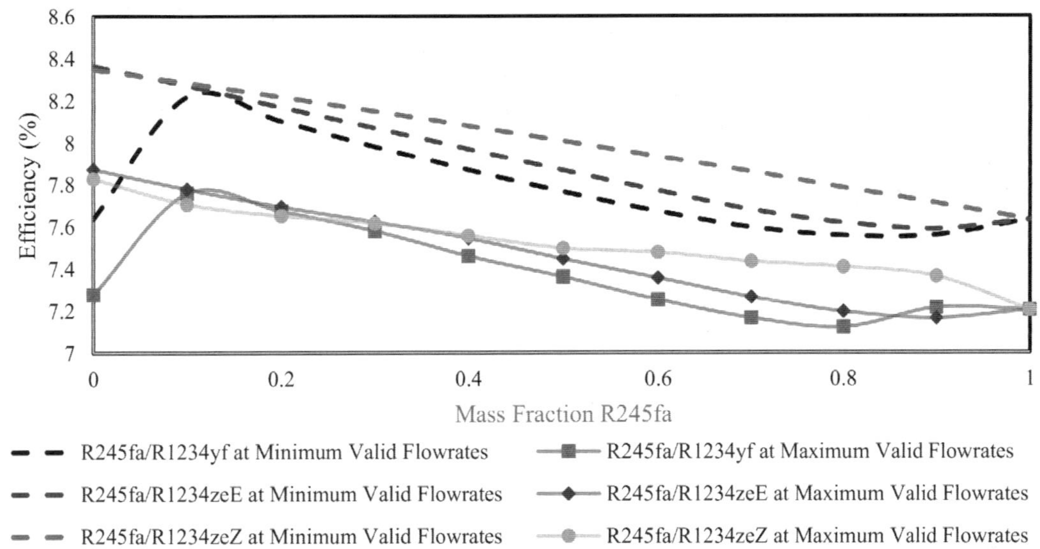

Figure 7. Efficiency (%) vs Mass Fraction R245fa

For the efficiency plots vs mass fraction R245fa at minimum-maximum valid mass flowrates, Figure 7 shows that the efficiencies of the OWF mixtures at minimum mass flowrate is consistently above that of the corresponding efficiencies at maximum flowrates. For R245fa/R1234yf, no pinch point can be observed, but the sudden increase in efficiency, can be observed in 0.1 mass fraction R245fa. R245fa/R1234zeZ at minimum flowrates yields the highest consistent efficiency with varying mass fractions of R245fa.

The sensitivity analysis has shown that at non-optimal conditions, R245fa/R1234yf blend is the most viable of the three OWF mixtures, followed closely by R245fa/R1234zeE and then R245fa/R1234zeZ.

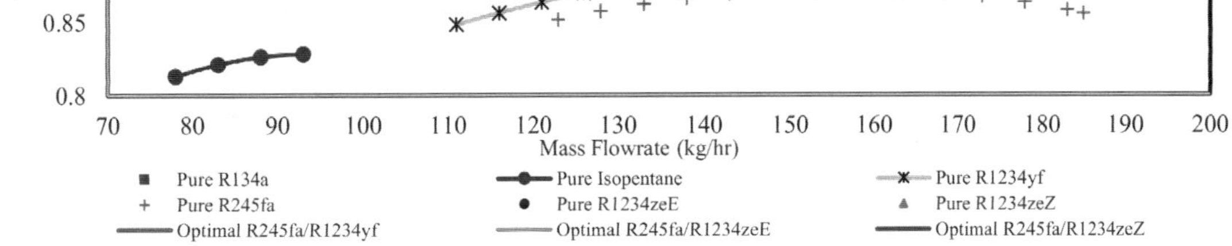

Figure 8. Turbine Power (kW) vs Mass Flowrate (kg/hr) of Optimal HFO-R245 Mixtures (in Mass Fraction R245fa: R245fa/R1234yf - 0.3589. R245fa/R1234zeE - 0.2678. R245fa/R1234zeZ - 0.2619), Pure, and Conventional Fluids

Figure 8 shows that for the mixture compositions, the optimal OWF mixtures are more effective for power generation than the pure components and conventional fluids. Specifically, R245fa/R1234zeE at optimal compositions yield greater than 1kW turbine power output for its valid mass flowrate range, followed by R245fa/R1234zeZ and R245fa/R1234yf at optimal compositions.

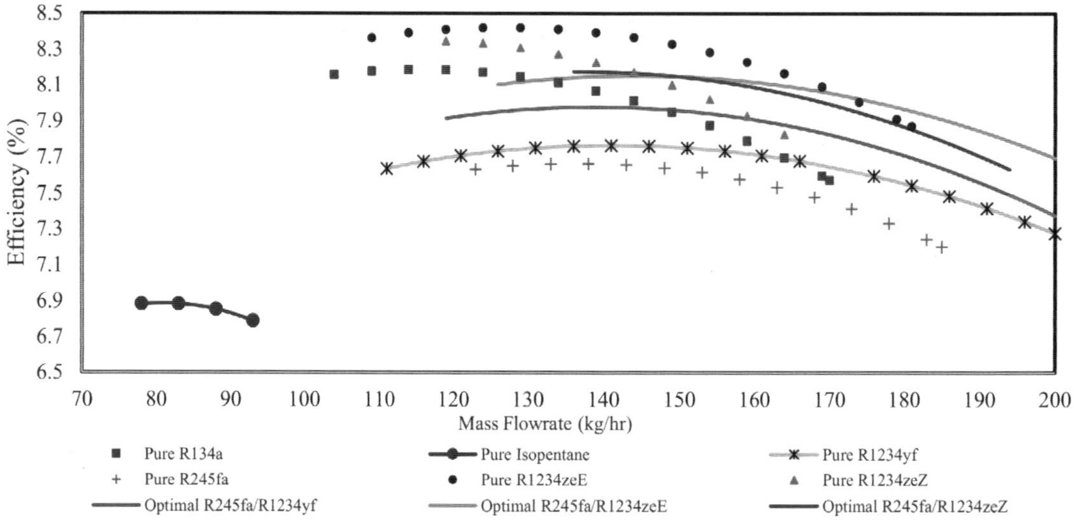

Figure 9. Efficiency (%) vs Mass Flowrate (kg/hr) of Optimal R245-HFO Mixtures (in Mass Fraction R245fa: R245fa/R1234yf - 0.3589. R245fa/R1234zeE - 0.2678. R245fa/R1234zeZ - 0.2619), Pure, and Conventional Fluids

The efficiencies of the optimal OWF mixtures, while not higher than the pure components and conventional fluids at its valid mass flowrates, are still viable as an alternative as these efficiencies are competitive with the pure and conventional fluids as observed in Figure 9. Particularly, the only OWF mixture components that are more efficient than the pure are R1234zeE and R1234zeZ, with mixtures' efficiencies reaching similar peaks as R134a.

Table 1 summarizes the global warming potentials (GWP) of the OWF mixtures, the pure and conventional fluids, and the OWF mixtures at optimal conditions. From the table, at optimal compositions, R245fa/R1234zeZ is the least potent in global warming (274.2 GWP), followed closely by R245fa/R1234zeE (281 GWP), and then R245fa/R1234yf (372.2 GWP). While it is expected that the GWPs of these mixtures will not be better than that of their pure HFO components, the performance benefits that the mixtures offer for a low-temperature ORC offsets the higher GWP values even at optimal compositions.

The optimal composition determined from the study can be traced in the temperature glide plot in Figure 10. However, it can be observed that these compositions have temperature glides that are not matching with the average cooling fluid temperature rise. The expected composition to have the optimal performances were not the resulting values from the optimization. The difference between the actual and expected results can be due to the other irreversibility[17] and the different objective function and boundary conditions utilized in the optimization[14].

Table 1. Global Warming Potentials (GWP) of (a) R245fa-HFO Mixtures, (b) Pure and Conventional Fluids, (c) R245fa-HFO Mixtures at Optimal Conditions

(a) R245fa-HFO Mixtures

R245fa composition	**OWF Mixture**		
	R245fa/R1234yf	**R245fa/R1234zeE**	**R245fa/R1234zeZ**
0.1	106.6	109.3	108.4
0.2	209.2	211.6	210.8
0.3	311.8	313.9	313.2
0.4	414.4	416.2	415.6
0.5	517	518.5	518
0.6	619.6	620.8	620.4
0.7	722.2	723.1	722.8
0.8	824.8	825.4	825.2
0.9	927.4	927.7	927.6

(b) Pure and Conventional Fluids

Component	**Pure Fluid GWPs**
R134a	1300
Isopentane	20
R245fa	1030
R1234yf	4
R1234zeE	7
R1234zeZ	6

(c) R245fa-HFO Mixtures at Optimal Conditions

Mixture	**Optimal Composition (Mass Fraction R245fa)**	**GWP at Optimal Composition**
R245fa/R1234yf	0.3589	372.2
R245fa/ R1234zeE	0.2678	281
R245fa/ R1234zeZ	0.2619	274.2

IOP Conf. Series: Materials Science and Engineering **1318** (2024) 012054 doi:10.1088/1757-899X/1318/1/012054

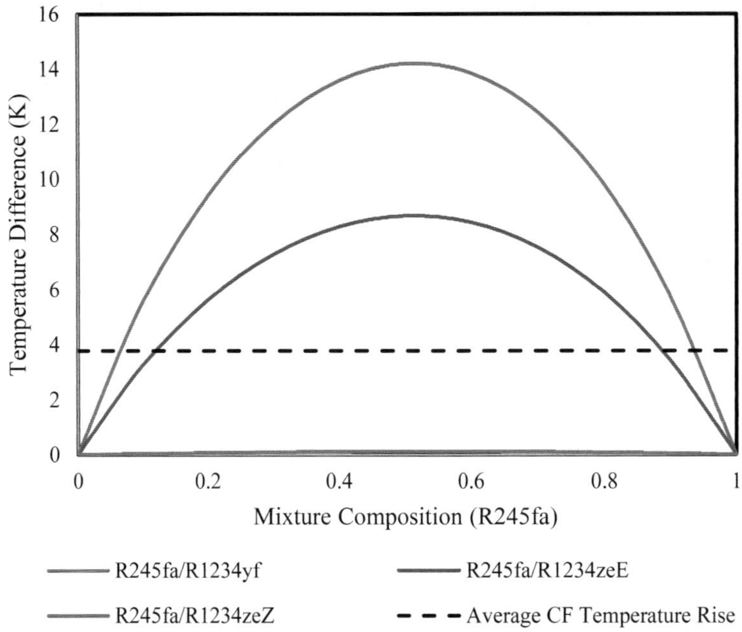

Figure 10. Temperature Glide of HFO-R245fa Mixtures

4. Conclusion and Recommendation

Optimal compositions of the OWF mixtures in terms of R245fa were determined to be R245fa/R1234yf: 0.3589; R245fa/R1234zeE: 0.2678; and R245fa/R1234zeZ: 0.2619. At these compositions, observable differences in power output vs pump outlet pressures are present and showed that the optimal mixtures have better turbine power output against its pure components and other conventional fluids. To supplement this viability, the OWF mixtures (excluding R245fa/R1234yf) reached the required 1 kW target power output. In terms of efficiency, it was determined that R245fa/R1234zeZ at optimal compositions had the highest peak efficiency (7.5545%), followed by R245fa/R1234zeE (7.4824%), then R245fa/R1234yf (7.4338%). In terms of GWP, R245fa/R1234zeZ was calculated to have the least GWP, followed by R245fa/R1234zeE, then R245fa/R1234yf.

R245fa/R1234zeZ was determined to be the most viable, with R245fa/R1234yf being the least preferable among the studied mixtures. It was then concluded that the HFO-245fa mixtures are competitive with conventional fluids and pure OWFs in terms of turbine power output, cycle efficiency, and GWP across valid mass flowrate ranges. Thus, these mixtures are viable working fluids in a low-temperature solar-powered ORC.

The proponents recommend that for future iterations of this study, the fluids undergo further optimization processes including exergy and economic analysis. A more comprehensive unit-by-unit analysis, particularly, more complex selection methods and analysis can also be employed to determine the effect of initial boundary errors on overall system performance. A levelized cost of energy (LCOE) analysis for the ORC system is also recommended to check its economic viability and cost-effectiveness in other applications. Furthermore, the researchers advocate for an experimental setting of the studied HFO-R245fa blends, along with other complex blends such as ternary mixtures.

References

[1] Aboelwafa O, Fateen S-EK, Soliman A, Ismail IM. A review on solar Rankine cycles: Working fluids, applications, and cycle modifications. Renew Sustain Energy Rev [Internet]. 2018;82:868–85. Available from: https://doi.org/10.1016/j.rser.2017.09.097

[2] Şen Z. Solar energy in progress and future research trends. Prog Energy Combust Sci [Internet]. 2004;30(4):367–416. Available from: https://doi.org/10.1016/j.pecs.2004.02.004

[3] Calise F, d'Accadia MD, Vicidomini M, Scarpellino M. Design and simulation of a prototype of a small-scale solar CHP system based on evacuated flat-plate solar collectors and Organic Rankine Cycle. Energy Convers Manag [Internet]. 2015;90:347–63. Available from: https://doi.org/10.1016/j.enconman.2014.11.014

[4] Giovannelli A. State of the art on small-scale concentrated solar power plants. Energy Procedia [Internet]. 2015;82:607–14. Available from: https://doi.org/10.1016/j.egypro.2015.12.008

[5] Tzivanidis C, Bellos E, Antonopoulos KA. Energetic and financial investigation of a stand-alone solar-thermal Organic Rankine Cycle power plant. Energy Convers Manag [Internet]. 2016;126:421–33. Available from: https://doi.org/10.1016/j.enconman.2016.08.033

[6] United Nations. Montreal Protocol on Substances that Deplete the Ozone Layer [Internet]. Treaties.un.org. 1987. Available from: https://treaties.un.org/doc/publication/unts/volume%201522/volume-1522-i-26369-english.pdf

[7] Department of Environment and Natural Resources. Revised Regulations on the Chemical Control Order for Ozone Depleting Substances (ODS) [Internet]. pod.emb.gov.ph. 2013. Available from: https://pod.emb.gov.ph/wp-content/uploads/2016/06/DAO-2013-25.pdf

[8] Bahrami M, Pourfayaz F, Kasaeian A. Low global warming potential (GWP) working fluids (WFs) for Organic Rankine Cycle (ORC) applications. Energy Rep [Internet]. 2022;8:2976–88. Available from: https://doi.org/10.1016/j.egyr.2022.01.222

[9] Xu W, Zhao R, Deng S, Zhao L, Mao SS. Is zeotropic working fluid a promising option for organic Rankine cycle: A quantitative evaluation based on literature data. Renew Sustain Energy Rev [Internet]. 2021;148(111267). Available from: https://doi.org/10.1016/j.rser.2021.111267

[10] Bamorovat Abadi G, Kim KC. Investigation of organic Rankine cycles with zeotropic mixtures as a working fluid: Advantages and issues. Renew Sustain Energy Rev [Internet]. 2017;73:1000–13. Available from: https://doi.org/10.1016/j.rser.2017.02.020

[11] Li Y-R, Du M-T, Wu C-M, Wu S-Y, Liu C. Potential of organic Rankine cycle using zeotropic mixtures as working fluids for waste heat recovery. Energy (Oxf) [Internet]. 2014;77:509–19. Available from: https://doi.org/10.1016/j.energy.2014.09.035

[12] Shalaby, S. M., Gadalla, M. A., el Sayed, A. R., Meliha, E. M., & Abosheiasha, H. F. (2022). Aspen Plus simulation of a low capacity organic Rankine cycle heated by solar energy. Energy Reports, 8, 416–421. https://doi.org/https://doi.org/10.1016/j.egyr.2021.11.109

[13] Loni R, Mahian O, Markides CN, Bellos E, le Roux WG, Kasaeian A, et al. A review of solar-driven organic Rankine cycles: Recent challenges and future outlook. Renew Sustain Energy Rev [Internet]. 2021;150(111410). Available from: https://doi.org/10.1016/j.rser.2021.111410

[14] Xu W, Zhao L, Mao SS, Deng S. Towards novel low temperature thermodynamic cycle: A critical review originated from organic Rankine cycle. Appl Energy [Internet]. 2020;270(115186). Available from: https://doi.org/10.1016/j.apenergy.2020.115186

[15] Qyyum MA, Khan A, Ali S, Khurram MS, Mao N, Naquash A, et al. Assessment of working fluids, thermal resources and cooling utilities for Organic Rankine Cycles: State-of-the-art comparison, challenges, commercial status, and future prospects. Energy Convers Manag [Internet]. 2022;252(115055). Available from: https://doi.org/10.1016/j.enconman.2021.115055

[16] Zhao L, Bao J. The influence of composition shift on organic Rankine cycle (ORC) with zeotropic mixtures. Energy Convers Manag [Internet]. 2014;83:203–11. Available from: https://doi.org/10.1016/j.enconman.2014.03.072

[17] Shahrooz M, Lundqvist P, Nekså P. Performance of binary zeotropic mixtures in organic Rankine cycles (ORCs). Energy Convers Manag [Internet]. 2022;266(115783). Available from: https://doi.org/10.1016/j.enconman.2022.115783
[18] Heberle F, Preißinger M, Brüggemann D. Zeotropic mixtures as working fluids in Organic Rankine Cycles for low-enthalpy geothermal resources. Renew Energy [Internet]. 2012;37(1):364–70. Available from: https://doi.org/10.1016/j.renene.2011.06.044

IOP Conf. Series: Materials Science and Engineering **1318** (2024) 012055 doi:10.1088/1757-899X/1318/1/012055

Comparison of single- vs dual-chamber microbial electrolysis cells performance for biohydrogen production

M T Isla-Cabaraban[1,2], S Paclijan[2], N P Logico[2], J M Dalagan[2], F A Gonzales[2], G M S Cabaraban[3], N P Tan[4]

[1] Department of Science and Technology Philippine Council for Industry, Energy, and Emerging Technology Research and Development (DOST-PCIEERD), Bicutan, Taguig City, Philippines

[2] Department of Chemical Engineering, Xavier University – Ateneo de Cagayan, Corrales Avenue, Cagayan de Oro City, Philippines

[3] Department of Electrical Engineering, Xavier University – Ateneo de Cagayan, Corrales Avenue, Cagayan de Oro City, Philippines

[4] Center for Advanced New Materials, Engineering, and Emerging Technologies (CANMEET), University of San Agustin, Department of Chemical Engineering, University of San Agustin, Iloilo City, Philippines

matheresa.cabaraban@pcieerd.dost.gov.ph

Abstract. In this study, fed batch single- and dual-chamber microbial electrolysis cells (MECs) were operated to compare differences in the electrohydrogenesis performance at 0.8 V applied voltage and hydraulic retention time (HRT) of 48 h. Results showed that current intensity, current density, COD removal efficiency, and H_2 production rate were higher in the dual-chamber MEC than in the single-chamber configuration. Electromethanogenesis and homoacetogenesis scavenged the H_2 from electrohydrogenesis thus reducing H_2 generation, as demonstrated by cathodic efficiencies less than 10%. It appears that homoacetogenesis overshadowed H_2 production in the dual-chamber MEC, as reflected by a coulombic efficiency greater than 100%, likely induced by a high H_2 partial pressure. These results portray MECs as an energy-efficient scheme for bioH_2 production in addition to removal of pollutants (COD). This study presented valuable information when deciding on suitable reactors and optimizing reactor design.

1. Introduction

The twin challenges of waste management and energy sustainability have amplified research efforts on innovative and efficient technologies that can simultaneously address both issues by transforming wastes into valuable energy carriers. Hydrogen (H_2) is a clean and efficient energy carrier that can potentially substitute fossil fuels and mitigate environmental pollution. However, the conventional methods of hydrogen production (e.g. methane or natural gas reforming, coal gasification, thermochemical water splitting, electrolysis) often make use of non-renewable resources, require very high processing temperatures [1,2], and emit greenhouse gases [3,4]. Hence, among the major challenges of the adoption of hydrogen as an energy carrier is finding sustainable production methods.

H_2 can also be produced from biological processes that are mostly operated at ambient conditions, hence are expected to be less energy intensive than the other H_2 production methods. These processes can utilize abundant organic resources, including industrial, agricultural, municipal or forestry wastes, as carbon sources. BioH_2 production processes include biophotolysis, photofermentation, dark fermentation, or biocatalyzed electrolysis (aka electrohydrogenesis). In biophotolysis, photoautotrophic microorganisms (e.g. microalgae, cyanobacteria) split water into molecular H_2 and O_2 [5–9]. Biophotolysis often requires the use of expensive catalysts (e.g. platinum and other noble metal components) to facilitate the H_2 production [10]. Photofermentation is a bioprocess in which organic compounds are oxidized under anaerobic, nitrogen-limited conditions to produce molecular H_2 [11–13] A constant and adequate supply of light as energy source is needed to drive the process, which can be challenging due to the intermittent nature of sunlight, making photofermentative H_2 production systems less reliable. Low light conversion efficiency results in low hydrogen yield, further inhibiting the application of photofermentation for H2 production [11]. Dark fermentation uses a wide range of organic substrates for producing bioH_2 utilizing specific microbes through the process of anaerobic fermentation [14–16]. Mixed cultures or microbial consortia are preferred, especially for industrial applications of complex substrates such as agro-industrial and food wastes [16]. However, the competing or negatively interacting microbiota can adversely impact overall bioH2 production, and the suppression or elimination of acid-producing (propionic acid or lactic acid) bacteria remains a challenge [17]. The maximum yield from fermentative bioH_2 production processes is also thermodynamically constrained to around 4 moles per mole of glucose, making them practically uncompetitive to other biofuels produced from the same biomass [18].

Bioelectrochemical systems offer an attractive option for sustainable biological H_2 production, out of which microbial electrolysis cell (MEC) is gaining interest due to its ability to concurrently remove organics and generate bioH_2 [19–21]. MECs are essentially modified microbial fuel cells where biodegradable materials are converted into H_2 through an electrolytic method referred to as electrohydrogenesis. Electrohydrogenesis offers the possibility of producing bioH_2 through the metabolic activities of exoelectrogenic microorganisms from a variety of substrates, ranging from pure substances (e.g. carbohydrates and proteins) to fermentation products (e.g. acetate and butyrate) to high-strength industrial and domestic wastewaters [18,22,23].

The exoelectrogenic microbes are present at the anode and oxidize the organic substrates to release the free electrons and protons. An external voltage is supplied across the MEC, stimulating the movement of electrons from the anode to the cathode, and either the diffusion of protons through the electrolyte to the cathode (single chamber MEC) or the transport of protons to the cathode (dual chamber MEC) where they are reduced by electrons transferred from the anode to form H_2. The electrodes are among the critical components that affect MEC performance. The anode provides the necessary area where electrochemically active microorganisms attach, and the cathode acts as the electron acceptor and carries out the hydrogen evolution reaction. As the anode supplies electrons from the microbial oxidation of the substrate to the cathode, anodic reaction therefore is rate-limiting to hydrogen production in MECs [24]. Anode materials do not need to be sophisticated materials but should have such features as large surface area and porosity that are favorable for microbial attachment, biocompatibility, and good conductivity [25,26]. The most used anodic materials in MEC are carbon cloth, carbon felt, carbon paper, carbon nanotubes, graphite felt, graphite brush, and graphite granules. Although carbon-based cathodes are preferred [21] due to their availability and low cost, hydrogen evolution reaction on such surfaces is kinetically slow and requires electrochemical catalyst such as Pt to improve the efficiency [21,25,26].

The presence of other bacteria can potentially reduce H_2 production through methanogenic competition. The most common inoculum pretreatment methods to avoid these competing microorganisms include physical (heat-shock, aeration, ultrasonication, microwave irradiation, freezing,

IOP Conf. Series: Materials Science and Engineering **1318** (2024) 012055 doi:10.1088/1757-899X/1318/1/012055

and infrared) and chemical (acid, alkali, inhibition with acetylene, chloroform, iodopropane, 2-bromo ethanesulfonic acid) processes [27,28].

The hydrogen production performance of MECs is also influenced by the reactor configuration [29], necessitating testing of this element to improve operating efficiency and energy recovery. To understand its application strategy, this study constructed and operated a single-chamber MEC and a dual-chamber MEC to compare the electrohydrogenesis and H_2 generation performance.

2. Materials and Methods

2.1. MEC construction

Two dual-chamber MEC (D-MEC) reactors were constructed of polypropylene were separated with a low-cost nonselective microporous polyethylene membrane (Rhinohide, size: 11 cm x 11.5 cm), while two single-chamber MEC (S-MEC) reactors were made of borosilicate glass. The components of (a) the S- MEC reactor and (b) the D-MEC reactor are shown in a schematic diagram in Figure 1. Each of the two D-MEC reactors had the same anodic and cathodic chambers of 432 mL, while the volume of each of the S-MEC reactors was at 345 mL. The anodes were made of carbon cloth (CC, size: 2.8 cm x 2.8 cm for each of the S-MEC and 5 cm x 5 cm for each of the D-MEC), while similarly sized cathodes were of SS 304 flat mesh (SM). The CC was pretreated sequentially with distilled water, isopropyl alcohol, hydrogen peroxide, and distilled water before use, and the SM with deionized water and sulfuric acid. The distance between the anode and cathode was 3.5 cm for all MEC reactors. An outlet was placed on top of the MEC reactors for gas collection in Tedlar bags and for gas volume determination.

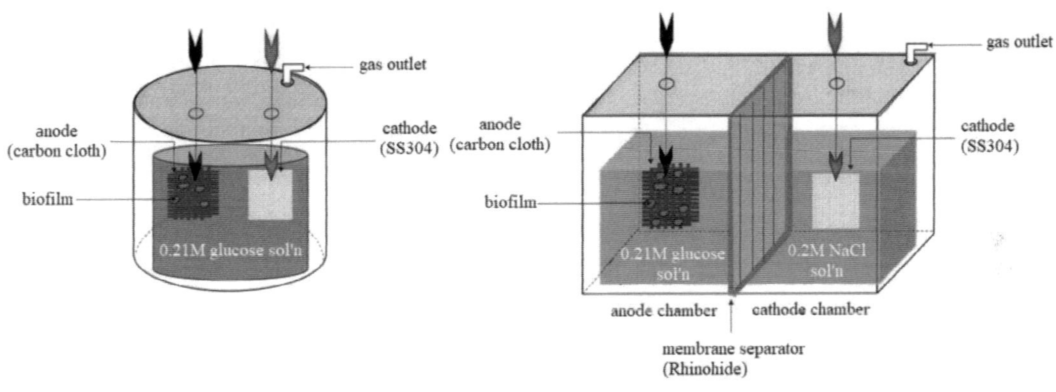

Figure 1. Reactor schematics for single-chamber MEC (left) and dual-chamber MEC (right)

2.2. MEC startup and operation

The entire S-MECs and the anode chambers of D-MECs were inoculated with anaerobic sludge from an upflow anaerobic sequential batch reactor treating fruit processing wastewater and were fed with synthetic wastewater. The composition of the synthetic substrate in the S-MECs included (per L of water): glucose, 52.689 g and phosphate buffer solution (PBS), 433.05 g. Meanwhile, the composition of the synthetic anolyte included (per L of water) in the D-MECs: glucose, 57.38 g and 567.92 g. The catholyte for the cathodes of the D-MECs was 0.2M NaCl solution. Chamber solutions were refreshed once every two days to maintain anodic hydraulic residence time (HRT) of all MECs equivalent to 48 h. After the 7- to 13-day initial cultivation following the sludge inoculation and biofilm formation on the CC, the chambers were sparged with nitrogen gas for 1.5 min before applying an external voltage set at 0.8 V to all MEC systems and the data collection was started. All MEC experiments were conducted in batch mode at room temperature (~30 °C).

IOP Conf. Series: Materials Science and Engineering **1318** (2024) 012055 doi:10.1088/1757-899X/1318/1/012055

2.3. Measurements

The equations for the anodic and cathodic electrochemical reactions in the MEC reactors associated with organic substrate degradation during wastewater treatment are shown below.

Anodic reaction: $C_6H_{12}O_6 + 6H_2O \rightarrow 6CO_2 + 24H^+ + 24e^-$

Cathodic reaction: $24H^+ + 24e^- \rightarrow 12H_2$

Supernatant water samples were collected from the chambers at the end of each batch for wastewater treatment assessment. The COD was measured using a modified open-reflux method. Gas production was quantified using the water displacement method and gas composition was determined by using a gas chromatograph equipped with thermal conductivity detector. The voltage (V) and current intensity (I) was monitored every hour using a digital multimeter (Sanwa Electric Instrument Co., Ltd., Japan). The resulting current density was estimated as I/A, where A is the surface are of the anode used.

2.4. Calculations

The effect of reactor configuration on MEC performance was investigated, based on cumulative hydrogen production, and calculated cathodic hydrogen recovery (r_{cat}), coulombic efficiency (CE), hydrogen yield (Y_{H_2}), hydrogen production rate (Q_{H_2}), and energy efficiency (η_E). r_{cat}, or the moles of H_2 recovered from the generated biogas during the MEC operation (n_{H_2}) relative to that possible based on the measured current (n_{CE}), was calculated as:

$$r_{cat} = \frac{n_{H_2}}{n_{CE}} \quad (1)$$

where n_{CE} is given by:

$$n_{CE} = \frac{\int_{t=0}^{t} I dt}{2F} \quad (2)$$

In Equation (2), 2 is generally used to convert the moles of electrons generated during the oxidation of the organic substrate into moles of H_2. F is Faraday's constant (96,485 C/mol e^-) and dt is the data recording interval period of 48 h. CE, an important metric of MEC performance, is the ratio between the electrons transported from the organic substrate to the cathode and the maximum electrons if all the substrates were degraded and converted to useful current. CE was calculated as:

$$CE = \frac{n_{CE}}{n_{th}} \quad (3)$$

Here, n_{th} refers to the theoretical number of moles of bioH_2 produced based on chemical oxygen demand (COD) and was calculated as [30]:

$$n_{th} = \frac{b_{H_2/S}\, v_L\, S}{M_S} \quad (4)$$

where $b_{H_2/S}$ is the maximum stoichiometric H_2 production possible from the organic substrate (12 mol H_2/mol $C_6H_{12}O_6$), v_L is the total volume of substrate in the reactor, S is the amount of organic substrate added (in terms of COD), and M_S is the molecular weight of the substrate (180.56 g/mol). H_2 yield was calculated as:

$$Y_{H_2} = \frac{Cumulative\ H2\ production\ in\ mL}{\Delta S} \quad (5)$$

and H_2 production rate as:

$$Q_{H_2} = \frac{Cumulative\ H2\ production\ in\ L}{H_2\ evolution\ period\ (d) \times bioreactor\ volume\ (L)} \quad (6)$$

Energy efficiency was calculated based on the electricity input, as follows:

$$\eta_E = \frac{E_{H_2}}{E_{in}} \times 100 \quad (7)$$

E_{H_2} is the energy produced when n_{H_2} is burned, and E_{in} is the electric energy provided by the external power calculated as described by Yan, et al [31]. The heat of combustion of H_2 is 285.83 kJ/mol [32,33].

IOP Conf. Series: Materials Science and Engineering **1318** (2024) 012055 doi:10.1088/1757-899X/1318/1/012055

3. Results and Discussion

Higher current intensities are required to better facilitate organic matter degradation, and hydrogen production rate had been shown to have a linear relationship with current density [34]. The boxplots in Figure 2 shows the variations in the current intensities. As can be seen, the mean current intensity was

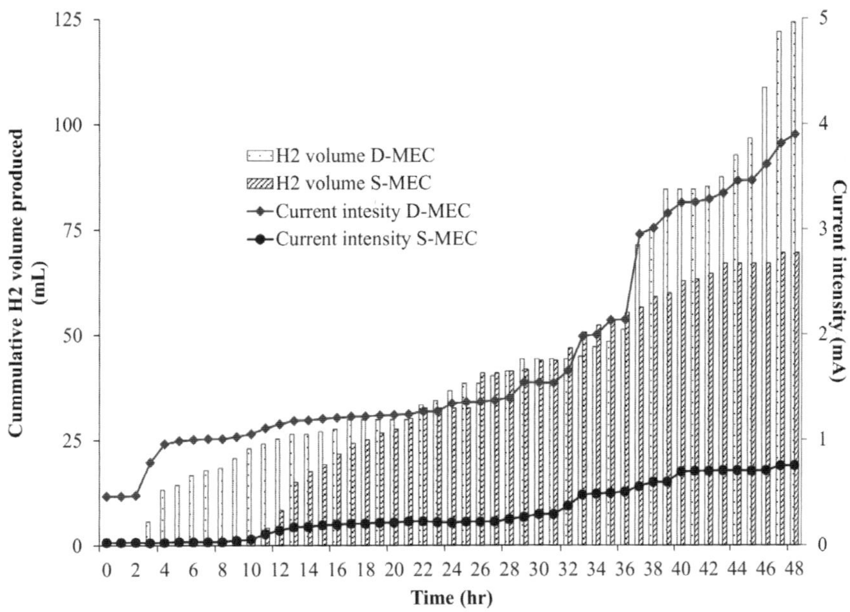

Figure 2. Biohydrogen production and current intensity evolution in single- and dual-chamber MECs

0.32±0.25 mA for S-MEC and it exhibited a slight increase to 1.46±0.76 mA for D-MEC, pointing to greater exoelectrogenic activity in the dual-chamber reactor. This likely explains the higher COD removal efficiency in D-MEC (14.67%) than in S-MEC (1.07%). The measured current in these two reactor configurations highly varied with time, as evidenced by the large standard deviations relative to the mean, suggesting that the process was highly unstable during the experimental period. A maximum of 1.56 A/m^2 was achieved in D-MEC, three times larger than the average current density calculated across the operation of 48 h. Cathodic specific area was three times greater in D-MEC (6 m^2/m^3) than in S-MEC (2 m^2/m^3), which may have explained the larger current density calculated in the former in comparison with the latter [35]. Electrohydrogenesis seemed to have occurred while a voltage of 0.8 was applied. Y_{H_2} and Q_{H_2} reached up to 470.33 mL/g COD and 0.10 L/L-d, respectively, for S-MEC and 49.05 mL/g COD and 0.144 L/L-d, respectively, for D-MEC. These Q_{H_2} values are considerably lower than other studies with glucose, although such values were obtained using a cathode catalyst (Pt) and with longer HRT (³50 h) [25]. In the present study, results revealed that the bioH_2 production of the D-MEC (3.55±4.30 mL/day) is significantly higher than that of the S-MEC (2.24±1.75 mL/d) (Figure 2).

Figure 3 shows the observed cumulative biogas production rate and yield for the MEC operation with glucose as substrate. There was a small amount of CH_4 evolution detected (S-MEC: $Y_{CH_4} = 0.049$ mL/g COD, $Q_{CH_4} = 0.006$ L/L-d; D-MEC: $Y_{CH_4} = 0.001$ mL/g COD, $Q_{CH_4} = 0.004$ L/L-d), suggesting that the methanogenic inhibition applied (heat shock at 100 °C for 30 minutes) was not able to completely suppress the occurrence of electromethanogenesis. Additionally, higher Q_{CH_4} in S-MEC suggests that more electrons flowed to CH_4 production in this reactor configuration, instead of to the

IOP Conf. Series: Materials Science and Engineering **1318** (2024) 012055 doi:10.1088/1757-899X/1318/1/012055

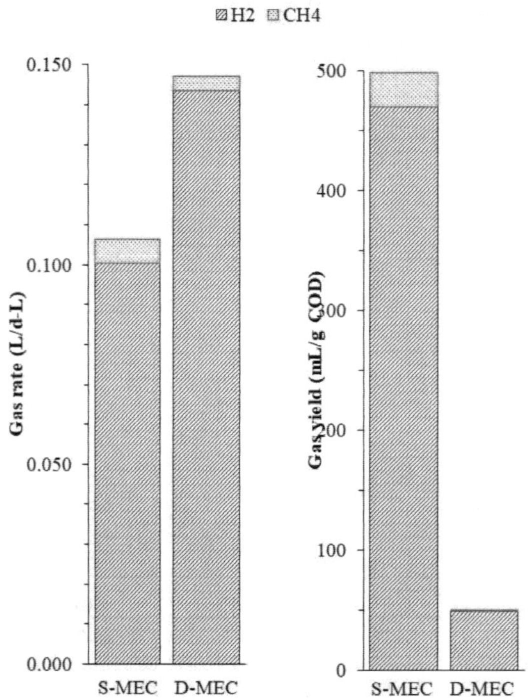

Figure 3. Cumulative biogas production rate (a) and yield (b) after 48 h of operation

external circuit, than in D-MEC. This may be due to electrotrophs present in the cathodic biofilm that are able to uptake electrons through direct extracellular electron transfer ($CO_2 + 8H^+ + 8e^- \rightarrow CH_4 + H_2O$) or utilize cathodic H_2 through hydrogenotrophic methanogenesis ($4H_2 + CO_2 \rightarrow CH_4 + 2H_2O$) [30].

The electrohydrogenesis performance of the two MEC systems were evaluated using CE, r_{cat}, and η_E. The CE of D-MEC exceeded 100%, likely resulting from H_2 recycling [30,36–38]. This phenomenon occurs when part of the H_2 formed at the cathode is converted by homoacetogenic bacteria into acetate ($4H_2 + 2CO_2 \rightarrow$ Acetate) [39–41], which in turn is oxidized by exoelectrogens at the anode (or anode-respiring bacteria) to generate current [42]. The homoacetogens may have permeated through the porous membrane separator and populated the cathode chamber of D-MEC. On the other hand, the CE of S-MEC was only at 32.77%, which contributed to a higher r_{cat} of 9.91% compared to 3.84% for D-MEC. The higher H_2 partial pressure in D-MEC may have induced acetate production, outweighing the formation of H_2 [41]. Nevertheless, these H_2 recovery efficiency values are still unsatisfactorily low, translating to at least 90% of the generated current not being captured as H_2. This may have been due to a substantial fraction of the electrons generated and converted to current from the oxidation of the organic substrate not having been effectively used for proton reduction at the cathode. This influenced the η_E (based on electricity input), which was lower in D-MEC (22.75%) than in S-MEC (51.93%). The incidence of H_2 recycling in D-MEC likely competed with electrohydrogenesis, thereby undermining r_{cat}, explaining to some degree the drop in η_E.

4. Conclusions

In this paper, we compared single-chamber and dual-chamber MECs in terms of bioH_2 production performance. It was found that MEC operated in either configuration attained ample energy efficiency (>20%) by bioH_2 production. Results showed that the H_2 concentration and current density of the dual-chamber MEC were higher than those of the single-chamber MEC. The stability of the single-chamber system was better, however H_2 production rate was greater in the dual-chamber system. BioH_2 was produced for 48 h, but the apparent modest cathodic efficiency (<10%) indicates significant H_2 losses accounting to biocathode formation that promoted electromethanogenesis, thus weakening the H_2 production rate. It is essential to identify the optimum anodic HRT to minimize electromethanogenesis and maximize cathodic efficiency, thereby maximizing hydrogen production. Since the hydrogen evolution reaction is a critical parameter in MEC development, prospective research should also be directed toward the development of low-cost and efficient cathode or catalyst needed to enhance cathodic efficiency. Electromethanogenesis was somewhat inhibited by the application of a Rhinohide membrane separator (i.e. dual-chamber MEC), but the high H_2 partial pressure proved to be beneficial for the homoacetogens, promoting H_2 recycling (or scavenging). Future research efforts on dual-chamber MECs should emphasize the replacement of Rhinohide with other suitable membrane material, e.g. natural polymer-based membranes like cellulose, agar, gelatine, etc. Finally, since the metabolic pathway for hydrogen production is highly impacted by the H_2 partial pressure, the continuous removal of H_2 that accumulate in the MEC headspace could reduce the partial pressure, thereby limiting H_2 scavenging by acetate formation and favoring greater H_2 production. One way to do this is to purge the reactor with an inert gas (e.g. N_2) to adjust the partial pressure [41]. However, this will dilute the H_2, subsequently increasing the cost for downstream separation.

References

[1] Do T N, Kwon H, Park M, Kim C, Kim Y T and Kim J 2023 Carbon-neutral hydrogen production from natural gas via electrified steam reforming: Techno-economic-environmental perspective *Energy Convers Manag* **279** 116758
[2] Szima S and Cormos C-C 2019 Techno – Economic assessment of flexible decarbonized hydrogen and power co-production based on natural gas dry reforming *Int J Hydrogen Energy* **44** 31712–23
[3] Moshrefi M M and Rashidi F 2014 Hydrogen production from methane by DC spark discharge: Effect of current and voltage *J Nat Gas Sci Eng* **16** 85–9
[4] Sazali N 2020 Emerging technologies by hydrogen: A review *Int J Hydrogen Energy* **45** 18753–71
[5] Javed M A, Zafar A M, Aly Hassan A, Zaidi A A, Farooq M, El Badawy A, Lundquist T, Mohamed M M A and Al-Zuhair S 2022 The role of oxygen regulation and algal growth parameters in hydrogen production via biophotolysis *J Environ Chem Eng* **10** 107003
[6] Li S, Li F, Zhu X, Liao Q, Chang J S and Ho S H 2022 Biohydrogen production from microalgae for environmental sustainability *Chemosphere* **291** 132717
[7] Jiménez-Llanos J, Ramírez-Carmona M, Rendón-Castrillón L and Ocampo-López C 2020 Sustainable biohydrogen production by Chlorella sp. microalgae: A review *Int J Hydrogen Energy* **45** 8310–28
[8] Sadvakasova A K, Kossalbayev B D, Zayadan B K, Bolatkhan K, Alwasel S, Najafpour M M, Tomo T and Allakhverdiev S I 2020 Bioprocesses of hydrogen production by cyanobacteria cells and possible ways to increase their productivity *Renewable and Sustainable Energy Reviews* **133** 110054
[9] Bozieva A M, Khasimov M K, Voloshin R A, Sinetova M A, Kupriyanova E V., Zharmukhamedov S K, Dunikov D O, Tsygankov A A, Tomo T and Allakhverdiev S I 2023 New cyanobacterial strains for biohydrogen production *Int J Hydrogen Energy* **48** 7569–81
[10] Aziz M, Darmawan A and Juangsa F B 2021 Hydrogen production from biomasses and wastes: A technological review *Int J Hydrogen Energy* **46** 33756–81
[11] Putatunda C, Behl M, Solanki P, Sharma S, Bhatia S K, Walia A and Bhatia R K 2023 Current challenges and future technology in photofermentation-driven biohydrogen production by utilizing algae and bacteria *Int J Hydrogen Energy* **48** 21088–109

[12] García-Sánchez R, Ramos-Ibarra R, Guatemala-Morales G, Arriola-Guevara E, Toriz-González G and Corona-González R I 2018 Photofermentation of tequila vinasses by Rhodopseudomonas pseudopalustris to produce hydrogen *Int J Hydrogen Energy* **43** 15857–69

[13] Ghosh S, Dairkee U K, Chowdhury R and Bhattacharya P 2017 Hydrogen from food processing wastes via photofermentation using Purple Non-sulfur Bacteria (PNSB) – A review *Energy Convers Manag* **141** 299–314

[14] D'Silva T C, Khan S A, Kumar S, Kumar D, Isha A, Deb S, Yadav S, Illathukandy B, Chandra R, Vijay V K, Subbarao P M V, Bagi Z, Kovács K L, Yu L, Gandhi B P and Semple K T 2023 Biohydrogen production through dark fermentation from waste biomass: Current status and future perspectives on biorefinery development *Fuel* **350** 128842

[15] Ahmad A, K R, Hasan S W, Show P L and Banat F 2023 Biohydrogen production through dark fermentation: Recent trends and advances in transition to a circular bioeconomy *Int J Hydrogen Energy*

[16] Villanueva-Galindo E, Vital-Jácome M and Moreno-Andrade I 2023 Dark fermentation for H2 production from food waste and novel strategies for its enhancement *Int J Hydrogen Energy* **48** 9957–70

[17] Dahiya S, Chatterjee S, Sarkar O and Mohan S V 2021 Renewable hydrogen production by dark-fermentation: Current status, challenges and perspectives *Bioresour Technol* **321** 124354

[18] Parkhey P and Gupta P 2017 Improvisations in structural features of microbial electrolytic cell and process parameters of electrohydrogenesis for efficient biohydrogen production: a review *Renewable and Sustainable Energy Reviews* **69** 1085–99

[19] Kadier A, Kalil M S, Abdeshahian P, Chandrasekhar K, Mohamed A, Azman N F, Logroño W, Simayi Y and Hamid A A 2016 Recent advances and emerging challenges in microbial electrolysis cells (MECs) for microbial production of hydrogen and value-added chemicals *Renewable and Sustainable Energy Reviews* **61** 501–25

[20] Brar K K, Cortez A A, Pellegrini V O A, Amulya K, Polikarpov I, Magdouli S, Kumar M, Yang Y H, Bhatia S K and Brar S K 2022 An overview on progress, advances, and future outlook for biohydrogen production technology *Int J Hydrogen Energy* **47** 37264–81

[21] Bora A, Mohanrasu K, Angelin Swetha T, Ananthi V, Sindhu R, Chi N T L, Pugazhendhi A, Arun A and Mathimani T 2022 Microbial electrolysis cell (MEC): Reactor configurations, recent advances and strategies in biohydrogen production *Fuel* **328** 125269

[22] Koul Y, Devda V, Varjani S, Guo W, Ngo H H, Taherzadeh M J, Chang J-S, Wong J W C, Bilal M, Kim S-H, Bui X-T and Parra-Saldívar R 2022 Microbial electrolysis: a promising approach for treatment and resource recovery from industrial wastewater *Bioengineered* **13** 8115–34

[23] Tang J H 2021 A Review of Suitable Substrates for Hydrogen Production in Microbial Electrolysis Cells *IOP Conf Ser Earth Environ Sci* **621** 012145

[24] Lim S S, Yu E H, Daud W R W, Kim B H and Scott K 2017 Bioanode as a limiting factor to biocathode performance in microbial electrolysis cells *Bioresour Technol* **238** 313–24

[25] Gautam R, Nayak J K, Ress N V., Steinberger-Wilckens R and Ghosh U K 2023 Bio-hydrogen production through microbial electrolysis cell: Structural components and influencing factors *Chemical Engineering Journal* **455** 140535

[26] Park S G, Rajesh P P, Sim Y U, Jadhav D A, Noori M T, Kim D H, Al-Qaradawi S Y, Yang E, Jang J K and Chae K J 2022 Addressing scale-up challenges and enhancement in performance of hydrogen-producing microbial electrolysis cell through electrode modifications *Energy Reports* **8** 2726–46

[27] Srivastava P, García-Quismondo E, Palma J and González-Fernández C 2023 Coupling dark fermentation and microbial electrolysis cells for higher hydrogen yield: Technological competitiveness and challenges *Int J Hydrogen Energy*

[28] Luo L, Sriram S, Johnravindar D, Louis Philippe Martin T, Wong J W C and Pradhan N 2022 Effect of inoculum pretreatment on the microbial and metabolic dynamics of food waste dark fermentation *Bioresour Technol* **358** 127404
[29] Luo S, Liu F, Fu B, He K, Yang H, Zhang X, Liang P and Huang X 2021 Onset Investigation on Dynamic Change of Biohythane Generation and Microbial Structure in Dual-chamber versus Single-chamber Microbial Electrolysis Cells *Water Res* **201** 117326
[30] Zhen G, Kobayashi T, Lu X, Kumar G, Hu Y, Bakonyi P, Rózsenberszki T, Koók L, Nemestóthy N, Bélafi-Bakó K and Xu K 2016 Recovery of biohydrogen in a single-chamber microbial electrohydrogenesis cell using liquid fraction of pressed municipal solid waste (LPW) as substrate *Int J Hydrogen Energy* **41** 17896–906
[31] Yan X, Wang B, Liang H, Yang J, Zhao J, Ndayisenga F, Zhang H, Yu Z and Qian Z 2022 Enhanced straw fermentation process based on microbial electrolysis cell coupled anaerobic digestion *Chin J Chem Eng* **44** 239–45
[32] Cheng S and Logan B E 2007 Sustainable and efficient biohydrogen production via electrohydrogenesis *Proceedings of the National Academy of Sciences* **104** 18871–3
[33] Ndayisenga F, Yu Z, Wang B and Zhou D 2023 Effects of the applied voltage on electroactive microbial biofilm viability and hydrogen production in a recalcitrant organic waste-fed single-chamber membrane-free microbial electrolysis cell performance *Chemical Engineering Journal* **469** 144002
[34] Rousseau R, Ketep S F, Etcheverry L, Délia M L and Bergel A 2020 Microbial electrolysis cell (MEC): A step ahead towards hydrogen-evolving cathode operated at high current density *Bioresour Technol Rep* **9** 100399
[35] Cotterill S E, Dolfing J, Jones C, Curtis T P and Heidrich E S 2017 Low Temperature Domestic Wastewater Treatment in a Microbial Electrolysis Cell with 1 m^2 Anodes: Towards System Scale-Up *Fuel Cells* **17** 584–92
[36] Zhou H, Xuanyuan X, Lv X, Wang J, Feng K, Chen C, Ma J and Xing D 2023 Mechanisms of magnetic sensing and regulating extracellular electron transfer of electroactive bacteria under magnetic fields *Science of The Total Environment* **895** 165104
[37] Montpart N, Rago L, Baeza J A and Guisasola A 2015 Hydrogen production in single chamber microbial electrolysis cells with different complex substrates *Water Res* **68** 601–15
[38] Wang Y, Xi B, Jia X, Li M, Qi X, Xu P, Zhao Y, Ye M and Hao Y 2021 Characterization of hydrogen production and microbial community shifts in microbial electrolysis cells with L-cysteine *Science of The Total Environment* **760** 143353
[39] Wang L, Singh L and Liu H 2018 Revealing the impact of hydrogen production-consumption loop against efficient hydrogen recovery in single chamber microbial electrolysis cells (MECs) *Int J Hydrogen Energy* **43** 13064–71
[40] Wang L, Chen Y, Long F, Singh L, Trujillo S, Xiao X and Liu H 2020 Breaking the loop: Tackling homoacetogenesis by chloroform to halt hydrogen production-consumption loop in single chamber microbial electrolysis cells *Chemical Engineering Journal* **389** 124436
[41] Hasibar B, Ergal I, Abreu Dias S, Bochmann G, Rittmann S K M R and Fuchs W 2020 Competing acetate consumption and production inside a microbial electrolysis cell *J Environ Chem Eng* **8** 103847
[42] Dhar B R, Elbeshbishy E, Hafez H and Lee H S 2015 Hydrogen production from sugar beet juice using an integrated biohydrogen process of dark fermentation and microbial electrolysis cell *Bioresour Technol* **198** 223–30

AUTHOR INDEX